Vol. 29. **The Analytical Chemistry of Sulfur and Its Compounds** (*in three parts*). By J. H. Karchmer
Vol. 30. **Ultramicro Elemental Analysis.** By Günther Tölg
Vol. 31. **Photometric Organic Analysis** (*in two parts*). By Eugene Sawicki
Vol. 32. **Determination of Organic Compounds: Methods and Procedures.** By Frederick T. Weiss
Vol. 33. **Masking and Demasking of Chemical Reactions.** By D. D. Perrin
Vol. 34. **Neutron Activation Analysis.** By D. De Soete, R. Gijbels, and J. Hoste
Vol. 35. **Laser Raman Spectroscopy.** By Marvin C. Tobin
Vol. 36. **Emission Spectrochemical Analysis.** By Morris Slavin
Vol. 37. **Analytical Chemistry of Phosphorus Compounds.** Edited by M. Halmann
Vol. 38. **Luminescence Spectrometry in Analytical Chemistry.** By J. D. Winefordner, S. G. Schulman and T. C. O'Haver
Vol. 39. **Activation Analysis with Neutron Generators.** By Sam S. Nargolwalla and Edwin P. Przybylowicz
Vol. 40. **Determination of Gaseous Elements in Metals.** Edited by Lynn L. Lewis, Laben M. Melnick, and Ben D. Holt
Vol. 41. **Analysis of Silicones.** Edited by A. Lee Smith
Vol. 42. **Foundations of Ultracentrifugal Analysis.** By H. Fujita
Vol. 43. **Chemical Infrared Fourier Transform Spectroscopy.** By Peter R. Griffiths
Vol. 44. **Microscale Manipulations in Chemistry.** By T. S. Ma and V. Horak
Vol. 45. **Thermometric Titrations.** By J. Barthel
Vol. 46. **Trace Analysis: Spectroscopic Methods for Elements.** Edited by J. D. Winefordner
Vol. 47. **Contamination Control in Trace Element Analysis.** By Morris Zief and James W. Mitchell
Vol. 48. **Analytical Applications of NMR.** By D. E. Leyden and R. H. Cox
Vol. 49. **Measurement of Dissolved Oxygen.** By Michael L. Hitchman
Vol. 50. **Analytical Laser Spectroscopy.** Edited by Nicolo Omenetto
Vol. 51. **Trace Element Analysis of Geological Materials.** By Roger D. Reeves and Robert R. Brooks
Vol. 52. **Chemical Analysis by Microwave Rotational Spectroscopy.** By Ravi Varma and Lawrence W. Hrubesh
Vol. 53. **Information Theory As Applied to Chemical Analysis.** By Karel Eckschlager and Vladimir Štěpánek
Vol. 54. **Applied Infrared Spectroscopy: Fundamentals, Techniques, and Analytical Problem-solving.** By A. Lee Smith
Vol. 55. **Archaeological Chemistry.** By Zvi Goffer
Vol. 56. **Immobilized Enzymes in Analytical and Clinical Chemistry.** By P. W. Carr and L. D. Bowers
Vol. 57. **Photoacoustics and Photoacoustic Spectroscopy.** By Allan Rosencwaig
Vol. 58. **Analysis of Pesticide Residues.** Edited by H. Anson Moye
Vol. 59. **Affinity Chromatography.** By William H. Scouten
Vol. 60. **Quality Control in Analytical Chemistry.** By G. Kateman and F. W. Pijpers
Vol. 61. **Direct Characterization of Fineparticles.** By Brian H. Kaye
Vol. 62. **Flow Injection Analysis.** By J. Ruzicka and E. H. Hansen

(*continued on back*)

Radiochemistry and Nuclear Methods of Analysis

CHEMICAL ANALYSIS

A SERIES OF MONOGRAPHS ON
ANALYTICAL CHEMISTRY AND ITS APPLICATIONS

Editor
J. D. WINEFORDNER
Editor Emeritus: **I. M. KOLTHOFF**

VOLUME 116

A WILEY-INTERSCIENCE PUBLICATION

JOHN WILEY & SONS, INC.

New York / Chichester / Brisbane / Toronto / Singapore

Radiochemistry and Nuclear Methods of Analysis

WILLIAM D. EHMANN

Professor, Department of Chemistry
University of Kentucky
Lexington, Kentucky

DIANE E. VANCE

Training Accreditation Coordinator, Analytical Laboratories
Westinghouse Savannah River Co.
Aiken, South Carolina

A WILEY-INTERSCIENCE PUBLICATION

JOHN WILEY & SONS, INC.

New York / Chichester / Brisbane / Toronto / Singapore

Library of Congress Cataloging in Publication Data:

Ehmann, William D.

Radiochemistry and nuclear methods of analysis / William D. Ehmann, Diane E. Vance.

p. cm.—(Chemical analysis, ISSN 0069-2883: v. 116)

"A Wiley-Interscience publication."

Includes bibliographical references and index.

ISBN 0-471-60076-8 (c)

1. Radiochemistry. 2. Nuclear chemistry. I. Vance, Diane E. II. Title. III. Series.

QD601.2.E34 1991 90-26336

541.3'8—dc20 CIP

Printed in the United States of America

10 9 8 7 6 5 4 3 2 1

To Nancy

To John

PREFACE

The fields of nuclear and radiochemistry have been dynamic and momentous areas of scientific research since the discovery of radioactivity by Henri Becquerel in 1896. After nearly a century, they remain vigorous fields of study as nuclear and radiochemists probe deeper into the subatomic world and develop new and exciting ways to use radioactive materials. The applications of radioactivity have played an essential role in many of the advances in analytical chemistry, biological science, medicine, materials science, pharmacology, geology, industrial process control, and environmental science. Many of the scientists who use radioactive materials are specialists in areas other than radionuclear chemistry, and their need for a background in radiochemistry differs from that needed by students planning a career in radionuclear chemistry. This point was developed more fully in the report *Training Requirements for Chemists in Nuclear Medicine, Nuclear Industry, and Related Areas* (Washington, DC: National Academy Press, National Research Council, 1988). One of the recommendations made in this report is that new curricula and textbooks be developed for radiochemistry lecture and laboratory courses for workers in other disciplines who require a basic understanding of radioactivity. Our book was already in progress at the time the report was published (1988), and we believe that it will serve the need identified above.

This text is based on the lectures for the radiochemistry course that has been taught by one of the authors (WDE) for 32 years at the University of Kentucky. Of the 30 or so students who take this course each year, only a few are graduate students in nuclear or radiochemistry. Most are undergraduate chemistry majors or graduate students from biology, health physics, pharmacy, geology, or medically related areas. The excellent comprehensive texts that serve the needs of the student majoring in nuclear or radiochemistry at the graduate level often seem formidable to students with less extensive backgrounds in mathematics, physics, and quantum mechanics, even when the students are highly motivated to learn. Our book is designed to serve two functions. We felt that there was a need for a radiochemistry textbook that would fill the gap between the one-chapter treatments in general chemistry and some undergraduate analytical chemistry textbooks and those available in the more comprehensive texts. In addition to serving as a textbook in radiochemistry, we have stressed applications in the broad field of radioanalytical chemistry. Hence, this volume should also serve as a useful introduction to the use of tracers and nuclear methods of analysis for analytical chemists.

Radiochemistry and Nuclear Methods of Analysis is suitable for a one-semester course in introductory radiochemistry and radioanalytical chemistry at the undergraduate or beginning graduate level. The book has been designed to be understandable to a reader who has had at least two semesters of general chemistry, but no previous exposure to the study of nuclear or radiochemistry. Exposure to introductory college-level courses in physics, analytical chemistry, and calculus would enhance understanding of certain sections, but is not a prerequisite. The book could also serve as an introduction to radiochemistry for professionals who find that they need some knowledge of the area for a work assignment. The nuclear industry and government national laboratories, for example, employ a great many chemists who have had little or no previous training in nuclear or radiochemistry. We hope this volume may prove to be of value in these on-the-job training programs.

Our aim in this book has been to give the reader a clear and concise introduction to the basic concepts of radiochemistry, and then to survey the broad applications of radioactivity in a variety of other fields. We have deliberately kept the book relatively brief, easily readable, and "user-friendly." Doing this imposes a limitation on the number of topics that can be incorporated into the text and on the level of detail that can be included. We first selected those topics that seemed essential for any radiochemistry course and covered these in Chapters 1–8. Chapters 9–14 deal mostly with applications, with the greatest emphasis on analytical chemistry. However, we have tried to alert the reader to applications in as many other fields as possible. Nuclear chemistry (including theoretical treatments of nuclear structure, models, reactions, and decay modes) is treated only briefly because these topics are thoroughly covered in a number of other excellent texts. We think that these advanced topics are taught more productively to students after they have had time to assimilate the basic concepts of radiochemistry and have had sufficient calculus and physical chemistry to comprehend the material. This intentional selection process may have resulted in the omission of material that other instructors will consider essential. The reading lists at the end of each chapter provide sources for further information. We have not cited original journal literature extensively in these lists, but instead have cited mostly monographs and reviews that are more likely to be in the libraries of educational institutions that might not have extensive collections of research journals. We think that these will generally be more comprehensible and thus more useful to the beginning student. To help direct the student's attention to specific points, we intentionally made figures generalized, instead of reproducing figures directly from the research literature. The more complex literature figures may sometimes distract the student from visualizing the overall trends. The exercises at the ends of each chapter range from simple to complex. The simple ones are intended to help the reader become familiar with the equations and concepts presented in the text and to gain confidence in manipulating the formulas. More complex problems are included to challenge the more advanced student. The exercises are concentrated in the earlier chapters, where most of the basic material is presented.

We have drawn heavily for background material and specific equations on the well-established comprehensive texts in nuclear and radiochemistry, such as *Nuclear and Radiochemistry,* 3d ed., by Friedlander, et al. (New York: Wiley, 1981) and *Nuclear Chemistry* by Choppin and Rydberg (Oxford: Pergamon, 1980). In addition, the excellent textbook by G. F. Knoll entitled *Radiation Detection and Measurement* (New York: Wiley, 1979 (1st ed.), 1989 (2d ed.)) provided much of the background material for our chapter on radiochemistry instrumentation. Our briefer text will not replace these detailed and more comprehensive works.

We would like to acknowledge the help of many colleagues during preparation of this book. Dave Potocik of the Savannah River Site Health Protection group read the chapters relating to health physics and made useful comments. John C. Vance of Humana Hospital in Lexington, Kentucky, and John Moll of Westinghouse Savannah River Co., Aiken, South Carolina, read the entire manuscript in great detail, and we especially appreciate their suggestions relating to the clarity of the text. We are grateful to colleagues at the University of Kentucky who read sections of the manuscript and offered technical suggestions and corrections and worked the problems; they include M. A. Lovell, M.T. McEllistrem, J. D. Robertson, L. Tandon, D. J. Van Dalsem, and S.W. Yates. While we have made every effort to eliminate errors in the text, it is quite possible that some have still escaped our notice. While acknowledging the assistance of others, the authors take sole responsibility for any errors, and would appreciate it if readers would inform us of any errors that they might find.

It has taken us a good deal longer to write this volume than we had anticipated. That is a reflection of how much radiochemistry has grown in the past 40 years. To discuss the many applications of nuclear methods it was necessary to explore the literature of many diverse fields. In the future, it is probable that few books will be written to cover the entire field of radiochemistry and its applications. Radiochemists will most likely become specialists and write books only in the subdisciplines of the field. This trend is already observed in the listing of new specialized books provided in the biennial review entitled "Nuclear and Radiochemical Analysis" in *Analytical Chemistry Fundamental Reviews* (Ehmann, Robertson, and Yates, *Anal. Chem.* **62**:50R (1990)). If this is so, we hope our book can continue to serve as an interdisciplinary introduction to this challenging and fascinating field of study. In recent years, nuclear studies have received more bad publicity than good, and students are often reluctant to enter into study of it. Perhaps this book will provide a doorway into radiochemistry that students can enter with relative ease. Once inside, they may discover for themselves the many positive contributions this science has made to our world, and the intellectual enjoyment that can be gained from its study. Our commitment to opening doors to students through education and training in radionuclear chemistry was the source of our efforts to write this book. We hope that it will prove useful to the student and to the radiochemistry community.

Finally, we want to express our sincere appreciation to our spouses, Nancy Ehmann and John Moll, for their help and understanding during this time. Perhaps now, many "lost weekends" devoted to wordprocessing and library searches can be recouped.

William D. Ehmann
Diane E. Vance

Lexington, Kentucky
Aiken, South Carolina
July 1991

CONTENTS

Radiochemistry and Nuclear Methods of Analysis

CHAPTER

1

INTRODUCTION TO RADIOCHEMISTRY

Nearly a century has passed since the discovery of the phenomenon of radioactivity by Henri Becquerel in 1896 (Fig. 1.1). Since that time, concepts of the basic structure of matter have changed dramatically. The phenomenon of radioactivity has assumed a great importance in daily life, with the potential for both constructive and destructive uses. The importance of the work in this area may be illustrated by noting that there have been 12 Nobel Prizes in the field of chemistry and 13 in the field of physics for studies that could be considered to be in the province of nuclear or radiochemistry (Table 1.1). Research in these areas is now directed toward a deeper understanding of basic nuclear structure and the development of new technologies for the application of radioactivity to a wide variety of problems.

Nuclear chemistry may be defined as the application of procedures and techniques common to chemistry to study the structure of the nucleus and to define the nature of the fundamental particles. **Radiochemistry** may be defined as the application of the phenomenon of radioactive decay and techniques common to nuclear physics to solve problems in the field of chemistry. It is with the latter that this book is mainly concerned, although there is an overlap in the subject matter and experimental techniques used in both fields.

This chapter outlines the history of the discovery of radioactivity, notes the later developments in the field through the 20th century, and briefly identifies some current research areas. During the historical development, many terms, concepts, and definitions that are used throughout the study of radiochemistry will be introduced.

1.1. THE DISCOVERY OF RADIOACTIVITY AND EVOLUTION OF NUCLEAR THEORY

The latter part of the 19th century and the first part of the 20th century were times of major advances in chemistry and physics in the understanding of atomic structure. During that period, scientists' views of the basic structure of matter were completely revolutionized.

Figure 1.1. Henri Becquerel (1852–1908). In 1896 he discovered the phenomenon of radioactivity by observing penetrating radiations emitted from various salts of uranium. It is of interest that his father, Edmund Becquerel, had also earlier studied the phosphorescence of uranium salts. (Courtesy of the American Institute of Physics, Niels Bohr Library; gift of Wm. G. Meyers.)

The story of the discovery of radioactivity began in 1895 with W. C. Röntgen and his work with cathode ray tubes (CRT). These tubes were used by several researchers of that era to study electrical phenomena. A typical CRT consists of an evacuated glass tube that has two electrodes sealed inside. An electrical potential is applied across the electrodes, and the streams of electrons passing between the electrodes, the *cathode rays,* can be studied. Röntgen allowed the cathode rays to impinge upon different metal targets inside the tube, and noticed that highly penetrating radiations were emitted. These he called **x rays** (Fig. 1.2). He also noted

Table 1.1. Some Nobel Laureates Who Have Made Significant Contributions[a] to the Fields of Radiochemistry and Nuclear Chemistry

	Prize in Chemistry	
1908	Ernest Rutherford	Modes of radioactive decay, Rn characterization
1911	Marie Sklodowska Curie	Discovery of Po and Ra
1921	Frederick Soddy	Concept of isotopes, change of Z in radioactive decay
1922	Francis W. Aston	Separation of stable isotopes of Ne, gaseous diffusion for separation of isotopes
1934	Harold C. Urey	Discovery of deuterium, cosmic abundances, evolution of the meteorites
1935	Frederic Joliot Irene Joliot Curie	Production of artificial radionuclides
1943	Georg Hevesy	Use of radioisotopes as tracers, development of neutron activation analysis (with H. Levi)
1944	Otto Hahn	Discovery of nuclear fission (with F. Strassman)
1951	Edwin M. McMillan Glenn T. Seaborg	Discovery of 1st transuranium elements, Np and Pu
1960	Willard F. Libby	^{14}C and ^{3}H dating methods
	Prize in Physics	
1901	Wilhelm C. Röntgen	Production and characterization of x rays
1903	Henri Becquerel Marie S. Curie, Pierre Curie	Discovery of radioactivity, identified as an atomic process
1935	James Chadwick	Discovery of the neutron
1938	Enrico Fermi	Search for transuranium elements, first controlled nuclear fission reactor
1939	Ernest O. Lawrence	Development of the cyclotron
1951	John D. Cockcroft Ernest T. S. Walton	Developed voltage-doubling accelerator, first accelerator-produced nuclear transformation
1961	Rudolf L. Mössbauer	Mössbauer spectrometry, isomer shifts in γ-ray emission
1963	Maria Goeppert-Mayer J. H. D. Jensen, E. P. Wigner	Fundamental nuclear structure studies
1968	Luis Alvarez	Discovered electron capture decay
1983	William F. Fowler	Nucleosynthesis of elements in stars
	Prize in Physiology or Medicine	
1977	Rosalyn S. Yalow	Radioimmunoassay

[a]The contributions listed may include some that were not specifically a part of the Nobel award. The year listed is that of the Nobel Prize, not of the contributions.

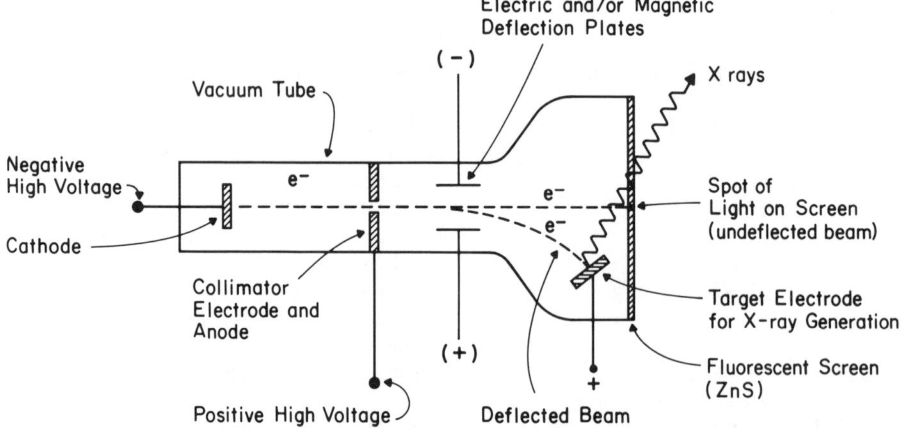

Figure 1.2. Cathode-ray tube. Electric and/or magnetic deflection plates can be used to deflect the electron beam (cathode rays) onto a target electrode to generate x rays characteristic of the target element.

similarities between x rays and sunlight, in that both could darken a photographic plate and cause fluorescence in certain minerals and salts.

The fluorescence of natural minerals had long been of interest to the French scientist Henri Becquerel. Both he and his father had studied the fluorescence of potassium uranyl sulfate [$K_2UO_2(SO_4)_2 \cdot 2H_2O$]. In some of these experiments, Becquerel had fastened crystals of $K_2UO_2(SO_4)_2 \cdot 2H_2O$ to photographic plates that had been covered with opaque materials, and exposed the crystals and the plate to sunlight for a time. Upon development of the plates, an image of the crystal could be seen. At first Becquerel concluded that the fluorescence induced in the crystal by the sunlight penetrated the opaque materials and exposed the photographic plate. However, in late February of 1896, the sun did not cooperate with the experimenter, and Becquerel was not able to expose the plates and crystals to sunlight for several days. He placed the photographic plates, with crystals attached, into a drawer. Several days later Becquerel developed the plates anyway, expecting to find only weak images of the crystals. Instead, he observed very intense silhouttes of the crystals. He reported these findings and stated, "these radiations, whose effects possess a strong analogy with the effects produced by the radiations studied by Lenard and Röntgen, might be invisible radiations emitted by phosphorescence, whose duration of persistence might be infinitely greater than that of the luminous radiations emitted by these substances." His understanding of the phenomenon was limited at that time, but what Becquerel had observed was, of course, the effect of uranium radioactivity.

Figure 1.3. Marie Sklodowska Curie (1867–1934) and Pierre Curie (1859–1906) in their laboratory. Mme. Curie, together with her husband, Pierre, and Henri Becquerel, received the Nobel Prize in physics in 1903 for studies of "radiation phenomena." She alone received the Nobel Prize in chemistry in 1911 for the discovery of polonium and radium. Photograph taken in their laboratory in 1896. (Courtesy of the American Institute of Physics, Niels Bohr Library.)

Marie and Pierre Curie (Fig. 1.3) pursued the study of "Becquerel rays" with other minerals. Many of these evidenced even greater activity than did the potassium uranyl sulfate crystals originally used by Becquerel. The Curies were convinced that this additional activity was due to the presence of an active substance other than the uranium and thorium which had already been identified. They performed a series of chemical extractions of the uranium ore pitchblende and succeeded in isolating a new active substance in the bismuth fraction of the separation. They proposed the name

polonium for this new substance, in honor of Mme. Curie's homeland, Poland. Further chemical investigations of pitchblende led to the discovery of a second, even more active substance, in the barium-containing fraction from their separation procedure. They proposed the name **radium** for this substance, and also used the term **radioactive** for the first time to describe the observed activity. The Curies then set themselves on the task of isolating a pure sample of radium from the pitchblende.

Working in a small, unheated shed, the Curies processed some two tons of pitchblende ore (75% U_3O_8) and eventually isolated a total of 100 mg of $RaCl_2$. This represented 25% of the total amount of the Ra that had actually been present in the ore—an impressive chemical feat by any standards!

While the Curies worked to isolate and characterize the chemical substances responsible for the phenomenon of radioactivity, other scientists concentrated on characterizing the radiations themselves and on a further elucidation of nuclear structure. Ernest Rutherford (Fig. 1.4) was an important investigator in this area. In 1903, he and Frederick Soddy postulated that radioactivity was not just a consequence of an atomic change that had previously taken place, but rather that the radioactive emissions were directly associated with the change. They characterized three types of radiations: **alpha (α), beta (β), and gamma (γ)** rays. The α rays were shown to be deflected by electric and magnetic fields in a direction opposite to that for cathode rays, and thus were positively charged. The charge-to-mass ratio (e/m) of the α particle was measured to be one-half that for a proton. The symbol e refers to the unit charge on an electron, and m is the mass of the particle in units of daltons, or atomic mass units (amu). In later work (1909), Rutherford and Royds were able to demonstrate spectroscopically that α particles were actually helium nuclei (^4He). The β rays behaved in the same manner as cathode rays, and so were thought to be negatively charged particles. They were originally estimated to have a mass only 1/1000 that of the hydrogen atom. The actual ratio is now known to be 1/1837.15. The γ rays were shown to be extremely penetrating and were unaffected by electrical or magnetic fields.

Early in this century, the atom was thought to consist of positive and negative particles (protons and electrons) evenly distributed throughout the atom. Rutherford saw that the particles from radioactive decay might be used to investigate this hypothesis. In 1911, Rutherford reported on the famous scattering experiments performed by H. Geiger and E. Marsden. In these experiments, they measured the degree of scattering of α and β particles as they passed through a gold foil of 0.00004 cm thickness. Rutherford concluded from the results of these measurements that the subatomic particles were not evenly distributed in the atom, but rather that the atom "consists of a central charge supposed concentrated at a point." The strong α

Figure 1.4. Ernest Rutherford (1871–1937) (*right*) talking with J. A. Ratcliffe in 1935. Rutherford received the Nobel Prize in physics in 1908 for his characterization of different modes of radioactive decay, studies of radon, and recognition that radioactive decay resulted in the transformation of atoms of uranium and thorium into atoms of other elements. (Courtesy of the American Institute of Physics, Niels Bohr Library: photograph by C.E. Wynn-Williams.)

and β deflections that had been observed were due to their interactions with this strong central field. Rutherford's hypothesis is now well established. The central portion of the atom, where the charge and mass are concentrated, is called the **nucleus**. The nucleus has a radius of approximately 10^{-13} to 10^{-12} cm, while the atom as a whole has a radius of about 10^{-8} to 10^{-7} cm. Rutherford's concept of a nuclear model for the atom opened the door for rapid development of new sophisticated theories of atomic structure.

Chemical studies of the phenomenon of radioactivity had revealed the existence of many apparently new chemical substances. The nature and true identities of these substances were unknown at first. It was recognized that many of the active substances originated from the known radioactive elements (thorium, polonium, radium, and uranium). Therefore, these substances were initially given names like "RaA" (now known as ^{210}Po), or "UX_2" (now known to be ^{234m}Pa) to indicate their source. Rutherford and Frederick Soddy, in their 1902 paper, proposed that these relatively short-lived "atom-fragments" were not really elements. They suggested the term "metabolon" to describe them. However, in papers in 1911 and 1913, Soddy and K. Fajans elucidated the true nature of the uranium and thorium decay products, and established the relationship of the decay products to specific elements in the periodic table. Soddy also discussed the nature of the three **decay chains** known at that time. These decay chains were successive transformations by radioactive decay of one element into another, beginning with ^{235}U, ^{238}U, or ^{232}Th as the **parent nuclei,** that is, the first naturally occurring radioactive substance in the chain. The numbers to the upper left of the elemental symbol are known as **mass numbers.** As will be discussed in more detail later, the mass number (A) is the sum of the neutrons (N) and protons (Z) in the nucleus of an atom. The products of the decays, the **daughters**, were themselves radioactive and so decayed to produce further daughter elements. Soddy and Fajans chemically characterized the daughters of the three natural chains and showed their similarities. For example, the "emanations" (gaseous products) of each of the three decay chains were all associated with Group VIIIA of the periodic table, the noble gases. All decayed by α emission in a short period of time.

Fajans also showed that α emission resulted in a jump of two spaces from right to left in a horizontal row of the periodic table, and that in β decay, the jump was one space, from left to right. Using these rules, he showed that all the radioelements could be placed into existing places in the periodic table. Thus, the phenomenon of radioactivity actually resulted in the changing of one element into another element, a process called **transmutation**. This realization led to another puzzle, because there did not appear to be enough places in the periodic table for all the species that appeared to be new elements. This problem was addressed in 1913, when Soddy proposed the existence of **isotopes**. Isotopes are forms of an element that have the same number of protons, but different masses, due to different numbers of neutrons in their nuclei. This means that these several different forms of an element could occupy the same block of the periodic table. The isotope hypothesis was confirmed in 1919 through Francis Aston's investigations of the isotopes of Ne with his newly developed mass spectrometer. In 1934, Harold Urey (Fig. 1.5) received the Nobel Prize in chemistry for the discovery of deuterium, a heavy stable isotope of the element hydrogen.

Figure 1.5. Harold C. Urey (1894–1981) received the Nobel Prize in chemistry in 1934 for the discovery and characterization of deuterium, a stable heavy isotope of hydrogen. In 1931, R. T. Birge and D. H. Menzel had suggested the presence of a heavy isotope of hydrogen in ordinary hydrogen to the extent of approximately one part in 4500, based on mass spectrometric measurements of hydrogen. At Columbia University, Urey speculated in 1931 that different isotopes of hydrogen should have different vapor pressures. To concentrate any heavy isotope that might be present, he allowed 4 L of liquid hydrogen to evaporate down to 1 mL. An optical spectrum of the residual liquid hydrogen was found to exhibit a spectral line in the position predicted for an isotope of hydrogen with a mass near 2 daltons. (Courtesy of The Joseph Regenstein Library, University of Chicago.)

Up to 1919, the observations made on radioactivity were done with substances that were naturally radioactive. In that year, however, Rutherford reported on the first experimentally induced nuclear transformation. He placed a source of α particles in a box equipped with a movable fluorescent ZnS detection screen. As expected, the α particles produced scintillations, or flashes of light, when they struck the screen, but Rutherford also found that he would still see scintillations even when the screen was moved far out of the range of the α particles. After a systematic study of this "anomalous effect in nitrogen," Rutherford concluded that the N atoms in the air were being disintegrated by collisions with the α particles and that H atoms were being emitted. The reaction that occurred was

$$^{4}\text{He} + {}^{14}\text{N} \longrightarrow {}^{17}\text{O} + {}^{1}\text{H} + Q$$

where Q refers to the energy change of the reaction. In nuclear shorthand, we can write the above equation as

$$^{14}\text{N}(\alpha, \text{p})^{17}\text{O}$$

where the first species outside the parentheses is the target nucleus, the first species inside the parentheses is the bombarding particle, the second species inside the parentheses is the smaller mass product, and the last species outside the parentheses is the larger mass product. Rutherford noted at the conclusion of this report on experimental transmutation, "if α particles—or similar projectiles—of still greater energy were available for experiment, we might expect to break down the nuclear structure of many of the lighter atoms." In 1929, E. O. Lawrence (Fig. 1.6) developed a device to do just that—the cyclotron.

In his Bakerian lecture in 1920, Rutherford proposed the existence of a third atomic particle in addition to the proton and electron. This particle, with a mass equivalent to 1 amu and an electrical charge of zero, would have important uses as a probe of nuclear structure. In the audience listening to this lecture was James Chadwick, an assistant to Rutherford. It was Chadwick who would prove the existence of this particle (the **neutron**) some 12 years later, in 1932. Chadwick's experimental setup consisted of a Po radiation source mounted behind a disk of pure Be. Alpha particles from the Po struck the Be, and very penetrating "radiation" was emitted. This radiation was capable of causing the ejection of high-energy protons from paraffin wax or other hydrogen-rich materials placed in the path of the radiations. Chadwick went on to demonstrate that this radiation had properties consistent with those proposed by Rutherford for the neutron.

The nuclear reaction that occurred to produce the neutrons was

$$^{9}\text{Be}(\alpha, \text{n})^{12}\text{C}$$

This reaction is still used today in laboratory isotopic neutron sources. In these sources, an α emitter (often Ra or Pu) is mixed with Be powder to

Figure 1.6. Ernest O. Lawrence (1901–1958) (*right*) proposed development of the cyclotron in 1929 and constructed the first working model in 1930 with M. S. Livingston (*left*). The first cyclotron, located at the University of California, produced 80-keV protons. (Lawrence Berkeley Laboratory photograph; courtesy of the American Institute of Physics, Niels Bohr Library.)

generate a relatively low yield of neutrons. The discovery of the neutron made possible many later advances in nuclear science. Because of its neutrality, the neutron could enter a nucleus without suffering Coulomb repulsion, so complicated accelerating devices were not needed for many neutron-induced reactions. Studies of the interactions of a neutron with nuclei could give information not only on nuclear reactions, but also on the fundamental structure of the nucleus itself.

Another subatomic particle, the **positron** (β^+), was discovered in 1932 by C. D. Anderson. This particle is an electron with a positive charge and had been predicted by Dirac in 1930 on theoretical grounds. Anderson used cloud-chamber photographs of cosmic-ray tracks to identify a curved track in the field of observation that was identical in radius of curvature to that of the electron (β^-), but opposite in direction. This track was attributed to the positron (Fig. 1.7). In 1934, Frederick Joliot and Irene Curie (daughter of Pierre and Marie Curie, Fig. 1.8) reported on the first artificial production of a radioelement. They irradiated several elements with α particles

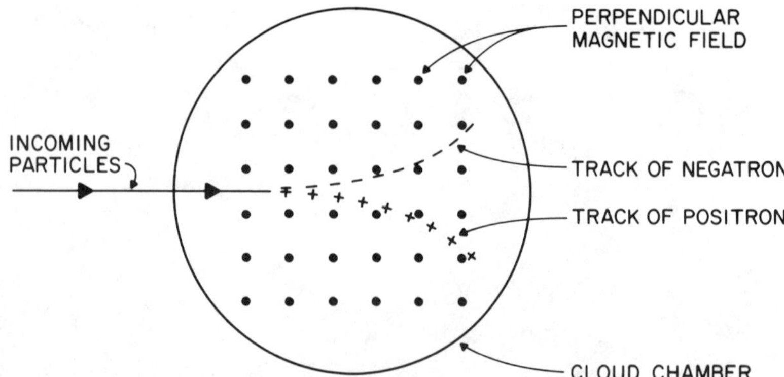

Figure 1.7. Discovery of the positron by C. D. Anderson. Positrons generated by cosmic rays follow a curved path in a magnetic field placed perpendicular to a cloud chamber detector. Negatrons follow a path with the same radius of curvature and same ionization density, but in the opposite direction.

Figure 1.8. Frederic Joliot (1900–1958) and Irene Joliot-Curie (1897–1956) bombarded thin foils of aluminum, boron, and magnesium with α particles emitted by polonium and observed the formation of the first "artificial" radionuclides. They jointly received the Nobel Prize in chemistry in 1935. Irene was the daughter of Mme. Marie Curie. (Original photograph from the French Embassy, Press Information Division, courtesy of the American Institute of Physics, Niels Bohr Library.)

from Po, and observed the formation of new radioactive isotopes. The first artifical **radioisotope**, ^{13}N, was formed by the reaction

$$^{10}B(\alpha, n)^{13}N$$

The ^{13}N, which they called "radionitrogen," decayed by emission of a positron. At first, production of other artificial radioisotopes was slow, but by the beginning of the 1990s over 2000 different radioisotopes of the elements had been produced.

The neutrons discovered by Chadwick were first used for analytical purposes in 1936 by Georg Hevesy and Hilde Levi. They noted that rare earth elements, notably Dy, became highly radioactive upon being bombarded with neutrons. They recognized the potential for both qualitative and quantitative identification of the elements present in a sample through determination of the "neutron absorbing power" and half-lives of the radioactive elements. (The half-life of a radioisotope is the time needed for one-half of a statistically large number of radioactive atoms to decay. See Chapter 5 for further discussion of half-life.) This method of analysis, called **neutron activation analysis (NAA),** is now an important technique for multielement analysis. It is further discussed in Chapter 9.

Neutrons were also used in 1939 by O. Hahn and F. Strassman (Fig. 1.9) to split an atom for the first time (**induced nuclear fission**). They bombarded U with neutrons and succeeded in demonstrating the presence of radioactive products that were of much lower mass than U. They postulated that the neutrons caused the U atoms to split, or fission, into two fragments. An example of a fission reaction would be

$$^{235}U + n \longrightarrow {}^{144}Ba + {}^{90}Kr + 2n$$

This phenomenon had actually been observed by Enrico Fermi (Fig. 1.10) in 1934. He had, however, incorrectly interpreted the experimental result and assumed that some of the products he observed were elements with atomic masses heavier than U, the **transuranium elements.** Unfortunately, he did not do the chemical separations needed to confirm this hypothesis. However, he was later awarded the Nobel Prize (1938) for his work on neutron absorption and the production of new radioactivities.

The first transuranium element was actually produced in 1940 by E. M. McMillan and P. H. Abelson. They irradiated ^{238}U with neutrons to produce ^{239}U, which subsequently underwent β^- decay to produce neptunium (Np):

$$^{238}U(n, \gamma)^{239}U$$

$$^{239}U \longrightarrow {}^{239}Np + \beta^-$$

The ^{239}Np then decayed by β^- emission to the next transuranium element, ^{239}Pu, but this element was not identified at that time, due to its long half-life

Figure 1.9. (*A*) Otto Hahn (1879–1968) (*right*) and Fritz Strassmann (1902–1980) (*left*). Hahn discovered nuclear isomers in 1921 (^{234}Pa and ^{234m}Pa). While working with Strassmann in Berlin in 1939, they observed that some of the radioactive products formed in the bombardment of uranium with neutrons were actually isotopes of elements near the middle of the periodic table. Lise Meitner and O. R. Frisch shortly thereafter helped provide the explanation that uranium atoms bombarded by neutrons split into two smaller nuclei of approximately equal size; that is, uranium undergoes fission. Hahn alone received the Nobel Prize in chemistry in 1944 for the discovery of nuclear fission. The photograph was taken in 1962 in Munich and shows some of their original counting equipment. (Courtesy of Dietrich Hahn, Ottobrunn, Germany.)

Figure 1.9. *(Continued)* (*B*) O. Hahn and L. Meitner in their counting room, University of Berlin, about 1909. (Courtesy of Archiv zur Geschichte der Max-Planck-Gesellschaft, Berlin.)

of 24 100 years. Plutonium was first identified in 1940 by J. W. Kennedy, G. T. Seaborg (Fig. 1.11), E. M. McMillan, and A. C. Wahl. They bombarded ^{238}U with deuterium nuclei (^{2}H or d) to form ^{238}Np, which then decayed to ^{238}Pu:

$$^{238}\text{U}(\text{d}, 2\text{n})^{238}\text{Np}$$

$$^{238}\text{Np} \longrightarrow {}^{238}\text{Pu} + \beta^-$$

$$^{238}\text{Pu} \longrightarrow {}^{234}\text{U} + \alpha \qquad \text{(half-life} = 87.7 \text{ years)}$$

The possibility of releasing large amounts of energy by the fission of atoms had occurred to many people. In the process of neutron-induced fission, more neutrons are released in a fission event than the one required for the next fission event. This excess of neutrons leads to the potential for a **chain reaction;** that is, the splitting of one atom could release two or three neutrons that could split more atoms, and so on. If many fissions took place in a short time, a large amount of energy would be released. In the 1940s, under the direction of Enrico Fermi, the first nuclear reactor for the controlled release of nuclear energy was built. On December 2, 1942, the reactor "went critical," that is, sustained a chain reaction, for the first

Figure 1.10. Enrico Fermi (1901–1954). During his early work in Rome in the 1930s, he studied radioactivities generated by bombarding uranium with neutrons, neutron absorption properties of the elements, and developed selection rules for β decay. He received the Nobel Prize in physics in 1938 for his studies of neutron-induced radioactivities and neutron absorption. However, he had erroneously identified several radioactivities produced by neutron bombardment of uranium as being due to the production of new transuranium elements. His "new elements" were later shown by Hahn and Strassmann to be fission products of uranium. After coming to the United States he did criticality experiments with uranium–graphite lattices at Columbia University in 1941, and in 1942 constructed the world's first controlled nuclear fission reactor under the Stagg Field football stadium at the University of Chicago. His "Chicago Pile 1 (CP1)" achieved criticality (a self-sustaining nuclear chain reaction) on December 2, 1942. His success was announced to government officials with the message "the Italian navigator has just landed in the new world." The famous physicist Leo Szilard, however, waited until all but he and Fermi had left the balcony over CP-1 that day and countered Fermi's enthusiasm by telling him privately that he thought "this day would go down as a black day in the history of mankind." (Courtesy of The Joseph Regenstein Library, University of Chicago.)

Figure 1.11. Glenn T. Seaborg (1912–) received the Nobel Prize in chemistry (with E. M. McMillan) for discovery and characterization of the chemistry of transuranium elements. Most of his work has been done at the University of California, Berkeley. The element plutonium (^{238}Pu) was first produced and identified by Seaborg and coworkers in 1940. (Courtesy of Dr. Seaborg and the University of California Lawrence Berkeley Laboratory.)

time. Three years later, during World War II (WWII), the first atomic bombs were unleashed on Japan.

After WWII, knowledge in both theoretical and applied areas of nuclear and radiochemistry grew rapidly. Table 1.2 lists some of these developments. Only the most significant advances are highlighted in the following brief topical review.

In 1949, Willard Libby (Fig. 1.12) at the University of Chicago reported the development of the radiocarbon dating method. This method is widely used for archaeological dating. Dating methods based on many other

Table 1.2. Important Developments in Nuclear Science after World War II.

1945	First tests and use of a nuclear fission bomb and the end of WWII
1949	Development of the radiocarbon dating method by Willard Libby, University of Chicago [*Radiocarbon Dating,* Chicago: Univ. of Chicago Press, 1955]
1951	First token use of a nuclear power reactor, Argonne National Laboratory
1952	First test of a hydrogen bomb
1955	First International Conference on Atomic Energy, Geneva
1955	Discovery of mendelevium (Md), Z = 101, Ghiorso et al., Berkeley [*Phys. Rev.* **98:**1518 (1955)]
1957	Establishment of the International Atomic Energy Agency, Vienna
1958	Discovery of nobelium (No), Z = 102, by Ghiorso et al., Berkeley [*Phys. Rev. Lett.* **1:**18 (1958)]
1959	First test of a nuclear rocket engine
1961	Discovery of the last element in the actinide series, lawrencium (Lw), Z = 103 by Ghiorso et al., Berkeley [*Phys. Rev. Lett.* **6:**473 (1961)]
1980	Evidence that neutrinos might have mass [*Science* **208:**697 (1980)]
1982	Discovery of elements 107 and 109, Darmstadt, Germany [*Naturwissenschaften* **70:**383 (1983)]
1983	Identification of two proton decay [*Phys. Rev. Lett.* **50:**404 (1983); *Phys. Lett.* **133B:**146 (1983)]
1984	Discovery of element 108, Darmstadt, Germany [*Z. Phys. A At. and Nucl.* **317:**235 (1984)]
1984	Identification of ^{14}C decay [*Nature(London)* **307:**245 (1984)]
1984	Identification of delayed triton emission from ^{11}Li [*Phys. Lett.* **146B:** 176 (1984)]
1985	Limits established for the neutrino mass [*Yad. Fiz.* **42:**1441 (1985)]
1986	Irradiation of fruits and vegetables to kill insects and bacteria and to slow ripening is approved in the United States by the FDA
1989	Evidence presented for "cold fusion" of deuterium to produce tritium and neutrons by electrolysis of heavy water [*J. Electroanal. Chem.* **261:**301 (1989)]

Figure 1.12. Willard Libby (1908–1980) developed the radiocarbon and tritium dating methods at the University of Chicago. He received the Nobel Prize in chemistry in 1960 for ^{14}C dating. (Courtesy of the Joseph Regenstein Library, University of Chicago.)

radioactive elements also have been developed. These are discussed in Chapter 12.

In the 1950s, nuclear reactors were increasingly used for both research and for electric power production. These reactors provided the means for production of many different radioisotopes for which applications were quickly found in areas such as biology, agriculture, industrial quality control, medicine, and geology. The neutrons produced in the reactors also were used as activating particles for neutron activation analysis techniques, and this field of elemental analysis expanded rapidly.

An area of active research today is the search for new elements. The last actinide element (Lw, number 103) was discovered in 1961 by Ghiorso et al. Three laboratories (the Lawrence Berkeley Laboratory at Berkeley,

California; the Joint Institute for Nuclear Research at Dubna, USSR; and the Institute for Heavy-Ion Research, GSI at Darmstadt in Germany) have done most of the work of discovering the elements past atomic number 103. The usual approach used in this research is to bombard heavy spherical nuclei ($Z = 94$, or greater) with a beam of medium mass nuclei from a particle accelerator. By 1989, elements through $Z = 109$ had been discovered and authenticated. Authentication of a new element requires proof of Z, usually by chemical properties, or by emission of characteristic x rays. The correlation of its decay with the decay of its radioactive daughters is required, but characterization of the mass number (A) is not necessary.

In the mid-1960s, theoretical calculations predicted the possible existence of stable or long-lived isotopes for nuclei with approximately 114 protons and 184 neutrons. These **superheavy elements** would form an **island of stability** beyond the identified unstable transuranium elements. Thus far, however, these unusually stable elements have not been experimentally produced. Whether this is because production techniques are inadequate or because theoretical predictions of stability are incorrect is not clear.

The three original modes of radioactive decay characterized by Rutherford and others are not the only ones that can occur. In addition to the α-, β-, and γ-decay modes, many heavier elements undergo **spontaneous fission decay.** Some nuclei may emit protons or neutrons promptly after β decay. Decay involving the simultaneous emission of 2 or 3 particles has also been observed. ^{11}Li, with 3 protons and 8 neutrons, has a very high decay energy, and both 2-neutron and 3-neutron emission have been observed. ^{22}Al, with 13 protons and 9 neutrons, decays by emission of 2 protons. Some nuclei may decay by emission of much heavier particles. The emission of ^{14}C from ^{223}Ra was discovered by H. J. Rose and G. A. Jones in 1984. And β^--delayed triton (^{3}H or t) emission from ^{11}Li was observed by M. Langevi et al. in 1984. Radioactive decay modes are discussed in more detail in Chapter 2.

In addition to theoretical advances, new practical uses for radioisotopes and radioactive decay phenomena are continually being developed. Their use as diagnostic, treatment, and research tools in medicine and in pharmacology is expanding rapidly. These uses are discussed in Chapters 10 and 11.

1.2. FORCES IN MATTER AND THE SUBATOMIC PARTICLES

1.2.1. Forces in Nature

There are four basic forces in nature: gravitational, electromagnetic, strong, and weak. **Gravitational forces** were the first to be mathematically described, by Isaac Newton in the 1600s. In comparison to the other three forces, gravity is very weak, but it can act over very large distances. The force of gravity is responsible for the motions of celestial bodies and the motion of the universe itself. Gravity does not seem to play an important role in the microscopic structure of matter.

Electrical and **magnetic** forces are familiar to all of us. They are the forces that exist between charged or magnetic bodies. It was an important advance in physics when J. Maxwell in the 1800s found a way to describe both electrical and magnetic forces in a unified theory of electromagnetism. Electromagnetic forces are the second strongest of the four forces and, like gravity, act over very large distances.

The strong and weak forces are important in nuclear interactions. The **strong force** is, as its name implies, the strongest of the four basic forces. It is some 10^{38} times stronger than gravity. However, the strong force acts only over very short distances, around 1 fm (10^{-13} cm). The strong force is responsible for holding the protons and neutrons together in the nucleus. The **weak force** is weaker than either the strong or electromagnetic forces, and acts over even shorter distances (0.001 fm). The weak force is involved in a variety of nuclear decay processes.

Physicists would like to develop a quantum theory that could unify all the basic forces, but this has not yet been accomplished. Successful quantum theories have been proposed that unify the electromagnetic and weak forces (the **electroweak** force), and that describe the interactions of the strong force (**quantum chromodynamics**). No single theory, however, has yet unified the electromagnetic, weak, and strong forces into a **grand unified theory (GUT).** There is no successful quantum theory of gravity, nor has anyone found a way to unify gravity with the other three forces into a **Theory of Everything (TOE).**

1.2.2. The Subatomic Particles

A whole netherworld of subatomic particles has been discovered since the 1950s. Further characterization of these particles continues to be an active area of research today.

The subatomic particles may be divided into the **elementary particles** and the **composite particles.** Elementary particles are those with no known internal structure; that is, they are not themselves made of other particles. There are three groups of elementary particles: the **leptons**, the **quarks**, and the **elementary vector bosons.** The six leptons are the **electron, muon** **(μ),** and **tauon (τ),** and the corresponding **neutrino** for each (ν_u, ν_e, and ν_τ). Quarks have not as yet been isolated as free particles. There are six **flavors** of quarks. They have been given the names up, down, strange, charmed,

Table 1.3. Chart of Subatomic Particles

I. Elementary Particles	II. Composite Particles (Hadrons)
A. Leptons	A. Baryons
1. electron, electron neutrino	1. protons and neutrons
2. tauon, tauon neutrino	2. many others
3. muon, muon neutrino	
B. Quarks	B. Mesons
1. up, down	1. pions
2. strange, charmed	2. kaons
3. bottom, top	3. many others
C. Elementary Vector Bosons	
1. photons	
2. gluons	
3. intermediate vector bosons	

bottom (or beauty), and top (or truth). The top quark has not been experimentally observed yet, but its existence is well supported theoretically. Flavor refers to a property that is related to the amplitude of the photon coupling constant. Each flavor of quark has three **colors**. Color is a type of polarization similar to charge. The elementary vector bosons are the particles that mediate the forces that exist between subatomic particles. There are three of these: The **photons** mediate electromagnetic interactions, the **gluons** mediate the strong force between quarks, and the **intermediate vector bosons** mediate the weak force.

Composite particles (the **hadrons**) are composed of elementary particles. There are over 100 of these known and no attempt will be made to list all of them. The hadrons may be broadly divided into the **baryons** and the **mesons**. The baryons consist of 3 quarks confined in a "bag," while the mesons consist of a quark–antiquark pair. Protons and neutrons are examples of baryons. The mesons mediate the strong force between hadrons. Table 1.3 summarizes the classes of fundamental particles.

Each of the subatomic particles also has its **antiparticle**. The antiparticle has the same mass and spin number as the particle, but differs in either electric charge, color, or flavor. For example, the antiparticle of the electron is the positron. A positron has the same mass and spin as the electron, but has a positive electrical charge instead of a negative one.

1.3. NUCLIDES AND NATURAL DECAY CHAINS

There are many symbols and terms used in radiochemistry that are important to know. An extensive compilation entitled, "Glossary of Terms Used in Nuclear Analytical Chemistry" has been published by the International Union of Pure and Applied Chemistry (see Reading List for this chapter). Some of these are defined in the following section.

1.3.1. Nuclides and Symbols

A **nuclide** is an atomic species characterized by specific values of the **atomic number** (*Z*, the number of protons in the nucleus of the atom) and the *mass number* (*A*, the sum of the number of protons and neutrons in the nucleus). If the nuclide is radioactive, then the term **radionuclide** is used. A nuclide is symbolized as ${}^{A}_{Z}X$, where X is the chemical symbol for the element. The value for *Z* is often omitted, because it is uniquely identified by the element's symbol.

The distinctions among mass number, nuclidic (isotopic) mass, and atomic mass (or atomic weight) should be noted. The mass number (*A*) is always a whole number, because it is simply the sum of the number of protons and neutrons in the nucleus. The **nuclidic** (or **isotopic**) **mass** (***M***) refers to the mass of an atom of a *given nuclide* relative to the mass of a ${}^{12}C$ atom, which is set equal to exactly 12 **daltons**, or atomic mass units (amu). The **atomic mass** (also called the **atomic weight** of an *element*) refers to the weighted average of the masses of all the naturally occurring isotopes in a sample of the element. Another term dealing with nuclide masses is the **mass excess,** often also called the **mass defect (Δ).** This term represents the difference between the nuclidic mass and the mass number, or $M - A$. This concept is discussed further in Chapter 3. The distinctions among these terms may be illustrated by using nitrogen as an example. The mass number for the nuclide ${}^{14}N$ is 14, because its nucleus contains 7 protons and 7 neutrons. The nuclidic mass of a ${}^{14}N$ atom is 14.003074 daltons. The atomic mass of the element nitrogen as it occurs in nature (99.63% ${}^{14}N$ and 0.37% ${}^{15}N$) is 14.0067 daltons. The mass defect ($M - A$) for ${}^{14}N$ would be $14.003074 - 14 = 0.003074$.

1.3.2. Classification of Nuclides

Nuclides may be classified in several ways. One is according to stability; that is, does the nuclide undergo radioactive decay (a radionuclide), or is it stable to decay (a stable nuclide)? In practice, five different groups of nuclides may be defined:

1. *Stable nuclides* include those for which no radioactive decay has been observed to date. There are now 264 of these stable nuclides. However, this number continues to decrease as more sophisticated methods

of detecting very low activity levels are developed. Examples of stable nuclides include ^{12}C, ^{14}N, and ^{16}O.

2. *Primary natural radionuclides* are those found now in nature which are radioactive and have persisted on Earth from the origin of the solar system. These radionuclides have very long half-lives. There are about 26 of these known. Examples include ^{238}U (half-life = 4.47×10^9 years), ^{40}K (half-life = 1.28×10^9 years), and ^{87}Rb (half-life = 4.8×10^{10} years).
3. *Secondary natural radionuclides* are those found in nature that have been produced by the decay of the primary natural radionuclides, but have half-lives too short for them to have survived from the origin of the solar system. About 38 of these are known, including ^{226}Ra (half-life = 1600 years), and ^{234}Th (half-life = 24.1 days), both of which are continuously produced by the decay of ^{238}U.
4. *Induced natural radionuclides* are those found in nature that are constantly being produced by the action of cosmic rays on the earth's atmosphere. Approximately 10 of these are known. Two well-known examples are ^{3}H (tritium, half-life = 12.3 years), and ^{14}C (half-life = 5730 years). Both of these are used in radioactive dating methods (see Chapter 12). ^{14}C is produced by the action of cosmic-ray-produced neutrons on ^{14}N in the atmosphere, in the $^{14}N(n, p)^{14}C$ reaction. Tritium (t) is produced in the atmosphere by the $^{14}N(n, t)^{12}C$ reaction.
5. *Artificial radionuclides* are those that are man-made and do not occur to any significant extent in nature. Approximately 2000 of these are known, including ^{60}Co, ^{137}Cs, and ^{24}Na.

It is also possible to classify nuclides based on their atomic numbers and mass numbers. Isotopes are nuclides that have the same Z, but different A. That is, they have the same number of protons, but differing numbers of neutrons. An example of a set of isotopes would be the four commonly used isotopes of carbon: $^{11}_{6}C$ (radioactive), $^{12}_{6}C$ and $^{13}_{6}C$ (both stable), and $^{14}_{6}C$ (radioactive). The three isotopes of H have special names: protium ($^{1}_{1}H$), deuterium ($^{2}_{1}H$), and tritium ($^{3}_{1}H$). **Isobars** are nuclides having the same mass number (A), but different numbers of protons (Z). An example of a pair of isobars would be $^{14}_{6}C$ and $^{14}_{7}N$. **Isotones** are nuclides that have the same number of neutrons (N), but a different number of protons (Z). An example of a set of isotones is $^{15}_{8}O$, $^{14}_{7}N$, and $^{13}_{6}C$, all of which have seven neutrons. **Nuclear isomers** are nuclides having the same Z and A, but different states of nuclear excitation. For example, the nuclear isomers of ^{234}Pa are designated as ^{234m}Pa and ^{234g}Pa. The m stands for **metastable state** and the g

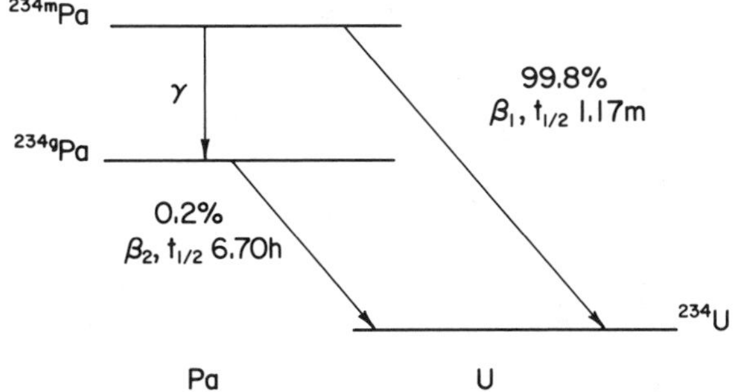

Figure 1.13. Decay scheme for ^{234}Pa; an example of nuclear isomers.

stands for **ground state.** This indicates that one isomer is in a nuclear excited state and the other in the lowest energy state for the specified nuclide. The decay from the metastable state to the ground state or any lower energy state is called an **isomeric transition.** This is discussed more in Section 2.3. The decay scheme for the isomers of ^{234}Pa is shown in Fig. 1.13.

1.3.3. Chart of the Nuclides

A **chart of the nuclides** is an important reference tool for the radiochemist. As its name implies, the chart contains spaces for all the isotopes of the elements, not just one box per element, as in the ordinary periodic table. A chart of the nuclides will have values for neutron number increasing left to right in the horizontal direction, and number of protons increasing upward in the vertical direction. Therefore, each horizontal row represents all the isotopes for an element, and each vertical row all the isotones for a given neutron number. Isobars are found along diagonal lines. A small section of a chart of the nuclides is shown in Fig. 1.14.

The individual boxes on the chart convey a variety of information, depending on the publisher of the chart. The shading of the box indicates the stability of the nuclide. Shaded or dark outlined boxes, for example, often indicate the stable nuclides. In some charts, the box may also use color schemes to indicate the neutron-absorption properties of stable nuclides, or the decay mode of the radionuclides. Other information that is often given would include the percent natural abundances of the stable isotopes, the nuclidic mass, modes of decay and half-lives (if radioactive), energies of emitted radiations, and spins and parities of the nuclear states. Typical boxes from a chart of the nuclides are shown in Fig. 1.15. Table 1.4 lists some of the symbols commonly used for particles in a chart of the nuclides.

		10	11	12	13	14	15	16	17	18	19	20
Z=17	Cl 35.453	β⁺ ε				Cl31 0.15s	Cl32 297ms	Cl33 2.51s	Cl34 32.2m	**Cl35 75.77**	Cl36 3x10⁵y	**Cl37 24.23**
Z=16	S 32.066				S29 0.19s	S30 1.18s	S31 2.55s	**S32 95.02**	**S33 0.75**	**S34 4.21**	S35 87.2d	**S36 0.02**
Z=15	P 30.974		P26 20ms		P28 270ms	P29 4.14s	P30 2.50m	**P31 100**	P32 14.3d	P33 25.3d	P34 12.4s	P35 47s
Z=14	Si 28.086	Si24 0.1ms	Si25 220ms	Si26 2.20s	Si27 4.14s	**Si28 92.23**	**Si29 4.67**	**Si30 3.10**	Si31 2.62h	Si32 100y	Si33 6.2s	Si34 2.8s
Z=13	Al 26.982	Al23 0.47s	Al24 2.07s	Al25 7.17s	Al26 7x10⁵y	**Al27 100**	Al28 2.25m	Al29 6.5m	Al30 3.69s	Al31 0.64s		
Z=12	Mg 24.305	Mg22 3.86s	Mg23 11.3s	**Mg24 78.99**	**Mg25 10.00**	**Mg26 11.01**	Mg27 9.45m	Mg28 21.0h	Mg29 1.3s	Mg30 0.33s	Mg31 0.25s	β⁻
	N=	10	11	12	13	14	15	16	17	18	19	20

Figure 1.14. Section of a typical chart of the nuclides. The numbers along the left side represent the atomic numbers (Z, the number of protons in the nucleus) and the weighted atomic masses for the elements as found in nature, relative to a ^{12}C atom = exactly 12.0 daltons. Each square represents a nuclide identified by chemical symbol and mass number (A). Vertical columns are labeled at the bottom with the number of neutrons in the nucleus (N). Directions for sequences of positron decays (β^+), electron capture decays (ε), and negatron decays (β^-) are indicated by arrows. Shaded squares represent stable nuclides and include numbers for the atomic abundances (atom percent in nature.) Other squares are for radionuclides with values given for their half-lives.

RADIONUCLIDE

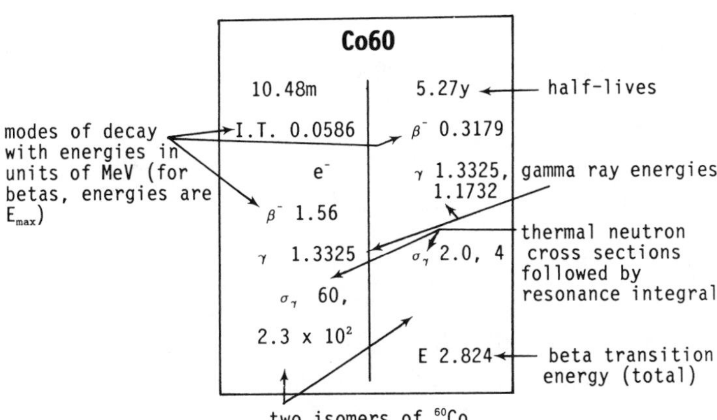

e^- = emission of a conversion electron
I.T. = decay via isomeric transition to the lower lying isomer of ^{60}Co
β^- = decay via emission of a negatron to ^{60}Ni
γ = emission of gamma rays following beta emission
E = total energy of the beta transition to ground state of the daughter

STABLE NUCLIDE

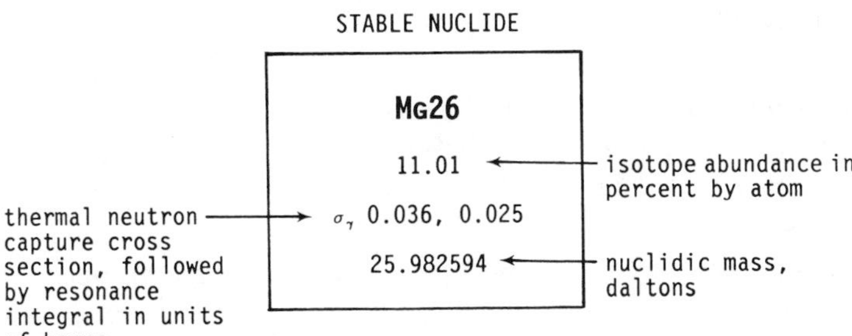

Figure 1.15. Information typically contained in a square of a chart of the nuclides for a radioactive and a stable nuclide.

Table 1.4. Symbols for Particles and Radiations

p	proton	^{1}H nucleus
n	neutron	Particle of zero charge and $A = 1$
d	deuteron	^{2}H nucleus
t	triton	^{3}H nucleus
α	alpha particle	^{4}He nucleus
e^-	electron	Electron from atomic orbitals
β^-	negatron	Negative electron emitted by the nucleus
β^+	positron	Positive electron emitted by the nucleus
γ	gamma ray	Electromagnetic energy (photon) emitted by the nucleus
x ray	x ray	Electromagnetic energy emitted as a result of electron transitions in the atomic orbitals
ν	neutrino	Emitted in β-decay processes (two types)

1.3.4. Natural Decay Chains

The existence of three naturally occurring decay chains has been noted earlier in this chapter. These chains originate from isotopes of uranium and thorium and culminate with isotopes of lead (Figs. 1.16–1.18). The three series are:

- ***The Thorium Decay Series (4n + 0).*** The parent primary natural radionuclide for this decay series is ^{232}Th. The series is given the designation **(4n + 0)** because the mass number of the parent, and of all the members of the decay chain, is evenly divisible by 4. The half-life of ^{232}Th is 1.4×10^{10} years, and the final daughter nuclide of the series is ^{208}Pb. There are six α particles and four β^- particles emitted in this series.
- ***The Uranium Decay Series (4n + 2).*** The parent of this series is ^{238}U and the final product is ^{206}Pb. There are eight α particles and six β particles emitted in the series. The half-life of the parent is 4.5×10^9 years.
- ***The Actinium Decay Series (4n + 3).*** The parent of this decay chain is ^{235}U, which has a half-life of 7.04×10^8 years. The end product is ^{207}Pb. Seven α particles and four β particles are emitted in the series.

It is possible to calculate the number of α and β particles emitted in the overall decay chain. Mass number changes in the natural decay chains will be due entirely to alpha emission. By subtracting the mass number (A) of the final daughter from the parent, the total change in mass number is obtained. Each α particle removes 4 **nucleons**, so the number of α particles

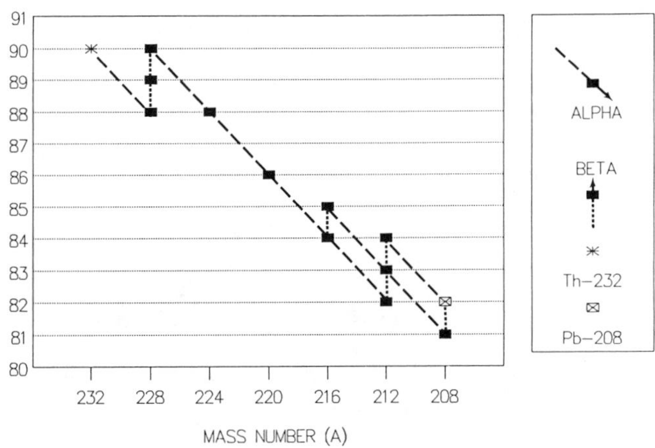

Figure 1.16. Thorium natural decay series.

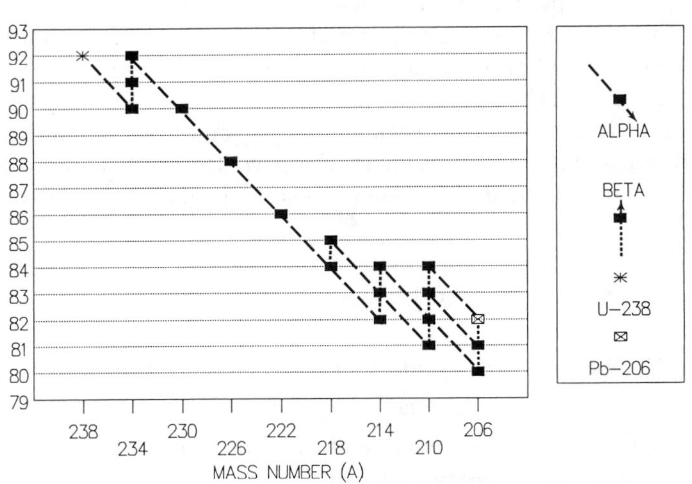

Figure 1.17. Uranium natural decay series.

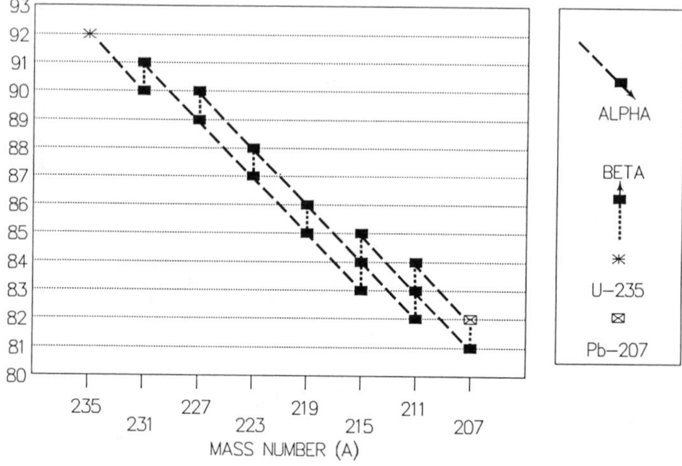

Figure 1.18. Actinium natural decay series.

emitted is calculated by dividing this mass difference by 4. The emission of β particles does not change the mass number, but does change the atomic number (Z). Emission of a negative electron will cause Z to increase by one. The emission of each α will decrease Z by 2 units. Thus, after calculating the number of alpha particles emitted, multiplication by 2 will give the number of protons removed from the parent. Although we *remove* two Z units in each α decay, we *add* one Z unit in each β^- decay. Therefore, the difference in Z between the parent and a given daughter in a chain must equal 2 times the number of α particles emitted minus the number of β particles emitted. An example of this calculation for the $4n + 0$ series is shown below:

$$^{232}_{90}\text{Th} \longrightarrow \longrightarrow \longrightarrow \cdots \, ^{208}_{82}\text{Pb}$$

Number of α particles lost: $(232 - 208)/4 = 24/4 = 6$
Number of Z units removed by α decay: $6 \times 2 = 12$
Difference in Z of parent and daughter: $90 - 82 = 8$
Number of β^- decays: $12 - 8 = 4$

1.3.5. An Extinct Natural Decay Chain

Another decay series may have existed early in the history of the solar system. It is the ***neptunium decay series (4n + 1).*** The neptunium decay series is not included among the naturally occurring decay chains because the parent, ^{237}Np, has a half-life of only 2.14×10^6 years. Thus, this chain would have completely decayed since the earth's formation, 4.7×10^9 years ago. Hence, the series is no longer found naturally on earth, but members of the series may be produced artificially. The final product of this chain is ^{209}Bi. Seven α particles and four β particles are emitted in the series.

In all of these decay chains, γ rays may also be associated with any of the decay steps. However, γ-ray emission does not affect the values of Z or A.

TERMS TO KNOW

Alpha (α)
Atomic mass
Atomic number
Baryons
Beta (β)
Cyclotron
Daughter
Decay chains
Electron
Elementary particles
Elementary vector bosons
Gamma (γ)
Hadrons
Isobars
Isomeric transition
Isotones
Isotopes
Leptons
Mass excess
Mass number
Mesons
Neutron
Neutron activation
Nuclear chemistry
Nuclear fission
Nuclear isomers
Nucleons
Nucleus
Nuclide
Parent
Positron
Proton
Quarks
Radioactive
Radiochemistry
Radionuclide
Transmutation
Transuranium

READING LIST

Armbruster, P., and Münzenberg, G., Creating superheavy elements. *Scientific American* **260**:66 (1989).

Arnikar, H. J., *Essentials of Nuclear Chemistry,* 2d ed. New York: Wiley, 1987. [general reference for this, and subsequent chapters]

Choppin, G. R., and J. Rydberg, *Nuclear Chemistry.* Oxford: Pergamon, 1980. [general reference for this, and subsequent chapters]

Close, F. E., *The Cosmic Onion: Quarks and the Nature of the Universe,* New York: American Institute of Physics, 1986. [subatomic particles]

Curie, E., *Madame Curie, a Biography.* New York: Doubleday, 1939. [history]

Ehmann, W. D., J. R. Robertson, and S.W. Yates, Nuclear and radiochemical analysis, *Anal. Chem. Fund. Rev.* **62**:50R (1990). [review of current literature in the field]

Fermi, L., *Atoms in the Family.* Chicago: University of Chicago Press, 1954. [history]

Friedlander, G., J.W. Kennedy, E. S. Macias, and J. M. Miller, *Nuclear and Radiochemistry,* 3d ed. New York: Wiley, 1981. [general reference for this, and all succeeding chapters]

International School of Subnuclear Physics (22nd, Erice, Sicily, A. Zichichi, ed.), *Quarks, Leptons, and Their Constituents,* New York: Plenum, 1988. [subatomic particles]

International Union of Pure and Applied Chemistry, Glossary of terms used in nuclear analytical chemistry—(Provisional), Prepared for publication by M. de Bruin, *Pure Appl. Chem.* **54**:1533 (1982). [nomenclature]

National Research Council (U.S.), Elementary Particle Physics Panel, *Elementary Particle Physics.* Washington, DC: National Academy Press, 1986. [forces and subatomic particles]

National Research Council (U.S.), Panel on Nuclear Physics, *Nuclear Physics,* Washington, DC: National Academy Press, 1986. [forces and subatomic particles]

Nuclides and Isotopes, 14th ed. San Jose, CA: General Electric Co., 1989. [nuclear data]

Oliphant, M., *Rutherford: Recollections of the Cambridge Days.* New York: Elsevier, 1972. [history]

Rafelski, J., and Jones, S. E., Cold nuclear fusion. *Scientific American* **257**:84 (1987).

Rhodes, R., *The Making of the Atomic Bomb.* New York: Simon & Schuster, 1986. [history of the discovery of nuclear fission and the Manhattan Project in World War II]

Rose, H. J., and G. A. Jones, *Nature(London)* **302**:245 (1984). [^{14}C decay]

Seaborg, G.T., ed., *Transuranium Elements: Products of Modern Alchemy,* Series on Benchmark Papers in Physical Chemistry and Chemical Physics, New York: Academic, 1978. [transuranium elements]

Seaborg, G.T., and W. D. Loveland, eds. *Nuclear Chemistry Benchmark Papers.* New York: Van Nostrand Reinhold, 1982. [history]

Segrè, E., *Enrico Fermi, Physicist.* Chicago: University of Chicago Press, 1970. [history]

Starke, K., The detours leading to the discovery of nuclear fission, *J. Chem. Ed.* **56**:771–775 (1979). [history]

Tuli, J. K., *Nuclear Wallet Cards.* Upton, NY: Brookhaven National Laboratory, 1990.

Vèrtes, A., and I. Kiss, *Nuclear Chemistry.* Amsterdam: Elsevier, 1987. [general reference for this, and subsequent chapters]

EXERCISES

1. Identify the stable nuclide that is the end product in the natural decay chain that includes the secondary natural radionuclide ^{214}Bi.
2. Calculate how many α particles and negatrons would be emitted in the decay of $^{238}_{92}U$ to $^{210}_{82}Pb$.
3. Calculate the number of neutrons, protons, and orbital electrons in the following atomic species: $^{235}_{92}U$ neutral atom, $^{14}_{6}C$ neutral atom, $^{40}_{19}K^{+}$ ion, $^{18}_{8}O^{2-}$ ion.
4. What information did Anderson derive from measuring the radius of curvature of a positron in a magnetic field?
5. Which of the following pairs are isotopes: ^{14}C and ^{14}N, ^{11}C and ^{14}C, ^{235}U and ^{238}U, ^{39}K and ^{40}Ca?
6. Identify which of the following nuclides are isotones: ^{50}Cr, ^{48}Ti, ^{50}V, ^{48}Sc, ^{46}Ca, ^{48}Ca.
7. With the assistance a chart of the nuclides, write the nuclear equation for the production of ^{238}Pu by α-particle bombardment of ^{235}U.
8. Use a chart of the nuclides to classify the following nuclides according to stability: **(a)** ^{4}He **(b)** ^{222}Rn **(c)** ^{232}Th **(d)** ^{67}Ga **(e)** ^{3}H
9. Identify the parent of the decay chain that includes the nuclide ^{234}Th.
10. Write the "shorthand" form for this nuclear reaction:

$$^{238}U + ^{12}C \longrightarrow ^{246}Cf + 4n$$

11. Provide the following data for chlorine: **(a)** M for ^{37}Cl **(b)** A for ^{37}Cl **(c)** atomic mass of naturally occurring Cl

CHAPTER

2

TYPES OF RADIOACTIVE DECAY

Radioactive decay is the spontaneous emission of particles or electromagnetic radiation from an atom due to a transition within its nucleus. The rate of decay of atoms of a particular radioactive nuclide is, in general, unaffected by the factors that control the rates of chemical reactions, such as temperature, pressure, physical state, and chemical nature. A rare exception to this generalization will be noted in the section on electron capture decay. Radioactive decay is always an exoergic process. A nucleus will be unstable to a particular mode of decay if the mass(es) of the product(s) of the decay are less than that of the parent. The usual conservation laws are followed in decay processes. Among these are conservation of total energy, linear and angular momentum, charge, and mass number.

The energy unit commonly used to measure the energy changes involved in radioactive decay is the electronvolt (eV). One electronvolt is the energy acquired by a particle of unit electrical charge (1.6×10^{-19} coulombs, or 4.8×10^{-10} electrostatic units, esu), when accelerated through a potential difference of one volt. Other common units are the keV (10^3 eV), MeV (10^6 eV), and GeV (10^9 eV). In nuclear work, calculations are usually done on a per-atom basis, rather than on a mole basis. Hence, electronvolt units are more convenient than the units of ergs or joules. The latter are commonly used in chemistry and atomic physics when dealing with mole quantities. The most commonly used relationships among electronvolts, ergs, and joules are

$$1 \text{ eV} = 1.60219 \times 10^{-12} \text{ erg} = 1.60219 \times 10^{-19} \text{ J}$$

$$1 \text{ MeV} = 1.60219 \times 10^{-6} \text{ erg} = 1.60219 \times 10^{-13} \text{ J}$$

As noted previously, a variety of natural decay modes exist. The three most common are alpha, beta, and gamma decay. Less common decay modes that also are discussed briefly in this chapter include spontaneous fission, delayed neutron and proton emission, two-proton decay, composite particle emission, and double β decay.

2.1. ALPHA-PARTICLE DECAY

In α decay, a nucleus emits a particle consisting of two protons and two neutrons. This is, of course, the nucleus of the ^{4}He atom. This emission results in a loss of 4 mass number units (A) and 2 atomic number units (Z) from the parent radionuclide (E), to produce a daughter nuclide (F). A general equation for α-particle decay is

$$^{A}_{Z}\mathrm{E} \longrightarrow {}^{A-4}_{Z-2}\mathrm{F} + {}^{4}_{2}\mathrm{He} + (\gamma \text{ rays})$$

Gamma rays may or may not be emitted subsequent to the α decay. An example of α decay is given in the following equation for the decay of ^{210}Po:

$$^{210}_{84}\mathrm{Po} \longrightarrow {}^{4}_{2}\mathrm{He} + {}^{206}_{82}\mathrm{Pb} + \gamma$$

The half-life of ^{210}Po is 138.4 days (d), and the energy of the α particle (E_α) emitted is 5.304 MeV.

Nuclei with masses greater than 150 are theoretically unstable to α emission, but most naturally occurring α emitters have Z greater than 83. A few lighter naturally occurring α emitters are found in the rare-earth region of the periodic table (e.g., ^{144}Nd, ^{147}Sm, and ^{148}Sm).

The measured half-life range for natural α emitters is very large, from microseconds to more than 10^{15} years, but the range in α-particle energies is quite small, typically 5–7 MeV for the heavier parent radionuclides and 1–2.5 MeV for the rare-earth α emitters. There is an inverse relationship between the α energies and their half-lives. This was first recognized by Rutherford in 1906, and later was formally expressed by Geiger and Nuttall. A simple statement of this relationship is known as the **Geiger–Nuttall rule:**

$$\log t_{1/2} \propto \frac{1}{\log E_\alpha} \tag{2.1}$$

Thus, α emitters that have higher energy α particles have shorter half-lives. For example, ^{216}Rn, which emits α particles with an energy of 8.05 MeV, has a half-life of 45 μs. ^{144}Nd, with an α-particle energy of only 1.83 MeV, has a half-life of 2.1×10^{15} years (y). This relationship is illustrated by looking at the α emitters in a single sequential decay chain, such as members of each of the ^{235}U, ^{238}U, and ^{232}Th natural decay chains. In these decay chains, the proportionality represented in the Geiger–Nuttall rule may be replaced by equations containing specific proportionality constants characteristic of each decay chain. The energies lost in α decay processes will, in many cases, exhibit an inverse relationship (Fig. 2.1) to the mass numbers for isotopes of any given element. For example, ^{222}Rn has $E_\alpha = 5.5$ MeV, and ^{214}Rn has $E_\alpha = 9.04$ MeV. Exceptions to this relationship occur at very stable nuclear configurations associated with the nuclear "magic numbers" (Chapter 3).

Alpha particles are emitted with discrete energies; that is, only α particles with a few specific energies are emitted by a given radionuclide. A plot of the number of α particles versus their energy would be expected to yield a simple line spectrum. The observed energy spectrum contains broadened peaks, due to the uncertainty principle and experimental factors. An energy spectrum for an α emitter is shown in Fig. 2.2. A simple **energy level diagram,** or **decay scheme** is shown in Fig. 2.3. In a decay scheme, a horizontal line for the parent radionuclide is drawn at the top of the diagram. Daughter nuclide energy levels are represented by horizontal lines drawn below the initial parent level. The relative vertical positions of the lines

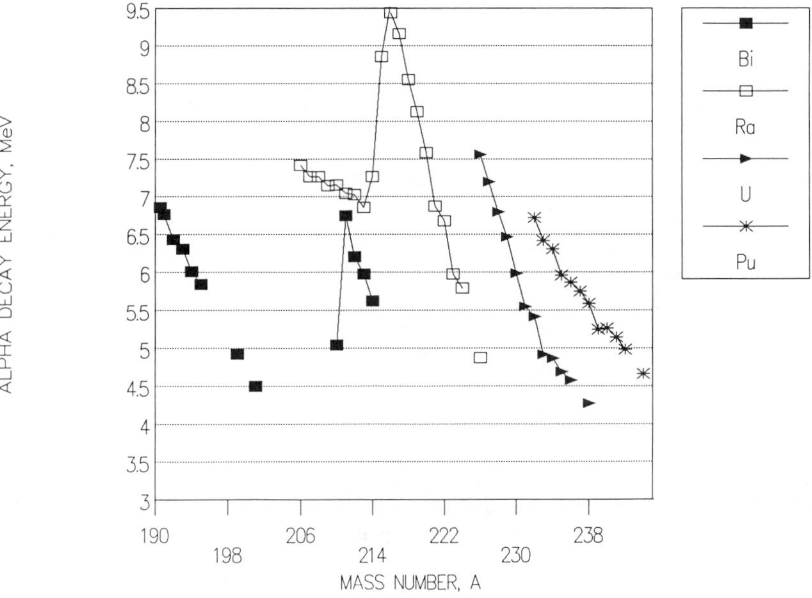

Figure 2.1. The relationship between α-decay energy and mass number among α-emitting isotopes of a given element. Where α decay energies have not been experimentally measured, no values are listed. Data are from E. Browne and R. B. Firestone, *Table of Radioactive Isotopes,* 1986. Deviations from the inverse relationship occur at closed neutron and proton "shells."

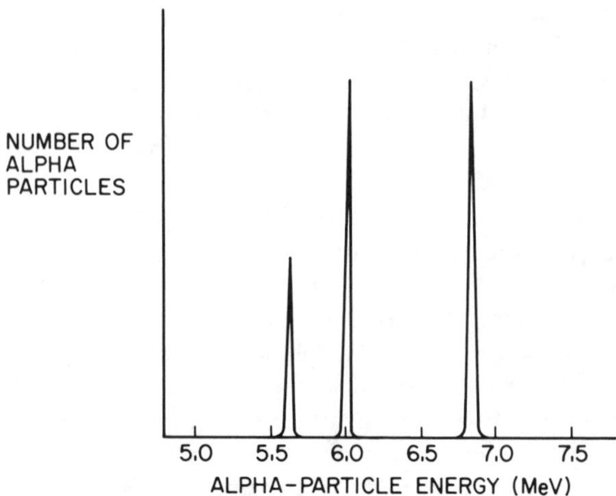

Figure 2.2. Energy spectrum of α particles emitted according to the decay scheme in Fig. 2.3. Ideally, a simple line spectrum would be produced with each line corresponding to the energy of the emitted α particle. In reality, the lines have a certain "width" due to the uncertainty principle and experimental factors.

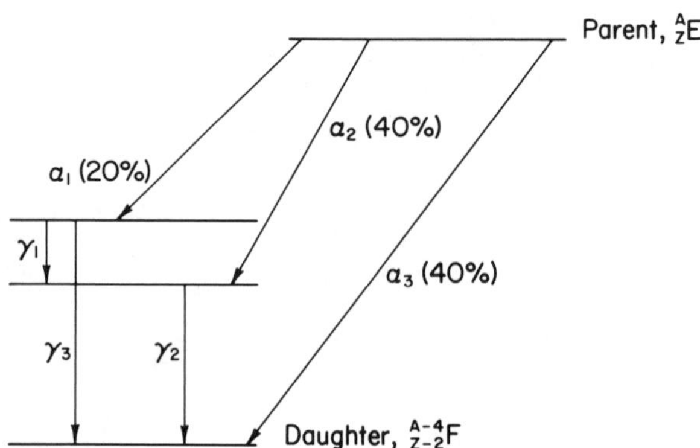

Figure 2.3. Decay scheme for a radionuclide undergoing α decay. Three α particles of different energies and intensities are emitted in this illustration. Gamma rays are emitted in transitions among the energy levels in the daughter nuclide.

can give qualitative information on energy level differences; that is, the lowest energy levels lie lowest in the diagram and a large spacing between the lines reflects a large energy gap. The Z value increases from left to right for each vertical column of energy levels. If a decay mode results in a decrease of atomic number, the diagonal line(s) representing decay to the

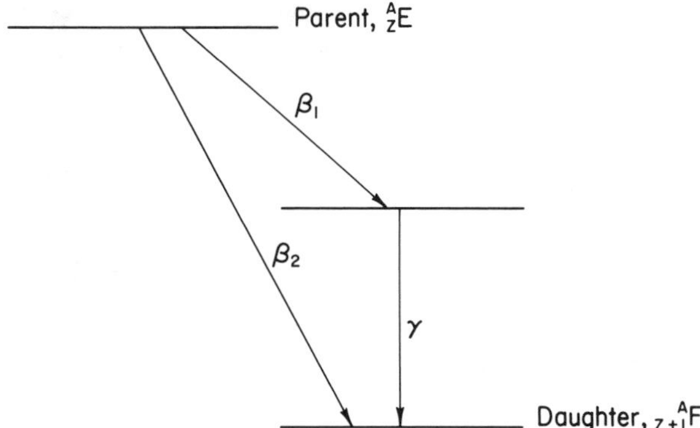

Figure 2.4. Decay scheme for a radionuclide undergoing negatron decay. Two β decay paths are shown. A γ ray is emitted from an excited state of the daughter in one decay path.

energy level(s) of the daughter are drawn to the left of the parent. This is the case for α decay (Fig. 2.3). If the decay results in an increase in atomic number, diagonal lines representing the decay are drawn to the right. The latter case is illustrated by β^- decay (Fig. 2.4).

It is possible for one radionuclide to experience several different modes of decay, or to have one mode populating several different energy levels of the daughter nuclide. In this case, diagonal lines in the decay scheme are drawn from the parent to each appropriate daughter, or level, and the percentage of all parent decays occurring by each given route is given. The half-life of the parent may also be included in the energy level diagram. In Fig. 2.3, the α decay of a parent to three different states of a daughter nucleus is shown. Daughter nuclides produced in excited states then decay to a lower energy state, or to the ground state, by emission of γ rays. The experimental α-particle spectrum shown in Fig. 2.2 has three peaks, one for each α transition that the parent undergoes. The height of each peak, or more precisely the baseline-corrected net area under a peak, is representative of the relative percentage of parent decays occurring by each branch.

The energy actually imparted to the α particle is not exactly equal to the total energy of the transition. This is because the daughter nuclide will recoil upon emission of the α particle. The **recoil energy** imparted to the daughter nucleus is a consequence of the need to conserve momentum in the system. This is analogous to the recoil of a rifle when a shot is fired. The total transition energy (E_{trans}) is given by the sum of the α-particle energy and the energy imparted to the recoil daughter nucleus (E_{recoil}):

$$E_{\text{trans}} = E_\alpha + E_{\text{recoil}} \tag{2.2}$$

The amount of energy imparted to the recoil particle (E_{recoil}) can be calculated in a simple manner. Both particles (the α and the recoil nucleus) have a kinetic energy (E) given by:

$$E = \tfrac{1}{2}mv^2 \tag{2.3}$$

where m is the mass of the particle and v is its velocity.

If both sides of this equation are multiplied by $2m$, we have

$$2mE = m^2v^2 = (mv)^2 \tag{2.4}$$

Momentum (p) is defined as

$$p = mv \tag{2.5}$$

and

$$p^2 = m^2v^2 = (mv)^2 \tag{2.6}$$

$$2mE = p^2 \tag{2.7}$$

Therefore, by the *law of conservation of momentum,*

$$p_\alpha = p_{recoil} \tag{2.8}$$

Using the symbols E_α and E_{recoil} for the two kinetic energies, we obtain

$$2m_\alpha E_\alpha = 2m_{recoil} E_{recoil} \tag{2.9}$$

Solving for the energy of the recoil nucleus gives

$$E_{recoil} = (m_\alpha/m_{recoil})E_\alpha \tag{2.10}$$

Example. Calculate the recoil energy and total transition energy for a 5.00-MeV α emitter with a daughter of mass number equal to 200 (assume for this calculation that the masses are exactly equal to mass numbers of the parent radionuclide and the α particle):

$$E_{recoil} = (4/200)(5.00) = 0.10 \text{ MeV}$$

$$E_{trans} = E_\alpha + E_{recoil} = 5.00 \text{ MeV} + 0.10 \text{ MeV} = 5.10 \text{ MeV}$$

2.2. BETA DECAY

A decay process is included under the heading of β decay if it results in a change in the Z of the nuclide with no change in A. There are three common processes that will be discussed under the heading of β decay: (1) negatron decay (β^-), (2) positron decay (β^+), and (3) electron capture decay (EC, or ε).

Unlike α decay, which occurs primarily among nuclides in specific regions of the periodic table, β decay is possible for certain isotopes of all elements. A single β-decay process will occur if the change of a proton to a neutron, or the change of a neutron to a proton in the nucleus of the parent will yield a daughter nuclide with greater stability (lower mass) than the parent radionuclide.

2.2.1. Negatron Decay

In negatron decay, negative electrons (β^- particles) are emitted from the nucleus. A general equation for the decay of radionuclide E to a daughter, F, which has a Z value one unit larger than that of E is

$$^{A}_{Z}\mathrm{E} \longrightarrow {}^{A}_{Z+1}\mathrm{F} + \beta^- + \bar{\nu} + (\gamma \text{ rays})$$

where $\bar{\nu}$ is an antineutrino, whose presence will be explained shortly. Some specific examples of negatron decay are

$$^{90}_{38}\mathrm{Sr} \longrightarrow \beta^- + \bar{\nu} + {}^{90}_{39}\mathrm{Y} \qquad t_{1/2} = 29.1 \text{ y}$$

$$^{32}_{15}\mathrm{P} \longrightarrow \beta^- + \bar{\nu} + {}^{32}_{16}\mathrm{S} \qquad t_{1/2} = 14.3 \text{ d}$$

Neutron-rich nuclides (those that have a large N/Z ratio) are most likely to undergo negatron decay.

Beta-decay processes show a wide range of both half-life and decay energies. The relationship between half-life and decay energy for β decay is not so simple as for α decay, because β decay rates depend on several other properties, such as spin and parity changes. It is broadly true, however, that larger decay energies are associated with shorter half-lives.

A simple decay scheme for a radionuclide undergoing negatron decay is shown in Fig. 2.4. In β^- decay, Z increases by 1 unit, so the daughter nuclide is to the right of the parent. It is important to note the distinct difference between α- and β-decay processes with respect to the energy spectra of the emitted particles (Figs. 2.2 and 2.5). The electrons emitted from the

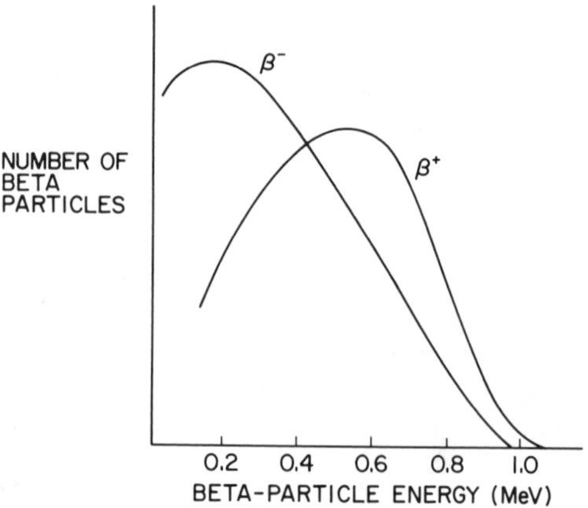

Figure 2.5. Energy spectra for negatrons and positrons in β decay. Each spectrum has a maximum β energy, E_{max}, which corresponds to the energy of the specific transition, corrected for recoil energy imparted to the daughter nuclide. The average negatron energy is approximately one-third of E_{max}, while the average positron energy is somewhat larger.

nucleus in β-decay processes do not have fixed (discrete) energies. In β decay (Fig. 2.5) there is a continuous distribution of particle energies, from 0 up to a maximum value, E_{max}. Spectra obtained for emission of negative (negatron decay) or positive electrons (positron decay) are similar. E_{max} represents the energy difference between the initial and final nuclear states, corrected for the relatively small recoil energy of the daughter nuclide.

In early studies of β decay, the observed β-particle spectrum appeared to violate laws of physics regarding conservation of angular momentum and energy. These two problems led W. Pauli, in 1930, to postulate that another particle must also be emitted in β decay. This particle, the **antineutrino** ($\bar{\nu}$), would have no charge, no magnetic moment, a zero or near zero rest mass, and a spin quantum number of $\frac{1}{2}$. The reason that an antineutrino is emitted in negatron decay, rather than a neutrino, is that leptons must always be conserved in nuclear reactions. Thus, if one begins a decay process with no leptons on the reactant side, the product side must have no leptons, or a lepton (the β^-) and an antilepton (the $\bar{\nu}$). In positron decay and electron capture decay, the extra particle emitted is called simply the **neutrino** (ν). The presence of the antineutrino would allow the conservation laws to be valid, and the continuous β-particle spectrum could be explained by assuming that, in a decay, the neutrino and the β-particle each carry off part of the total energy of the nuclear transition and momentum is also conserved:

$$E_{\text{trans}} = E_{\text{negatron}} + E_{\text{antineutrino}} + E_{\text{recoil}} \tag{2.11}$$

Neutrinos (*note:* the term neutrino is commonly used to describe both the neutrino and the antineutrino) proved to be quite elusive, and it was not until 1953 that F. Reines and C. L. Cowan, Jr., detected them by observing their interactions with protons to form a neutron and a positron:

$$\bar{\nu} + \mathrm{p} \longrightarrow \mathrm{n} + \beta^+$$

In their experiment, a tank of water containing $^{113}CdCl_2$ was surrounded by detectors and placed next to a nuclear fission reactor. A few of the antineutrinos emitted in the fission-product decay processes occurring in the reactor would interact with the hydrogen in the water to produce neutrons, as shown above. The neutrons would then be captured by ^{113}Cd, which has a very high probability of interacting with neutrons. This interaction results in the prompt emission of a very high-energy γ ray. The positrons (β^+) produced simultaneously in the reaction would interact with a neighboring extranuclear electron in an **annihilation process** that would produce two 0.511-MeV γ rays (more correctly called annihilation photons). Simultaneous detection (a triple coincidence event) of these three γ rays (the very high-energy ^{113}Cd prompt capture γ and the two annihilation photons) would serve to demonstrate that an antineutrino had interacted with a proton. Reines and Cowan observed a count rate of 2.88 events/h above background, which was sufficient to experimentally demonstrate the existence of the antineutrino. As previously noted in Table 1.3, there are actually three classes of neutrinos (electron, tauon, and muon) and each has its antiparticle. There is much current interest in neutrino research, especially in the fields of astrophysics and cosmology (Chapter 13).

2.2.2. Positron Decay

In this type of β decay, positive electrons (β^+) are emitted from the nucleus. A general equation for the decay of parent radionuclide E to daughter F which has a Z one unit less than that of the parent is

$${}^{A}_{Z}\mathrm{E} \longrightarrow {}_{Z-1}^{A}\mathrm{F} + \beta^+ + \nu + (\gamma)$$

Although positron decay results in a decrease in Z of one unit, no change occurs in A. In this decay mode, neutrinos (ν) are emitted. Some examples of positron decay are

$${}^{22}_{11}\mathrm{Na} \longrightarrow {}^{22}_{10}\mathrm{Ne} + \beta^+ + \nu \qquad t_{1/2} = 2.605\ \mathrm{y}$$

$${}^{11}_{6}\mathrm{C} \longrightarrow {}^{11}_{5}\mathrm{B} + \beta^+ + \nu \qquad t_{1/2} = 20.3\ \mathrm{min}$$

The decay scheme for positron decay is similar to that for negatron decay (Fig. 2.4), except that the decay lines to the daughter are drawn to the left of the column containing the parent, because Z decreases by 1 in the decay. The β^+ energy spectrum, like that for negatron decay, is continuous. The transition energy in this case is split between the positron and the neutrino. The energy spectrum for the positrons emitted will in most cases show fewer low-energy particles than a negatron spectrum with the same maximum energy. This is due to the acceleration provided by the repulsive nuclear Coulomb effect on the emitted positrons.

Positron decay occurs primarily among proton-rich nuclides (i.e., the N/Z ratio is small). This situation occurs rarely among the heavier nuclides, but is more common among the lighter nuclides. In positron decay, a proton is converted to a neutron in the parent nucleus, a positive electron is emitted by the nucleus, and an orbital electron originally present in the parent atom is lost to form the neutral daughter atom. In the original Dirac theory this process could be visualized as the creation of a positron–electron pair in the nucleus from the available transition energy. The energy equivalent to the mass of one electron is 0.511 MeV, so a minimum of 1.02 MeV is needed to create the masses of two electrons. Therefore, positron decay is possible only when the energy of the transition is greater than 1.02 MeV.

The fate of a positron after emission from the nucleus is conversion to pure energy in a process called **positron annihilation.** After the positrons have slowed to energies comparable to that of surrounding extranuclear electrons, they will interact with the negative electrons and the two interacting electron masses will be converted to energy. Usually, this energy is emitted in the form of two 0.511-MeV photons emitted at 180° to each other. Less common processes of positron annihilation include the formation of one, three, or zero annihilation photons. Which of these events actually occurs depends on the spin orientations of the positron–electron pair. If the spins are parallel, a triplet state is formed (spin quantum number of 1). If the spins are antiparallel, a singlet state forms (spin quantum number of 0). The number of photons emitted must be even for the singlet state and odd for the triplet state, because of parity conservation requirements. The probability of 2-photon versus 3-photon annihilation is approximately 372:1.

It is possible that a positron–electron pair will form a very short-lived bound state similar to an atom, called a **positronium (Ps)** "atom." The likelihood of formation of Ps is related to the medium in which the e^+–e^- pair is found. Positronium may be considered as a light "isotope" of hydrogen, with the positron substituting for the nuclear proton. The form of Ps in which the spins of the e^+ and e^- are parallel is called **ortho Ps;** antiparallel spins result in **para Ps.** Ortho Ps has a lifetime of about 10^{-7} s; the para form has a lifetime of only about 10^{-10} s. In spite of their short lifetimes, the chemistry of Ps atoms has been investigated extensively and their reactions have been characterized.

2.2.3. Electron Capture (EC or ε)

Electron capture is a decay mode in which an orbital atomic electron is captured by the excited nucleus. The general equation for the decay of parent radionuclide E to daughter nuclide F which has a Z one unit less than that of the parent is

$$ {}^{A}_{Z}\mathrm{E} \longrightarrow {}^{A}_{Z-1}\mathrm{F} + \text{x rays or Auger electrons} + \nu + \text{inner bremsstrahlung} + (\gamma) $$

Inner bremmstrahlung is a continuous spectrum of very low-intensity electromagnetic energy that is emitted in all β-decay processes. It constitutes some of the energy normally attributed to emission of neutrinos. **Auger electrons** are relatively low-energy atomic orbital electrons that may be emitted as an alternative to x-ray emission.

Some examples of EC are

$$ {}^{172}_{71}\mathrm{Lu} \longrightarrow {}^{172}_{70}\mathrm{Yb} + \text{x rays} + \text{Auger electrons} + \nu \qquad t_{1/2} = 6.70\ \mathrm{d} $$

$$ {}^{188}_{78}\mathrm{Pt} \longrightarrow {}^{188}_{77}\mathrm{Ir} + \text{x rays} + \text{Auger electrons} + \nu \qquad t_{1/2} = 10.2\ \mathrm{d} $$

It is important to note that EC produces the same daughter nuclide that would be produced by positron decay; that is, Z is decreased by one unit with no change in A. Electron capture decay is the only possible decay mode to this daughter when the transition energy is less than 1.02 MeV. This is because positron emission requires at least this amount of available energy ($2m_e$). When both modes of decay are energetically feasible, EC is favored over positron emission in cases where (1) the transition energy is relatively low, and (2) the Z of the parent is high. The latter effect is explained by noting that high-Z elements have a greater electron population density in the field of the nucleus, and greater electron density enhances the probability of capture by the nucleus. Capture of a K electron ($n = 1$ principal atomic quantum level) is most common, because these electrons have the highest probability distribution (population density) in the nuclear region. This process is termed **K capture.** L capture is also observed and becomes more probable for the heavier elements. Capture of higher-quantum-level orbital electrons may occur, but these processes are less probable. Electron capture processes were not discovered until 1938, perhaps because this decay does not produce any easily detectable unique nuclear radiation. As noted earlier, electron capture processes can be an exception to the rule that decay processes are not measureably affected by chemical factors. Because electron capture involves extranuclear electrons, the chemical state of a radioactive atom can affect decay rates slightly. For example, a difference of 0.08% in the electron capture decay rates of a given amount of ^{7}Be as a free metal, or in chemical combination as the BeF_2 salt, have been observed. For ^{90m}Nb, there is a 3.6% difference in EC decay rates between Nb metal and the NbF_5 complex.

Neutrinos emitted in EC decay are different from those emitted in other β-decay processes in that they are **monoenergetic**. They do not have a distribution of energies like neutrinos emitted in positron decay. Their energy is equal to the transition energy minus the binding energy of the electron captured by the nucleus. This binding energy is quite small compared to the transition energy and can usually be ignored. Neutrinos cannot readily be detected, so the EC process is most often detected by means of the inner bremsstrahlung radiation that is emitted. This radiation is emitted as the electron in the process of being captured is accelerated toward the nucleus. Determination of the upper limit of this inner bremsstrahlung energy distribution can be used to directly measure the EC transition energy.

The other two products in electron capture decay, the x rays and the Auger electrons, are the results of atomic electron rearrangements following nuclear deexcitation. The capture of an orbital electron by the nucleus results in a vacancy in the K or L atomic shell. This vacancy is quickly filled by another orbital electron falling from a higher shell. The potential energy lost by this electron as it falls can either be emitted as x rays or can cause the ejection from the atom of another electron in a higher shell. These secondary electrons of relatively low energy (much less than typical energies of electrons emitted in negatron decay) are the Auger electrons. There are, therefore, two competing processes that may take place subsequent to EC in an atom:

- ***Fluorescence yield*** (ω) is a measure of the fraction of the EC vacancies filled with subsequent emission of x rays.
- ***Auger yield*** is a measure of the fraction of EC vacancies filled with emission of Auger electrons.

The fluorescence yield generally increases with increasing Z, as illustrated in Fig. 2.6.

2.3. GAMMA DECAY

In γ decay, electromagnetic radiation is emitted as a nucleus undergoes transitions from its higher excited state(s) to lower excited state(s), or its

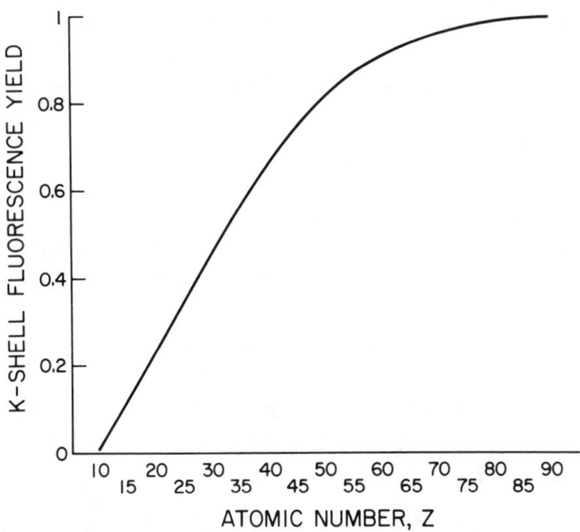

Figure 2.6. Variation of fluorescence yield with Z in electron capture.

ground state. The general equation for γ decay, where E^* signifies the higher energy state, is

$$ {}^{A}_{Z}\mathrm{E}^* \longrightarrow {}^{A}_{Z}\mathrm{E} + \gamma $$

Some examples of γ decay are

$$ {}^{110m}_{47}\mathrm{Ag} \longrightarrow \gamma + {}^{110}_{47}\mathrm{Ag} \qquad t_{1/2} = 249.8\ \mathrm{d} $$

$$ {}^{115m}_{49}\mathrm{In} \longrightarrow \gamma + {}^{115}_{49}\mathrm{In} \qquad t_{1/2} = 4.5\ \mathrm{h} $$

Note that γ decay results in no changes in either Z or A.

The excited state of the nucleus and the lower-energy daughter state that is reached as a result of the γ emission are referred to as nuclear isomers only when the half-life of the excited state is long enough to be easily measurable (milliseconds or greater). When this is the case, the γ decay is described as an isomeric transition (IT). Otherwise, the terms "metastable state" or "excited state" are used to describe species in energy states above the ground state.

Three modes of γ decay will be discussed: (1) pure γ-ray emission, (2) internal conversion (IC), and (3) pair production (PP).

2.3.1. Pure Gamma-Ray Emission

Pure γ decay is illustrated by the decay of ^{234m}Pa to ^{234g}Pa in Fig. 1.13. The γ rays emitted by a nucleus in a γ-decay process are monoenergetic for each transition between energy levels and typically range in energy from 2 keV to 7 MeV. These γ energies are very close to the energies of the transitions between the nuclear quantum states. A small amount of recoil energy is imparted to the daughter nucleus, but this is usually small enough relative to the energy of the γ ray that it may be ignored.

2.3.2. Internal Conversion (IC)

In this decay mode, the excited nucleus deexcites by transferring its energy to an orbital electron, which is then ejected from the atom. No γ ray is emitted. The general equation is

$$^{A}_{Z}\text{E}^{*} \longrightarrow {}^{A}_{Z}\text{E} + \text{IC electrons} + \text{x rays} + \text{Auger electrons}$$

The IC electrons are monoenergetic. They have an energy equal to the transition energy of the nuclear levels involved minus the binding energy of the atomic electron.

$$E_{\text{IC electron}} = E_{\text{trans}} - \text{BE}_{\text{atomic electron}} \tag{2.12}$$

Differences in the energies of the IC electrons emitted are thus equivalent to the atomic electron energy level spacings, and can be used to define the atomic number of the parent radionuclide. Figure 2.7 illustrates this process.

Because IC decay results in a vacancy in an atomic orbital, the processes of x-ray emission and Auger electron emission will also occur, as described earlier for EC decay. The x rays emitted following IC are characteristic of both parent and daughter, since both are isomers of the same element. This is in contrast to the EC process, where the x rays are characteristic only of the daughter nuclide. Internal conversion and pure γ emission are competing processes. The ratio of the fraction of decays occurring by IC to the fraction occurring by pure γ emission is the **internal conversion coefficient (α):**

$$\alpha = \frac{\text{Fraction of decays occurring by } \gamma \text{ emission}}{\text{Fraction of decays occurring by IC}} \tag{2.13}$$

The values of α may vary from 0 to infinity. The internal conversion coefficient varies with transition energy, atomic number, and change in nuclear spin.

$$E_{IC\ electron} \dashrightarrow E_{trans} - B.E._{atomic\ electron}$$

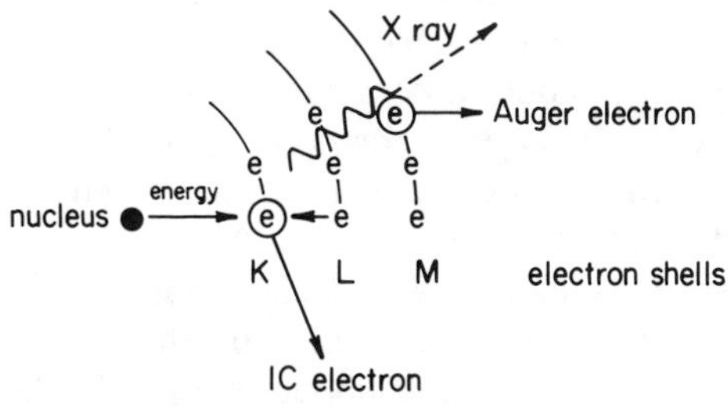

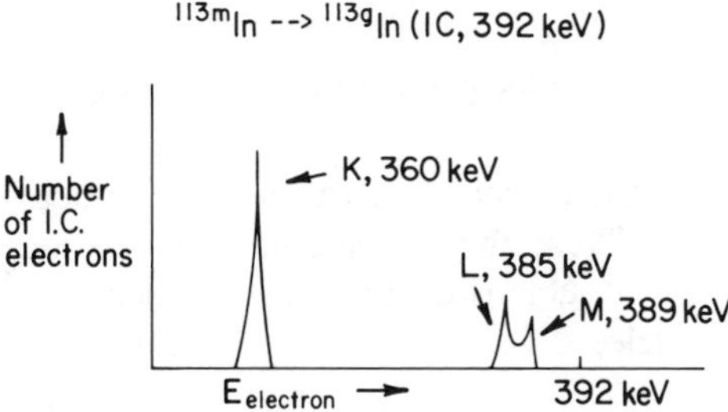

Figure 2.7. Internal conversion decay.

2.3.3. Pair Production (PP)

For nuclear transitions with energies greater than 1.02 MeV, pair production is an alternative, although uncommon, mode of decay. In this process, the energy of the transition is used to create an electron–positron pair and then eject the pair from the nucleus. The total kinetic energy given to the pair is equal to the difference between the transition energy and the 1.02 MeV needed to create the pair. A general equation for this process is

$$^{A}_{Z}E^{*} \longrightarrow {}^{A}_{Z}E + \beta^{-} + \beta^{+}$$

The positron will eventually undergo annihilation, as discussed previously. An example of a nuclide that decays by PP is ^{16m}O

$$^{16m}O \longrightarrow {}^{16}O \qquad E_{trans} = 6.05\ MeV \quad t_{1/2} = 7 \times 10^{-11}\ s$$

2.4. BRANCHING DECAYS AND DECAY SCHEMES

Nuclides may decay in ways that are much more complex than those shown here. The same nuclide may be able to undergo several kinds of decay. For example, ^{252}Cf undergoes both α decay and spontaneous fission in a ratio of 97:3, and ^{185}Hg decays by both α emission and by electron capture. The relative amounts of each of these decay modes are given by the **branching ratios** (see the section on α decay).

Actual energy level diagrams and decay schemes for most radionuclides are far more complex than those used in the examples. These decay schemes also provide information about the specific energies above the ground state for the various energy levels and about nuclear properties such as spin quantum numbers and parities of the nuclear states involved.

2.5. LESS COMMON DECAY MODES

Alpha, beta, and gamma decay are the most common decay modes, but several others exist. These include spontaneous fission, delayed neutron and proton emission, double-β decay, 2-proton decay, and decay by emission of heavier particles.

2.5.1. Spontaneous Fission Decay

Spontaneous fission is a naturally occurring decay process in which a nucleus breaks into two fragments, with emission of 2–3 neutrons. This process should be distinguished from **neutron-induced fission,** which requires capture of a neutron to initiate the fission process. An example of a spontaneous fission *decay process* is

$$^{252}_{98}\text{Cf} \longrightarrow {}^{98}_{38}\text{Sr} + {}^{152}_{60}\text{Nd} + 2 \text{ neutrons}$$

Fission products is the term used to describe the nuclides formed in a fission process. Each fissionable nuclide can produce a wide range of fission products and those listed here are only examples. An example of a *neutron-induced fission reaction* is

$$^{235}_{92}\text{U} + {}^{1}\text{n} \longrightarrow {}^{140}_{56}\text{Ba} + {}^{94}_{36}\text{Kr} + 2 \text{ neutrons}$$

Spontaneous fission is more common for heavier nuclei. For the transuranium elements, it sometimes can compete significantly with alpha decay.

Fission in Nature

An interesting case of a fission chain reaction in nature was seen at the Oklo Mine in Gabon, Africa. In 1972, French scientists were analyzing U ore samples from the mine. Normally, the ^{235}U isotope is found to be 0.7202% of the total uranium abundance. However, in some of the Oklo samples, ^{235}U abundances were found to be only 0.7171%. Analyses of other samples later showed abundances as low as 0.296% ^{235}U. Scientists speculated that perhaps the depletion of ^{235}U had occurred as a result of a "fossil fission reactor" that operated several million years ago.

The age of the rock containing the U ore was determined to be 1.74×10^9 y. In a rock this old, the original ^{235}U content would have been approximately 3% of the total uranium present (the $t_{1/2}$ of ^{235}U is 7.04×10^8 y). This is close to the isotopic enrichment of uranium used in a modern power reactor. To achieve enough mass to sustain the chain reaction the ore would have had to have been approximately 20% uranium. Samples have been found at Oklo that do contain 20–60% uranium. Spontaneous fission of ^{238}U, or cosmic-ray neutrons could provide the "trigger" neutrons required to initiate the neutron-induced fission of ^{235}U. ^{235}U has a much higher probability of undergoing neutron-induced fission than does ^{238}U. This is the reason the ^{235}U isotope is enriched by gaseous diffusion, or other processes, for use in nuclear reactors and in the early nuclear bombs. The ^{235}U fission events provided several additional neutrons to propagate a nuclear chain reaction by inducing more than one secondary fission event for each initial or parent fission event.

Other conditions must have been present for the chain reaction to have taken place. The neutrons that induce the fission must be slowed to lower or **thermal energies** (near the energy of gas molecules at room temperature, $E_{mean} \approx 0.04$ eV). In modern reactors, water is a common **moderator** used to reduce fission neutron energies by multiple-collision reactions with the hydrogen atoms in water. The soil at Oklo is a clay containing 5% water. This is about the same water/U ratio as is used in a modern reactor. Also, some elements have a very high probability of capturing neutrons, and thus stopping the chain reaction. These elements, such as V, were found to be in very low abundance in the Oklo region soils. Other elements with high neutron-capturing potentials, such as B, Nd, and Gd, are present at extremely low levels at Oklo. The small amount of these elements originally in the Oklo soil were probably "burned out" during the early operation of the reactor.

Besides the depletion of ^{235}U, the absence of "poisons" like V, and the altered amounts of B, Nd, and Gd, there is other experimental evidence that points to an operating natural nuclear reactor at Oklo. There is an abnormally high abundance of isotopes of nuclides that are commonly produced in fission processes. Among these are isotopes of trapped Kr and Xe gases which are known to be fission products produced in high yield from ^{235}U fission. There is an absence of volatile organic compounds in the rocks of the area, due to radiolysis reactions induced by γ rays and heating. The local rocks show signs of having been strongly heated, but there is no evidence of local volcanic activity. Finally, there are unusual isotopic ratios for certain elements, like Nd, that have isotopes with varying abilities to capture neutrons. Those isotopes that had a high probability of neutron capture are depleted relative to the other isotopes, indicating the presence of an intense flux of neutrons.

The Oklo site is of interest to environmentalists, because it has implications with respect to radioactive waste disposal problems. Other sites of natural fission reactors are being sought, but Oklo is the only known fossil nuclear reactor site at the time of this writing.

2.5.2. Delayed-Neutron Emission

Fission products, such as the ^{140}Ba and ^{94}Kr shown previously, are most often neutron-rich species, and will undergo β^- decay. After β^- decay, the nucleus is usually in an excited state, but this state of excitation is not high enough for particle emission, so γ decays will occur to return the daughter nuclide to its ground state. In a few cases, however, the β-decay process leaves the daughter nucleus in a high enough state of excitation that particle emission from the excited nuclide is possible. Because the nuclei are neutron-rich, a neutron is the most likely particle to be emitted. These **delayed neutrons** are emitted very rapidly after the β decay, so their half-life is that of the β decay. There are approximately 100 delayed-neutron emitters known, most of which are fission products.

An example of a delayed-neutron emitter is

$$\underset{\text{Precursor}}{^{87}\text{Br}} \longrightarrow \underset{\text{Emitter}}{^{87}\text{Kr}} \longrightarrow \underset{\text{Product (stable)}}{^{86}\text{Kr}} + \text{n} + \beta^-$$

The product of a delayed-neutron emission often is one that has a "magic number" of neutrons. *Nuclear magic numbers* are nuclear configurations that have unusual stability (see Chapter 3 for more on the magic numbers). Delayed-neutron emitters are very important in the engineering control of nuclear reactors. The rate of delayed-neutron emission following irradiation of uranium ores with reactor or ^{252}Cf neutrons is used for online monitoring of uranium content.

2.5.3. Delayed-Proton Emission

This decay mode is similar in concept to that of delayed neutron emission, but occurs less often. Proton-rich precursors are often produced by nuclear reactions employing charged-particle bombardment. The product nuclide then decays either by immediate proton emission, or by positron decay, followed by proton emission. An example of the process is

$$\text{Production of precursor:} \quad {}^{54}\text{Fe(p, 2n)}{}^{53}\text{Co}$$

$$\text{Decay by proton emission:} \quad {}^{53}\text{Co} \longrightarrow {}^{52}\text{Fe} + \text{p}$$

As with delayed neutron emission, the stable daughters are often magic number nuclei.

2.5.4. Double-Beta Decay

Earlier it was stated that a nucleus is unstable to a given decay if the mass(es) of the product(s) of the decay are less than that of the parent. Calculations involving the masses of certain nuclides, like ^{130}Te and ^{82}Se, show that these nuclides are stable to ordinary β decay, but are unstable toward 2-β decay. This process is known as **double-beta decay,** and involves the *simultaneous*, not sequential, emission of two β particles. The equations would be

$$^{130}_{52}\text{Te} \xrightarrow{2.5 \times 10^{21}\text{ y}} {}^{130}_{54}\text{Xe} + 2\,\beta^-$$

$$^{82}_{34}\text{Se} \xrightarrow{1.4 \times 10^{20}\text{ y}} {}^{82}_{36}\text{Kr} + 2\,\beta^-$$

These decays have been observed, but there is still question about how they actually take place. One possibility is the emission of 2 β^- and 2 $\bar{\nu}$ simultaneously. This process would not be expected to occur easily, partly because of the number of particles involved. A different process has been postulated which involves the emission of 2 β^- particles, but no antineutrinos. This process would occur through a **virtual intermediate state;** that is, a state that is very short-lived. First, a β^- decay would occur, producing a β particle and a virtual antineutrino. This antineutrino would then be immediately reabsorbed and the second β^- decay would occur. A similar process in which a nucleus absorbs a neutrino and reemits a positron has actually been observed. However, this mechanism would be possible only if a neutrino and an antineutrino were identical particles. The half-life for this mechanism would be expected to be much shorter than that for the 2-β, 2-antineutrino mechanism. The measurements on the ^{82}Se double-beta decay by Elliott et al. (1987) which show a half-life of about 10^{20} y offer more support for the 2-β, 2-antineutrino process.

2.5.5. Two-Proton Decay

In the early 1970s, Goldanski proposed the existence of a **two-proton decay** process, which is similar in principle to double-beta decay. The double-proton decay process would be most likely to occur among lighter nuclides that are very proton-rich. In 1983, this decay was actually observed for ^{22}Al and ^{26}P by Cable et al. and Honkanen et al.

2.5.6. ^{14}C and Other Cluster Emission Decay

In 1984, Rose and Jones found that one decay out of 10^9 in ^{223}Ra occurs by emission of a ^{14}C nucleus with an energy of 29.8 MeV, instead of by α emission. Emission of a particle this massive in a decay process had never been observed, except for the formation of fission products in spontaneous fission decay. ^{222}Ra and ^{224}Ra have also been observed to undergo ^{14}C decay. Emission of ^{24}Ne by ^{232}U has also been reported. Beta-delayed triton emission has been identified in the decay of ^{11}Li, a nuclide that had been shown earlier to undergo β-delayed two- and three-neutron emission. Emissions of these **cluster particles** are clearly rare processes. However, it would not be unexpected if the number of decay processes eventually proves to be much larger than was believed only a few years ago.

TERMS TO KNOW

Alpha decay
Annihilation
Antineutrino
Auger electrons
Auger yield
Beta decay
Branching decay
Bremmstrahlung
Cluster emission decay
Coincidence detection
Daughter nuclide
Decay chain
Decay scheme
Delayed-n emission
Delayed-p emission
Double-beta decay
E_{max}
Electron capture
Electronvolt
Energy level diagram
Energy spectrum
Fission products
Fluorescence yield
Fossil fission reactor
Gamma decay
Geiger–Nuttall Rule
IC coefficient
Induced fission
Internal conversion
K capture
Moderator
Momentum
Negatron decay
Neutrino
Pair production
Parent nuclide
Positron decay
Positronium
Radioactive decay
Recoil energy
Spontaneous fission
Transition energy
Two-proton decay

READING LIST

Bahcall, J. N., R. Davis, Jr., and L. Wolfenstein, Solar neutrinos: a field in transition. *Nature(London)* **334**:487 (1988). [the solar neutrino problem]

Barwick, S.W., P. B. Price, and J. D. Stevenson, Radioactive decay of ^{232}U by ^{24}Ne emission. *Phys. Rev. C* **31**:984 (1985). [^{24}Ne decay]

Browne E., and R. B. Firestone, *Table of Radioactive Isotopes,* V. S. Shirley, ed. New York: Wiley, 1986. [decay schemes]

Cable, M. D., J. Honkanen, R. F. Parry, S. H. Zhou, Z.Y. Zhou, and J. Cerny, Discovery of beta-delayed two-proton radioactivity: ^{22}Al. *Phys. Rev. Lett.* **50**:404 (1983). [two-proton decay]

Elliott, S. R., A. A. Hahn, M. K. Moe, Direct evidence for two-neutrino double-beta decay in ^{82}Se. *Phys. Rev. Lett.* **59**:2020 (1987). [double-beta decay]

Honkanen, J., M. D. Cable, R. F. Parry, S. H. Zhao, Z.Y. Zhao, and J. Cerny, Beta-delayed two-proton decay of ^{26}P. *Phys. Lett.* **133B**:146 (1983). [two-proton decay]

Hopke, P. K., Extranuclear effects on nuclear decay rates. *J. Chem. Ed.* **51**:517 (1974).

Langevin, M., C. Detraz, M. Epherre, D. Guillemaud-Mueller, B. Jonson, and C. Thibault, Observation of β-delayed triton emission. *Phys. Lett.* **146B**:176 (1984). [β-delayed triton emission]

Moe, M. K., and Rosen, S. P., Double-beta decay. *Scientific American,* **261**:48 (1989).

Reines, F., The early days of experimental neutrino physics. *Science* **203**:11 (1979). [history of neutrino detection]

Rose, H. J., and G. A. Jones, A new kind of natural radioactivity. *Nature(London)* **307**:245 (1984). [^{14}C decay]

EXERCISES

1. A radionuclide undergoes decay by isomeric transition to emit a 1.24-MeV γ ray. How much energy, in joules, would be emitted by the decay of 1.00 mol of this radionuclide?

2. Write the complete nuclear equations for the decay of the following radionuclides:

(a) the α decay of ^{238}Pu

(b) the negatron decay of ^{60}Co

(c) the positron decay of ^{11}C

(d) the electron capture decay of ^{208}Bi

3. Describe how delayed-neutron emission could be used for on-line measurement of the uranium content of ores.

4. By reference to the recent literature, describe the discovery of any new decay modes not mentioned in this text.

5. Discuss why neutrinos are so difficult to detect.

6. How can we distinguish a radionuclide undergoing negatron decay from one undergoing decay by internal conversion (IC), since both emit electrons?

7. Radioactive ^{222}Rn is known to accumulate in the basements of buildings and may consititute a health risk.

(a) What is the parent primary natural radionuclide that produces ^{222}Rn?

(b) Why are none of the intervening radionuclides between parent and the ^{222}Rn daughter of serious health concern?

(c) Write the nuclear equation for the decay of ^{222}Rn.

(d) How many negatrons and α particles are emitted in the chain of decay from ^{222}Rn to the stable nuclide daughter at the end of the decay chain?

8. The nuclides listed below will decay either by α emission, negatron emission, positron emission, or electron capture. Predict a reasonable decay mode for each nuclide, and explain your reasoning. Compare your prediction with the actual decay mode as found in a chart of the nuclides.

(a) ^{243}Am **(b)** ^{17}F **(c)** ^{49}Ca **(d)** ^{181}Ir

9. Calculate the recoil energy and total transition energy for the α emitted by ^{225}Ac. Find needed data in a chart of the nuclides.

10. Sketch the decay scheme for the decay of the following nuclides. Include the types and energies of the decay modes when possible.

(a) ^{77}As **(b)** ^{40}K **(c)** ^{220}Rn

11. Use a chart of the nuclides to locate one nuclide that decays by:

(a) delayed neutron emission

(b) spontaneous fission

(c) delayed proton emission

(d) isomeric transition

12. ^{237}Np is the parent of an extinct natural decay chain that ended at ^{209}Bi. What is the longest half-life daughter radionuclide in this decay chain?

CHAPTER

3

NUCLEAR CHEMISTRY AND MASS–ENERGY RELATIONSHIPS

Nuclear physicists and chemists have learned much about the properties and structure of the nucleus and its component nucleons. The first part of this chapter summarizes some of this information, including descriptions of nuclear size, shape, density, forces, momentum and spin, magnetic moments, and parity. Because these topics fall principally into the realm of nuclear chemistry rather than radiochemistry, the discussion is kept brief. The interested reader may consult the nuclear chemistry references listed at the end of the chapter for more detailed treatment of these topics. The second part of the chapter explains the mass–energy relationships important in radioactive decay and in nuclear reactions.

3.1. DESCRIPTION OF THE NUCLEUS

The nucleus of an atom consists of protons and neutrons, known collectively as **nucleons**. Most of the naturally occurring nuclei in their ground states are approximately spherical in shape, although truly spherical nuclei are rare. The ellipsoidal deviations from the spherical shape are described by a quantity known as the **electric quadrupole moment.** Prolate (football-shaped) and oblate (lens- or disk-shaped) ellipsoidal nuclear shapes have nonzero values of the electric quadrupole moment. These shapes are illustrated in Fig. 3.1.

The radius of a heavy nucleus is on the order of 10^{-12} cm, or about 10 ***fermis*** (1 fermi = 1 femtometer = 1 fm = 10^{-13} cm). This is only a small portion of the radius of the entire atom, which is typically on the order of $1\text{–}2 \times 10^{-8}$ cm. The radius of a nucleus may be estimated by

$$R = R_0 A^{1/3} \tag{3.1}$$

where R is the radius, A is the mass number, and R_0 is a constant. The value of R_0 varies, depending on the experimental methods used to measure it. The value of R_0 determined from α-decay systematics is $\approx$1.3 fm; from fast neutron scattering experiments is $\approx$1.4 fm; and from charged-particle

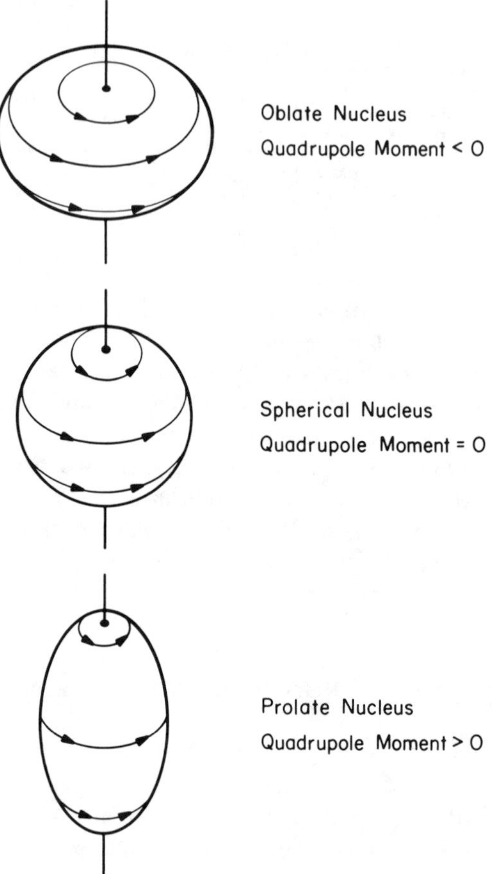

Figure 3.1. Shapes of the nucleus. All even-Z, even-N nuclei are approximately spherical in their ground states. Nuclei with odd Z and/or N have shapes that are spheroidal in their ground states. The deformation is indicated by the quadrupole moment of the nucleus. A negative quadrupole moment signifies a shape similar to a flattened spinning toy top (an oblate spheroid). A positive quadrupole moment is associated with a cigar-shaped spheroid spinning around its long axis (a prolate spheroid).

scattering is 1.2–1.6 fm. The variable experimental values obtained by charged-particle scattering are due, in part, to the effect of Coulomb forces between the incoming charged particle and the positively charged target nucleus. For most calculations, the value of 1.4 fm may be used without serious error. As may be inferred from the different R_0 values, the term "nuclear radius" may have several meanings. The radius may refer either to the

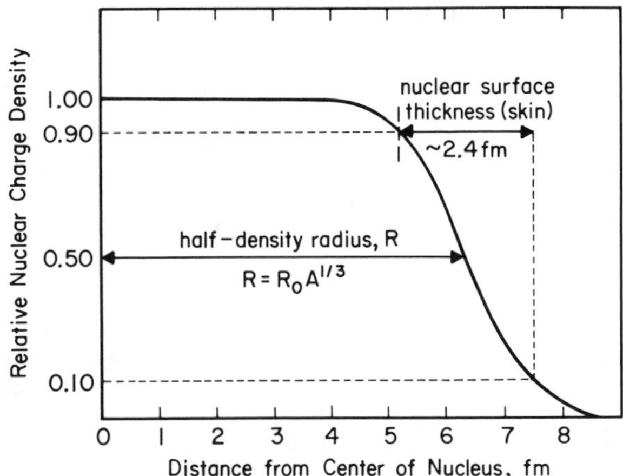

Figure 3.2. The charge distribution in a nucleus. The region of the nucleus where the relative charge density changes from 0.9 to 0.1 is known as the nuclear surface, or "skin." This skin has a thickness of approximately 2.4 fm. The most commonly used radius for a nucleus is the *half-density radius.* It is the distance from the center of the nucleus to the point where the nuclear charge density has decreased to one-half of its value at the center of the nucleus.

charge distribution, the matter distribution, or a radius representing actual penetration of the nuclear potential energy barrier. Each of these can be measured in different ways. A graph illustrating the charge distribution in a typical nucleus is shown in Fig. 3.2.

From Eq. 3.1, it can be inferred that the volume of a nucleus is approximately proportional to its mass. The densities of all nuclei are, therefore, essentially constant. Simple calculations will show that the density of a nucleus, about 10^{14} g/cm^3, is much greater than that of ordinary matter (gases at atmospheric pressure, $\approx 10^{-3}$ g/cm^3; liquids and solids up to 22.6 g/cm^3). Nucleons are held together in the nucleus by the strong force (see Chapter 1). The strong force or interaction of hadrons is the greatest in magnitude of the four basic forces, but acts only over very short distances ($\approx$1 fm). The strong force has **exchange character.** This means that the force is maintained by the exchange of particles between the nucleons. For the strong force, the exchange particle is the **pi meson,** or **pion**. Nuclear forces are not central field forces; there is no central attractive particle to which all the other particles are attracted. It is difficult to determine a mathematical function to describe the nuclear forces, so they are usually represented by a **nuclear potential** of some type. The shape and nature of this potential depends upon the type of experiment used to assess it.

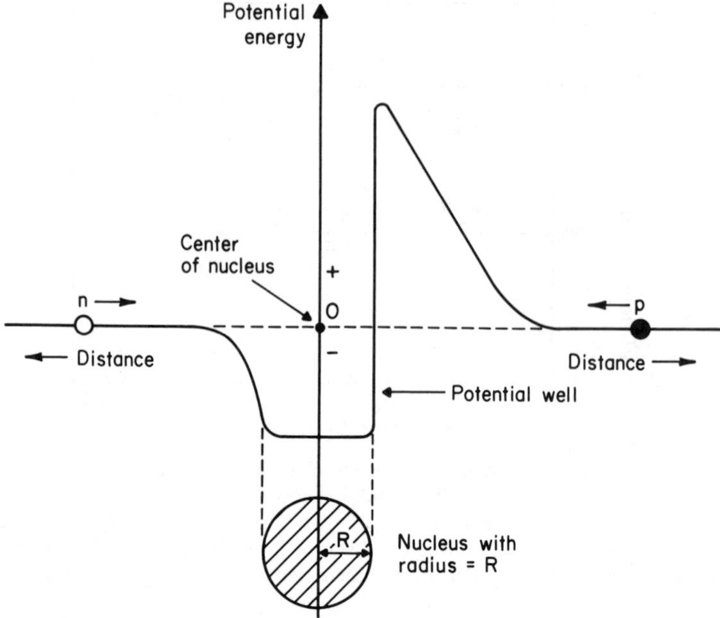

Figure 3.3. A representation of the nuclear potential experienced by both a neutron and a proton approaching a nucleus with radius = R. The proton, being a positively charged particle, experiences a Coulomb repulsive force as it approaches the positively charged nucleus. This creates a potential energy "barrier" that the proton must either overcome, or penetrate through (quantum-mechanical tunneling) before it can enter the target nucleus.

Figure 3.3 shows the nuclear potential experienced by a proton and a neutron as they approach the nucleus. The horizontal axis of this graph represents the distance from the center of the nucleus, and the vertical axis represents potential energy. The shapes of these potentials show that nuclear forces are very short range, falling off quickly to zero just outside the nuclear radius, while Coulomb repulsive forces are important at very small distances of nucleon separation ($\approx$0.4 fm). Figure 3.3 shows that as a proton approaches the nucleus it will first experience a repulsive force due to Coulomb interactions with the nuclear protons, while the neutron experiences no such force. At around 2.4 fm, the proton and neutron both experience the attraction of the strong force, and fall into the potential energy well at the bottom of the curve. The depth of this potential well is on the order of 25–30 MeV. An accurate knowledge of the nuclear potential is important for further understanding of many nuclear phenomena.

3.2. NUCLEAR PROPERTIES

3.2.1. Angular Momentum and Nuclear Spin

The **total angular momentum** of a nucleus is attributable to both the intrinsic spin of the nucleons and to their orbital motion within the nucleus. Neutrons and protons both have an intrinsic spin (s) of $\pm\frac{1}{2}$. The orbital angular momentum (ℓ) of the nucleons is quantized, and has integer values. The total angular momentum of a single nucleon (j) is equal to the sum of the spin and orbital angular momenta:

$$j = \ell + s = \ell + \left(\pm\frac{1}{2}\right) \tag{3.2}$$

For the nucleus as a whole, the total angular momentum (I, the *nuclear spin*) is equal to the vector sum of the angular momenta of all the nucleons:

$$I = \sum j \tag{3.3}$$

For even-A nuclei, I will be either 0 or integral; for odd-A nuclei, I will be half-integral. Nuclei that have even numbers of both protons and neutrons (**even–even nuclei**) always have a ground state spin equal to 0.

3.2.2. Magnetic Moment (μ)

Any moving electrically charged object gives rise to a **magnetic moment** (**μ**), defined as

$$\mu = (\text{pole strength}) \times (\text{distance between poles}) \tag{3.4}$$

The magnetic moment of a nucleus is due to the orbital motion of the proton, the spin of the proton, and the spin of the neutron. It is not surprising that the proton contributes to the magnetic moment, because it is a charged particle. It is unexpected, however, for the neutron to have a magnetic moment, because it has zero net charge. The existence of a measured magnetic moment for the neutron and also one for the proton that differs from predicted values are indications that these two nucleons have an internal structure (each is composed of three quarks). A separation of charge within the neutron must lead to its measured magnetic moment of approximately -1.91 nuclear magnetons (μ_N, the nuclear magneton $= 5.051 \times 10^{-24}$ ergs/

gauss). The negative sign for the μ of the neutron indicates that the spin and magnetic moment of the neutron are not parallel. Nuclei with a nuclear spin of zero have a zero magnetic moment. This is because, when all spins are paired, the individual magnetic moments of the nucleons cancel out. Nuclei that have nonzero magnetic moments can be detected using nuclear magnetic resonance (NMR) techniques.

3.2.3. Parity and Symmetry

Nuclei are also characterized by their parity and by the type of statistics they obey. **Parity** is a quantum number that is related to symmetry properties of the wave function and is not related to any physical quantity. Parity is said to be either odd or even (− or +), according to what happens to the sign of the wave function when the signs of the space coordinates are changed. Odd A nuclides for which the orbital angular momentum, ℓ, of the unpaired nucleon is even will have even parity, and those for which ℓ of the unpaired nucleon is odd will have odd parity. Nuclides that have no unpaired nucleons will have even parity. The parity of the nucleus is normally represented together with the spin. For example, the symbol $\frac{7}{2}^-$ for ^{59}Co indicates that the nucleus has a nuclear spin (I) of $\frac{7}{2}$ and a negative parity.

Statistical approaches are needed to describe the behavior of large numbers of particles. There are two types of statistical laws that are applied to subatomic particles: **Fermi–Dirac** and **Bose–Einstein.** These two systems are based on the symmetry of the wave function describing the particles. If the coordinates of two particles in a system can be interchanged, and the wave function of the system is not altered in any way as a result of the interchange, the system has a symmetrical wave function and obeys Bose–Einstein statistics. The particles that obey these laws are called **bosons**. If the wave function is altered after the interchange of the coordinates of two particles, the wave function is antisymmetric and the system obeys Fermi–Dirac statistics. The latter particles are called **fermions**. The **Pauli exclusion principle,** which states that no two particles may have the same set of quantum numbers, applies to fermions but not to bosons. Referring to Table 1.3, all leptons and all quarks are fermions. The hadrons are divided: The baryons are fermions, but the mesons are bosons.

3.3. MODELS OF NUCLEAR STRUCTURE

Several models have been proposed to describe nuclear properties and reactions. Because none is completely satisfactory to explain all experimental observations, different models are used for interpreting various nuclear phenomena. The most useful descriptions of nuclear structure are the shell, Fermi gas, liquid drop, collective, and optical models.

3.3.1. Shell Model (Single-Particle Model)

The shell model was developed independently by M. Mayer and by O. Haxel, J. H. Jensen, and H. E. Suess. Mayer, Jensen, and Wigner were awarded the Nobel Prize in 1963 for this and other fundamental nuclear structure studies. This model was based on the observation that nuclei with certain numbers of protons or neutrons were found to be especially stable. These numbers (2, 8, 20, 28, 50, 82, and 126) are called **nuclear magic numbers.** To explain this observation, the shell model assumed that nucleons arranged themselves in discrete energy levels within the nucleus analagous to those for electrons in the orbitals of the atom. The most stable configuration would occur as the various nuclear energy levels for the protons and neutrons were completely filled. There are several experimental observations that support the idea that magic numbers of nucleons have an especially stable configuration.

- Nuclei that have magic numbers of nucleons have a high natural abundance. This is especially true for those isotopes that have a magic number of both protons and neutrons. These nuclei are said to be **doubly magic.** For example, V has two stable isotopes, ^{50}V and ^{51}V. The ^{50}V isotope, with a magic number of neutrons (28), accounts for 99.75% of naturally occurring V. The other isotope, with 29 neutrons, represents only a minor portion of the naturally occurring V. An example of a doubly magic nucleus is ^{16}O, with 8 protons and 8 neutrons. This nuclide accounts for 99.756% of the naturally occurring oxygen and is the most abundant nuclide in the earth's crust. There are two other stable isotopes of oxygen, ^{17}O and ^{18}O, that have only the single proton magic number of 8. Doubly magic ^{4}He is the second most abundant nuclide in our visible universe (hydrogen is the most abundant). Another way to illustrate this point is to look at a plot of nuclidic abundances in the solar system versus mass number (Fig. 3.4). The peaks on this plot, representing elements present in high abundance, are often magic number nuclei.
- The three naturally occurring decay chains all decay to a stable isotope of lead, a nuclide with a magic number of protons (82). The end product of the 4n + 0 chain, ^{208}Pb, is doubly magic, with 82 protons and 126 neutrons.

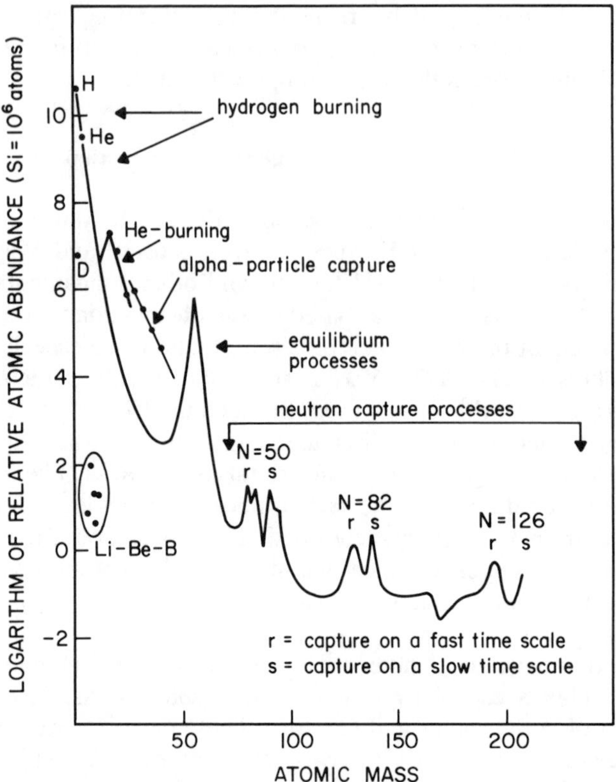

Figure 3.4. The atomic abundances of the elements in the solar system relative to silicon being set equal to 10^6 atoms. The abundance peaks in the neutron capture region of the curve are due principally to nuclear magic number effects. Major processes contributing to the nucleosynthesis of the elements in stars are noted for the various regions of the abundance curve (see Chapter 13). (Adapted from Burbidge et al., *Rev. Mod. Phys. 29*:547 (1957).)

- Nuclides with magic numbers of nucleons often have a low neutron capture **cross section;** that is, they have a very low tendency to capture an additional neutron. This would be analagous to the lack of chemical reactivity of the noble gases (He, Ne, Ar, etc.), which have a filled outer shell of electrons. Conversely, nuclei that are just one nucleon short of a magic number have a very high probability of capturing a neutron. The probability of capturing a neutron is measured in a unit called the **barn** (which is further discussed in Chapter 4). As an example, the neutron capture cross section for the magic number nuclide ^{116}Sn, with 50 protons, is 0.006 barns, which is a very low value.

However, the cross section for ^{115}In, which is just one proton short of the magic number 50, is 2600 barns, a quite high value.

- On the chart of the nuclides, bold lines mark the magic numbers of protons and neutrons. Examination of the chart of the nuclides will show that at positions where there are magic numbers of either protons or neutrons or both, there are often many stable nuclides (gray or shaded boxes). Tin, with 50 protons, has 10 stable isotopes, the most of any element. Antimony, with 51 protons, has only two stable isotopes.
- In the previous chapter, delayed neutron and proton decay was discussed. It was noted that this decay mode often results in the production of a magic number nuclide. For example, the decay of ^{87}Br by β emission results in a nucleus of ^{87}Kr, which has 51 neutrons. The rapid emission of a neutron from the ^{87}Kr results in ^{86}Kr, which has a magic number of neutrons (50).
- There are often "islands of isomerism" at atomic numbers just below the magic numbers. Recall that nuclear isomers are excited states of a nucleus that have a measurable half-life. Nuclides just below the magic numbers often have high nuclear spin quantum numbers (I), due to the presence of unpaired nucleons. Nuclear transitions in which the spin change between levels is large do not take place quickly, so half-lives are long and many isomers exist. For example, ^{113}Cd has 48 protons, just two protons short of a magic number. The nuclear spin for the first excited level of ^{113}Cd is $\frac{11}{2}$, while the spin for the ground state is $\frac{1}{2}$. The change in spin for the transition from the first excited state to the ground state is thus $(\frac{11}{2} - \frac{1}{2}) = \frac{10}{2} = 5$. This is a large value, so the half-life for this isomeric transition would be expected to be long. In fact, it is measured to be 14.1 years.

The shell model of the nucleus explains the magic number observations very well. It can also be used to predict with accuracy the spins and parities of odd-A nuclides, and accounts for their deviations from the spherical shape. It is most useful for descriptions of nuclei in their ground states.

3.3.2. Fermi Gas Model

This is a statistical model of the nucleus, which means that the nucleus is considered to be a collection of a statistically large number of particles, and the motions and interactions of the individual nucleons are not dealt with. The nuclear forces are expressed as a nuclear potential, or nuclear well. The nucleons are thought to be in the lowest possible energy states within the potential well. Each nucleon "gas" may be characterized by the kinetic energy of the highest filled level, called the **Fermi level.** A diagram of the

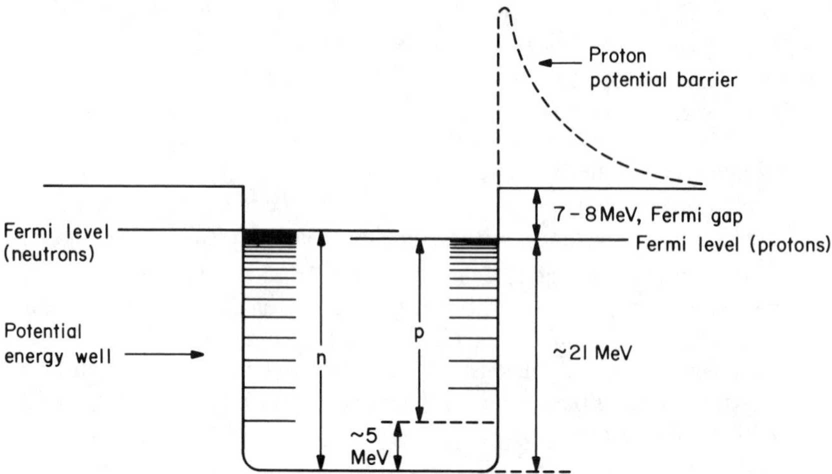

Figure 3.5. A representation of the Fermi potential for a medium to heavy element. The relative positions of the Fermi levels for neutrons and protons will vary, depending on the specific nuclide. However, the two levels are always very close to each other.

clear potential for the Fermi gas model is shown in Fig. 3.5. Note that the well is deeper for neutrons than for protons. This is because protons, unlike neutrons, experience repulsive Coulomb forces. Nuclear excitation will result in some nucleons being raised to levels above the Fermi levels.

The Fermi gas model of the nucleus is useful for the study of nuclear reactions and some decay transitions. It can also be used to characterize some thermodynamic properties of excited nuclei, such as nuclear entropy and temperature. In addition, the model gives good estimates of the momentum distribution of nucleons. The model fails, however, when used for predictions of the properties of low-lying states observed in radioactive decay.

3.3.3. Liquid Drop Model

The liquid drop model was proposed in the late 1930s by N. Bohr and J. Wheeler. In this model, the nucleus is treated as a drop of liquid. Like the Fermi gas model, this is a statistical model in which the properties of the individual nucleons are not considered.

The liquid drop model has been very useful to explain excited-state behavior and to provide a model for the mechanisms of low-energy reactions and fission processes. It also provides the basis for the semiempirical binding energy equation by Weiszäcker, which is discussed later in this chapter.

However, the liquid drop model is not able to explain shell and pairing effects, and is not equally applicable to all nuclei.

3.3.4. Optical Model (Cloudy Crystal Ball Model)

This model was proposed in the 1940s. The nucleus is viewed as a "cloudy crystal ball" that can reflect, refract, absorb, or transmit incoming particles. Mathematical techniques common in optics are used to describe these phenomena that may occur during nuclear reactions.

The optical model is best for explaining the results of the scattering of incoming particles by a nucleus. It does not adequately predict the results for inelastic scattering or for reactions in which a particle is absorbed by a nucleus.

3.3.5. Collective Model

This model was developed by A. Bohr and B. Mottelson in the early 1950s. It tries to incorporate features of both the shell and liquid drop models by considering the nucleus as a whole and the motions of individual outer nucleons.

This model has been successful in explaining the observed rotational energy levels of nuclei far from magic numbers. It also gives good predictions of the cross sections for Coulombic excitation and for certain types of gamma transitions. In addition, it can explain the observed magnetic moments, quadrupole moments, and isomeric transitions observed for nuclei far away from magic numbers

3.4. MASS–ENERGY RELATIONSHIPS

Early in this century, Einstein developed the equation describing the equivalence of mass and energy:

$$E = mc^2 \tag{3.5}$$

We know that, in any process, the total amount of matter and energy is constant, and that matter and energy are interconvertible. In ordinary chemical processes, the conversion of matter to energy is so small that it is not measurable. This is not true, however, for nuclear processes, which involve much larger energy changes than those of chemical reactions. In this part of the chapter, mass–energy relationships for nuclei and for nuclear reactions are discussed.

3.4.1. Mass–Energy Equivalence

In nuclear and radiochemistry, the masses of particles are sometimes expressed as their energy equivalents, instead of in daltons. The amount of energy equivalent to 1 dalton may be easily calculated, as shown below:

$$1 \text{ dalton} = (1 \text{ g/mol})/(6.022 \times 10^{23} \text{ units/mol}) = 1.66058 \times 10^{-24} \text{ g}$$

Using Einstein's equation $E = mc^2$, we have

$$E = (1.66058 \times 10^{-24} \text{ g})(2.99792 \times 10^{10} \text{ cm/s})^2$$
$$E = 1.49245 \times 10^{-3} \text{ ergs} = 1 \text{ dalton}$$

Using the relationship given in Chapter 2, we can convert ergs to MeV:

$$1.49245 \times 10^{-3} \text{ erg } (1 \text{ MeV}/1.60219 \times 10^{-6} \text{ erg}) = 931.5 \text{ MeV} = 1 \text{ dalton}$$

This relationship can be used to interconvert mass and energy values.

Example. Calculate the energy, in MeV, that is released in the decay of one neutron into a proton and a negatron.

$$^{1}\text{n} \longrightarrow {}^{1}\text{p} + \beta^{-} + \bar{\nu}$$

The mass of a neutron is 1.0086641 daltons. The mass of a proton and an electron = the mass of a H atom, 1.0078250 daltons.

$$\text{Energy of the transition} = (1.0086641 - 1.0078250)\ 931.5 \text{ MeV}$$
$$= 0.7816 \text{ MeV}$$

This is the energy released in the decay of one neutron to a proton and an electron.

Example. Calculate the amount of energy, in J/mol, that is released by the nuclear decay of 1.00 mol of a pure 1.00-MeV γ-ray emitter.

The decay of one atom releases 1.00 MeV of energy, so the decay of 1.00 mol (6.02×10^{23} atoms) will release 6.02×10^{23} MeV.

$$(6.02 \times 10^{23} \text{MeV/mol})(1.602 \times 10^{-13} \text{ J/MeV}) = 9.64 \times 10^{10} \text{ J/mol}$$

Also,

$$(9.64 \times 10^{10}\ \text{J/mol})(1\ \text{cal}/4.184\ \text{J}) = 2.30 \times 10^{10}\ \text{cal/mol}$$
$$= 23.0 \times 10^{6}\ \text{kcal/mol}$$

The latter example serves to illustrate the great difference between chemical and nuclear reactions in energy losses. Typical chemical reaction energies are often about 20–50 kcal/mol (≈80–200 kJ/mol), while the energy released by the γ-emitter in the above example is $\approx 10^6$ times greater.

The term mass excess (or mass defect) was mentioned briefly in Chapter 1. This quantity, symbolized by Δ, is equal to the difference between the nuclidic mass and the mass number:

$$\Delta = M - A \tag{3.6}$$

The value of the mass excess is given frequently in tables of the nuclides, usually as the energy equivalent. A sample calculation for the conversion of mass excess to the energy equivalent is given below.

Example. Calculate the mass excess, in MeV, for ^{14}C. The nuclidic mass of ^{14}C is 14.00324 daltons.

$$\Delta = M - A$$
$$\Delta = 14.00324 - 14$$
$$\Delta = 3.24 \times 10^{-3}\ \text{daltons}$$
$$(3.24 \times 10^{-3}\ \text{daltons})(931.5\ \text{MeV/dalton}) = 3.02\ \text{MeV}$$

Mass excess values may be either positive or negative and can be used directly for energy calculations from nuclear equations.

Example. Repeat the calculation done above for the energy released in the decay of a neutron to a proton and a negatron, given that $\Delta_n = +8.071$ MeV and $\Delta_H = +7.289$ MeV:

$$\Delta_n - \Delta_H = 8.071 - 7.289 = 0.782\ \text{MeV}$$

3.4.2. Energy Changes in Nuclear Reactions

Nuclear reactions, whether spontaneous or induced, result in changes in energy. The ***Q* value** of a nuclear reaction refers to the amount of energy released or absorbed in the reaction. The **binding energy** is simply the Q value for the specific reaction involving the formation of a nucleus from its component nucleons. These two quantities are analagous to the thermodynamic quantities of the enthalpy of a reaction (ΔH) and the enthalpy of formation (ΔH_f).

The Q value for a nuclear reaction is calculated by subtracting the masses of all the products of the reaction from the masses of all the reactants and converting the Δ(mass) to energy units. As seen above, it is also possible to use the mass excess values in place of the actual masses.

$$Q = \left(\sum \text{masses}_{\text{reactants}} - \sum \text{masses}_{\text{products}}\right)(931.5\ \text{MeV/dalton}) \quad (3.7)$$

where mass is in daltons, or

$$Q = \sum \Delta_{\text{reactants}} - \sum \Delta_{\text{products}} \quad (3.8)$$

where the Δ values are in energy units. A positive value of Q means that the masses of the products are less than those of the reactants, energy has been released, and the reaction is, therefore, exoergic. Conversely, a negative Q value indicates an endoergic reaction. The nuclidic masses found in handbook tabulations include the masses of the orbital electrons. As noted previously for positron decay, attention is required for electron balance in decay equations.

Example. Calculate the energy change, in MeV, for the nuclear reaction below. Is the reaction exoergic or endoergic?

$$^{31}\text{P} + \text{n} \longrightarrow {}^{28}\text{Al} + \alpha$$

$$Q = [\Delta(^{31}\text{P}) + \Delta(^{1}\text{n})] - [\Delta(^{28}\text{Al}) + \Delta(^{4}\text{He})]$$

$$Q = [(-24.441) + (+8.071)]_{\text{MeV}} - [(-16.851) + (+2.424)]_{\text{MeV}}$$

$$= -1.94\ \text{MeV}$$

Since Q is negative, the reaction is endoergic and will not proceed unless energy is provided externally.

The nuclear process in which a nucleus is formed from its components is illustrated by the following general equation:

$$Zm_{\text{H}} + Nm_{\text{n}} \longrightarrow \text{a neutral atom of } {}^{A}\text{E} + \text{energy}$$

where m_H is the mass of a ^1H atom, and m_n is the mass of a neutron, in daltons. The energy released in this reaction is called the nuclear binding energy (BE) and can be calculated as follows:

$$E_{MeV} = (\text{Masses of reactants} - \text{Masses of products})\ 931.5\ \text{MeV/dalton}$$

and,

$$BE_{MeV} = [(Zm_H + Nm_n) - M]\ 931.5\ \text{MeV/dalton} \qquad (3.9)$$

A similar equation could be written using Δ values in units of MeV:

$$BE_{MeV} = (Z\Delta_H + N\Delta_n) - \Delta_{\text{product nuclide}} \qquad (3.10)$$

The total binding energy gives an indication of the stability of a nucleus. The addition of nucleons to a nucleus always results in an increase in the total binding energy of the system, due to the strong forces that exist between hadrons. However, the best indicator of nuclear stability is the binding energy per nucleon: BE/A. The higher the value of BE/A, the more stable is the nuclide. Values of BE/A for the stable nuclides with $A > 11$ range from approximately 7.5 to 8.8 MeV/nucleon, with an average value of $\approx$8.0. The stable nuclide ^{56}Fe has the highest BE/A (approximately 8.8 MeV/nucleon). Therefore, on the basis of thermodynamics, ^{56}Fe is the most stable of all nuclides. Based solely on energetics, it could be said that the ultimate fate of the universe would be the conversion of all matter into ^{56}Fe! Two graphs of BE/A versus A for the stable nuclides, one for the mass range up to about 200 and one for the low mass region, are shown in Fig. 3.6 and 3.7. In Fig. 3.7, it is noted that the lighter magic number nuclei lie on "spikes" higher than their neighbors, due to their enhanced nuclear stability. Also, nuclides with equal numbers of protons and neutrons have higher BE/A values than their neighbors, reflecting enhanced stability due to a pairing effect among nucleons.

From the BE/A curves it can be seen why energy is released in both fission and fusion processes. Fission is the breakup of a large nucleus to form middle mass nuclides, which have a higher BE/A than the original heavy nucleus; hence, residual energy is released. Fusion is the putting together of small nuclei (e.g., deuterium and tritium) to make heavier nuclei, which have higher BE/A values, again releasing residual energy. The change in BE/A is greatest per unit mass for the fusion process. This is illustrated by the fact that the slope is much steeper in the approach toward ^{56}Fe for low mass fusion than it is in the approach toward ^{56}Fe from the high mass region where fission occurs.

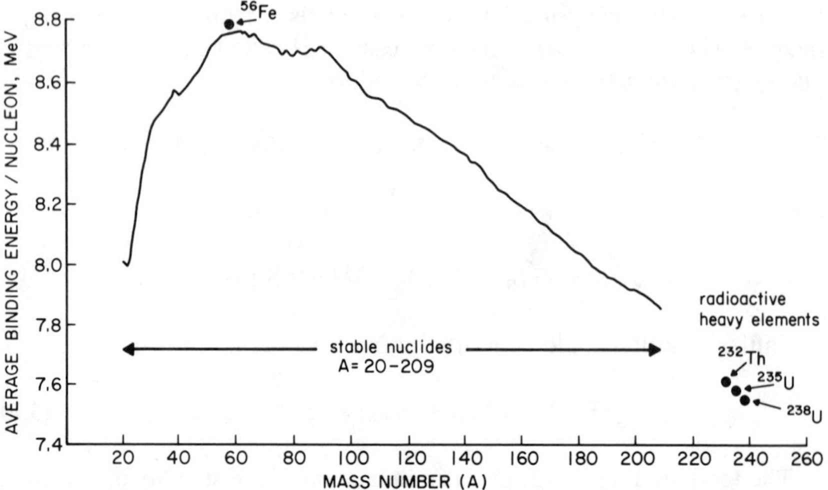

Figure 3.6. The average binding energy per nucleon (in MeV) for stable nuclides with A = 20–209. The line is drawn through odd-A nuclides to minimize variations due to even–even and odd–odd pairing effects for even-A nuclides. The even–even nuclide, ^{56}Fe, is the most thermodynamically stable nuclide. BE/A values are also shown for three primary natural radionuclides.

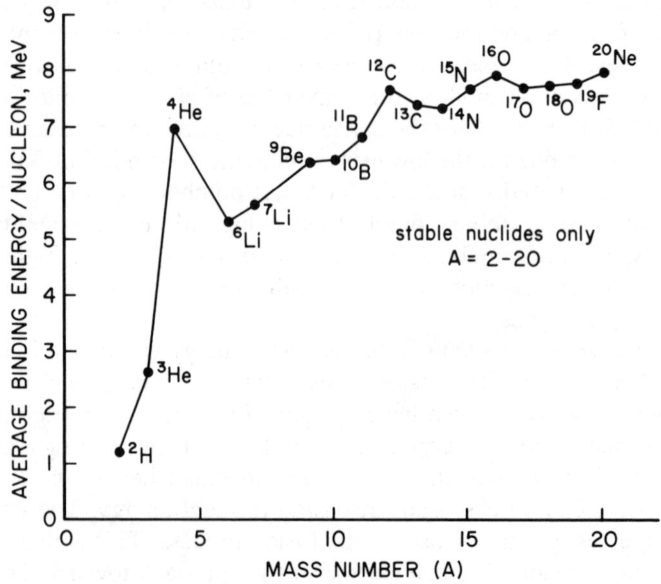

Figure 3.7. The average binding energy per nucleon (units of MeV) for stable nuclides with A = 2–20. The enhanced stability of even–even nuclides can be seen.

It should be emphasized that the BE/A values plotted in the figure for the stable nuclides are the averages for all nucleons in a given nucleus. However, not all the nucleons in a given nucleus are bound equally. The surface nucleons, for example, are surrounded by attractive neighbors on only one hemisphere and are, therefore, less tightly bound than the average. Interior nucleons are more tightly bound. We can use these ideas to illustrate a point mentioned earlier in the section on α decay: that this decay mode occurs most commonly (hence is an exoergic process) among the heavier nuclides with Z greater than that of Bi ($Z = 83$).

Example. Show, using binding energy considerations, that α decay should be energetically feasible for elements between bismuth ($Z = 83$) and uranium ($Z = 92$).

We must first determine the amount of energy needed to "form" an α particle inside the nucleus:

$$\begin{aligned} BE_\alpha &= (2\Delta_H + 2\Delta_n - \Delta_{He}) = [2(7.289) + 2(8.071) - 2.424] \\ &= 28.3 \text{ MeV} \end{aligned}$$

Emitted α's typically have an energy of 5 MeV, so the energy freely available for the detachment of the α would be only $28.3 - 5 = 23.5$ MeV.

Using the binding energy equation (Eq. 3.9), the BE for ^{209}Bi is found to be 1640 MeV and its average BE/A is $1640/209 = 7.8$. Based on this average BE/A it would thus appear that for emission (detachment) of an α particle, $(4 \times 7.8) = 31.2$ MeV of energy would be required, but only 23.5 MeV is available from the formation of the particle. However, 7.8 MeV is the average BE/A, and the nucleons on the surface of the nucleus would be bound *less* tightly. The addition of 29 more nucleons to ^{209}Bi would form ^{238}U. The BE of ^{238}U can be calculated from Eq. 3.9 to be 1800 MeV. The difference in BE beween these two nuclides is 160 MeV. The addition of the 29 nucleons increased the BE by 160 MeV, so the BE/A for these last added nucleons is $160/29 = 5.5$ MeV/nucleon. Hence, to detach 4 of these "outside" nucleons would require only $(4 \times 5.5) = 22.0$ MeV. This much energy is available from formation of the α particle, so α decay is energetically feasible in this region. A similar calculation for much lower Z regions would generally indicate that α decay was not energetically possible.

Nuclides with $A > 100$ are also unstable with respect to spontaneous fission and this decay process can compete with α decay. However, high

Coulomb barriers restrict *measurable* spontaneous fission decay rates to nuclides with $A > 230$.

3.4.3. Energy Changes in Radioactive Decay

Alpha Decay

The calculation of the Q value for α decay is straightforward using Eq. 3.7 or 3.8. The parent nuclide is the *reactant*, and the alpha particle and the daughter nuclide are the *products*.

Example. Calculate the Q value for the α decay of ^{252}Cf.

$$^{252}\text{Cf} \longrightarrow {}^{248}\text{Cm} + \alpha$$

$$Q = \Delta_{\text{Cf}} - (\Delta_{\text{Cm}} + \Delta_{\text{He}})$$

$$Q = 76.030 - (2.424 + 67.388) = 6.22 \text{ MeV}$$

Remember, Δ values are based on atomic masses (with orbital electrons).

Negatron Decay

In the general equation for negatron decay (Chapter 2), the product side of the reaction shows a negatron and the daughter nuclide. However, the mass of the β^- is not included in calculation of the Q value. The reason for this is that the daughter nuclide will, when initially formed, be deficient 1 electron (because Z has increased by 1), but it will quickly pick up an electron to become a neutral atom. Because the nuclidic masses or Δ values normally tabulated represent *atomic* masses, which include the masses of the orbital electrons, the β^- particle mass (an electron) is already accounted for by the atomic mass (or Δ) of the daughter nuclide for the calculation of Q.

Example. Calculate the Q value for the negatron decay of ^{32}P to ^{32}S.

$$^{32}_{15}\text{P} \longrightarrow {}^{32}_{16}\text{S} + \beta^- + \bar{\nu} + Q$$

$$Q = \Delta(^{32}\text{P}) - \Delta(^{32}\text{S}) = -24.305 - (-26.016) = 1.711 \text{ MeV}$$

Positron Decay

The electron balance situation for positron decay is different from that for negatron decay. Here, one electron mass is lost through emission of the positron. In addition, the daughter nuclide would initially be formed with one extra orbital electron, because the atomic number of the daughter is one unit lower than that of the parent. This electron is lost to form the neutral daughter atom. Thus, 2 electron masses must be accounted for in the calculation of Q values for positron decay.

Example. Calculate the Q value for the positron decay of ^{26}Al.

$$^{26}_{13}\mathrm{Al} \longrightarrow {}^{26}_{12}\mathrm{Mg} + \beta^+ + \text{atomic electron} + \nu + Q$$

$$Q = \Delta(^{26}\mathrm{Al}) - [\Delta(^{26}\mathrm{Mg}) + 2\ (\text{energy equivalent of the } m_e)]$$

$$Q = -12.210 - (-16.214) - 2(0.511) = 2.982\ \mathrm{MeV}$$

3.4.4. Closed-Cycle Decay for Mass–Energy Calculations

In all of the mass–energy calculations illustrated in previous sections, the necessary nuclidic masses (or Δs) were known, so the calculations were straightforward. However, experimentally determined nuclidic masses of many nuclides, especially short-lived radioactive species, are not available. Therefore, indirect or empirical approaches are used to calculate needed nuclidic masses or energy values.

One of the indirect methods used for this purpose is the **closed-cycle decay diagram.** Four different nuclides are placed at the corners of a box. One of these nuclides is the parent of all the others. Two independent decay paths lead to a final daughter through two intermediate daughters. An example of one of these diagrams is

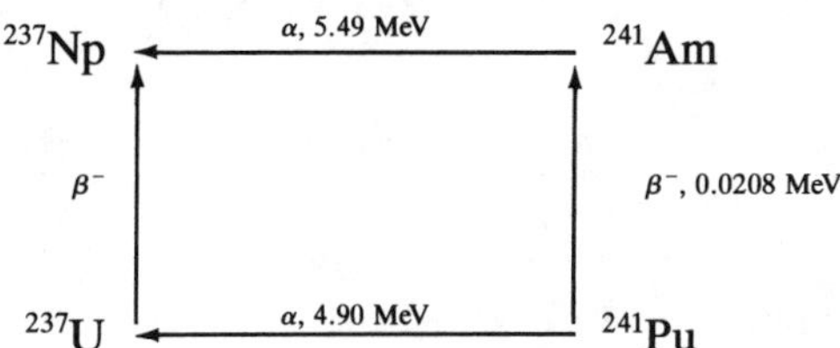

In the cycle shown above, ^{241}Pu is the parent. It decays by α emission to ^{237}U, which in turn undergoes β^- decay to ^{237}Np. The ^{241}Pu also decays by β^- emission to ^{241}Am, which then forms ^{237}Np by α emission. The overall energy change from the parent ^{241}Pu to the daughter ^{237}Np must be the same regardless of the path taken. Therefore, we can write

$$Q_{\mathrm{Pu}\to\mathrm{U}} + Q_{\mathrm{U}\to\mathrm{Np}} = Q_{\mathrm{Pu}\to\mathrm{Am}} + Q_{\mathrm{Am}\to\mathrm{Np}}$$

If the energies and masses of three of the four sides of the cycle are known, it is obvious that we can calculate the fourth.

Example. Using the decay cycle shown above, calculate the Q-value for the decay of ^{237}U to ^{237}Np. The energies for the other three sides are shown on the diagram.

Using the energies given in the diagram above,

$$Q_{\mathrm{U}\rightarrow\mathrm{Np}} = Q_{\mathrm{Pu}\rightarrow\mathrm{Am}} + Q_{\mathrm{Am}\rightarrow\mathrm{Np}} - Q_{\mathrm{Pu}\rightarrow\mathrm{U}}$$

$$Q_{\mathrm{U}\rightarrow\mathrm{Np}} = 0.0208 + 5.49 - 4.90 = 0.61 \text{ MeV}$$

Mass calculations may be done similarly.

3.4.5. Semiempirical Binding Energy Equation

A semiempirical equation for calculation of binding energies of nuclides for which the nuclidic mass was not known was developed by C. F. von Weizsäcker in 1935. The equation is based on the liquid drop model of the nucleus, and expresses the binding energy as the sum of five terms that are functions of only Z and A. The constants in the equation were determined by fitting the empirical equation with known binding energies. For nuclei with $A > 40$, the agreement between actual values obtained from use of experimental nuclidic masses in the regular BE equation and the predicted values is often better than 1%.

The equation as modified by W. Myers and W. Swiatecki is

$$\begin{aligned} \text{BE (MeV)} &= c_1 A\{1 - k[(N - Z)/A]^2\} - c_2 A^{2/3}\{1 - k[(N - Z)/A]^2\} \\ &\quad - c_3 Z^2 A^{-1/3} + c_4 Z^2 A^{-1} + \delta \end{aligned} \tag{3.11}$$

It is possible to associate a physical meaning to each term, and illustrate how the term affects BE:

Volume Energy Term ($+c_1 A\ \{1 - k[(N - Z)/A]^2\}$)

This term accounts for the direct dependence of BE on the number of nucleons present in the nucleus. Each nucleon added to the nucleus increases its total binding energy, because hadrons are subject to the strong force. Hence, the term has a positive value. The term in { } is called the **symmetry term.** Binding energy is greatest when there are equal numbers of neutrons and protons. This term accounts for the reduction in BE when the numbers of protons and neutrons are not equal, and the term equals zero when $Z = N$. The constant c_1 has a value of 15.677 MeV, and the constant k in the symmetry term has a value of 1.79.

Surface Energy Term $(-c_2 A^{2/3} \{1 - k[(N - Z)/A]^2\})$

In an earlier section, it was noted that nucleons on the surface of the nucleus experience less attractive force because they are not completely surrounded by other nucleons. Therefore, surface nucleons contribute less to the total binding energy than interior nucleons and the surface energy term is negative. The volume of the nucleus is proportional to A, so the surface area is proportional to $A^{2/3}$. The symmetry term appears here too, although it does not affect the calculated binding energy values very much. It is included so that if the symmetry term approaches zero ($N >> Z$), then the surface energy term also goes to zero. The value of the constant c_2 is 18.56 MeV.

Coulomb Energy Term $(-c_3 Z^2 A^{-1/3})$

In addition to the attractive strong force experienced by all the nucleons, the protons in the nucleus experience repulsive forces due to electrostatic (Coulomb) interactions. This reduces the binding energy, and this term is also negative. The Z^2 dependence indicates that this term becomes more significant for higher Z nuclides. These nuclides will, therefore, have more neutrons than protons. The value of c_3 is 0.717 MeV.

Diffuse Boundary Correction Term $(+c_4 Z^2 A^{-1})$

The term describing the lowering of binding energy due to electrostatic repulsion between protons in the nucleus assumed that there is an even distribution of charge in the nucleus. This is not truly the case, so the diffuse boundary term actually serves as a slight positive correction for the Coulomb energy term. The value of c_4 is 1.211 MeV.

Pairing Energy Term (δ)

Nuclei that have even numbers of both protons and neutrons exhibit higher stability than do nuclides that have an odd number of protons and/or neutrons. This can be illustrated by the chart below, showing approximate numbers of stable nuclides according to odd-even characteristics:

	Even Z	***Odd Z***
Even N	159	50
Odd N	53	4

Notice that there are more even–even nuclides than all the other types combined. Even–even nuclei make up about 80% of the nuclides in the earth's crust, with ^{16}O, ^{24}Mg, ^{28}Si, ^{40}Ca, ^{48}Ti, and ^{56}Fe being especially abundant. On the other hand, there are only four known stable **odd–odd** nuclides; that is, nuclides that have an odd number of both protons and neutrons. These four nuclides are ^{2}H, ^{6}Li, ^{10}B, and ^{14}N. The **even–odd** and **odd–even** combinations are nearly equal in occurrence. The contribution of odd or even numbers of nucleons to binding energy is described by the pairing energy term. This term is positive for even–even nuclei, because the pairing of protons and neutrons provides added stability. The term is negative for the odd–odd nuclei, and zero for odd–even nuclei. The actual values of δ are $+11/A^{1/2}$ for even–even nuclei, and $-11/A^{1/2}$ for odd–odd nuclei. Nuclei that are even–odd or odd–even have a δ value of 0.

3.4.6. Nuclear Energy Surface Diagrams

The semiempirical equation may also be rewritten as a quadratic function of Z. If a plot is made of the equation, with binding energy on the vertical axis and Z on the horizonal axis, a single parabola will be obtained for constant values of odd A and two parabolas for constant values of even A. In these plots the value for binding energy is increasing in a *downward* direction. These parabolas are called **isobaric binding energy parabolas** or **nuclear energy surface diagrams.** The term *surface diagram* is used since these parabolas are slices through a three-dimensional plot of binding energy, atomic number, and neutron number that has an undulating surface. Such three-dimensional plots of the binding energy "surface" illustrate that the stable nuclei lie in a "valley of stability" that follows the line of beta stability as represented on the chart of the nuclides (Fig. 3.8). The plots obtained for odd-number or even-number isobar families are not identical, as indicated in Figures 3.9 and 3.10.

For an odd-A set of isobars (Fig. 3.9), the isobar at the bottom of the parabola has the largest total binding energy and BE/A, so it is most stable of the set. The other isobars, up either side of the parabola, will spontaneously decay toward it by beta decay processes. The isobars with a Z value greater than that of the most stable isobar will undergo positron decay or electron capture. Those with a lower Z value will undergo negatron decay. The single parabola is characteristic of all odd-A isobar families.

Figure 3.10 shows a nuclear energy surface diagram for an even-A set of isobars. Here, not one but two parabolas are seen. This is due to the differing stability of even–even and odd–odd nuclei, both of which result in an even-A nucleus. The even–even isobars, being more stable, form the lower of the two parabolas shown, while the odd–odd isobars constitute the upper parabola. For the odd–odd set, only one nuclide is stable, and that is the one at the bottom of the valley. If Fig. 3.10, represented, for example,

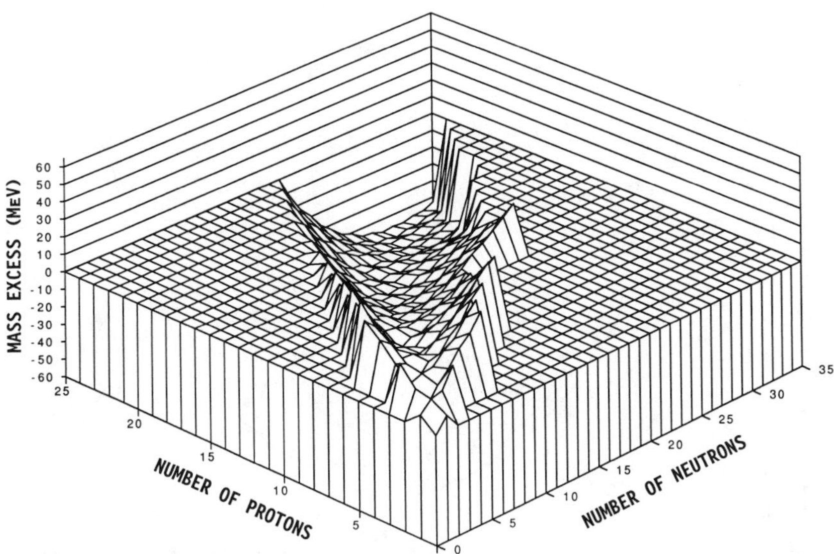

Figure 3.8. A three-dimensional representation of the valley of beta stability for the lighter elements. Mass excess values for each Z and N location are plotted at line intersections on the vertical axis. The lower the mass excess value, the greater is the BE/A. Hence, the stable nuclides lie at the bottom of the valley. Beta decays (either positron, electron capture or negatron) within sets of isobars eventually lead to the formation of the stable nuclides in the valley. In this representation, unknown isobars beyond the edge of the "valley" are assigned the default mass excess value of zero. This emphasizes the change from positive mass excess values for many low-Z nuclides to all negative mass excess values for heavier nuclides. (Plot courtesy of D. J. Van Dalsem.)

the $A = 64$ isobar set, the most stable odd–odd nuclide would be the $^{64}_{29}Cu$ isobar. For the even–even set, however, there are often two or more nuclides near the bottom of the lower parabola that can exhibit β-decay stability. These are called **stable isobaric pairs.** Decay between stable isobaric pairs is favored energetically, because one will inevitably have a higher BE/A than the other. However, decay of one to the other would require a double-β decay process, where two β particles are emitted simultaneously. As has already been noted, this is a very rare process and only a few examples of double-β decay have been observed. Using the example of the $A = 64$ isobar set again, the decay of ^{64}Zn (63.92914 amu) to ^{64}Ni (63.92796 amu) would require the emission of 2 β^+ particles. Although energetically feasible in many cases, double-β decay has been experimentally observed only for the double-negatron decay of $^{82}_{34}Se$ to $^{82}_{36}Kr$ ($t_{1/2} \approx 1.4 \times 10^{20}$ y).

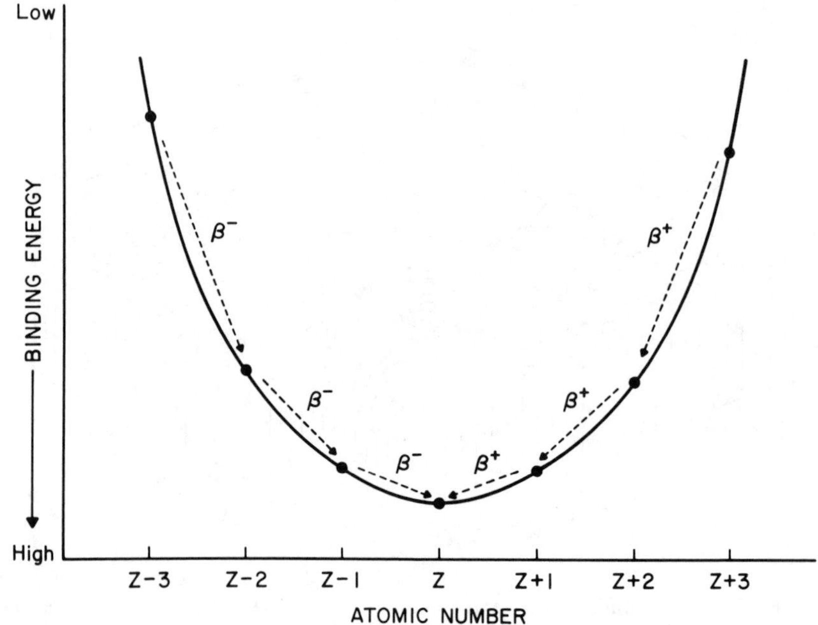

Figure 3.9. Binding energy parabola for a set of odd-*A* isobars. Electron capture can be an alternative to positron emission on all binding energy parabolas.

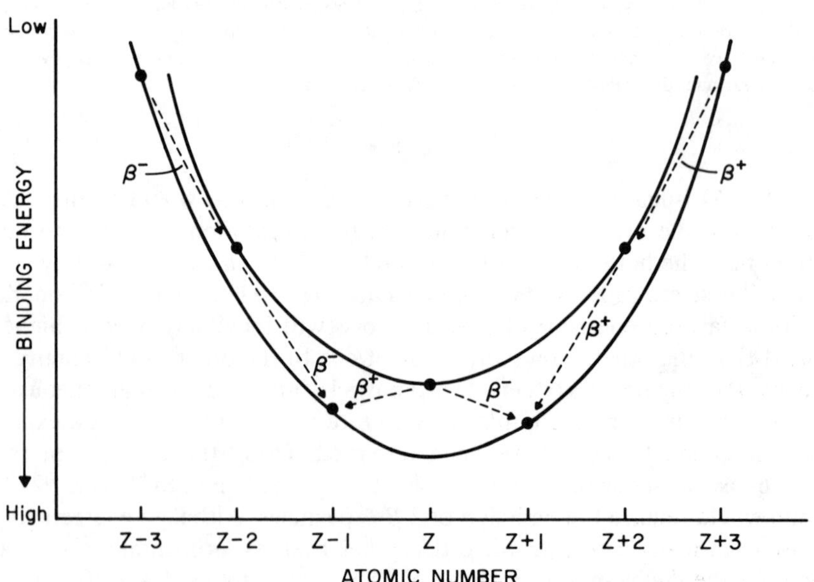

Figure 3.10. Binding energy parabolas for a set of even-*A* isobars.

The nuclear energy surface diagrams can be used to illustrate an observation made many years ago by J. Mattauch. He had noted that isobars differing by unity in their values of Z would not both be stable with respect to β decay. This might be expected, because the conversion of a proton to a neutron, or the reverse, would proceed spontaneously to form the isobar with the highest BE/A. It is unlikely that two isobars would have identical BE values. For example, the isobars ^{73}Ga, ^{73}Ge, and ^{73}As have Z values of 31, 32, and 33, respectively. According to Mattauch's Rule, they cannot all be stable. If Fig. 3.9 represented the $A = 73$ isobar set and if Ga, Ge, and As are represented by $Z - 1$, Z, and $Z + 1$, respectively, then we can see that the Ge isobar is the only stable nuclide and that Ga and As isobars will decay to Ge by β^- and E.C. decay, respectively. Another example of Mattauch's rule is the element Pm. Examination of a chart of the nuclides shows that all of the Pm isotopes near the line of β stability (hence most likely to be stable) are adjoined by stable isobars of other elements. One can predict, therefore, that there are no stable isotopes of Pm, and this is the case. Although initially the isobaric pairs ^{50}V–^{50}Cr, ^{123}Te–^{123}Sb, and ^{113}Cd–^{113}In were thought to violate Mattauch's rule, the first listed members of each pair have now been determined to undergo β decay processes. No exceptions to the rule are now known.

The half-lives of the β decays along the nuclear energy surfaces can be estimated by systematics. For the odd-A isobars, half-lives are short for nuclides with atomic numbers far from that of the one stable nuclide. As the line of β stability is approached, the half-lives generally become longer. For example, for the β^- decay to ^{133}Cs in the $A = 133$ isobar set, the following series of decays occurs:

$$^{133}_{50}\mathrm{Sn} \xrightarrow[\beta^-]{1.44\ \mathrm{s}} {}^{133}_{51}\mathrm{Sb} \xrightarrow[\beta^-]{2.48\ \mathrm{min}} {}^{133}_{52}\mathrm{Te} \xrightarrow[\beta^-]{12.4\ \mathrm{min}} {}^{133}_{53}\mathrm{I} \xrightarrow[\beta^-]{20.8\ \mathrm{h}} {}^{133}_{54}\mathrm{Xe}$$

$$\xrightarrow[\beta^-]{5.245\ \mathrm{d}} {}^{133}_{55}\mathrm{Cs}\ (\text{stable})$$

Note the increase in half-life as the stable ^{133}Cs is approached.

For even A isobars, the half-lives increase by alternate isobars as the line of β stability is approached, due to the nucleon pairing effect. Isobars on the even–even line have longer half-lives than those on the odd–odd line. For example, for the $A = 90$ isobars, the decay of ^{90}Br to ^{90}Zr occurs as follows (even–even isobars on the top line; odd–odd isobars are on the bottom line):

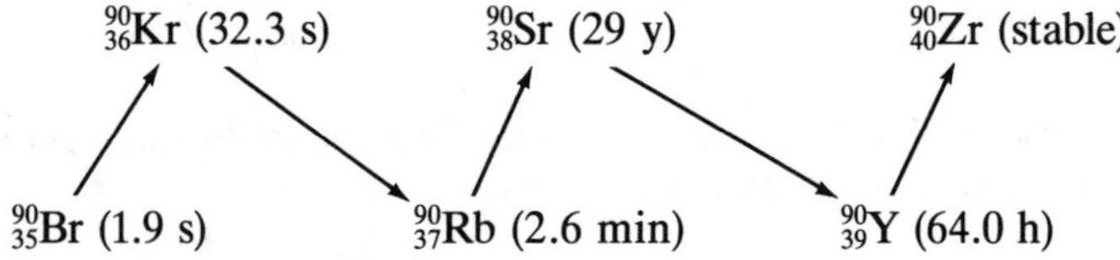

TERMS TO KNOW

Angular momentum
Barn
Binding energy
Bosons
Closed-cycle decay
Collective model
Coulomb energy term
Diffuse boundary correction term
Doubly magic
Electric quadrupole moment
Exchange character
Fermi gas model
Fermi level
Fermions
Isobaric pairs
Line of β stability
Liquid drop model
Magic numbers
Magnetic moment
Mass excess (defect)
Mattauch's rule
Nuclear energy surface diagram
Nuclear potential
Nucleus
Nuclidic mass
Odd–even rules of stability
Optical model
Pairing energy term
Parity
Q value
Semiempirical energy equation
Shell model
Symmetry
Symmetry term
Volume energy term

READING LIST

Arnikar, H. J., *Essentials of Nuclear Chemistry,* 2nd ed., New York: Wiley, 1987. [properties of nucleons and nuclei]

Burbidge, E. M., G. R. Burbidge, W. A. Fowler, and F. Hoyle, Synthesis of the elements in stars, *Rev. Mod. Phys.* **29**:547 (1957). [nucleosynthesis and the abundances of the elements]

Casten, R. F., *Nuclear Structure from a Simple Perspective,* New York: Oxford University Press, 1990. [fundamentals of nuclear structure]

Fermi, E., *Nuclear Physics,* rev. ed. Chicago: University of Chicago Press, 1950. [nuclear models]

Jensen, J. H. D., The history of the theory of the structure of the nucleus. *Science* **147**:1419 (1965). [history]

Mayer, M. G., The shell model. *Science* **145**:999 (1964). [nuclear structure]

Myers, W. D., and Swiatecki, W. J., Nuclear masses and deformations. *Nucl. Phys.* **81**:1 (1966).

Preston, M. A., and R. K. Bhaduri, *Structure of the Nucleus.* Reading, MA: Addison-Wesley, 1975. [nuclear structure]

Segré, E., *Nuclei and Particles.* Reading, MA: Benjamin, 1977. [nuclear structure]

EXERCISES

1. Show by calculation in which direction the following reaction would be energetically spontaneous: 2 $^{30}Si \leftrightarrow {}^{60}Ni$.

2. Assume that all the energy output of the sun is measured as 2.00 calories per minute per square centimeter at the earth's surface. The mean earth–sun distance is 1.49×10^8 km. How many kilograms of ^{4}He does the sun produce each year, if it is assumed that energy production is largely by the following nuclear reaction?

$$4\ {}^1\mathrm{H} \longrightarrow {}^4\mathrm{He} + 2\ \beta^+$$

3. ^{13}N decays by pure positron emission (no γ ray is emitted).

(a) Write the balanced nuclear equation for the decay.

(b) Calculate $E_{\max}$ for the positrons emitted, using mass excess (delta) tables.

(c) How could this decay mode be distinguished from negatron decay and electron capture decay?

4. ^{147}Sm is a rare-earth α particle emitter.

(a) What prediction would you make as to the energy of the α particles emitted and the half-life of the radionuclide as compared to α particle emitters in the transuranium region?

(b) Given that the energy of the α particles emitted is 2.23 MeV, calculate the recoil energy imparted to the daughter nuclide in units of kilojoules (kJ)/mole.

(c) How does the recoil energy calculated above compare to typical chemical bond energies?

5. The decay of ^{238}U to form ^{234}Th by α decay releases 4.1×10^{11} joules (J) of energy per mole of ^{238}U atoms. Given that the nuclidic masses of $^{238}\mathrm{U} = 238.0508$ amu and an α particle $= 4.0026$ amu, calculate the mass of ^{234}Th in units of g/mol.

6. (a) Calculate both the total binding energy (units of MeV) and the binding energy per nucleon for ^{204}Pb.

(b) What is the binding energy added by one additional neutron, if added to ^{204}Pb to make ^{205}Pb?

7. Solve the following problems using a chart of the nuclides:

(a) List two nuclides with magic neutron number 82.

(b) List two nuclides with doubly magic numbers.

(c) Which isotope of Ac is a radioactive daughter of ^{235}U?

(d) ^{18}F is a radionuclide used in positron emission tomography (PET) studies of glucose metabolism. The radionuclide may be made by bombarding ^{20}Ne with deuterons in a cyclotron. Write the equation for this nuclear reaction, including all reactants and products.

8. Calculate the radius in units of fermis (fm) of the nucleus of a ^{238}U atom, as it would be determined by a fast neutron scattering experiment.

9. Why do the Fermi levels of neutrons and protons differ?

10. There is a large abundance peak in Fig. 3.4 for elements in the vicinity of Fe. These elements are produced by an "equilibrium process" involving a variety of simple nuclear reactions. Describe two factors that contribute to the high cosmic abundance values for elements in this region.

11. **(a)** Calculate the density, in g/cm^3, of a nucleus of ^{19}F and of ^{239}Pu. (Determine radius of the nucleus using Eq. 3.1; assume the nuclei are spherical ($V = \frac{4}{3}\pi r^3$); assume the mass number, in g, of the nuclide contains Avogadro's number of atoms.)

(b) Compare the nuclear densities for ^{19}F and ^{239}Pu. Do they differ significantly?

12. Which of the following nuclides would be most likely to decay via delayed neutron emission? Give a reason for your answer.

(a) ^{89}Sr **(b)** ^{86}Sr **(c)** ^{90}Sr

13. Calculate Q values for the following decays. Obtain decay mode and energy information from a chart of the nuclides or nuclear data tables.

(a) ^{39}Ca **(b)** ^{72}Zn **(c)** ^{217}Rn

14. Calculate the binding energy for ^{26}Al, using the semiempirical binding energy equation.

CHAPTER

4

NUCLEAR REACTIONS

Nuclear reactions are interactions between an incoming fundamental particle, photon, neutrino, or multinucleon nucleus, and a target nucleus. The reaction may result in the scattering of the projectile, the excitation of the target nucleus, or excitation followed by nuclear transformations of the target to another nuclide by gain or loss of subatomic particles. Detailed studies of nuclear reaction energetics, mechanisms, and models are in the province of nuclear chemistry and nuclear physics. In this chapter some practical aspects of nuclear reactions that are of most interest and use to the radiochemist are examined.

4.1. TYPES OF NUCLEAR REACTIONS

4.1.1. Scattering Reactions

Two broad groups of particle-induced nuclear reactions may be identified. In **scattering** reactions, an incoming particle or *projectile* is deflected, or scattered, away from the target nucleus. This interaction results in a change of direction for the particles involved, but no change in the identity of either the projectile or the target nucleus. Scattering reactions are either elastic or inelastic. In **elastic scattering** the projectile–target system has essentially the same total kinetic energy before and after the interaction. Small kinetic energy losses, such as those to atomic or molecular excitation, may alter the energy balance slightly. The general notation for elastic scattering is

$$\mathrm{A(a, a)A}$$

where A is the target nucleus and a is the incoming particle. In a strict sense, elastic scattering is not a nuclear reaction, but merely a "billiard-ball" scattering event such as that employed by Rutherford in calculating the size of the nucleus. Rutherford backscattering is a method of elemental surface analysis based on measurement of the energies of elastically scattered particles (see Chapter 11).

In **inelastic scattering,** the incoming particle loses energy to the target nucleus, causing excitation of nuclear energy levels. The notation for inelastic scattering is

$$A(a, a')A^*$$

where a′ refers to the scattered projectile, and A* is the same target nuclide, but in a postreaction nuclear excited state. In inelastic scattering the excited nucleus of nuclide A* will later de-excite by gamma transition(s) and the identity of the target nuclide will be preserved. Scattering reactions have been used to study nuclear structure, including nuclear size, density, strong force interactions, and the nuclear energy level structure. Scattering reactions are of less interest to analytical chemists than reactions that involve transformation of the target nuclide into a different nuclide by absorption and emission of subatomic particles or radiation by a nucleus.

4.1.2. Other Reactions

A variety of particles are used to induce nuclear reactions in a target. Some of the more common ones are neutrons, protons, deuterons, and α particles. There are many possible outcomes that may result from the interaction of an incoming particle with a target nucleus. These different reaction paths are known as **reaction channels.** For example, consider the reaction of a deuteron (d) with ^{59}Co. Some possible reaction channels are summarized below.

Reaction	***Reaction Type***
^{59}Co(d, d)^{59}Co	elastic scattering
^{59}Co(d, d′)^{59}Co*	inelastic scattering
^{59}Co(d, γ)^{61}Ni	radiative capture
^{59}Co(d, p)^{60}Co	stripping reaction
^{59}Co(d, n)^{60}Ni	stripping reaction
^{59}Co(d, ^{3}He),^{58}Fe	pickup reaction
^{59}Co(d, α)^{57}Fe	pickup reaction

Figure 4.1 shows other common reactions that occur with various types of projectiles. However, just because a balanced nuclear reaction equation can be written does not mean that the reaction will actually occur. The occurrence of the reaction at a measurable rate will depend on such factors as the energy of the incoming projectile and a factor expressing the probability of the interaction. The second column in the list of possible reaction channels above gives a name for some types of outcomes that occur frequently. These mechanisms are further discussed later in this chapter.

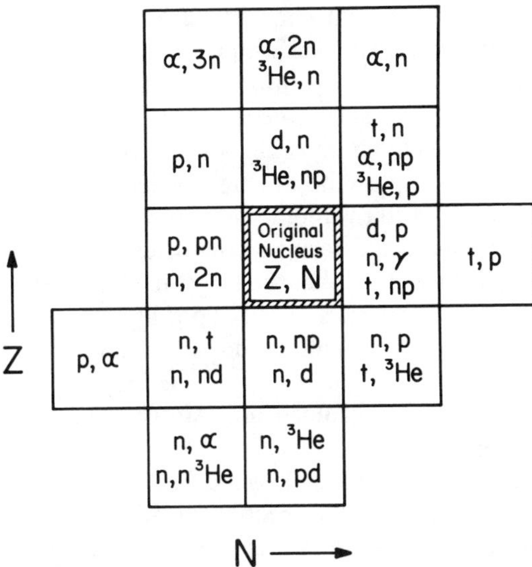

Figure 4.1. Positioning of products of selected particle-induced nuclear reactions (<25 MeV) on a chart of the nuclides.

4.2. ENERGETICS OF NUCLEAR REACTIONS

Energy, mass number (except for very high energy reactions), and momentum (both linear and angular) are conserved quantities in nuclear reactions. The amount of energy released or absorbed in a reaction is defined as the *Q value* for the reaction (see Chapter 3). If the value of Q is positive, the reaction releases energy and is exoergic. If Q is negative, the reaction requires energy and is endoergic. Most kinds of induced nuclear reactions fall into the latter group. In this section, the energy actually needed to bring about a given nuclear reaction will be calculated. An example for this calculation is the first induced nuclear reaction, discovered by Rutherford early in this century:

$$^{14}\text{N} + {}^{4}\text{He} \longrightarrow {}^{1}\text{H} + {}^{17}\text{O} + Q$$

The value of Q for this reaction is -1.19 MeV, so the reaction is endoergic. However, if this reaction were attempted using α particles of 1.19 MeV energy, the reaction would not proceed at a measureable rate. The reason for this is that not all of the energy of the incoming particle is effectively used in bringing about the reaction. Some of the incident particle energy is also used up in conserving momentum of the system, and extra energy above the Q value may be required to overcome the repulsive Coulomb barrier that exists between the incident charged particle and the nucleus. Therefore, corrections must be made for both of these factors to determine the actual incident energy required to initiate an endoergic reaction.

The minimum amount of energy needed to bring about appreciable reaction is called the **threshold energy.** This energy is equal to the Q value for the reaction, if negative, plus the amount of additional energy needed to conserve momentum and to overcome the Coulomb barrier. The presentation here will be based on calculation of the threshold energy in the *laboratory system,* where it is assumed that the target nucleus is at rest and is struck by a moving projectile. In physics, calculations are often done in a *center of mass system,* where a separate momentum correction is not required.

4.2.1. Momentum Correction

Momentum, both linear and angular, is a conserved quantity in nuclear reactions. In the simple case of a projectile impact at the center of a target nucleus, no angular momentum is transferred from the projectile to the target and we need only consider transfer of linear momentum. To calculate the amount of the projectile energy that goes into conserving linear momentum, consider the reaction of x (a bombarding particle) on y (the stationary target nuclide). Assume that x has a velocity v_x before impact, and that the target has 0 velocity before impact. After impact, the initial compound nucleus, z, has a velocity v_z. (It should be noted that the compound nucleus may later dissociate into other final products).

$$[\mathrm{x} \xrightarrow{v_x} \mathrm{y}] \longrightarrow [\mathrm{z} \xrightarrow{v_z} \ ]$$

Before impact After impact

The kinetic energy (E_x) of the incident particle is

$$E_x = \tfrac{1}{2}m_x v_x^2 \tag{4.1}$$

The kinetic energy (E_z) of the compound nucleus is

$$E_z = \tfrac{1}{2}m_z v_z^2 = \tfrac{1}{2}(m_x + m_y)v_z^2 \tag{4.2}$$

By conservation of momentum ($p = mv$),

$$m_x v_x = (m_x + m_y)v_z \tag{4.3}$$

and rearranging gives

$$v_x = (m_x + m_y)v_z/m_x \tag{4.4}$$

Squaring this expression and substituting it into Eq. 4.1 gives

$$v_x^2 = (m_x + m_y)^2 v_z^2/m_x^2 = 2E_x/m_x \qquad (4.5)$$

Solving for E_x gives

$$E_x = (m_x + m_y)^2 v_z^2/2m_x = \tfrac{1}{2}(m_x + m_y)^2 v_z^2/m_x \qquad (4.6)$$

Dividing E_z by E_x gives

$$\frac{E_z}{E_x} = \frac{\frac{1}{2}(m_x + m_y)v_z^2}{\frac{1}{2}(m_x + m_y)^2 v_z^2/m_x} = \frac{m_x}{m_x + m_y} \qquad (4.7)$$

Equation 4.7 represents the fraction of the incoming particle energy that ends up as kinetic energy of the compound nucleus to achieve conservation of momentum. Another way to look at this is to consider that the part of the particle energy available to cause the reaction is

$$[m_y/(m_x + m_y)]E_x \qquad (4.8)$$

Thus, the incident energy actually needed to initiate a reaction for which Q is negative would be

$$E_{\text{incident}} = [(m_x + m_y)/m_y]\,|Q| \qquad (4.9)$$

For the (α, p) reaction on ^{14}N, the incident energy required would be

$$E_{\text{incident}} = [(14 + 4)/14](1.19) = 1.5 \text{ MeV}$$

However, if the reaction were tried with 1.5-MeV α particles, it would still not proceed at a significant reaction rate. Another correction is needed.

4.2.2. Coulomb Barrier Correction

The incoming α particle must also overcome a Coulomb barrier (the repulsion between two particles of like charge, see Fig. 3.3) to effect the desired reaction. As the α particle approaches the nucleus, it experiences repulsion up to a certain distance, where it begins to feel the attractive nuclear strong force. The height of the Coulomb barrier (V^0) is given by

$$V^0 = \frac{Z_x Z_y e^2}{R_x + R_y} \qquad (4.10)$$

where Z_x and Z_y are the charges on the projectile (x) and nucleus (y), e is the charge on an electron, and R_x and R_y are the radii of the projectile (x) and target (y) nuclei. If R values are given in cm and e in esu, then V^0 is in units of ergs. If R values are given in fermis, and $e^2 = 1.44$ MeV-fm, then we can write

$$V^0(\text{MeV}) = \frac{1.44\, Z_x Z_y}{R_x + R_y} \tag{4.11}$$

Equation 3.1 states that the radius of a nucleus is approximately $1.4A^{1/3}$ fm. Therefore, for the (α,p) reaction on ^{14}N, the Coulomb barrier is

$$V^0 = \frac{1.44\,(2)\,(7)}{1.4[(14)^{1/3} + (4)^{1/3}]} = 3.6 \text{ MeV}$$

It is important to note that the barrier exists both for the incoming projectile and the emitted particle. Therefore, barrier calculations for nuclear reactions are based on either the projectile in, or the particle out, whichever has the highest charge. The value for the radius ($1.4A^{1/3}$ fm) is useful for most reactions except for proton-induced reactions on medium to heavy nuclei, where the proton radius is taken to be zero (a point charge). The V^0 values for medium to heavy elements are about 12 MeV for proton-induced reactions, and 24 MeV for α-induced reactions. **Quantum-mechanical tunneling** is always possible, so some reactions do occur at lower energies. The threshold energy in the laboratory system for a nuclear reaction is the projectile energy required for the reaction to occur with a high probability. It is calculated by applying the momentum correction either to the reaction's Q value (if negative), or to the Coulomb barrier, whichever is greater. Threshold energies are usually tabulated as positive values, even though they represent energies required for reaction. In the above example, the Coulomb barrier (3.6 MeV) is larger than the absolute value of Q (1.19 MeV), so the threshold energy in the laboratory system for the reaction of α particles on ^{14}N is

$$3.6 \text{ MeV } [(4 + 14)/14] = 4.6 \text{ MeV}$$

4.3. CROSS SECTIONS FOR NUCLEAR REACTIONS

In previous chapters, the concept of reaction **cross section,** the probability that a given nuclear reaction will occur, was alluded to briefly. The symbol for cross section is the Greek letter sigma (σ). Figure 4.2 can be used to help envision the concept of cross section. For a nuclear reaction to occur,

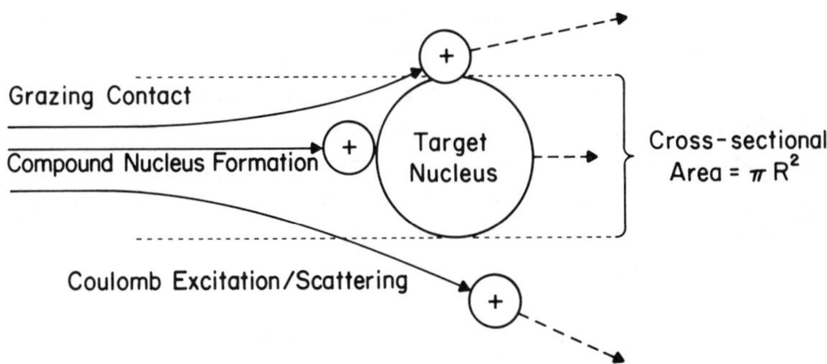

Figure 4.2. A representation of the concept of target nucleus cross-sectional area as a measure of reaction probability. Contact and probably a reaction occur when the edge of the incident particle intersects the circle representing the maximum cross-sectional area (πR^2) of the target nucleus. Grazing contacts are those where the projectile and target just touch, energy is transferred, and then the incident particle–target nucleus amalgamation separates anew. Grazing contact reactions can be important for reactions involving heavy incident particles. Coulomb excitation of the target nucleus and particle scattering occur when the incident projectile passes close to, but does not contact, the target nucleus. More direct contacts at moderate projectile energies often lead to the formation of an excited compound nucleus, which may then de-excite by emission of radiations and/or particles that are different than the incident projectile. (After R. Kaufmann and R. Wolfgang, *Proc. 2d Conf. Reactions between Complex Nuclei*. New York: Wiley, 1960.)

the projectile and the target nucleus must come close enough together to interact. This distance will vary for different types of reactions. For scattering reactions, there need not be any actual contact between the projectile and the nucleus. For other types of reactions, the projectile and the nucleus might have to come into direct contact. In the latter case, the probability of reaction with a given projectile could, as a first approximation, be related to the size of the target nucleus, that is, to the cross-sectional area presented to the projectile. This area calculated for a spherical nucleus with radius R is called the **geometrical cross section of the target nucleus:**

$$\sigma_{\text{geometric}} = \pi(R_{\text{target}})^2 \tag{4.12}$$

Example. Calculate the probability for a reaction of a fast neutron with a nucleus of $A = 100$, using only the concept of geometrical cross section of the target nucleus.

$\sigma = \pi R^2$ and R = target nucleus radius $= 1.4A^{1/3}$ fm.

For A = 100,

$$R = 1.4(100)^{1/3} = 6.5 \text{ fm} = 6.5 \times 10^{-13} \text{ cm}$$

$$\sigma = \pi(6.5 \times 10^{-13})^2 = 1.3 \times 10^{-24} \text{ cm}^2$$

The cross section for most nuclei, calculated in this simple way, would be on the order of 10^{-24} cm^2. More accurately, a true collisional cross section is based on an interaction zone with a radius that is equal to the sum of the radii of the projectile and the target nucleus, not just on the radius of the target nucleus. Notice that the units for cross section are those of area, and cm^2 is the most commonly used unit. Another traditional and still commonly used unit of cross section is the **barn (b).** Relationships among some common cross section measurements are

$$1 \text{ barn} = 1 \times 10^{-24} \text{ cm}^2 = 100 \text{ fm}^2$$

$$1 \text{ mb} = 1 \times 10^{-3} \text{ b} = 10^{-27} \text{ cm}^2$$

$$1\ \mu\text{b} = 1 \times 10^{-6} \text{ b} = 10^{-30} \text{ cm}^2$$

Cross sections may be defined for specific reaction channels, or for the overall reaction probability of the target nucleus. It is apparent that simple target nucleus geometrical cross-section calculations ignore many factors that could be important in estimating the probability of a nuclear reaction, such as projectile energy, projectile size, and stability of the target nucleus. Actual reaction cross sections do vary widely with the energy and identity of the incoming particle, and with the characteristics of the target nucleus. For example, cross sections for reactions with slow neutrons as projectiles are often much larger than the geometric cross section, while cross sections for neutron reactions with magic number nuclei are a great deal smaller than expected on the basis of simple geometrical cross section.

Reaction cross-section values may be calculated using quantum-mechanical considerations, or they can be experimentally determined. Various methods of calculation can be found in nuclear physics texts. Here, only experimental methods will be discussed.

4.3.1. Measurement of Cross Section

In the experimental determination of cross section, the rate of a particular reaction, the number of projectiles, and other parameters are measured and the cross section calculated from these values. The calculations differ depending on the relative thickness of the target and its effect on the intensity of the incoming particles.

Thin Targets

Thin targets are defined as those in which the incident particles experience no significant attenuation. The intensity (particles/s) of incoming particles passing through a given area of the target is known as the **flux**

density, or **fluence rate** (particles cm^{-2} s^{-1}). In a thin target, the flux density of projectiles is assumed to be constant throughout the target. Reactions in thin targets can involve either a well-defined beam of particles incident on one target surface (usually charged particles), or a homogeneous flux of particles incident on the target from all directions (usually neutrons). In either case, the assumptions made for subsequent calculations are the following:

1. There is no appreciable decay of the product nuclide during the irradiation period.
2. The number of target atoms is not greatly depleted during the course of the irradiation (i.e., there is no significant **burn-up**).
3. The cross section is constant throughout the target, because there is no significant change in the incident particle energy.

The first assumption is true for products that have half-lives that are long with respect to the time of irradiation. The second and third assumptions are generally valid with moderate irradiation times and thin targets.

For the case of a beam of particles incident on the target, the number of nuclear reactions induced in the target per unit time (R) is given by

$$R = Ixn'\sigma \tag{4.13}$$

where I = flux = incident particles/s

x = target thickness (cm)

n' = number of target nuclei/cm^3

σ = cross section (cm^2)

Particle beam experiments often are done with charged particles accelerated to relatively high energies. The cross sections for charged-particle reactions are generally no larger than a few barns, and are often much smaller.

For the case in which the target is embedded in a uniform flux of particles, as a target in a nuclear reactor, Eq. 4.13 for the rate of reaction (R) can be rewritten as follows:

$$R = n\phi\sigma \tag{4.14}$$

where n = the total number of target nuclei

ϕ = flux density = incident particles cm^{-2} s^{-1}

σ = cross section (cm^2)

Another form of the above equation is

$$N_{\text{products}} = N_{\text{target}}\phi\sigma t \tag{4.15}$$

where N_{products} = number of product nuclei produced

N_{target} = number of target atoms exposed to the incident particles

t = time of exposure of target to the incident particles, the irradiation time

These equations can be used to calculate cross sections.

Example. A thin sample of 1.00 μg of natural cobalt was irradiated in a thermal neutron flux density of 1.00×10^{13} n cm^{-2} s^{-1} for 10^3 s. 2.00×10^9 atoms of ^{60}Co were produced. Calculate the cross section (in barns) for the capture of neutrons by natural cobalt (100% natural abundance ^{59}Co) to produce radioactive ^{60}Co.

$$\begin{aligned}
N_{\text{product}} &= N_{\text{target}}\phi\sigma t \\
N_{\text{target}} &= [1.00 \times 10^{-6}\ \text{Co}/(58.9\text{g}\ ^{59}\text{Co/mol})] \\
&\quad \times (6.02 \times 10^{23}\ \text{atoms/mol}) \\
&= 1.02 \times 10^{16}\ \text{atoms}\ ^{59}\text{Co} \\
2.00 \times 10^{9}\ \text{atoms}\ ^{60}\text{Co} &= (1.02 \times 10^{16}\ \text{atoms}\ ^{59}\text{Co}) \\
&\quad \times (1.00 \times 10^{13}\ \text{n cm}^{-2}\ \text{s}^{-1})\sigma(10^{3}\ \text{s}) \\
\sigma &= (1.96 \times 10^{-23}\ \text{cm}^2)\ (1\ \text{b}/10^{-24}\ \text{cm}^2) \\
&= 19.6\ \text{b}
\end{aligned}$$

Thick Targets

In thick targets, the particle beam is significantly attenuated. Therefore, the flux of particles coming out of the target is less than the incident flux. For neutrons, there are few targets with thicknesses of a few millimeters, or less, that need to be considered as "thick." This is because neutrons have no charge and pass through matter freely. However, charged particles interact to a much greater extent with matter, so many charged-particle reactions will require calculations based on the following "thick" target considerations.

The attenuation in a target, dI, is given as

$$dI = I_0 - I \tag{4.16}$$

The amount of attenuation of the incident beam depends on the thickness of the target, the number of target nuclei, the intensity of the incident beam, and the cross section.

$$-dI = In'\sigma(dx) \tag{4.17}$$

where $-dI$ = differential attenuation of the beam intensity

n' = number target nuclei/cm^3

x = distance into the target (cm)

I = beam intensity (particles/s)

σ = cross section (cm^2)

Integration of the above equation will give

$$\int_{I_0}^{I_x} \frac{dI}{I} = -n'\sigma \int_0^x dx \tag{4.18}$$

$$\ln I_x - \ln I_0 = -n'\sigma x \tag{4.19}$$

$$\ln(I_x/I_0) = -n'\sigma x \tag{4.20}$$

$$I_x = I_0 e^{-n'\sigma x} \tag{4.21}$$

It is assumed in the use of this equation that the cross section is constant as the energy of the incident particle decreases as it passes deeper into the target. This is not always true. It is also assumed that the number of target nuclei is not significantly depleted during irradiation, which is most often true.

Example. Cadmium has a very high cross section for neutrons, and even thin sheets may significantly attenuate a neutron beam. Calculate the cross section (in barns) for Cd, given that a 0.0204-cm thickness of Cd reduces the intensity of a beam of thermal neutrons to exactly 0.1 of its original intensity. The density of Cd metal is 8.64 g/cm^3, and the atomic mass of Cd is 112.4 daltons.

$$\begin{aligned} n' &= (8.64 \text{ g/cm}^3)[6.02 \times 10^{23} \text{ atoms/(112.4 g/mol)}] \\ &= 4.63 \times 10^{22} \text{ atoms/cm}^3 \end{aligned}$$

$$I_x/I_0 = e^{-n'\sigma x}$$

$$0.1 = e^{-(4.63E22)(\sigma)(0.0204)} \qquad (\textit{Note: } 4.63E22 = 4.63 \times 10^{22})$$

$$\sigma = 2.44 \times 10^{-21} \text{ cm}^2 = 2.44 \times 10^3 \text{ b}$$

The methods described above for determination of cross sections measure total reaction cross sections. Often, it is the cross section for one particular reaction channel that is of interest. Appropriate experimental conditions, such as use of enriched isotope targets, would allow for determination of individual cross sections. The number of particles that have interacted to produce a given product radionuclide can be calculated from the activity of the product (Chapter 5). This number of product atoms can be set equal to $(I_0 - I)$ and the cross section for the particular reaction channel calculated from

$$I_0 - I = I_0(1 - e^{-n'\sigma x}) \tag{4.22}$$

4.3.2. Excitation Functions

Variations in cross sections for different projectiles at varying energies may be examined by looking at excitation functions. **Excitation functions** are plots of reaction cross section against energy of the incident particle for a specific reaction.

Neutron Excitation Functions

An excitation function for a neutron-induced reaction is shown in Fig. 4.3. Several features of this plot are worth consideration. The first is the general trend toward lower cross section with increasing neutron energy. This means that neutrons with very low energy have the greatest likelihood of inducing a nuclear reaction in a target nucleus. For the lower-energy neutrons (several hundredths of an eV to several hundred eV), the decrease in σ with increasing energy is roughly proportional to $1/v_n$, where v_n is the velocity of the neutron. This is referred to as the **one-over-*v* law.** Superimposed on this smoothly dropping curve are sharp peaks. These peaks, known as **resonance peaks,** mark neutron energies for which sharp increases in reaction probability are seen. These particular energies would correspond to the nuclear energy level spacings in the target that are being excited by the incoming neutron. An equivalent situation exists in atomic absorption spectroscopy where absorption of photons occurs at energies required to promote electrons to higher atomic energy levels. The analytical importance of the resonance region is discussed further in Chapter 9 in the section on epithermal neutron activation analysis.

The cross sections for slow-neutron reactions are sometimes quite large, as much as 10^4 or even 10^5 b. These are, of course, much larger values than would be expected using the simple cross-section approximation (about 1 b)

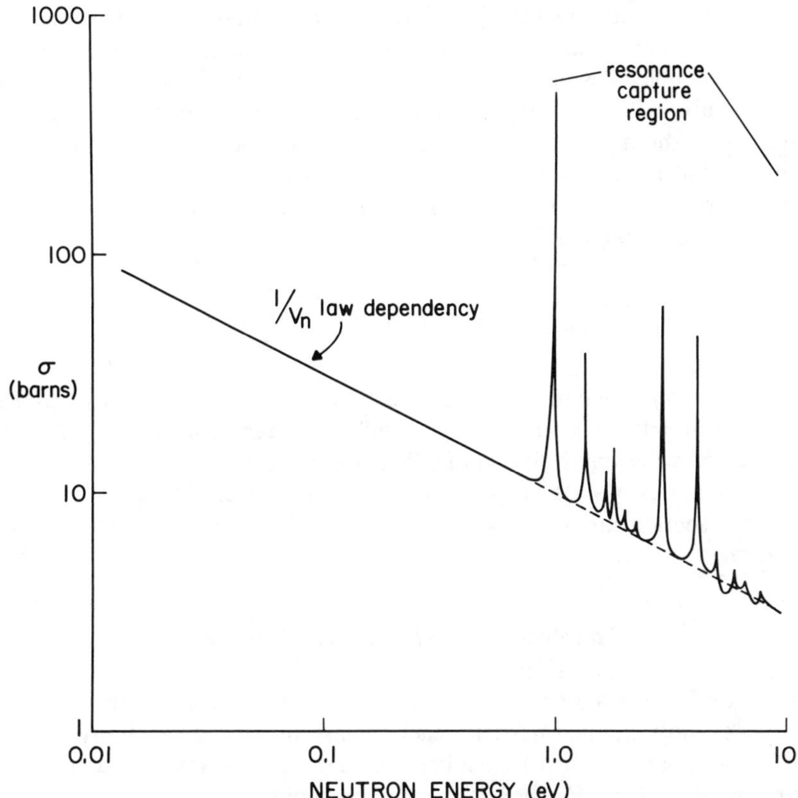

Figure 4.3. An excitation function for a neutron-induced reaction. Neutron capture cross sections (σ) are plotted against low incident neutron energies. Resonance peaks have small widths at half-height ($\approx$0.1 eV).

described earlier. The explanation for this observation requires the use of quantum-mechanical concepts.

Just as electromagnetic radiation has both a wave and particle nature, so particles have a wave component. The **de Broglie wavelength** (λ) of a particle is given by

$$\lambda = h/mv \tag{4.23}$$

where h = Planck's constant = 6.63×10^{-34} J · s

m = particle mass (kg)

v = particle velocity (m/s)

Small particles with low velocities have the most pronounced wave nature. Low-energy, low-velocity neutrons with energies approximately equal to that of gas molecules at room temperature have large wavelengths. These neutrons are called **slow** or **thermal neutrons.** The greater the de Broglie wavelength, the higher is the likelihood of an interaction between the neutron and the target nucleus. Thermal neutrons will, therefore, have high capture cross sections. It can be shown that the cross section for thermal neutrons is approximated by

$$\sigma = \pi (R + \lambdabar)^2 \tag{4.24}$$

where the simple geometrical cross section of the target nucleus (πR^2) is increased by adding $\lambdabar$ to the target nucleus radius. The term $\lambdabar$ refers to the neutron wavelength divided by 2π; that is, $\lambdabar = \lambda/2\pi$. As the neutron energy increases, the de Broglie wavelength decreases. At higher energies, the cross section approaches the value of πR^2 and exhibits little further variation.

Charged-Particle Excitation Functions

Excitation functions for charged-particle reactions have a very different appearance. Most charged-particle reactions are **threshold reactions;** that is, the incident particle must have a certain amount of energy before any reaction can take place. For neutrons, the threshold energy may be zero, because there is no Coulomb barrier to overcome and a large positive Q value will provide energy for an exiting charged particle to overcome its potential energy barrier. The threshold energy requirement for charged-particle induced reactions implies that, at lower energies, there will be little chance of a reaction occurring. However, some reactions at energies below the threshold energy can occur through quantum-mechanical tunneling. The probability of reaction will dramatically increase as the projectile reaches the threshold energy. It would be expected, therefore, that charged-particle cross sections would initially increase with increasing energy, instead of decreasing, as was the case for neutron excitation functions. Figure 4.4 shows the excitation function for several proton-induced reactions on a given target. The previous predictions are confirmed. Little reaction occurs until the threshold energy is reached. From that point, the cross section increases steadily with increasing energy, until it reaches a maximum. These maxima are reached just before the threshold energy for a competing reaction

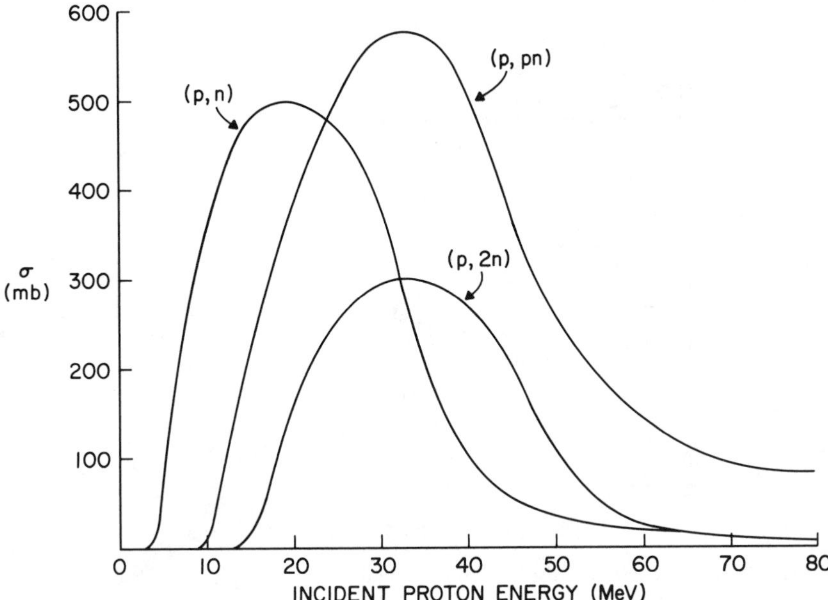

Figure 4.4. Excitation functions for several common proton-induced reactions on the same low-Z target nuclide.

channel is reached. After a maximum value is reached, the cross section for the reaction with the lower threshold energy will decrease, and that for the next higher threshold reaction will increase until it, too, reaches a maximum. Hence, the excitation function for a charged-particle reaction will show a series of curves, one for each different type of reaction that might occur. Cross sections for charged-particle reactions are usually much lower than those for neutron-induced reactions, often only fractions of a barn. Figure 4.5 compares the values for the cross sections of neutron and charged-particle-induced reactions. The horizontal line represents the geometrical cross section of πR^2. For neutrons, cross sections are highest for low energies, and decrease with increasing energy to approach πR^2. For charged-particle reactions, cross sections are lowest for low-energy particles, and gradually increase to approach the value of πR^2.

Table 4.1 summarizes some typical reactions that can occur on nuclides with intermediate and heavy mass numbers for various projectile energies.

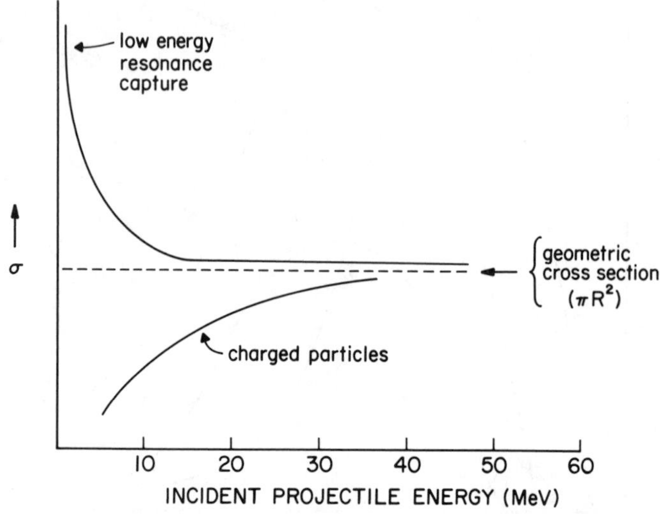

Figure 4.5. A comparison of the variations in neutron and charged-particle cross sections with incident projectile energy.

4.4. REACTION MECHANISMS

In this section, some of the mechanisms through which nuclear reactions involving incident particles are thought to occur are discussed. For projectiles with less than 50 MeV energy, there are two mechanisms that predominate: compound-nucleus formation and direct interactions.

4.4.1. Compound-Nucleus Formation

In the **compound-nucleus mechanism,** the projectile strikes the target nucleus and is assimilated into it to form a **compound nucleus.** The projectile adds its kinetic energy and binding energy to the target nucleus, so the compound nucleus is in an excited state. The energy brought in by the projectile is transferred to the other nucleons through numerous collisions until a state of statistical equilibrium is reached. At this time, the original particle is indistinguishable from the other nucleons, and the compound nucleus has no "memory" of how it was formed. If sufficient energy is present, eventually one nucleon, or a composite of nucleons, will acquire enough energy to overcome nuclear binding forces and escape from the

Table 4.1. Common Light Particle Reactions with Medium and Heavy Mass Target Nuclei

Incident Particle Energy		Medium Mass Targets	Heavy Mass Targets
Low, (0–1 keV)	n	(n, n_{el}), (n, γ)	(n, n_{el}), (n, γ)
Intermediate (1–500 keV)	n	(n, n_{el}), (n, γ)	(n, n_{el}), (n, γ)
	p	(p, n), (p, γ), (p, α)	
	α	(α, n), (α, γ), (α, p)	
	d	(d, p), (d, n)	
High (0.5–10 MeV)	n	(n, n′), (n, n_{el}), (n, p), (n, α)	(n, n′), (n, n_{el}), (n, p), (n, γ)
	p	(p, n), (p, p′), (p, α)	(p, n), (p, p′), (p, γ)
	α	(α, p), (α, n), (α, α′)	(α, p), (α, n), (α, γ)
	d	(d, p), (d, n), (d, pn), (d, 2n)	(d, p), (d, n), (d, pn), (d, 2n)
Very high (1–50 MeV)	n	(n, 2n), (n, n′), (n, n_{el}), (n, p), (n, np), (n, 2p), (n, α), ≥3 particles	(n, 2n), (n, n′), (n, n_{el}), (n, p), (n, np), (n, 2p), (n, α), ≥3 particles
	p	(p, 2n), (p, n), (p, p′), (p, np), (p, 2p), (p, α), ≥3 particles	(p, 2n), (p, n), (p, p′), (p, np), (p, 2p), (p, α), ≥3 particles
	α	(α, 2n), (α, n), (α, p), (α, np) (α, 2p), (α, α′), ≥3 particles	(α, 2n), (α, n), (α, p), (α, np), (α, 2p), (α, α′), ≥3 particles
	d	(d, p), (d, 2n), (d, np), (d, 3n), (d, d′), (d, t), ≥3 particles	(d, p), (d, 2n), (d, np), (d, 3n), (d, d′), (d, t), ≥3 particles

Source: After J. Blatt and V. Weisskopf, *Theoretical Nuclear Physics.* New York: Wiley, 1952, Chapter 9.

Note. Coulombic excitation for intermediate, high, and very high energy charged-particle reactions, charged-particle elastic scattering reactions, and induced fission reactions which may be important for the heaviest elements are not listed.

Abbreviations: el, elastic scattering; (n, n′), (p, p′), inelastic scattering.

nucleus. This process is called **evaporation**, because it is analagous to the way in which a molecule in the liquid phase acquires sufficient energy to escape into the vapor phase. The particle emitted from the compound nucleus bears no relationship to the original projectile, because a state of equilibrium had been reached before emission. If residual energy remains after one nucleon or composite unit has been evaporated, more may be emitted, until there is too little energy left to cause particle emission. The nucleus may then further de-excite via a gamma transition.

There are two distinct steps in the compound-nucleus-reaction mechanism:

1. Capture of the projectile to form a compound nucleus
2. Decay of the compound nucleus via particle evaporation and γ-ray emission

An important identifier of the compound-nucleus mechanism is the time separation of the capture and decay steps. The lifetime of the compound nucleus (10^{-14} to 10^{-20} s) is long as compared to the time it takes an energetic particle to cross the nucleus ($\approx 10^{-23}$ s), and the equilibrium state has time to develop. Once this state is reached, decay is related only to the amount of energy added to the nucleus, and not to the way the energy was added in the first place.

Neutron-capture reactions are good examples of processes that proceed via compound-nucleus formation. For thermal neutrons, the compound nucleus does not usually have enough energy for particle emission to take place. Therefore, emission of a γ ray is the primary means of de-excitation. These (n, γ) reactions are examples of **radiative capture** reactions, and are very important in activation analysis. Neutrons with higher energies do excite the compound nucleus sufficiently to cause particle emission. Therefore, reactions such as (n, n'), (n, p), and (n, α) are common with the 14-MeV neutrons produced by the $d(t, n)\alpha$ reaction in a small accelerator used as a neutron generator. Reactions induced by charged particles also may proceed via compound-nucleus formation. The charged particles need higher energies than do the neutrons to penetrate the nucleus (because they must penetrate the Coulomb barrier). After barrier penetration, this energy again becomes available to aid in particle detachment, so charged-particle-induced reactions will almost always result in particle evaporation.

4.4.2. Direct Interactions

In contrast to compound-nucleus formation, **direct interactions** do not result in the assimilation of the projectile into the nucleus. Instead, the incoming particle interacts directly with the nucleons at the periphery of the nucleus in a very fast process ($\approx 10^{-22}$ s) that involves the prompt emission of a nucleon. Direct interactions happen in only one step, with no time for equilibration between the initial interaction and resulting nuclear change. In direct interactions, therefore, the nucleus does have a "memory" of how it was formed, and the decay modes are related to the means by which the nucleus was originally excited. Direct interactions may occur by either **knock-on** or **transfer** processes.

In the knock-on mechanism, a collision between the projectile and an outer nucleon results in the direct ejection of one, or more particles. This type of process is most commonly observed for projectiles with higher energies (50 MeV and above).

Transfer mechanisms can be divided into **stripping** and **pickup** reactions. Their names are descriptive of the processes that occur. In stripping, the target nucleus takes one of the nucleons away from the composite incident particle. In a pickup reaction, the projectile takes a nucleon away from the target nucleus. Transfer reactions are most often observed for particles with energies less than 50 MeV.

The deuteron-induced reactions used as examples in Section 4.1 illustrate transfer processes. In the pickup reactions, the deuteron gained either a proton (d, ^{3}He) or another deuteron (d, α) from the ^{59}Co target. The stripping reactions show the loss of either a proton (d, n) or a neutron (d, p) from the deuteron to the ^{59}Co.

Deuteron reactions have been studied and used for radionuclide production since the early days of accelerator use. Even then, it was well known that the (d, p) reaction channel is much more probable than the (d, n) channel for low- or moderate-energy deuteron-induced reactions. This was explained by J. R. Oppenheimer and M. Phillips, who suggested that as the deuteron approached the target, it oriented itself so the proton was away from and the neutron closer to the target nucleus. The neutron, which experiences no Coulomb repulsion, is attracted by the nuclear forces in the target nucleus and is stripped from the weakly bound deuteron, leaving the proton as the "emitted" particle. This is called the **Oppenheimer–Phillips process.** This phenomenon made possible the production of radionuclides with small accelerators capable of producing particle energies that were below the calculated threshold energies, based on Coulomb barrier considerations.

4.5. SPECIAL NUCLEAR REACTIONS

4.5.1. Neutron-Induced Fission

In Chapter 2 the process of spontaneous fission as a mode of radioactive decay was introduced. A more important reaction from an economical and social viewpoint is that of **neutron-induced fission.** In this process, a neutron is captured by a heavy nuclide to form a compound nucleus. This nucleus then splits into two smaller fragments and emits two or three free

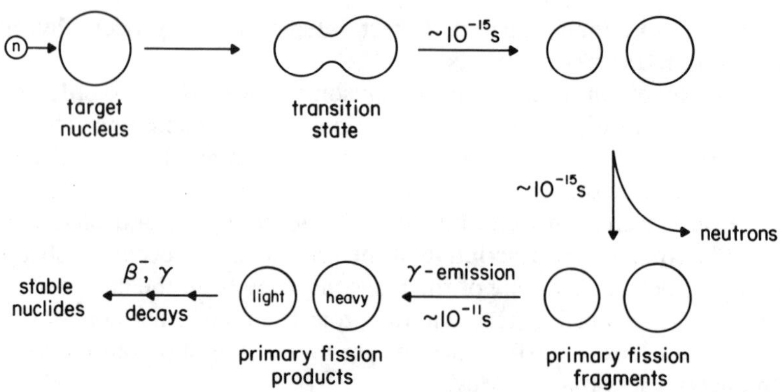

Figure 4.6. A representation of neutron-induced fission. Fission of heavy elements (e.g., actinides) induced by low-energy neutrons is principally asymmetric, leading to fission products with masses that typically differ by about 50 mass units.

neutrons, along with about 200 MeV of energy per nucleus fissioned. Fission may also be induced by charged particles, like protons, deuterons, or ^{3}He, but it is neutron-induced fission that will be discussed here.

Figure 4.6 gives a schematic illustration of how neutron-induced fission is thought to occur. In the target nucleus, there are two competing forces at work: a binding force that holds the nucleons together, and a Coulomb force that tends to push them apart. In a stable nucleus, the binding force is stronger. For low-mass nuclides, the binding force is much larger than the repulsive force, so they are not subject to fission. For heavier nuclides, the two forces are more equal in magnitude, and it is possible to shift the balance in favor of the repulsive forces by addition of a neutron. The neutron provides some internal excitation energy that deforms the nucleus into a transition state. If the deformation is severe enough, the Coulomb forces take over and split the nucleus into two fragments. These fission fragments are repelled away from each other with considerable kinetic energy, due to the Coulomb repulsive forces. The fragments are neutron-rich, so they will decay by negatron emission, often followed by a gamma transition, until a favorable N/Z ratio is attained and a stable product results. The time scale for these events is shown in Fig. 4.6.

The fission of a given nuclide will not always yield the same fission fragments. Figure 4.7 shows a **mass yield curve** for the neutron-induced fission of a heavy nuclide such as ^{235}U or ^{239}Pu. Plotted on the vertical axis is the **cumulative fission yield,** which gives the percentage of fission events that result in a product nuclide of mass A. Several features of this plot should be noted. The shape of the plot shows two large maxima with a valley

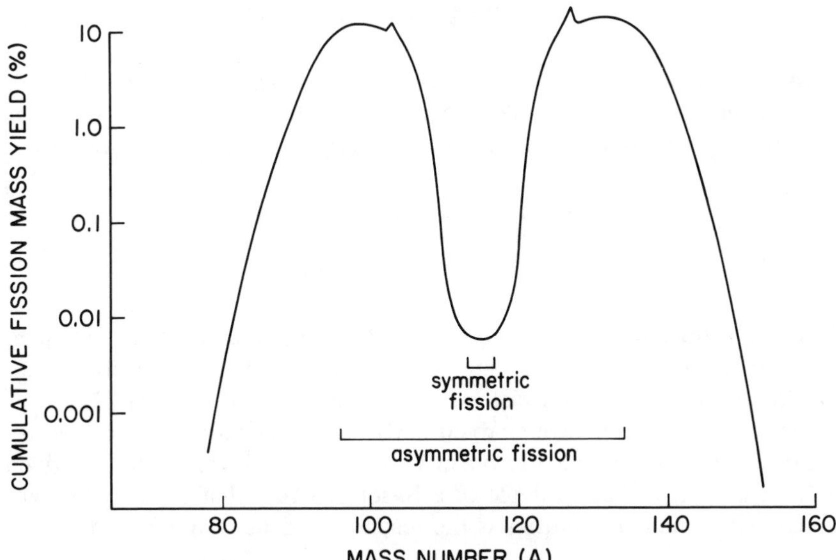

Figure 4.7. . A fission mass yield curve for the low-energy neutron-induced fission of a heavy-mass nuclide. At low neutron energies, asymmetric fission predominates. At high energies, neutrons and charged particles will produce a single-humped fission yield curve indicating that symmetric fission is favored. Closed shell (magic number) effects contribute to the spikes near the tops of the two maxima. Because fission is ordinarily binary (two fission fragments are produced), the total mass yield over all values of A is 200%. Ternary fission also occurs, but is rare compared to binary fission.

between them. The valley occurs at $A \approx 117$. This mass would result from a symmetric distribution of nucleons between the fission fragments. In fission induced by low-energy neutrons, then, an even split of nucleons between fragments is much less likely to occur than an uneven, or asymmetric, distribution. The two maxima occur at masses near $A = 95$ and $A = 138$, so fragments of this size are most likely to occur after fission. Superimposed on the maxima are two spikes, at $A = 100$ and $A = 134$. These spikes are due to shell effects, where a magic number of nucleons is reached. The appearance of the mass yield curve is altered as excitation energy increases. The likelihood of symmetric fission is greater for higher excitation energies, so the depth of the valley decreases.

Two nuclides that have a high cross section for fission by thermal neutrons are ^{235}U and ^{239}Pu. Both of these have applications in nuclear power generation and weapons production. The ^{238}U isotope is not readily fissionable with thermal neutrons, but can undergo fast neutron-induced fission.

4.5.2. Fusion

Fusion reactions are those in which two lighter nuclei are joined to form heavier nuclides, with the release of energy. Most useful fusion reactions take place between low-Z nuclei. Some examples of fusion reactions are

$$^{2}\text{H} + \text{p} \longrightarrow {}^{3}\text{He} + \gamma + Q$$

$$^{2}\text{H} + {}^{2}\text{H} \longrightarrow {}^{3}\text{He} + n + Q$$

Fusion reactions have not been observed to occur spontaneously on earth, but they do take place in the sun and the stars. The fusion reactions that are the source of the sun's vast energy are discussed in Chapter 13.

Fusion reactions involve charged particles, so they are all threshold reactions. They can be made to occur if the particles involved can be given sufficient energy. One example of a fusion reaction that is easily brought about using small accelerators is the reaction of deuterium and tritium:

$$^{2}\text{H} + {}^{3}\text{H} \longrightarrow {}^{4}\text{He} + n + Q$$

This reaction is the basis for the Cockcroft–Walton 14-MeV neutron generator, used in fast neutron activation analysis. This generator is further described in Chapter 14.

Fusion reactions like those above that take place using accelerators are not self-sustaining reactions; they require the continual input of energy to continue. A self-sustaining, but uncontrolled, fusion reaction takes place in the hydrogen or fusion bomb. A more desirable use of self-sustaining fusion, however, would be as a power source. Fusion has several advantages over fission as a source of energy. There is a practically unlimited supply of deuterium available from the world's oceans. The products of d–d and d–t fusion reactions that are of most interest for controlled fusion reactors are not radioactive, and do not pose the great problems of waste disposal that exist for fission reactors.

There are two major technological problems that need to be overcome to achieve controlled fusion. The first is the question of how to heat the gases to the temperature needed to sustain the reaction (about 10^8 K). At these temperatures, the material is actually a **plasma**, a very hot completely ionized gas. The second problem lies in confining the plasma, because no known materials could withstand the tremendous temperatures. One possible way to do this would be with magnetic fields, utilizing what is known as the "pinch effect." In 1989, several research groups claimed to have achieved "cold fusion" by bringing deuterium nuclei close together in a palladium cathode during electrolysis of heavy water. At this writing, cold fusion has not been satisfactorily confirmed by the international scientific community.

4.5.3. Heavy-Ion Reactions

Heavy ions are usually considered to be those with masses greater than that of ^4He; for example, $^{12}C^{6+}$, $^{40}Ar^{8+}$, $^{84}Kr^{30+}$, and $^{235}U^{50+}$. These ions differ from the lighter ions in their mass, in the high charge they possess, and in their small de Broglie wavelengths. The investigation of heavy ion reactions is a relatively recent undertaking, because the technology to accelerate these ions to an energy at which they could react was developed only in the 1960s and 1970s. One reason for the interest in heavy ions is that they will undergo a great variety of useful reactions. Heavy ion reactions are also the only practical route available to form the proposed stable superheavy elements, at around $N = 184$ and $Z = 114$. Due to their high charges, heavy ions experience a formidable Coulomb barrier to penetration of the nucleus. In fact, this large charge can excite a target nucleus by electromagnetic interactions without barrier penetration, a process known as **Coulomb excitation.** These interactions have provided abundant information on nuclear energy levels.

Heavy ions can react by the same mechanisms as the lighter ions, including scattering, compound nucleus formation, and direct interactions. In addition, however, heavy ions undergo a process that is different from any of these, called a **deeply inelastic reaction.** This is illustrated in Fig. 4.2. If the incoming heavy ion strikes the target nucleus near its center, a compound nucleus mechanism results. If the projectile is farther away from the nucleus, only scattering and Coulomb excitation of the target nucleus takes place. However, at a pathway intermediate between those two, the projectile will strike the target nucleus and form a very short-lived ($\approx 10^{-22}$ s) and distorted intermediate. A fragment then breaks off from the intermediate, leaving the nucleus with a considerable amount of excitation energy.

4.5.4. Photonuclear Reactions

Photonuclear reactions are nuclear reactions that are induced by electromagnetic radiation and result in the emission of one or more nuclear particles. Like charged-particle-induced reactions, they are threshold reactions, with a specific threshold energy for the various reaction channels. The energy imparted to the target nucleus is ordinarily low, so the most common reactions involve emission of only one or two nucleons, for example, (γ, n), (γ, p), and $(\gamma, 2n)$.

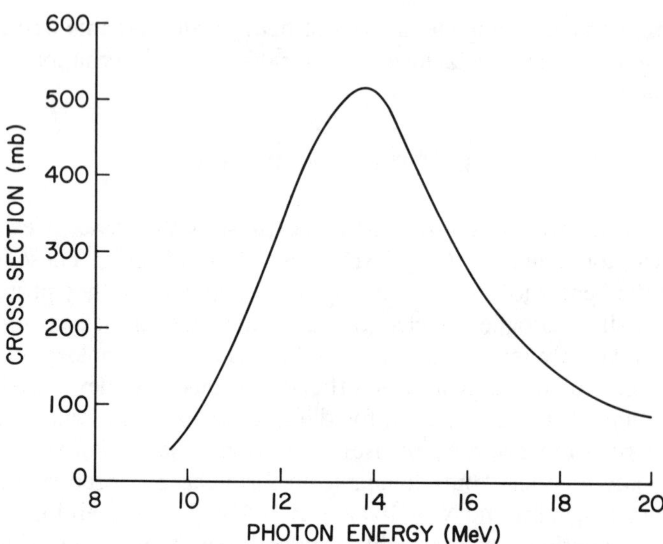

Figure 4.8. An excitation function for a photonuclear reaction showing a giant resonance.

Cross sections for photonuclear reactions are small, usually on the order of millibarns. Excitation functions for photonuclear reactions show an interesting feature called the **giant resonance,** illustrated in Fig. 4.8. In this figure, the cross section for the (γ, n) reaction has a threshold of approximately 7 MeV. The cross section then gradually increases with increasing photon energy, reaching a broad maximum at around 15 MeV for heavy nuclides and 25 MeV for light nuclides. This maximum is the giant resonance. This curve has the same general features for all nuclei.

Photonuclear reactions have some limited application in activation analysis. Electron accelerators called **synchrotron light sources** are used to produce the high-energy photons required. The high cost of the facility and limited time allocation permitted for existing facilities severely limit the use of photon-induced reactions for analytical purposes.

TERMS TO KNOW

Barn (b)	Cross section	Excitation function
Cold fusion	de Broglie wavelength	Fission yield
Compound nucleus	Deeply inelastic reaction	Flux density
Compound-nucleus reaction	Direct interactions	Fusion
Coulomb barrier	Elastic scattering	Giant resonance
Coulomb excitation		Heavy ions
		Induced fission

Inelastic scattering
Knock-on reaction
Laboratory system
Momentum correction
Nucleon evaporation
Oppenheimer–Phillips process
Photonuclear reaction
Pickup reaction
Q value
Radiative capture
Reaction channel
Resonance peak
Rutherford backscattering
Stripping reaction
Synchrotron
Thermal neutrons
Threshold energy
Threshold reactions

READING LIST

Gross, R. A., *Fusion Energy.* New York: Wiley, 1984. [the fusion process]

Hodgson, P. E., *Nuclear Reactions and Nuclear Structure.* Oxford: Clarendon, 1971. [nuclear reaction theory]

Margaritondo, G., *Introduction to Synchrotron Radiation.* New York: Oxford University Press, 1988. [synchrotron light sources]

Satchler, G. R., *Introduction to Nuclear Reactions.* New York: Wiley (Halsted), 1980. [basic principles of nuclear reactions]

Vandenbosch, R., and J. R. Huizenga, *Nuclear Fission,* New York: Academic, 1973. [the fission process]

EXERCISES

1. Consider the nuclear reaction ^{156}Gd (α, n) ^{159}Dy.

(a) Calculate the Q value (in units of MeV) for this reaction, using a table of mass excesses.

(b) Calculate the height of the potential energy barrier for this reaction.

(c) Calculate the threshold energy in the laboratory system required for significant production of ^{159}Dy.

2. How thick a sheet of Gd metal would be required to reduce the intensity of a beam of thermal neutrons by a factor of 1000? Atomic mass of Gd = 157.25 daltons, thermal neutron absorption cross section for Gd = 49.0×10^3 b, density of Gd = 7.90 g/cm^3.

3. It is found that it takes a 2.36-cm-thick piece of Pd metal foil (density 12.0 g/cm^3) to reduce the intensity of a beam of thermal neutrons to exactly 10^{-3} of its original intensity. Calculate the neutron capture cross section of Pd metal in units of barns.

4. How many grams of ^{204}Tl would be produced if exactly 0.750 g of natural Tl metal were irradiated with thermal neutrons for a period of 20.0 h? Assume the production reaction is ^{203}Tl(n, γ)^{204}Tl with a reaction cross section of 11.4 b and that the target is "thin." The flux density of neutrons (ϕ) = 1.00×10^{13} n cm^{-2} s^{-1}, and the half-life of ^{204}Tl = 3.78 y.

5. ^{252}Fm is produced by the following reaction:

$$^{244}_{94}\text{Pu} + ^{12}_{6}\text{C} \longrightarrow ^{252}_{100}\text{Fm} + 4^{1}_{0}\text{n}$$

Assuming an R_o value of 1.4 fm, calculate the height of the Coulomb energy barrier (in MeV) for this reaction. If the Coulomb barrier rather than the Q value is the controlling factor in determining the threshold energy for this reaction, calculate the threshold energy required for appreciable reaction in the laboratory system.

6. Consider the reaction ^{39}K(p, n)^{39}Ca. ^{39}Ca is a pure positron emitter with an E_{max} = 5.49 MeV.

(a) Calculate the height of the potential energy barrier for this reaction, assuming an R_o value of 1.4 fm.

(b) Show how it is possible to calculate Q for the reaction without a knowledge of the masses (or mass excesses) of ^{39}K or ^{39}Ca.

(c) Calculate the minimum energy in the laboratory system required to initiate this reaction without Coulomb barrier penetration.

7. Photons from a synchrotron light source have been used to measure concentrations of a number of light elements in human tissue samples. In this technique a radionuclide with a relatively short half-life is produced by photon irradiation of the sample. Write nuclear equations for photonuclear reactions that would be practical to determine the elements O, C, and N.

8. Classify the following reactions, using terms given in this chapter:

(a) ^{141}Pr(γ, n)^{140}Pr

(b) ^{197}Au(n, γ)^{198}Au

(c) $^{235}\text{U} \rightarrow ^{135}\text{Xe} + ^{98}\text{Sr} + 2\text{n}$

(d) ^{192}Os(n, n)^{192}Os

(e) $^{12}\text{C} + ^{12}\text{C} \rightarrow ^{24}\text{Mg} + \gamma$

9. **(a)** Calculate a simple geometrical cross section for ^{86}Kr and ^{135}Xe.

(b) Use a chart of the nuclides to look up the actual neutron capture cross sections for ^{86}Kr and ^{135}Xe.

(c) Are the calculated and tabled values in good agreement? Explain.

10. **(a)** "Cold neutrons" have very low energies, around 0.005 eV. Calculate the de Broglie wavelength for a neutron with this energy.

(b) Use this value to calculate the approximate cross section for the reaction of cold neutrons with ^{10}B.

11. Reactions with thermal and cold neutrons generally result in radiative capture rather than particle emission. Explain.

12. Tritons (^{3}H or t), as well as deuterons, can exhibit the Oppenheimer–Phillips effect in reactions. Based on this observation, predict which reaction below is most likely to occur and explain why: **(a)** $^{12}C(t,p)^{14}C$ **(b)** $^{12}C(t,n)^{14}N$ **(c)** $^{12}C(t,2n)^{13}N$

13. Induced natural radionuclides are formed in the atmosphere by the action of primary or secondary cosmic rays. These primary and secondary cosmic ray particles are predominantly high-energy protons and neutrons. Write nuclear reactions on target nuclides that would be abundant in the atmosphere or in the earth's surface that might produce the following induced natural radionuclides: **(a)** ^{36}Cl **(b)** ^{14}C **(c)** ^{32}Si **(d)** ^{39}Ar

14. The intensity of a beam of neutrons at the beam port of a ^{252}Cf irradiation facility is 1.57×10^4 neutrons/s. It is desired to put a door on the port that will reduce the beam intensity to 0.100 neutrons/s. What thickness of cadmium metal (density 8.60 g/cm^3, total absorption cross section 2.40×10^3 b) would be required for this door?

CHAPTER

5

RATES OF NUCLEAR DECAY

Radioactive decay is a statistically random process that follows the laws of first-order chemical kinetics. In almost any application of radioactivity, the amounts of radioactivity present and the length of time activity levels will endure must be known. In this chapter, a simple mathematical treatment of the laws of radioactive decay is presented, including equations for both decay and formation of radioactive species. The units used for measurement of radioactivity are introduced, and experimental methods used to determine half-lives are explained. More complicated cases of radioactive decay are also discussed, including chain decays and the formation and decay of radioactive materials through bombardment in a neutron flux.

5.1. RATES OF RADIOACTIVE DECAY

Radioactive decay is a random process. The probability that a given atom will decay in a certain interval (Δt) is independent of its history; it is a function only of the time of observation. When observing a sample consisting of a very large number of radioactive atoms, it is not possible to predict which atom will decay next, but the decay characteristics of the entire sample can be described. The rate of decay is called the **activity (*A*).** The activity represents the number of parent nuclides that decay per unit time. Often, units of **disintegrations per second (dps)** are used. The equation for activity may be written as

$$A = -\frac{dN}{dt} \tag{5.1}$$

where N = the number of parent atoms, and t is time. The activity is directly proportional to the number of parent atoms originally present, so we can also write

$$A = -\frac{dN}{dt} = \lambda N \tag{5.2}$$

where λ is the **decay constant.** Equation 5.2 has the form of the rate equation for a first-order reaction. Integration of this equation gives

$$\int_{N=N_0}^{N=N} \frac{dN}{N} = -\lambda \int_{t=0}^{t=t} dt \tag{5.3}$$

and,

$$\ln N - \ln N_0 = -\lambda t \tag{5.4}$$

$$\ln \frac{N}{N_0} = -\lambda t \tag{5.5}$$

$$N = N_0 e^{-\lambda t} \tag{5.6}$$

where N_0 refers to the original number of atoms, and N is the number of atoms left after time t has passed. Because the activity is proportional to the number of atoms, we can also write

$$A = A_0 e^{-\lambda t} \tag{5.7}$$

Equations 5.6 and 5.7 will be used frequently, so you should be thoroughly familiar with them and be able to use them effectively.

Example. A radionuclide often used in medical procedures is ^{99m}Tc. It undergoes gamma decay with a half-life of 6.01 h. The decay constant has a value of 0.115 h^{-1}. Calculate the amount of time it would take for the activity injected into a patient to be reduced to 0.1% of the original level.

$$A = A_0 e^{-\lambda t}$$

$$A/A_0 = 0.1\% = 0.001$$

$$0.001 = e^{-(0.115/\mathrm{h})\,(t)}$$

Taking the natural logarithm of both sides gives

$$\ln(0.001) = -(0.115\ \mathrm{h}^{-1})t$$

$$t = 60.1\ \mathrm{h}$$

5.1.1. Half-life and Average Life

The term half-life was defined in Chapter 1 as the time needed for half of a statistically large number of radioactive atoms in a sample to decay.

Therefore, after one half-life, $N = \frac{1}{2}N_0$. By substituting this relationship into Eq. 5.6 a relationship between the half-life ($t_{1/2}$) and the decay constant can be obtained:

$$\frac{1}{2}N_0 = N_0 e^{-\lambda t_{1/2}} \tag{5.8}$$

Taking natural logs of both sides and canceling the N_0 gives

$$\ln\frac{1}{2} = e^{-\lambda t_{1/2}} \tag{5.9}$$

$$t_{1/2} = 0.693/\lambda \tag{5.10}$$

Equation 5.10 can be used to calculate either the half-life or the decay constant if the other one is known.

Another parameter that describes the decay rate of a sample of radioactive atoms is the **average lifetime (τ).** This refers to the average time needed for an atom to decay. It can be shown that τ is slightly longer than the half-life, and is given by

$$\tau = 1/\lambda \tag{5.11}$$

The average lifetime is not used as frequently as the half-life. However, if only a few decay events are recorded for a newly produced radionuclide, an average life may be initially calculated and then converted to half-life by use of the above equation.

Example. The half-life for ^{117}Cd is 149 minutes (min). Calculate the decay constant for ^{117}Cd.

Equation 5.10 may be used to solve this problem:

$$149 \text{ min} = 0.693/\lambda$$

$$\lambda = 0.693/149 \text{ min} = 4.65 \times 10^{-3} \text{ min}^{-1}$$

Example. Calculate the decay constant for a radioactive sample of ^{37}Ar, if it takes 100 days for 86.3% of the atoms to decay.

If 86.3% of the atoms have decayed, then 100% − 86.3% = 13.7% of the atoms are left at the end of the 100-day decay period. Thus, $N/N_0 = 0.137$, and $t = 100$ days.

Equation 5.6 may be used to solve the problem:

$$N/N_0 = 0.137 = e^{-\lambda(100\ \mathrm{d})}$$

$$\ln 0.137 = -\lambda(100\ \mathrm{d})$$

$$\lambda = 1.99 \times 10^{-2}\ \mathrm{d}^{-1}$$

5.2. UNITS OF RADIOACTIVE DECAY

The activity of a radioactive sample may be expressed in a variety of ways. The expressions of activity are always based on the disappearance of the parent, rather than on the appearance of a product. The reason for this is that the parent may decay by several paths (see the discussion of branching decays later in this chapter), so monitoring the appearance of a product may not be representative of the total activity of the parent. The SI unit of radioactivity is the **becquerel (Bq).** It is equal to 1 dps. An older activity unit that is still widely used is the **curie (Ci).** The curie is defined to be exactly 3.7×10^{10} dps. Originally, the curie was defined as the measured activity of 1 g of ^{226}Ra, which does have an activity of approximately this value. The usual metric prefixes are also applied to the curie; for example, millicurie (mCi) = 10^{-3} Ci, microcurie (μCi) = 10^{-6} Ci, and picocurie (pCi) = 10^{-12} Ci. To the beginning student of radiochemistry, these activity units often have little functional relevance. To help gain some concept of the relative levels of activity designated by the Curie unit, the following generalizations may be made:

- *Curie Level.* Radioactive samples possessing activity at the Ci level are considered to be very "hot;" that is, highly radioactive. Special government licensing is required to obtain and handle radioactive samples with this level of activity. Laboratories would have to be specially equipped with "hot box" shielding for high-energy β- and all γ-ray emitters. Containment would be required for low-energy β-particle emitters and for α-particle emitters. Materials with curie levels of activity would be most likely to be found in nuclear facilities such as power plants, medical facilities, or radionuclide production facilities. They would not be normally found in chemical research laboratories.
- *Millicurie Level.* A fairly common working level of radioactivity for tracer studies would be at the mCi level (3.7×10^7 dps). A number of medical procedures, such as those involving ^{99m}Tc injection for PET scanning, involve mCi levels of activity. Government licensing and

specially equipped laboratories are also required for working with these activity levels, but the laboratory requirements are much simpler than for Ci levels. The mCi is the most common unit by which most radionuclides are sold.

- *Microcurie Level.* At the 1.0-μCi level (3.7×10^4 dps) exposure risks to personnel are considered to be minimal for most commonly used radioactive tracers. Heavy metal α-particle emitters such as plutonium (Pu) are an exception. Radionuclides with this level of activity may often be obtained without a governmental license, and no special laboratory facilities are necessary. For example, the radioactive materials found in the home smoke detectors contain activities at or below the μCi level.

Example. Calculate the natural activity of Nd, in Ci/g, given the following data: atomic mass of Nd = 144.24 daltons, half-life of ^{144}Nd = 2.10×10^{15} y, isotopic abundance of ^{144}Nd = 23.8% of natural Nd.

Equation 5.2 can be used to solve the problem. The decay constant is easily found from knowledge of the half-life. Because the answer is to be given in Ci, which has time units of seconds, the half-life and decay constant must be converted to seconds:

$$(2.10 \times 10^{15}\ \text{y})(3.1536 \times 10^{7}\ \text{s/y}) = 6.62 \times 10^{22}\ \text{s}$$

$$\lambda = 0.693/t_{1/2} = 0.693/6.62 \times 10^{22}\ \text{s} = 1.05 \times 10^{-23}\ \text{s}^{-1}$$

Now the number of radioactive atoms in a one-gram sample must be calculated. The isotopic abundance of the radioactive ^{144}Nd is 23.8%, which means that only this fraction of the sample will be ^{144}Nd atoms:

$$\begin{aligned} \text{N}(^{144}\text{Nd}) &= (1.00\ \text{g Nd}/144.24\ \text{g Nd mol}^{-1})(6.02 \times 10^{23}\ \text{atoms/mol}) \\ &\quad \times (0.238) \\ &= 9.93 \times 10^{20}\ \text{atoms of}\ ^{144}\text{Nd} \end{aligned}$$

$$A = \lambda N = (1.05 \times 10^{-23}\ \text{s}^{-1})(9.93 \times 10^{20}) = 1.04 \times 10^{-2}\ \text{dps/g}$$

$$\begin{aligned} 1.04 \times 10^{-2}\ \text{dps} \times (1\ \text{Ci}/3.7 \times 10^{10}\ \text{dps}) &= 2.82 \times 10^{-13}\ \text{Ci/g} \\ &= 0.282\ \text{pCi/g} \end{aligned}$$

Example. Calculate the mass, in grams, of 1.0 mCi of pure ^{32}P, given the following data: $t_{1/2}$ = 14.28 d, M ^{32}P = 31.9998 daltons. If ^{32}P costs \$40/mCi, calculate the cost of ^{32}P in dollars/pound.

Equation 5.2 is used for this problem. The decay constant is calculated from the half-life. Units must be converted to seconds, to be consistent with the Ci unit of activity:

$$(14.28\ \text{d})(8.640 \times 10^4\ \text{s/d}) = 1.23 \times 10^6\ \text{s}$$

$$\lambda = 0.693/1.23 \times 10^6\ \text{s} = 5.63 \times 10^{-7}\ \text{s}^{-1}$$

$$A = 1.0\ \text{mCi} = 1.0 \times 10^{-3}\ \text{Ci}\ (3.7 \times 10^{10}\ \text{dps/Ci}) = 3.7 \times 10^7\ \text{dps}$$

$$N = A/\lambda = (3.7 \times 10^7\ \text{dps})/(5.63 \times 10^{-7}\ \text{s}^{-1}) = 6.57 \times 10^{13}\ \text{atoms}$$

$$\text{Mass} = [(6.57 \times 10^{13}\ \text{atoms})/(6.02 \times 10^{23}\ \text{atoms})] \times 31.9998\ \text{g}$$

$$= 3.49 \times 10^{-9}\ \text{g}$$

$$= 3.49 \times 10^{-9}\ \text{g}\ (1\ \text{lb}/454\ \text{g}) = 7.69 \times 10^{-12}\ \text{lb}$$

cost: (\$40/mCi)/(7.69×10^{-12} lb ^{32}P mCi^{-1}) = \$$5.2 \times 10^{12}$/lb ^{32}P = 5.2 trillion dollars/lb!

5.3. BRANCHING DECAY

The existence of **branching decay** has already been noted in Chapter 2. The parent nuclide decays by two (or possibly more) separate decay modes, as shown in the general diagram below:

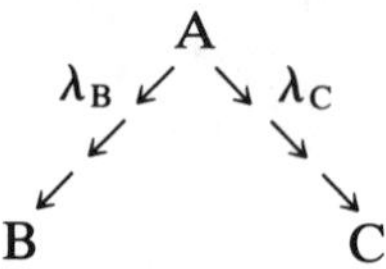

Branching decay occurs most often for odd–odd nuclei, such as ^{64}Cu and ^{40}K, and for heavier nuclides. Examination of Figs. 1.16–1.18 will demonstrate the existence of branching α–β decays in the naturally occurring decay chains. ^{64}Cu and ^{40}K may decay either by negatron emission or by electron capture. Figure 5.1 shows the decay scheme for ^{40}K.

Each decay branch is independent of the other, and will have its own characteristic half-life and decay constant. These are called **partial decay**

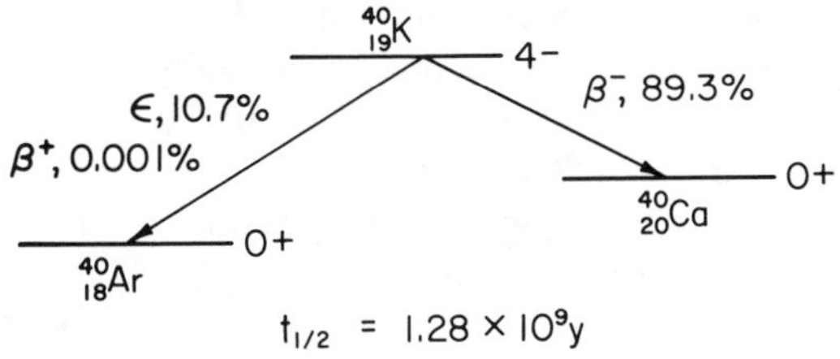

Figure 5.1. The decay scheme for ^{40}K, illustrating branching decay.

constants. The total activity of A is equal to the sum of the activity of each branch. Using Eq. 5.2 for activity we can write

$$A_t = \lambda_B N_A + \lambda_C N_A \tag{5.12}$$

$$A_t = N_A(\lambda_B + \lambda_C) = N_A \lambda_t \tag{5.13}$$

The overall decay constant is the sum of the individual decay constants.

5.4. EXPERIMENTAL METHODS FOR DETERMINATION OF HALF-LIFE

The half-life is an important characteristic of a nuclide that must be experimentally determined. The range in the half-lives of known radionuclides is very large, from $<10^{-18}$ s to $>10^{22}$ years. Due to this enormous range, no single technique is sufficient to measure all of them. For convenience, we will divide the half-lives into four groups: long (several years and up); medium (seconds to a few years); short (seconds to milliseconds); and very short (less than milliseconds). These divisions are arbitrary, and the lines drawn among them are not sharp.

5.4.1. Long Half-lives

Two methods for the measurement of long half-lives will be discussed: the specific activity method and the solid-state nuclear track detectors.

Specific Activity Method

To determine the half-life of a radionuclide in the simplest manner, the change in the activity of a nuclide over a period of time would be measured. The point at which the activity has dropped to half of its original value is the half-life. The experimental difficulty with long-lived nuclides is that the measured activity will remain essentially constant within error limits imposed by the counting statistics over periods of weeks, months, or years. This makes it impossible to monitor the change of activity over time. However, a quantity that can be directly determined and related to half-life is the **specific activity (SA)** of the substance. The SA is the activity per unit mass:

$$\text{SA} = \text{disintegrations s}^{-1}\,(\text{unit mass})^{-1} \tag{5.14}$$

Recalling Eq. 5.2, the activity (A) is equal to

$$A = -\frac{dN}{dt} = \lambda N = \frac{0.693}{t_{1/2}}N$$

The number of atoms present can be calculated using mole relationships. However, because a sample may contain isotopes that are not active, the fraction of the atoms that are radioactive must also be known. The **isotopic abundance (IA)** of the nuclide of interest can be used to calculate the number of atoms of a specific radionuclide in a naturally occurring mixture of isotopes:

$$N = (\text{grams of element/atomic mass})\,(6.02 \times 10^{23})\,(\text{IA}) \tag{5.15}$$

Substituting this into Eq. 5.2 gives

$$A = (0.693/t_{1/2})\,(\text{g/atomic mass})\,(6.02 \times 10^{23})\,(\text{IA}) \tag{5.16}$$

Using the definition of specific activity, SA = A/g, and rearranging the equation to solve for the half-life, we can write

$$t_{1/2} = \frac{(0.693)\,(6.02 \times 10^{23})\,(\text{IA})}{(\text{SA})\,(\text{atomic mass})} \tag{5.17}$$

To determine the half-life using Eq. 5.17, we need only measure the absolute activity of the sample and its mass. Mass is determined in the usual ways (analytical balance, mass spectrometer, etc). Determination of absolute activity is more difficult. The **absolute activity** is the true total disintegration rate of the sample. In most routine measurements of radioactivity, all of the emitted radiation is not detected. This is because a sample normally emits radiation isotropically (i.e., equally in all directions), but a typical detector arrangement will monitor only a fraction of this total. The

counting rate recorded by the detection system will not be equal to the absolute activity in this case. There are three ways to determine the absolute activity: use of calibrated standards, 4π counters, and use of coincidence techniques.

Calibrated standards are radioactive materials that have a well-defined level of activity. Such standard materials are produced by several organizations, such as the National Institute of Standards and Technology (NIST) in the United States, the International Atomic Energy Agency (IAEA), and the Commission of the European Communities (BCR).

To determine absolute activity, the calibrated standard is counted first, using exactly the same apparatus that will be used to count the sample whose half-life is to be measured. The measured activity of the standard can be used to calculate the **efficiency** of the detector. The efficiency refers to the proportion of decays that are actually observed by the detector system. For example, if the standard activity were 1 μCi (3.7×10^4 dps), and the detector system counted 3.7×10^2 dps, the efficiency factor for the counting system would be 0.01, or 1% efficient. This means that only 1% of the activity emitted by the standard is actually counted by the system. The unknown sample can then be counted, and the observed counts divided by the efficiency to obtain the absolute activity. Half-life calculations can then be performed using Eq. 5.17.

When using this method, it is important that the thickness of the standard and sample be comparable. The reason for this is that radiation emitted by a sample may be stopped within a sample, if it is thick enough. This is called **self-absorption.** If the sample were much thicker than the standard, significant amounts of radiation might be lost through self-absorption, and the correction would not be valid. It is also necessary that the detector system be configured in exactly the same way for both standard and sample. This means that both sample and standard should be counted on the same detector, with the same electronics, and in exactly the same position relative to the detector. It would not be possible to compare two activities if one sample were located much closer to the detector than the other.

A **4π counter** refers to one that is constructed in such a way that virtually 100% of all decays from the sample are detected. This kind of detector system can determine absolute activity directly. A typical arrangement of a 4π counter is shown in Fig. 5.2. A thin sample is placed into a closed chamber with detectors arranged all around the sample. Any radiation emitted by the sample will encounter one of the detectors, and the total activity will be recorded.

It will be apparent that this type of absolute activity determination assumes that all radiation travels far enough to reach the detector, and also that the radiation is not so energetic that it can escape the chamber

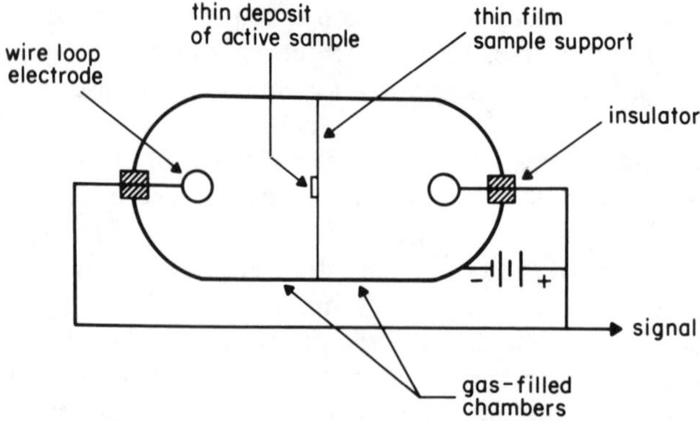

Figure 5.2. Schematic of a 4π, alpha or beta, gas-filled detector.

completely. Therefore, this type of activity determination is used most often for β particles. Gamma rays are too penetrating, so a significant fraction of them will escape from the chamber. Sample self-absorption and absorption in the sample support film may result in less than 100% efficiency in counting alpha particle emitters.

Coincidence techniques can be used to determine absolute activity for those cases where two coincident radiations are emitted by the sample. A generalized example of this is shown below:

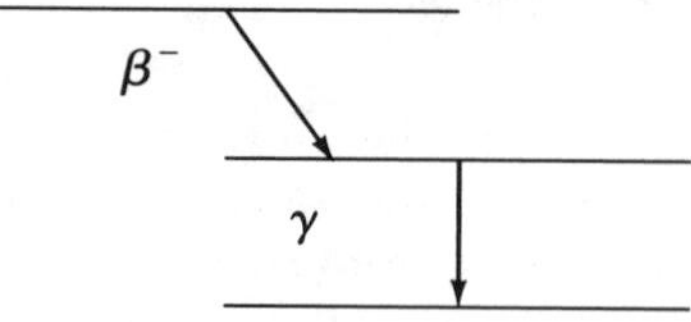

In this example, the sample emits a β^- and a γ **in coincidence,** that is, virtually simultaneously. The β–γ combination is the most commonly encountered, although others can be used. It is assumed in using the coincidence method that there is no **angular correlation** between the β and the γ; that is, there is no fixed relationship between the directions in which the β and the γ are emitted. A detector system, as represented in Fig. 5.3, is used. The β–γ emitter is placed between two detectors, one for β detection and one for γ detection. The counting rates for the β's (R_β) and the γ's (R_γ) are given by

$$R_\beta = \varepsilon_\beta R_0 \quad \text{and} \quad R_\gamma = \varepsilon_\gamma R_0 \tag{5.18}$$

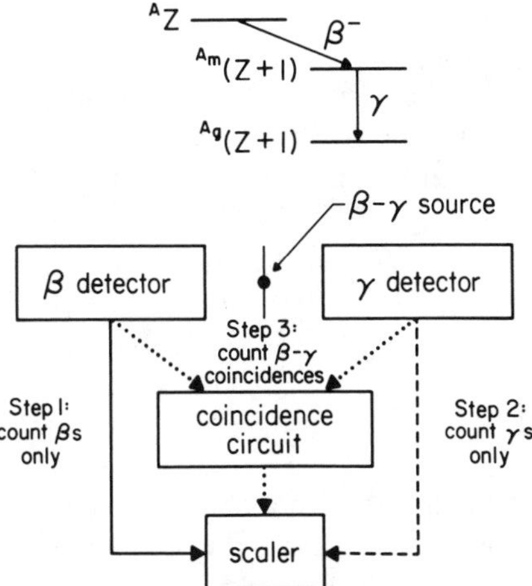

Figure 5.3. Experimental setup for the determination of absolute activity by a coincidence technique.

where ε refers to the efficiency of the detectors for the β and γ rays, and R_0 is the absolute activity of the source. The electronics are set up in such a way that a coincidence signal will be obtained only when a β and a γ are emitted and detected simultaneously. The probability of this happening is the product of the individual probabilities, so the coincidence counting rate, $R_{\beta\gamma}$, can be written as

$$R_{\beta\gamma} = \varepsilon_\beta \varepsilon_\gamma R_0 \tag{5.19}$$

If the two equations in Eq. 5.18 are multiplied and divided by Eq. 5.19, the efficiencies cancel out. Solving for the absolute activity gives the relation

$$R_0 = \frac{R_\beta R_\gamma}{R_{\beta\gamma}} \tag{5.20}$$

Solid State Nuclear Track Detectors

Solid-state nuclear track detectors (SSNTD) are materials that respond to radiation with the formation of some type of track. They can be made of photographic emulsions, plastic, glass, cellulose, or a variety of other materials. If a photographic emulsion is dipped into a solution of a long-lived nuclide, some measureable mass of the nuclide will be absorbed into the emulsion. The decay of the nuclide will produce tracks. The number of tracks per unit area per time can then be used to give an indication of the half-life.

5.4.2. Medium Half-lives

The most common way to determine the half-life in the minute to year range is by direct measurement of the change in activity over time. The method is called the **simple decay curve method,** or the **normal method of timed counts.**

The equation for radioactive decay (Eq. 5.7) can be transformed into one that has the form of a straight line ($y = mx + b$) by taking the natural logarithms:

$$\ln A = \ln A_0 - \lambda t \tag{5.21}$$

A plot of $\ln A(y)$ against $t(x)$ gives a straight line with a slope of $-\lambda$ and a y intercept of $\ln A_0$. The half-life can be obtained from the slope and the relationship between λ and $t_{1/2}$ given in Eq. 5.10. Figure 5.4 shows an example of a plot of $\ln A$ against t.

Example. Calculate the half-life of a radionuclide, given the following data:

Time (s):	60	180	300	420	540	660	780
Activity (d/s):	1360	910	694	463	351	232	140
$\ln A$:	7.22	6.81	6.54	6.14	5.86	5.45	4.94

The simple decay curve method is used to determine the half-life. A plot of $\ln A$ against t is made, similar to that shown in Fig. 5.4. The slope of this line (determined using a least-squares procedure) is -0.00304 and the intercept is 7.42.
To calculate the half-life:

$$\lambda = -0.00304 = -0.693/t_{1/2}$$

$$t_{1/2} = 228 \text{ s}$$

The original activity can also be calculated:

$$\ln A_0 = 7.42 \qquad A_0 = 1669 \text{ dps}$$

The simple decay curve method gives best results if a pure radionuclide source is used. Therefore, in some cases, a separation procedure may have to be performed before the half-life determination. There are alternatives to actual chemical separation, however. When the components of a mixture emit different types of radiation, it may be possible to devise a way to

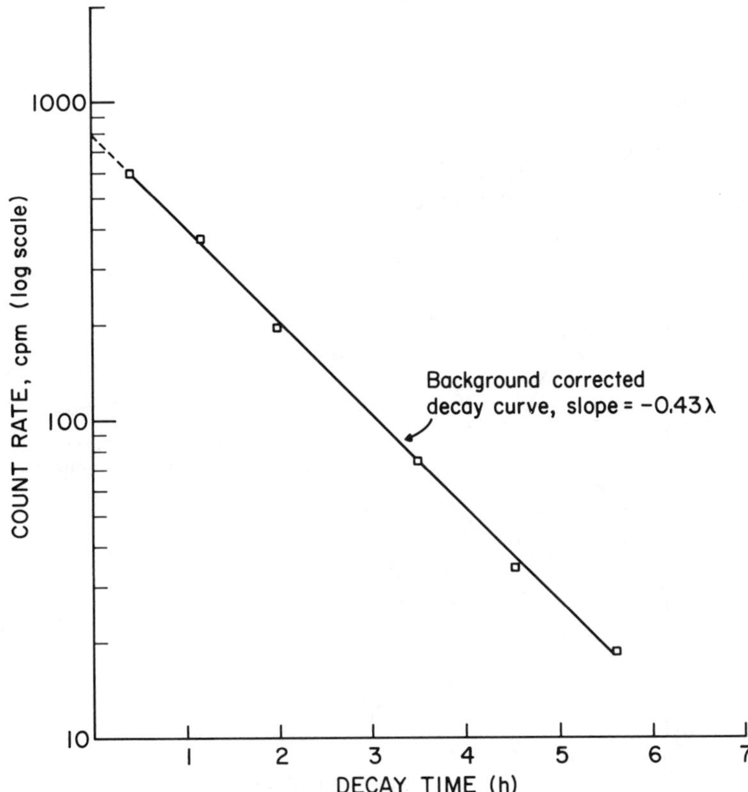

Figure 5.4. A decay curve for a single radionuclide.

count only the desired activity. If the desired radiation is more penetrating than an interferent, absorbers could be placed between the sample and the detector to eliminate the unwanted activity.

The simple decay curve method can be used to resolve individual activity levels for mixtures of two radionuclides, but is rarely used for mixtures of three or more radionuclides. For this resolution to be practical, the half-lives of the two active species should differ by at least a factor of two. The activity of the mixture is measured over time and plotted, as is done for a single radionuclide. However, this plot will differ in appearance from that for the single species. An example of a decay curve from a mixture of activities is shown in Fig. 5.5. Notice that the curve is initially not linear, until the activity due to the shorter half-life radionuclide becomes negligible. The linear portion of the decay curve will be due to the longer-lived species. The graph must then be resolved into two components, one for the

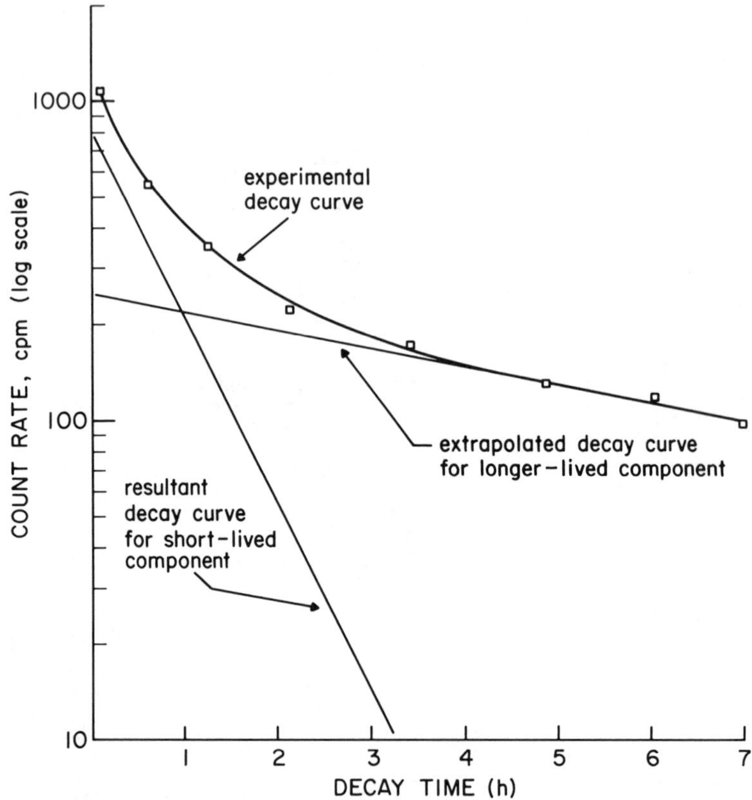

Figure 5.5. A decay curve for a source containing two radionuclides with different half-lives.

longer-lived component and one for the shorter-lived component. Extrapolation of the tail of the curve back to the y axis will give the decay line for the longer-lived species. Point-by-point subtraction of this line from the original curve will give the decay line for the shorter-lived species. These separate decay lines can then be used to determine the half-life of each component, as shown in the previous example.

This type of decay-curve analysis of two activities can be accomplished instrumentally by using a **multichannel analyzer (MCA)** (see Chapter 8) in what is called the **multiscaling mode.** An MCA is an instrument that records the activity sensed by a radiation detector. When used in the multiscaling mode, the analyzer counts all signals for a fixed time and records the total counts. The vertical display on the MCA represents the number of counts. The horizontal line is divided into units called **channels**. In the multiscaling mode, the channels are the predetermined time intervals. For example, suppose the unit is set to count for one-second intervals. Each channel on the MCA would represent one second. The instrument would determine the total number of events for the first second of counting and record it in the first channel. Then counting would begin for the next second of time, and the activity recorded. This process is repeated for any desired length of time. The result looks very much like any plot of exponential decay. An example of the use of this technique is in the determination of N and P in biological tissues using 14-MeV neutron activation analysis. Both N and P are activated by 14-MeV neutrons, N by the $^{14}N(n,2n)^{13}N$ reaction, and P via the $^{31}P(n,2n)^{30}P$ reaction. Both are positron emitters. However, their half-lives are different: 9.97 min for ^{13}N and 2.50 min for ^{30}P. Thus, a sample may be irradiated using 14-MeV neutrons and then counted. The net activity is recorded on an MCA using the multiscaling mode, and the decay curve analyzed as described above.

5.4.3. Short Half-lives

The determination of short half-lives can often be accomplished using the decay-curve method just described, with some modifications. Many of the short-lived nuclides in the chart of the nuclides are not naturally occurring and must be produced at the time of the half-life determination. A major experimental challenge is to get these nuclides to the counting system quickly enough to measure the activity before it decays away. An MCA used in multiscaler mode, as described above, can be useful in this regard if a transfer system is available that can get the sample to the detector rapidly enough. A variety of imaginative methods have been devised to accomplish this task, two of which will be described here as examples. Both of these methods rely on the fact that the nuclear reaction that initially forms the radionuclide of interest will result in the recoil of the nuclide, due to the conservation of momentum. At least some of these atoms will recoil completely out of the target, and these can be captured and counted.

Figure 5.6 shows a block diagram of a commonly used method to transfer short-lived reaction products to a counting station. The target is bombarded by a beam of particles to induce the nuclear reaction. Some of the reaction products recoil off the target, and are flushed from the area by a jet of helium gas. The gas drives the active species through a nozzle to a foil impactor, which catches the product nuclides. Here, the radioactive atoms are counted.

The determination of the half-life of ^{254}No also was accomplished using recoil effects. Figure 5.7 shows the experimental apparatus. ^{12}C was used to bombard a target of ^{246}Cm. The products of the reaction are ^{254}No and

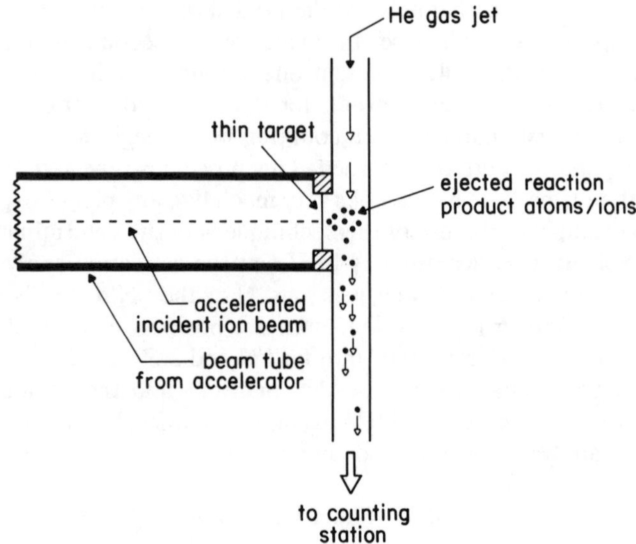

Figure 5.6. Collection of nuclear reaction products using a helium jet.

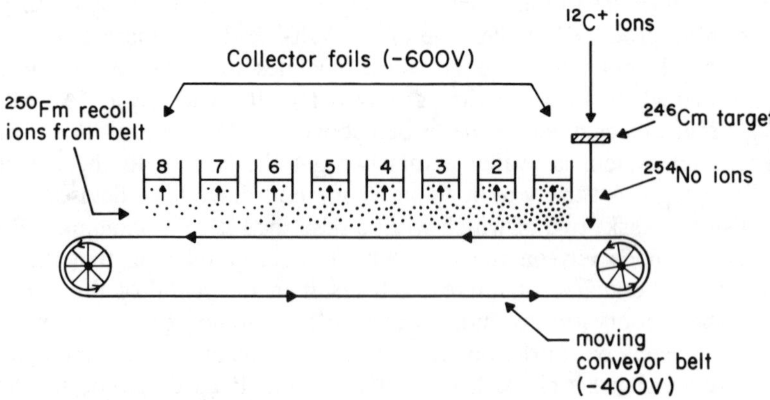

Figure 5.7. Schematic representation of a method employing a conveyer belt that was used to determine the half-life of ^{254}No.

some neutrons. Some of the ^{254}No produced in the reaction recoiled out of the target. The products of heavy ion reactions usually have high positive charges, so they may be easily accelerated. The ^{254}No atoms were attracted toward a conveyer belt held at -400 V, where they were collected. The belt moved continually and carried the newly formed ^{254}No away. The ^{254}No decayed by α emission, producing ^{250}Fm, which also decays by α emission. The nuclear reactions are shown below:

$$^{254}\text{No} \longrightarrow {}^{250}\text{Fm} + \alpha$$

$$^{250}\text{Fm} \longrightarrow {}^{246}\text{Cf} + \alpha$$

As the ^{254}No decays, some of the ^{250}Fm atoms will recoil off the conveyor belt. A series of catcher foils, at -600 V, are set up down the line from the conveyor belt. The ^{250}Fm atoms resulting from the ^{254}No decay will be attracted to and deposited on the foils. More ^{250}Fm will be deposited on the first foil than on the last foil, because the ^{254}No is decaying rapidly as the belt moves. The foils containing the ^{250}Fm atoms can be removed and counted. If a plot is then made of the log of the Fm activity versus foil number, the half-life of the ^{254}No can be determined from the slope of the plot. Thus, we are actually using the activity of the daughter to determine the half-life of the parent.

5.4.4. Very Short Half-lives

For the very short half-lives, the time needed for even the fastest sample transfer and the intrinsic timing limitations of electronic devices preclude direct measurement of the change of activity of the nuclide. Three indirect methods for determinations of these half-lives will be mentioned.

Method of Delayed Coincidences

In the **method of delayed coincidences,** the interval between the formation and the decay of the very short-lived state is measured. The formation of the state is determined by detecting the event that forms the isomer. For example, a nuclide may undergo β decay to an excited state of the daughter nucleus. Detection of the β would mark the formation of the excited state, while detection of the γ would mark the decay of the state. Figure 5.8 illustrates the experimental setup. Two separate detectors monitor the formation and decay events. The signal from the detector recording the formation event is passed through an electronic device that can delay it for a variable amount of time before it is allowed to pass to the coincidence unit. To record a signal, the coincidence unit must receive signals from both detectors simultaneously. If the delay time is initially set to 0, no true coincidences will be observed, because the two events (formation and decay) do, in fact, occur at different times. As the delay time is increased, the coincidence counting rate rises until the delay time is equivalent to the most

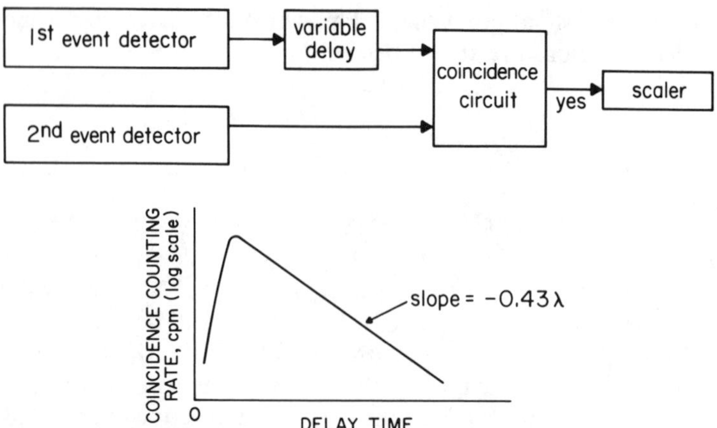

Figure 5.8. Block diagram representation of the method of determining short half-lives by the method of delayed coincidences.

probable lifetime of the short-lived radionuclide. Continued increases in the delay time result in a decrease in recorded coincidence events. A plot of the coincidence counting rate against the delay time is also shown in Fig. 5.8. The linear decreasing part of the curve can be used to determine the half-life of the second event. The method of delayed coincidences can be used either for radioactive decays, or for products formed in a nuclear reaction. It is useful for half-lives down to about 10^{-11} s.

An example of the application of delayed coincidences to a half-life determination is the decay of ^{24}Ne. The equations are

$$^{24}\text{Ne} \longrightarrow {}^{24m}\text{Na} + \beta^- \quad (1.98 \text{ Mev})$$

$$^{24m}\text{Na} \longrightarrow {}^{24g}\text{Na} + \gamma \quad (0.47 \text{ MeV})$$

In this example, the 1.98 MeV β^- from the ^{24}Ne marks the formation of the ^{24m}Na. The half-life for the isomeric transition of ^{24m}Na to ^{24g}Na would be the event to be measured. The half-life of this latter decay has been determined to be $\approx$20 ms.

Doppler Shift Method

The **Doppler shift** is the change in the wavelength of emitted radiation from a body as the body moves toward or away from the detecting device. A commonly cited example of the Doppler shift is the change in the pitch of a railroad engine whistle as it approaches, or moves away from, an observer. The lifetimes of very short-lived γ-ray emitters may be measured by noting the Doppler shift in the γ-ray energy. The nuclear reaction forming a radioactive species will impart momentum to it. The movement given to the nucleus in the momentum transfer is what causes the Doppler shift in energy. The magnitude of the Doppler shift is a function of the velocity of the original reaction product. The velocity of the product in a given medium as a function of time can often be determined experimentally. Therefore, a given Doppler shift determines a given velocity of the product. This velocity can then be related to the interval between formation of the product in the medium and emission of the γ ray. This time reflects the average lifetime for the product, which can then be used to calculate a value for the half-life of the transition.

Energy Width of Excited States

It can be demonstrated that quantum states that are not stable do not have exact energies. Therefore, a decay from a given nuclear energy state to another state does not have an exactly defined energy. The uncertainties in the energies and the half-lives of the states are governed by the Heisenberg uncertainty principle, as shown in this equation:

$$\Delta E \Delta t \geq h/2\pi \tag{5.22}$$

The uncertainty in the energy is called the **width (Γ)** of the state. The width of the state is related to the average lifetime (τ) by

$$\Gamma = \frac{h/2\pi}{\tau} \tag{5.23}$$

The measurement of the width of an excited state can be used to calculate the average lifetime and the half-life of the state.

5.5. ESTIMATION OF HALF-LIFE FROM THEORY AND SYSTEMATICS

In Chapter 2 it was noted that relative estimates of half-life can be made for α and β emitters based on the energy of the decay. For α emitters, this relationship is expressed by the Geiger–Nuttall rule (Eq. 2.1), which describes an inverse relationship between the energy of an α decay and its half-life. Figure 5.9 shows a Geiger–Nuttall plot, which plots the logarithm of the half-life against α energy for radionuclides in the ^{232}Th natural decay series. The half-lives decrease steadily as the energy of the emitted α increases.

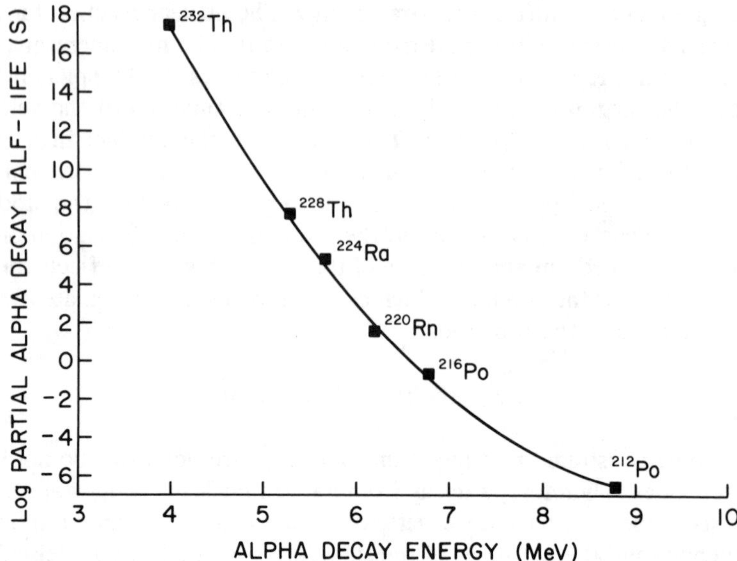

Figure 5.9. The Geiger–Nuttall relationship for α-particle emitters in the ^{232}Th natural decay chain.

The theory of β decay is more complex than that for α decay, so estimation of half-lives is not so straightforward. The probability of β decay is related to the energy of the transition, the atomic number of the parent, and the spin and parity changes involved in the decay from one nuclide to another. In the **Fermi theory of beta decay,** a parameter called the ***ft* value** is used to indicate the probability of β decay. In this parameter, f is a factor that depends on atomic number and E_{max}, and t is the representation used for half-life. This ft value is called the **comparative half-life** of the transition and is related to the distribution of energy between the electron and the neutrino emitted in beta decay. The ft value essentially factors out the effects of energy and atomic number, and is related only to the properties of the initial and final nuclear states involved in the beta transition. Hence, transitions having the same change in nuclear spin and parity have the same ft values.

Transitions that have low ft values are highly probable, and so have short half-lives. These are called **allowed** transitions. A few β decays occur especially easily, and these are called **superallowed**. The best example of nuclides that experience superallowed β transitions are those between **mirror nuclei,** that is, those with mirror numbers of protons and neutrons. For example, ^{38}Cl and ^{38}Ar are mirror nuclei. The radionuclide ^{38}Cl has 17 protons and 18 neutrons, while stable ^{38}Ar has 18 protons and 17 neutrons. Beta

Table 5.1. Comparative Half-Lives for Beta Decay

Transition Type	$\Delta\ell$	ΔJ	$\Delta\pi$	log *ft*
Super-allowed	0	0	no	≈3
Allowed	0	0, ±1	no	3–6
First forbidden	1	0, ±1, ±2	yes	6–10
Second forbidden	2	±2, ±3	no	10–14
Third forbidden	3	±4	yes	17–20

transitions that have large *ft* values do not have a very high probability of occurrence and are called **forbidden** transitions. Table 5.1 shows a summary of transition types and *ft* values. Values of log *ft* are ordinarily tabulated because *ft* values have a wide range. Notice that the greater the change in nuclear spin, momentum, and parity between the initial and final states, the larger is the log *ft* value, and the longer the half-life will be. If the spins and parities of the initial and final states are known, log *ft* values can be obtained from tables similar to Table 5.1. With the *ft* value and a value for *f* that can be calculated from equations that depend on Z and E_{max}, an estimate of the half-life (t) of the β transition may be obtained. Nomographs such as those published by S. A. Moszkowski that relate *ft* value, E_{max}, and half-life are also commonly used in making these half-life estimates.

Another method of estimating β half-life has already been discussed in Chapter 3. Examination of the chart of the nuclides will show that nuclei farther away from the valley of beta stability will generally have shorter half-lives than those nearer the valley. Beta decay half-lives for odd-A isobars generally increase regularly as the line of beta stability is approached. For even-A isobars, the trend of longer half-life as the line of beta stability is approached occurs in separate trends for the odd–odd and the even–even nuclides.

5.6. GROWTH OF RADIOACTIVE PRODUCTS IN A DECAY CHAIN

Up to now, only decay processes where the radioactive parent produced a stable daughter nuclide have been considered. Often, however, a parent will produce one or more radioactive daughters, which will themselves decay into stable or active "granddaughters." Calculations of the activities of the species involved in these decay chains are more involved than those for simple decay, because the daughter is simultaneously being produced and decaying. Two cases of chain decay will be considered: a parent radionuclide with only one radioactive daughter, and a parent radionuclide producing multiple radioactive daughters.

5.6.1. Parent with a Single Radioactive Daughter

A general equation for the case of a parent yielding a single radioactive daughter is

$$\mathrm{A} \xrightarrow{\lambda_A} \mathrm{B} \xrightarrow{\lambda_B} \mathrm{C}$$

A specific example would be the decay of ^{35}P to ^{35}Cl:

$$^{35}\mathrm{P} \xrightarrow{t_{1/2}=47\,\mathrm{s}} {}^{35}\mathrm{S} \xrightarrow{t_{1/2}=87.2\,\mathrm{d}} {}^{35}\mathrm{Cl} \quad \text{(stable)}$$

It is necessary to develop an equation to describe the activity of the daughter after some decay time, t, given the initial activity of the parent and the half-lives of the parent and daughter. The parent nuclide decays at a rate given by

$$\frac{dN_A}{dt} = -\lambda_A N_A \tag{5.24}$$

Because the daughter, B, is formed from the decay of the parent, A, the rate of decay of A is equal to the rate of formation of daughter B:

$$\frac{dN_B}{dt} = \lambda_A N_A \tag{5.25}$$

However, the radioactive daughter, B, is also decaying, so the loss of the daughter atoms must be accounted for using the decay equation:

$$\frac{dN_B}{dt} = -\lambda_B N_B \tag{5.26}$$

The overall change in B is the sum of its formation and decay:

$$\frac{dN_B}{dt} = \lambda_A N_A - \lambda_B N_B \tag{5.27}$$

Assuming that the original amount of the parent, A, is known, we can use Eq. 5.24 to get an expression for the number of atoms of A at some time t:

$$N_A = N_A^0 e^{-\lambda_A t} \tag{5.28}$$

Substituting this value for N_A into Eq. 5.27 gives

$$\frac{dN_B}{dt} = \lambda_A(N_A^0 e^{-\lambda_A t}) - \lambda_B N_B \tag{5.29}$$

This can be rearranged to give a first order linear differential equation:

$$\frac{dN_B}{dt} + \lambda_B N_B - \lambda_A N_A^0 e^{-\lambda_A t} = 0 \tag{5.30}$$

The solution of this equation will yield an expression that can be used to calculate the number of daughter atoms present at any time t:

$$N_B = \frac{\lambda_A}{\lambda_B - \lambda_A} N_A^0 (e^{-\lambda_A t} - e^{-\lambda_B t}) + N_B^0 e^{-\lambda_B t} \tag{5.31}$$

The activity of B can easily be calculated using the relationship: $A = \lambda N$.

There are two special cases where Eq. 5.31 can be greatly simplified. Both are situations where the half-life of the daughter is considerably shorter than that of the parent. The difference in half-life results in the eventual occurrence of an equilibrium condition between the activities of the parent and daughter activity. These two special cases are called transient equilibrium and secular equilibrium.

Transient Equilibrium

This equilibrium condition arises when the half-life of the parent is approximately 3–10 times longer than that of the daughter, but is short enough so there is a measurable decay of the parent during the experiment. After a decay time of at least 3–5 half-lives of the daughter B has elapsed, the term $e^{-\lambda_B t}$ in Eq. 5.31 will become very small and can be ignored. This equation then simplifies to

$$N_B = \frac{\lambda_A}{\lambda_B - \lambda_A}(N_A^0 e^{-\lambda_A t}) \tag{5.32}$$

The term in parentheses is equal to N_A, so we can also write

$$N_B = \frac{\lambda_A}{\lambda_B - \lambda_A} N_A \tag{5.33}$$

Rearranging this equation gives

$$\frac{N_B}{N_A} = \frac{\lambda_A}{\lambda_B - \lambda_A} \tag{5.34}$$

In this case of **transient equilibrium,** the ratio of the daughter atoms to the parent atoms becomes constant after several half-lives of the daughter have passed. Activities are proportional to the numbers of atoms, so it is also true that the ratio of the activities of the parent and daughter becomes constant:

$$\frac{A_B}{A_A} = \frac{\lambda_B}{\lambda_B - \lambda_A} \tag{5.35}$$

Figure 5.10 shows graphically the case of transient equilibrium. Note that the activity of the daughter is greater than that of the parent when transient equilibrium is achieved and that the ratio of the two activities is then constant.

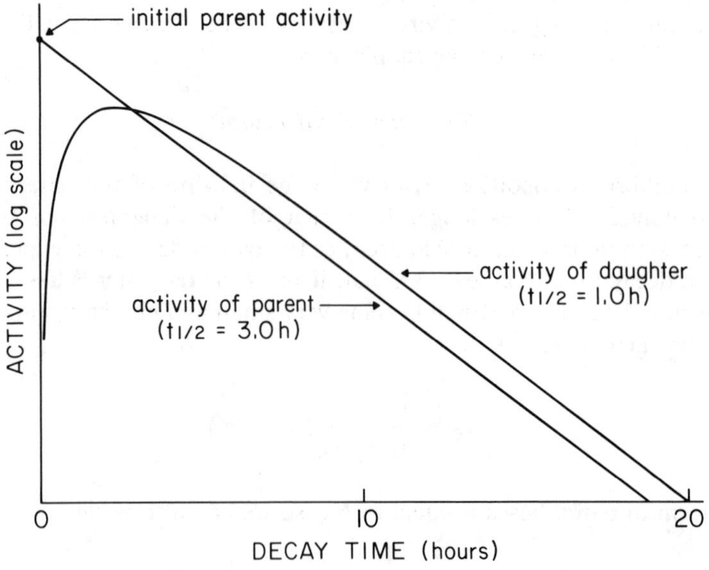

Figure 5.10. Parent and daughter activities as a function of decay time in transient equilibrium.

Secular Equilibrium

If the half-life of the parent is at least 10 times greater than that of the daughter and if there is essentially no change in the activity of the parent during the experimental observation, then the condition of secular equilibrium is attained. For decay times that are long with respect to the half-life of the daughter, the $e^{-\lambda_B t}$ term becomes very small, and can be ignored. In addition, the decay constant of the long-lived parent will be very small compared to that of the daughter. Therefore, we can assume that $(\lambda_B - \lambda_A) \approx \lambda_B$. Eq. 5.34 then simplifies to

$$\frac{N_B}{N_A} = \frac{\lambda_A}{\lambda_B} \quad \text{or} \quad N_B\lambda_B = N_A\lambda_A \tag{5.36}$$

Because $A = N\lambda$, we can also write

$$A_A = A_B \tag{5.37}$$

This situation where the activities of the parent and daughter become equal after a time is called **secular equilibrium.** A graph of secular equilibrium is shown in Figure 5.11. Note that the activity of the parent A is

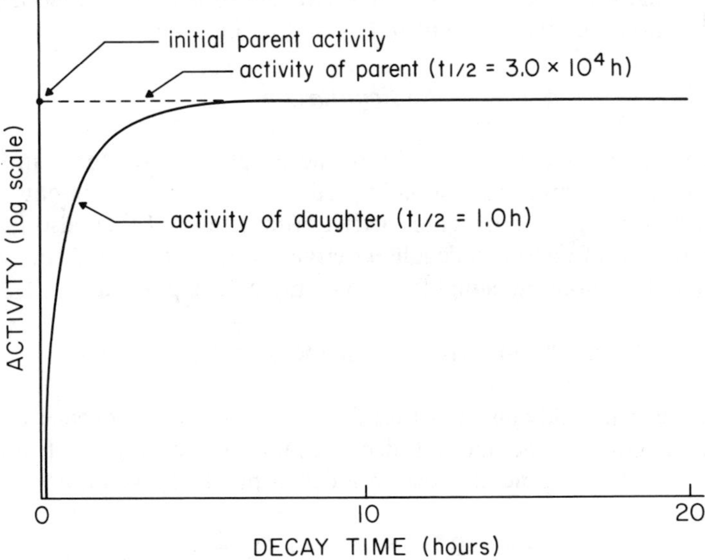

Figure 5.11. Parent and daughter activities as a function of decay time in secular equilibrium.

essentially constant, and that the activity of the daughter B increases from zero up to a maximum value, equal to that of the parent. The gross activity is the sum of the parent and daughter activities.

Earlier in this chapter, the difficulty of directly determining the half-life of a long-lived nuclide was discussed. Secular equilibrium (Eq. 5.37) can often be employed to calculate the half-life of a long-lived species that decays to a short-lived daughter radionuclide. An example is the decay of ^{238}U to ^{234}Th:

$$^{238}\text{U} \xrightarrow{t_{1/2}=4.5\times10^{9}\,\text{y}} {}^{234}\text{Th} \xrightarrow{t_{1/2}=24.1\,\text{d}} {}^{234}\text{Pa}$$

^{234}Th is a β emitter whose activity is easily measured. ^{238}U is an α emitter whose low specific activity in a bulk sample would be difficult to measure. Measurement of ^{238}U activity would additionally suffer from sample self-absorption of the α particles. A pure sample of ^{238}U could be chemically separated and then allowed to decay until conditions for secular equilibrium are reached. This would require about 5 half-lives of the ^{234}Th daughter, or about 120 days. After this decay time, the ^{234}Th could be chemically extracted from the sample, and the activity of the lower mass, higher specific activity β-emitting sample measured. Equation 5.37 shows that this activity is also equal to the activity of the ^{238}U. If the activity of the ^{238}U and the mass of the original ^{238}U sample is known, then the half-life of ^{238}U can easily be calculated using Eq. 5.17.

No Equilibrium

For the case in which the half-life of the daughter is greater than that of the parent, no equilibrium situation will ever occur. The parent activity approaches zero as the daughter activity first rises and then also falls. No constant ratio of parent to daughter activity over an extended time is ever attained. Therefore, no simplification of Eq. 5.31 is possible.

5.6.2. Parent with Multiple Radioactive Daughters

There are many radionuclides that decay to produce a sequence of radioactive daughters. The natural decay chains are examples of this (see Figs. 1.16–1.18). A general equation for multiple decay would be

$$\text{A} \xrightarrow{\lambda_A} \text{B} \xrightarrow{\lambda_B} \text{C} \xrightarrow{\lambda_C} \text{D} \xrightarrow{\lambda_D} \text{etc.}$$

As before, we need an equation to calculate the activity of any daughter nuclide at any time, t. Daughter C will be used as the example of how this is done. It is easy to see that the activity of C at some time t would equal the difference between its rate of formation and its rate of decay:

$$\frac{dN_C}{dt} = -\lambda_C N_C + \lambda_B N_B \tag{5.38}$$

This equation is similar to the one written earlier for the case of one radioactive daughter, but because the expression for N_B is more complex than the one for N_A, it is more complicated:

$$\frac{dN_C}{dt} = -\lambda_C N_C + \lambda_B \left[\frac{\lambda_A}{\lambda_B - \lambda_A} N_A^0 (e^{-\lambda_A t} - e^{-\lambda_B t}) + N_B^0 e^{-\lambda_B t} \right] \tag{5.39}$$

The expression for the next daughter, D, would be still more complicated. The **Bateman equation,** developed by H. Bateman, permits calculation of the activity of any member of a decay chain, when two assumptions are made. First, we assume that only the parent is present at $t = 0$ and that no two decay constants in the chain are equal. When this is the case, the number of any daughter atoms may be calculated using this equation:

$$N_N = C_A e^{-\lambda_A t} + C_B e^{-\lambda_B t} + C_C e^{-\lambda_C t} + \cdots C_N e^{-\lambda_N t} \tag{5.40}$$

where

$$C_A = \frac{\lambda_A \lambda_B \lambda_C \cdots \lambda_{N-1}}{(\lambda_B - \lambda_A)(\lambda_C - \lambda_A) \cdots (\lambda_N - \lambda_A)} N_A^0 \tag{5.41}$$

$$C_B = \frac{\lambda_A \lambda_B \lambda_C \cdots \lambda_{N-1}}{(\lambda_A - \lambda_B)(\lambda_C - \lambda_B) \cdots (\lambda_N - \lambda_B)} N_A^0 \tag{5.42}$$

In calculating the constants (C_N), the numerator is the product of all the decay constants except the last one. In the denominator, the decay constant of interest is subtracted from all other decay constants in the series and the results of these subtractions are all multiplied together. The value of the fraction expressed by these decay constants is then multiplied by the original number of parent atoms to get the value of the constant.

Example. A sample of ^{222}Rn has an activity of 1.00 μCi. Calculate the activity of the daughter ^{214}Pb atoms after 1.00 h has elapsed. The decay is

$$\begin{array}{llllll} ^{222}\text{Rn} & \xrightarrow{\hspace{2em}} & ^{218}\text{Po} & \xrightarrow{\hspace{2em}} & ^{214}\text{Pb} & \xrightarrow{\hspace{2em}} \\ t_{1/2} & 3.82\text{ d} & & 3.11\text{ min} & & 26.8\text{ min} \\ \lambda, \text{s}^{-1} & 2.10\times10^{-6} & & 3.71\times10^{-3} & & 4.31\times10^{-4} \end{array}$$

The number of atoms of the parent nuclide, ^{222}Rn, is

$$1.00\ \mu\text{Ci} = 3.7 \times 10^4 \text{ dps}$$

$$\text{N}(^{222}\text{Rn}) = \frac{A}{\lambda} = \frac{(3.7 \times 10^4 \text{ dps})}{(2.10 \times 10^{-6} \text{ s}^{-1})} = 1.76 \times 10^{10} \text{ atoms}$$

Using Eq. 5.40 and the related equations, we can first calculate the values of the constants for each nuclide:

$$\text{C}(^{222}\text{Rn}) = \frac{(2.10 \times 10^{-6})(3.71 \times 10^{-3})(1.76 \times 10^{10})}{(3.71 \times 10^{-3} - 2.10 \times 10^{-6})(4.31 \times 10^{-4} - 2.10 \times 10^{-6})}$$
$$= 8.62 \times 10^7$$

$$\text{C}(^{218}\text{Po}) = \frac{(2.10 \times 10^{-6})(3.71 \times 10^{-3})(1.76 \times 10^{10})}{(2.10 \times 10^{-6} - 3.71 \times 10^{-3})(4.31 \times 10^{-4} - 3.71 \times 10^{-3})}$$
$$= 1.13 \times 10^7$$

$$\text{C}(^{214}\text{Pb}) = \frac{(2.10 \times 10^{-6})(3.71 \times 10^{-3})(1.76 \times 10^{10})}{(2.10 \times 10^{-6} - 4.31 \times 10^{-4})(3.71 \times 10^{-3} - 4.31 \times 10^{-4})}$$
$$= -9.75 \times 10^7$$

Using these constants in Eq. 5.40 gives (E-6 equals 10^{-6}, etc.)

$$N(^{214}\text{Pb}) = 8.62 \times 10^7\, e^{-(2.10E-6/\text{s})(3600\,\text{s})} + 1.13 \times 10^7\, e^{-(3.71E-3/\text{s})(3600\,\text{s})}$$
$$+ [-9.75 \times 10^7\, e^{-(4.31E-4/\text{s})(3600\,\text{s})}] = 6.49 \times 10^7 \text{ atoms } ^{214}\text{Pb}$$

The activity of ^{214}Pb is

$$A(^{214}\text{Pb}) = \lambda N = (4.31 \times 10^{-4})(6.49 \times 10^7) = 2.80 \times 10^4 \text{ dps}$$

or

$$2.80 \times 10^4 \text{ dps } [1.00 \text{ Ci}/(3.7 \times 10^{10} \text{ dps})] = 7.6 \times 10^{-7} \text{ Ci}$$

5.7. GROWTH OF PRODUCTS IN A NEUTRON FLUX

In previous sections, activities of species that are undergoing or are formed by spontaneous radioactive decay have been considered. Often the radionuclide of interest is being produced by a nuclear reaction involving incident charged particles, photons, or neutrons. It is necessary to consider the equations that will allow calculation of the activity levels of product radionuclides produced in this way. As an example, equations for the growth of nuclides produced by neutron bombardment are discussed.

A nuclear fission reactor produces large numbers of neutrons. The neutron flux density (ϕ) is the number of neutrons incident on a unit area of the target per unit time (neutrons $cm^{-2}\,s^{-1}$). If a sample is placed in a nuclear research reactor, radiative capture reactions will occur:

$$A(n, \gamma)B \xrightarrow{\lambda_B} C$$

The target atoms (A) capture neutrons to form the radioactive species B, which then decays to product C. The rate of formation (R) of B is given by

$$R = n\phi\sigma \tag{5.43}$$

where n is the number of atoms of target present, which is assumed to be constant. The flux density is also assumed to remain constant throughout the target. The atoms of B are also decaying, at a rate equal to $-\lambda_B N_B$, so the net change in the number of B atoms is given as

$$\frac{dN_B}{dt} = n\phi\sigma - \lambda_B N_B \tag{5.44}$$

$$\int_{N_B=0}^{N_B=N_B} \frac{dN_B}{n\phi\sigma - \lambda_B N_B} = \int_{t=0}^{t=t} dt \tag{5.45}$$

There is a standard form integral for this expression:

$$\int \frac{dx}{a + bx} = \frac{1}{b} \ln(a + bx) \tag{5.46}$$

Substituting yields

$$\frac{1}{-\lambda_B} \ln(n\phi\sigma - \lambda_B N_B) - \frac{1}{-\lambda_B} \ln(n\phi\sigma) = t \tag{5.47}$$

and simplifying gives

$$n\phi\sigma - n\phi\sigma e^{-\lambda_B t} = \lambda_B N_B = A_B \tag{5.48}$$

Then

$$A_B = n\phi\sigma(1 - e^{-\lambda_B t}) \tag{5.49}$$

where A_B is the absolute activity of the sample at the end of the irradiation. Equation 5.49 is the basic equation of activation analysis and is discussed further in Chapter 9. It should be noted that after an irradiation period equivalent to 3–5 times the half-life of the radioactive product, activity levels of the product approach a saturation level equal to $n\phi\sigma$. Therefore, additional irradiation time does not significantly increase the activity level of the product radionuclide.

TERMS TO KNOW

Activation analysis
Activity (A)
Allowed transition
Average lifetime (τ)
Bateman equation
Becquerel (Bq)
Branching decay
Coincidence techniques
Comparative half-life
Curie (Ci)
Decay constant (λ)
Decay curve
Detector efficiency
Doppler shift
Energy width
Forbidden transition
4π counter
Half-life ($t_{1/2}$)
Isotopic abundance (IA)
Mirror nuclei
Multi-scaling
Neutron flux density
Nuclear track detectors
Partial decay constant
Radiative capture
Radioactive decay
Secular equilibrium
Self-absorption
Specific activity (SA)
Super-allowed transition
Transient equilibrium

READING LIST

Bateman, H., The solution of a system of differential equations occurring in the theory of radioactive transformations. *Proc. Cambridge Philos. Soc.* **15**:423 (1910). [the Bateman equation]

Chase, G. D., and J. L. Rabinowitz, *Principles of Radioisotope Methodology*, 2d ed. Minneapolis, MN: Burgess, 1962. [experimental methods of half-life determination]

Greenland, P.T., Seeking non-exponential decay, *Nature* **335**:298 (1988). [no deviations found for exponential decay law]

Moszkowski, S. A., Rapid method of calculating log (ft) values. *Phys. Rev.* **82**:35 (1951). [comparative half-lives]

Rubinson, W., The equations of radioactive transformations in a neutron flux. *J. Chem. Phys.* **17**:542 (1949). [activation analysis equation]

EXERCISES

1. Plutonium (^{238}Pu) was carried in the *Voyager* space probes to Jupiter as a power source. ^{238}Pu has also been used as a heat source to power heart pacemaker batteries. In a pacemaker this power system can have a life up to 20 years. Calculate the activity in units of curies (Ci) for the 0.25 g of pure ^{238}Pu used in the pacemaker battery. What properties of ^{238}Pu make it a good choice for human implants?

2. A sample of 0.250 g of a pure radionuclide with a mass number of 244 was observed to have an absolute activity of 4.45 microcuries (μCi). Calculate the half-life of this radionuclide and with the aid of a chart of the nuclides tentatively identify this radionuclide.

3. Consider the decay series A $\rightarrow$ B $\rightarrow$ C $\rightarrow$ D, where the half-lives of A, B, and C are 3.45 h, 10.0 min, and 2.56 h, respectively. We first prepare some pure radionuclide A and exactly 2.75 h after this preparation we measure the activity of daughter C. What would the activity of daughter C be (in Bq) after the 2.75-h decay of pure A, if we started with 7.35×10^7 becquerels of pure A (nuclidic mass of A = 158.9 daltons)?

4. A 1.00-mg Au wire was irradiated with neutrons at a flux of 1.00×10^{13} n cm^{-2} s^{-1}. The wire was counted immediately upon removal from the reactor using a detector with a counting efficiency of 20%. How long would the sample need to be irradiated to produce a measured activity of 1.00×10^6 cps? The thermal neutron capture cross section for the ^{197}Au(n, γ)^{198}Au reaction is 98.7 b.

5. Calculate the activity (in mCi) of a medical ^{60}Co source containing 1.00 mg of pure ^{60}Co.

6. The half-life of ^{137}Cs is 30.17 y. Calculate the decay constant for ^{137}Cs, in units of y^{-1} and s^{-1}.

7. Calculate the activity, in dps and Ci, expected for a 1.00 mg ^{252}Cf source that is 10.0 years old. The half-life of ^{252}Cf is 2.64 y.

8. The presence of high-energy β^- emission from ^{32}P is an interferent in some analytical situations. Calculate the time needed for the ^{32}P activity in a sample to be reduced to 0.1% of its original level.

9. The description of one household smoke detector states that it contains 1 μCi of ^{241}Am.
   (a) Calculate the mass, in grams, of the ^{241}Am present in the smoke detector.
   (b) How long will it take to reduce the activity of the ^{241}Am from 1.0 to 0.50 μCi?

10. Use a chart of the nuclides to identify two nuclides that undergo branching decay.

11. A detector used to count an NIST standard reference material records a counting rate of 500 cps. The certified value for the standard is 1200 dps.

(a) Calculate the efficiency of the detector.

(b) The same detector is now used to count a sample, and records an activity of 350 cps. What is the absolute activity of the sample?

12. The following data were collected for a half-life determination:

Activity (background corrected) (cpm)	Time (min)
34 730	1.0
16 235	3.0
9 029	5.0
5 169	7.0
2 576	9.0
1 157	11.0
755	13.0
362	15.0
68	17.0

Use the simple decay curve method to determine the half-life and the original activity of the sample.

13. ^{137}Cs decays via β^- emission to ^{137}Ba. An experiment is begun with 5.00×10^6 Bq of pure ^{137}Cs. Calculate the activity due to ^{137}Ba after a decay period of 5.00 min.

14. A sample contains a mixture of ^{239}Pu and ^{240}Pu in unknown proportions. The activity of the mixed sample was found to be 4.35×10^7 dpm for a sample of 0.125 mg of Pu. Calculate the weight % of each Pu isotope present.

15. In the decay chain $A \rightarrow B \rightarrow C_{\text{stable}}$ the half-lives of A and B are 17.75 and 43.9 min, respectively. If we start with pure A, how long a decay period would be required for the activity of B to become equal to the activity of A?

16. A sample of a radionuclide whose half-life is 13.85 min was removed from a nuclear reactor at 9:45 a.m. and counted from 9:50 a.m. to 10:10 a.m. The total number of counts recorded by a detection system for this counting period was 12 500. Calculate the counting rate of the sample at 9:45 a.m., when it was removed from the reactor.

17. The half-life of ^{82}Se is approximately 1.4×10^{20} years. This is one of the longest half-lives that has been directly measured in the laboratory. How would you undertake to measure this half-life and what specific experimental problems might you encounter?

18. ^{218}Po decays with a half-life of 3.10 min to ^{214}Pb, which in turn decays with a half-life of 26.8 min to ^{214}Bi. Assuming we have a source of pure ^{218}Po at the start of our experiment, what decay time will be required for the activity of ^{214}Pb to reach its maximum value?

CHAPTER

6

INTERACTIONS OF RADIATION WITH MATTER

An understanding of the ways radiation interacts with matter is a prerequisite for a good understanding of many topics discussed in later chapters. The unique interaction characteristics of different particles and radiations will affect the severity of physiological damage during exposure of living systems to radiation. Topics relating modes of interaction to health physics are discussed in Chapter 7. Different modes of interaction will require diverse types of detection systems, as discussed in Chapter 8. Many applications of radioactivity in the fields of pharmacology and medicine also require a knowledge of the reaction mechanisms of radiation with matter.

When used without qualification in this chapter, the term radiation includes both electromagnetic radiation and the energetic particles emitted by nuclei undergoing radioactive decay or a nuclear reaction.

6.1. MODES OF INTERACTION

There are five basic ways in which radiation interacts with matter: ionization, kinetic energy transfer, molecular and atomic excitation, nuclear reactions, and radiative processes.

- **Ionization** is the removal of an atomic electron from an absorber atom to form an **ion pair** consisting of a negative electron and a more massive positive ion. **Primary ionization** is initiated directly by the incident radiation. **Secondary ionization** is produced subsequently by the ions created in the primary ionization event. The amount of energy needed to form an ion pair varies with the type of absorbing medium. Approximate values for α-particle interactions with some common absorbing media, in units of eV/ion pair, are air = 35, helium = 43, xenon = 22, germanium = 2.9. These values are higher than the first ionization potentials for the elements, because other degrees of freedom, as mentioned below, may also receive energy in the interaction. However, the relative values for the monatomic gases are consistent with the ratios of their first ionization potentials, for example, 24 eV for He and 12 eV for Xe.

- **Kinetic energy transfers** are interactions that impart kinetic energy to ion pairs above the amount required to form the pair. Kinetic energy transfers may also occur due to elastic collisions between the incoming radiation and the nuclei of the absorber.
- **Molecular excitation and atomic electron excitation** are modes of interaction that may occur even when the energy transferred is less than the absorber ionization energy. As the atomic electrons fall back to lower energy levels, x rays and Auger electrons will be emitted. Molecular excitation occurs through translational, rotational, and vibrational processes, and also through electronic excitation. The molecular excitation energy is eventually dissipated by bond rupture, luminescence, or evolution of heat.
- **Nuclear reactions** of incoming radiations with nuclei of absorber atoms can be important modes of interaction. This is especially true for high-energy charged particles and neutrons.
- **Radiative processes** are those in which electromagnetic energy is released by decelerating high-velocity particles. Those of interest here are Čerenkov and bremsstrahlung radiation production.

In this chapter, radiations from decay processes or nuclear reactions are divided into two groups: charged and uncharged. Charged particles include both heavy (e.g., α, fission fragments, emitted nuclear reaction products) and light (e.g., β^-, β^+) varieties. Uncharged radiations include γ-rays, x rays, neutrinos, and neutrons. These two groups of radiations have characteristic differences in their interactions with matter. Charged radiations interact primarily with the electrons of the atoms in the absorbing medium through a series of many small energy-loss events. These events will result in the formation of ion pairs, kinetic energy transfers, and atomic or molecular excitation. Therefore, these radiations lose energy in an incremental and rather predictable manner. Uncharged radiations, in contrast, interact less frequently, or not at all, with the absorber electrons while dissipating their energy. Instead of a series of small sequential interactions, the uncharged radiations often undergo one or only a few major interaction events in which all their energy is lost. These interactions are not so predictable as those for charged particles. It is possible for uncharged radiations to pass through a large amount of matter with a low probability of interaction. This is not the case for charged particles. Hence, uncharged radiations are much more penetrating than charged particles of the same energy.

6.2. HEAVY CHARGED-PARTICLE INTERACTIONS

Heavy charged particles may be considered to include those with masses greater than one dalton. In radioactive decay processes, these particles have a positive charge. The heavy charged particles interact with matter primarily through ionization. For high-energy particles, nuclear reactions may also be important. The heavy charged particles are much more massive than the electrons with which they are interacting. Therefore, they are not significantly deflected by these interactions and the path of the particle through an absorber is relatively straight. Rarely, however, elastic collisions with the absorber nuclei will result in significant deviations of individual particles from a straight path (Rutherford scattering).

Alpha particles are the most commonly encountered heavy charged particles in radiochemistry. They will be used as examples of charged-particle interactions in the following discussion.

6.2.1. Range Relationships for Heavy Charged Particles

The high likelihood of interaction of positively charged particles with absorber electrons combined with their straight interaction path suggest that the particles will have a fairly well-defined **range** in a given absorber. The range of the particle may be stated in a variety of units, all of which reflect the distance traveled through the absorber until the particle is "stopped" (i.e., reaches the ambient average kinetic energy of the absorber atoms). The value of the range for a given particle in a specific absorber will show some variations because the energy loss events occur randomly and because the actual amount of energy lost per interaction varies. These variations result in an effect known as **range straggling.**

Because of straggling, more precise definitions of the range are needed. Figure 6.1 illustrates these definitions. This figure shows the proportion of α particles detected after passage through various thicknesses of an absorbing medium. The **mean range** (R_m) is the absorber thickness that will stop one-half of the incident α particles. Notice that the line indicating the α particles does not drop straight down to the x axis on the right side, but rather "tails off" at the end of the range. This is the straggling effect mentioned above. If the straggling is ignored, and the linear portion at the end of the plot is extrapolated to zero, the **extrapolated range** (R_e) is obtained. The extrapolated range and the mean range are related by this equation:

$$R_e = 1.1R_m \tag{6.1}$$

The mean range is most often used in tables and calculations.

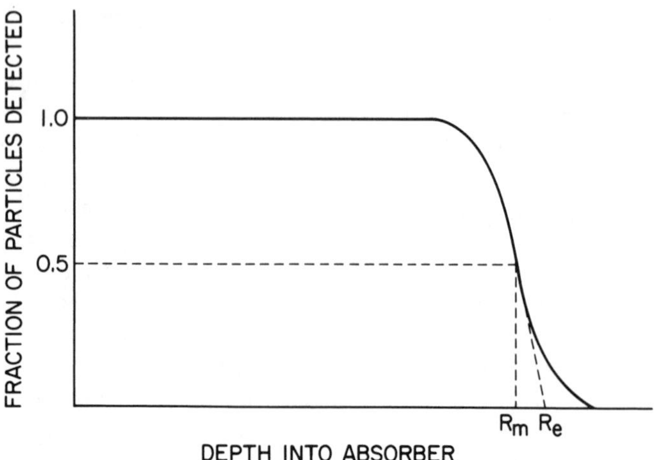

Figure 6.1. Absorption of α particles in matter. R_m is the mean range and is the value most often found in data compilations. R_e is the extrapolated range and is approximately 10% greater than R_m.

The range of a charged particle may be expressed in linear units, such as mm or cm. However, the density of the absorbing medium, which can vary with temperature and pressure, will affect the linear range. Therefore, another unit that takes into account the absorber density can be used to advantage. These units of thickness are called **mass thickness** units and are often given in mg/cm². The range in mass thickness units can be easily calculated from a knowledge of the density of the absorber and the linear range of the particle:

$$R\ (\text{mg/cm}^2) = R\ (\text{cm}) \times \text{Absorber density}\ (\text{g/cm}^3) \tag{6.2}$$

The advantage of using mass thickness units is that they are relatively independent of the temperature and pressure.

The range of any charged particle, expressed in either of the ways discussed above, is affected by several factors. These include

1. The Z or A of the absorber
2. The energy of the charged particle
3. The mass of the charged particle
4. The electronic charge on the particle

Charged particles will generally have shorter ranges in high-Z materials, due to the larger number of electrons with which they can interact. Higher-energy particles will have longer ranges than low-energy particles, because more interactions are required to dissipate the particle energy. A larger charge on the particle will result in shorter ranges, because multiple charges will increase the Coulomb interactions with the absorber electrons. A particle with a greater mass will have a shorter range than a particle of lesser mass with the same kinetic energy. This can be explained by an increased interaction time, because a massive particle travels more slowly than a small particle for a given kinetic energy.

There are several empirical equations that can be used to calculate the range of α particles of a given energy in air or other absorbers. One useful relationship is

$$R = a(E_\alpha)^b \tag{6.3}$$

where a and b are constants, R is the range in linear units, and E_α is the α energy in MeV. The constant b usually has a value of 1.5–1.8. For air at STP, Eq. 6.3 becomes

$$R_{\text{air}}(\text{cm}) = 0.31E_\alpha^{3/2} \tag{6.4}$$

The constants will vary with absorber type, temperature, pressure, and type of charged particle. Using mass thickness units, the α range in air is given by

$$R_{\text{air}}(\text{mg/cm}^2) = 0.40E_\alpha^{3/2} \tag{6.5}$$

Figure 6.2 shows a plot of the range of α particles in aluminum. The range of protons of the same energy would be four times greater, because range is inversely related to the square of the charge on the particle.

A general equation for the ranges of protons, deuterons, and α particles in the 0.1- to 1000-MeV energy range, when Z of the absorber is greater than 10, is

$$\frac{R_Z}{R_{\text{air}}} = 0.90 + 0.0275Z_{\text{abs}} + (0.06 - 0.0086Z_{\text{abs}}) \log \frac{E_{\text{MeV}}}{m_{\text{p}}} \tag{6.6}$$

where Z_{abs} refers to the Z of the absorber, E_{MeV} is the energy of the particle in MeV, and m_{p} is the mass of the particle in daltons. The ranges here are expressed in mass thickness units, mg/cm^2. The equation can be altered for absorbers of Z less than 10 by substituting these values for the first two terms: for hydrogen, 0.3; for helium, 0.82; for $Z = 3$–9, 1.0. For α particles, Eq. 6.6 simplifies to

$$R_Z(\text{mg/cm}^2) = 0.173E_\alpha^{3/2}A_{\text{abs}}^{1/3} \tag{6.7}$$

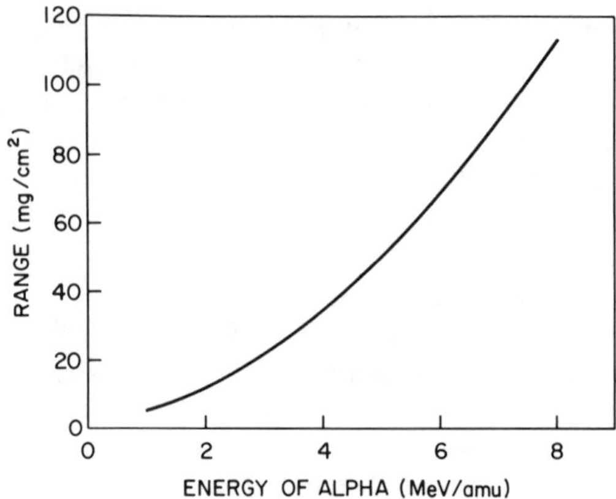

Figure 6.2. Range of α particles in aluminum (mg/cm^2) as a function of particle energy per amu of the particle. A value of 2 MeV/amu on this plot corresponds to an actual alpha particle energy of 8 MeV. It is common to express particle energies in units of MeV/amu in range-energy plots.

Example. Calculate the range, in both mass thickness and linear units, of a 5.0-MeV α particle in aluminum. The density of Al is 2700 mg/cm^3. The range for an α particle in an absorber of $Z > 10$ is given by Eq. 6.7. Substituting into this equation gives

$$R_Z = 0.173\ (5.0\ \text{MeV})^{3/2}\ (27)^{1/3}$$

$$R_Z = 5.8\ \text{mg/cm}^2 \times \frac{1}{2700\ \text{mg/cm}^3} = 0.0021\ \text{cm} = 0.021\ \text{mm of Al}$$

Often, absorbers are not pure elements, but rather are compounds or alloys. The range of the charged particle in these complex absorbers can be calculated using the following equation:

$$\frac{1}{R_c} = \frac{w_1}{R_1} + \frac{w_2}{R_2} + \cdots + \frac{w_n}{R_n} \tag{6.8}$$

In this equation, w is the weight fraction of each element in the absorber, R is the range of the particle in the element, in mass thickness units, and R_c is the range in the complex absorber.

Example. Calculate the range of 20.0-MeV α particles in polyethylene, $-(CH_2)_x-$. This is a complex absorber, so Eq. 6.8 must be used along with Eq. 6.7 to calculate the ranges in the individual elements:

$$
\begin{aligned}
R_H &= 0.173\,(20.0)^{3/2}\,(1)^{1/3} = 15.5 \text{ mg/cm}^2 \\
R_C &= 0.173\,(20.0)^{3/2}\,(12)^{1/3} = 35.4 \text{ mg/cm}^2 \\
w_H &= (2 \times 1.0)/(12.0 + 2.0) = 0.143 \\
w_C &= (1 \times 12.0)/(12.0 + 2.0) = 0.857 \\
1/R_{\text{polyethylene}} &= (0.857/35.4) + (0.143/15.5) = 0.0335 \\
R_{\text{polyethylene}} &= 29.9 \text{ mg/cm}^2
\end{aligned}
$$

For charged particles other than the α, different equations for range are needed. However, development of appropriate empirical equations for all charged particles would be a tedious task. What is often done, therefore, is to relate the range of more massive particles to a simple particle like the proton. An equation for this is

$$R_1 = \frac{(M_1)(\xi_2)^2}{(M_2)(\xi_1)^2} R_2 \tag{6.9}$$

where $\xi_{1,2}$ refers to the electrical charge on particles 1 and 2, and M equals the masses of the particles, in daltons. The energies are related by

$$E_2 = \frac{M_2}{M_1} E_1 \tag{6.10}$$

where $E_{1,2}$ refer to the energies of the particles.

Example. Estimate the range in Al metal of a 160-MeV electron-stripped ^{16}O ion, given a table of proton ranges in Al. Assume that particle 1 is oxygen and particle 2 is hydrogen.

$$
\begin{aligned}
R_{\text{oxygen}} &= (16/1)\,(1^2/8^2) = 1/4 \text{ of a proton's range,} \\
\mathrm{E}_{\text{proton}} &= (1/16)\,160 = 10 \text{ MeV} \\
R_{\text{proton}} \text{ of } 10 \text{ MeV} &= 171 \text{ mg/cm}^2 \text{ from a table of proton ranges in Al} \\
R_{\text{oxygen}} &= (1/4)\,171 = 43 \text{ mg/cm}^2
\end{aligned}
$$

For fission products, $E \approx 100$ MeV, $M \approx 100$ daltons, and $\xi \approx 20$. Thus, the range of fission products in air is only about 2–3 cm. This is about the same range as a 4-MeV α particle.

6.2.2. Stopping Power

In addition to the range of the charged particles, the rate of their energy loss in the absorber is also of interest. This rate of energy loss is called the **stopping power (S),** or the **specific ionization,** and is given by

$$S = -\frac{dE}{dx} \tag{6.11}$$

The stopping power is directly related to the particle's charge and mass, and inversely related to its energy. A plot describing the specific ionization of α particles is shown in Fig. 6.3. This is commonly called a **Bragg curve.** It plots the specific ionization of a particle as a function of distance traveled into the absorber. Examination of this curve shows that the rate of energy loss increases slowly at first, then reaches a peak near the end of the range. The reason for the higher specific ionization near the end of the path of an α particle is that it is moving slower, and has a greater interaction time near a given absorber atom. Therefore, it is more likely to produce an ion pair.

Another concept that was developed many years ago is that of **relative stopping power ($\mathbf{RSP_{abs}}$).** Although this concept is no longer widely used, it does illustrate the shielding effectiveness of various materials, as compared to air. The relative stopping power is defined by the relationship

$$\mathrm{RSP_{abs}} = \frac{R_{\mathrm{air}}}{R_{\mathrm{abs}}} \tag{6.12}$$

where R_{air} and R_{abs} are the *linear* ranges (cm) of the particle in air at STP and in another absorber, respectively. Some relative stopping power values for a 7-MeV α particle are

Absorber	RSP	Actual Range
Air	1	$\approx 5.7 \times 10^{-2}$ m
Water, biological tissue	800	$\approx 7 \times 10^{-5}$ m
Al metal	1735	$\approx 3.3 \times 10^{-5}$ m

6.3. BETA-PARTICLE INTERACTIONS

The interactions of light charged particles with matter are more complex than those of heavy charged particles. In radioactive decay, we are primarily concerned with β^- and β^+ particles. Beta particles interact with matter

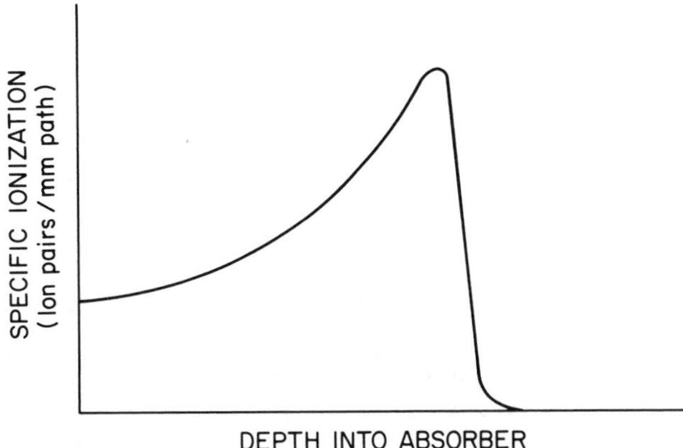

Figure 6.3. The specific ionization (ion pairs/mm of path) for α particles as a function of depth of penetration into an absorber (the Bragg curve).

via ionization and atomic and molecular excitation, as do the heavy charged particles. However, beta interactions with matter also include backscatter and the emission of bremsstrahlung and Čerenkov radiations. These latter three modes of interaction are not significant for heavy particles.

The smaller mass of the β particle means that its velocity is greater than a heavier particle with the same energy. Therefore, there are differences in the degree of ionization produced by heavy and light particles. Both types lose about equal amounts of energy during formation of an ion pair (about 35 eV/ion pair formed in air). However, the specific ionization is much lower for β particles, due to their higher velocities and the resulting reduced time available for interaction. The specific ionization for β particles is greatest near the end of the path, when the kinetic energy of the particle has been significantly reduced. This was also seen in the Bragg curve for α particles. Beta particles may lose a large fraction of their energy in a single collision with absorber electrons, because their masses are the same. (This would not be true for β particles with very high energies because these would have relativistic mass increases; see next paragraph.) Beta particles are also easily deflected, so they will be likely to scatter erratically away from the original beam path.

Another difference between the light and heavy particles is that, above 1 MeV, the energy of a β becomes significantly **relativistic**. This refers to the fact that an object's mass varies with its motion. This effect is not perceptible with objects in the everyday world, moving at relatively slow velocities, but for small particles like the electron very high velocities are possible. As the velocity and kinetic energy of a particle increase, so does the mass. This is called a **relativistic mass increase.** The relativistic mass of a particle is given by

$$M_{\text{rel}} = \frac{M_{\text{rest}}}{(1 - v^2/c^2)^{1/2}} \tag{6.13}$$

The mass of a particle at rest is called its **rest mass.**

6.3.1. Range Relationships for Beta Particles

In Chapter 2 it was noted that β particles are not monoenergetic, but rather show a continuum of energies up to some maximum value (Fig. 2.5). Therefore, the range of β particles is also a continuum, up to a maximum range. This maximum range corresponds to the maximum energy, that is, to a decay event in which the β receives all the decay energy, and no neutrino (ν) is emitted. The maximum range for a β will be greater than the range for a heavier particle of the same energy because of the greater velocity of the β. For example, the 0.157-MeV β^- from ^{14}C has a range of 30 mg/cm^2 in Al, and the 1.7-MeV β^- from ^{32}P has a range of 800 mg/cm^2 in Al. By contrast, the range for a 2-MeV α in Al is only about 10 mg/cm^2.

Empirical equations have been developed to interconvert maximum energies and ranges for β particles. The **Glendenin equations** are well-known examples of these:

$$R_{\max} = 407(E_{\max})^{1.38} \qquad \text{for } \beta \text{ energies 0.15–0.8 MeV} \tag{6.14}$$

$$R_{\max} = 542E_{\max} - 133 \qquad \text{for } \beta \text{ energies 0.8–3.0 MeV} \tag{6.15}$$

The ranges here are in aluminum, in units of mg/cm^2, and the energies are in MeV. Graphical representations of the range–energy relationship are also used. An example of a range–energy curve for β^- particles in Al is shown in Fig. 6.4.

Experimental determination of β ranges is accomplished by placing successively greater thicknesses of absorber between the β source and the detector, until the observed count becomes constant with further additions of absorber. In practice, the actual range cannot be determined directly in this way because the β activity drops to very low levels at higher absorber thicknesses. Therefore, graphs of counting rates versus absorber thickness are plotted using observed counting rate data and then extrapolated to a zero count rate.

Figure 6.5 shows an experimental setup for determination of maximum β ranges. The sample is placed at a fixed distance from the detector, and Al absorbers of increasing thickness are placed between the source and detector. The absorbers are placed close to the detector window to lessen

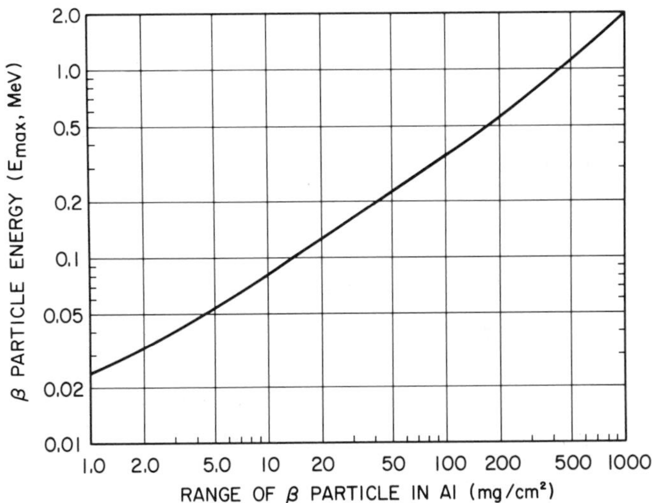

Figure 6.4. Maximum range of β particles in Al (mg/cm^2) as a function of maximum β-particle energy (E_{max}, MeV). (Adapted from G. Friedlander et al., *Nuclear and Radiochemistry,* 3d ed. New York: Wiley, 1981, p. 223.)

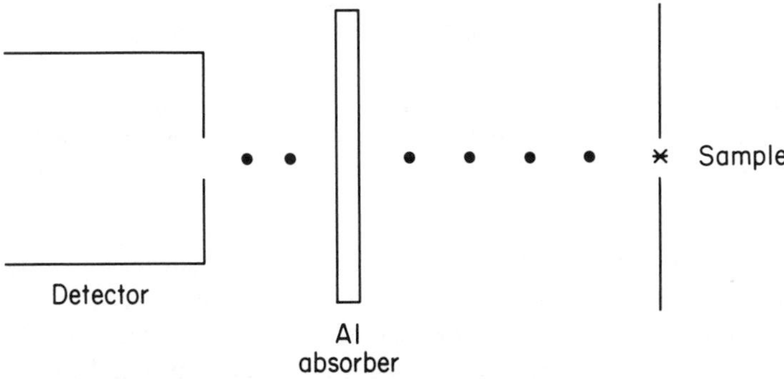

Figure 6.5. Experimental setup for the determination of β-particle maximum ranges. The Al absorbers are placed close to the detector window to minimize scattering losses. The detector is usually a Geiger–Müller or proportional counter.

scattering losses. The counting rate of the sample is measured for each absorber thickness. There are actually three absorbing media between the source and detector. They are the Al absorber, air, and the window of the detector. The sum of all three of these must be included in recording the total absorber thickness. A plot of log count rate (background corrected) against absorber thickness for a pure β emitter is shown in Fig. 6.6*a*. A vertical line drawn tangent to the curve as it approaches zero net counting rate provides an estimate of maximum β range. A similar plot for a β–γ

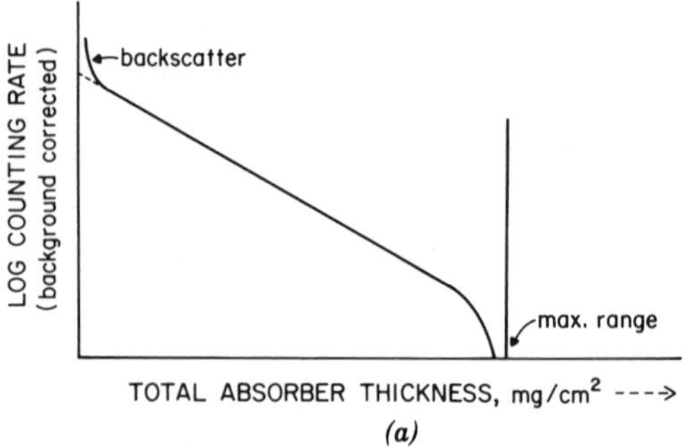

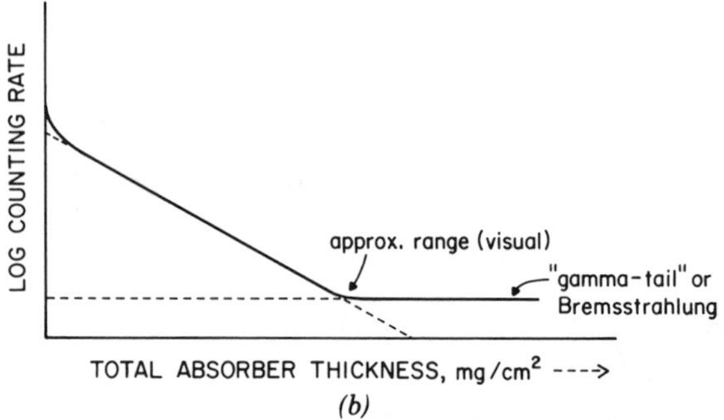

Figure 6.6. (*a*) Log count rate versus total absorber thickness for a pure negatron emitter. (*b*) Log count rate versus total absorber thickness for a source emitting both β particles and γ rays, or a high-energy beta emitter with associated bremsstrahlung radiation.

emitter is shown in Fig. 6.6*b*. Absorber thicknesses that totally absorb the β particles will not stop the γ rays. Therefore, it can be seen in Fig. 6.6*b* that the counting rate ultimately levels off at a rate that is greater than the ambient background. This is due to the γ radiation emitted. To determine the counting rate due only to the betas, the straight "γ tail" is extrapolated back to zero absorber thickness. The values from the extrapolated line are subtracted point by point from the gross counting rates at each absorber thickness. This process yields an absorption curve equivalent to that for a pure β emitter (Fig. 6.6*a*). This new curve can then be used to estimate the maximum β range.

6.3.2. The Feather Method

An alternative method for determination of maximum β ranges is the **Feather method.** This method is based on the assumption that absorber thicknesses that reduce the counting rates of a known and unknown sample by identical amounts are then identical fractions of the two maximum ranges. Counting rates must be corrected for any γ tail and for ambient background. For example, suppose that a pure β emitter is known to have a maximum β range of 200 mg/cm^2. An Al absorber of 50 mg/cm^2 thickness reduces the counting rate of this known β emitter to one-half of its original value. The Al absorber is 50/200, or 0.25 of the maximum β range for the known. Now, an unknown β emitter is measured and it is found that an absorber of 100 mg/cm^2 is needed to reduce the counting rate of the unknown to one-half of its original value. According to the Feather assumption, this absorber must be 0.25 of the maximum β range for the unknown. Therefore, R_{max} for the unknown is 100/0.25, or 400 mg/cm^2. In practice, the method is repeated at different fractional decreases in the counting rates, using data derived from the known and unknown absorption plots. The calculated maximum β range values are plotted as a function of each fractional range. This plot is extrapolated to 1.0 (100%) of the range to obtain the **Feather range.**

Example. Calculate the Feather range for an unknown β emitter, given the following experimental data.

Absorber added (mg/cm^2)	Fraction Range	Observed Count Rate (cpm)	Fraction of Activity	Calculated Range (mg/cm^2)
Known (range = 500 mg/cm^2, original counting rate = 10^5 cpm)				
50	0.10	0.50×10^5	0.5	
100	0.20	0.20×10^5	0.2	
150	0.30	0.10×10^5	0.1	
200	0.40	0.06×10^5	0.06	
250	0.50	0.04×10^5	0.04, etc.	
Unknown (range = ?, original counting rate = 10^4 cpm)				
21	0.10	0.50×10^4	0.5	210
39	0.20	0.20×10^4	0.2	195
57	0.30	0.10×10^4	0.1	190
76	0.40	0.06×10^4	0.06	189
94	0.50	0.04×10^4	0.04	188, etc.

The calculated ranges are plotted as a function of fraction of range and extrapolated to full range as shown in Fig. 6.7. The extrapolation yields a Feather range of approximately 183 mg/cm^2.

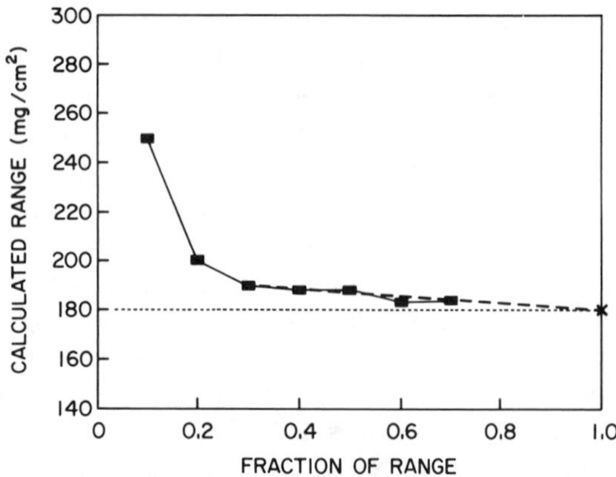

Figure 6.7. Extrapolation of calculated Feather ranges at different fractional ranges to obtain the estimated maximum range of beta particles.

6.3.3. Bremsstrahlung Radiation

This mode of interaction with matter occurs most often for high-energy β particles. When a β is near a nucleus, Coulomb interactions cause the acceleration of the electron away from its original path. This acceleration results in a loss of energy, which is emitted in the form of electromagnetic radiation called **bremsstrahlung radiation.** Bremsstrahlung radiation consists of a continuous spectrum of photon energies in the x-ray region, with a maximum energy equal to that of the β particle. (This is not identical to the inner bremsstrahlung radiation discussed in Chapter 2.) The intensity of bremsstrahlung radiation is inversely related to the square of the particle mass, so the heavy charged particles do not cause significant amounts of this radiation to be emitted. Bremsstrahlung emission is most likely to occur in high-Z absorbers. An equation describing the relative amounts of energy loss for β particles by bremsstrahlung radiation versus ionization is

$$\frac{\text{Energy loss by bremsstrahlung}}{\text{Energy loss by ionization}} = \frac{E_e Z_{abs}}{800} \tag{6.16}$$

where E_e is the energy of the beta in MeV, and Z_{abs} is the atomic number of the absorber.

The bremsstrahlung emitted by β particles may be either an interferent or a useful form of radiation. When detection of γ rays is the goal, the presence of bremsstrahlung radiation raises the general background level and makes gamma detection more difficult. In other cases, the bremsstrahlung radiation is purposely generated and used analytically for an x-ray excitation source or for medical applications.

6.3.4. Čerenkov Radiation

The interaction that results in emission of Čerenkov radiation occurs only for charged particles that have very high velocities. If such a particle enters a given medium traveling faster than light can travel in that medium, a type of "shock wave" is produced as the particle slows down to the speed of light. The energy lost by the particle as it slows down is emitted as electromagnetic radiation in the blue-white region of the spectrum. This radiation is called **Čerenkov radiation.** The blue glow seen around the core of a "swimming pool"-type reactor is Čerenkov radiation. The fission processes occurring in the core result in the production of large numbers of high-energy β particles. Some of the betas are traveling faster than the speed of light in water as they come out of the reactor core into the pool, and, as they are slowed, the Čerenkov radiation is emitted. The interactions responsible for Čerenkov emission are independent of particle mass, and depend only on the velocity and charge of the particle. There are analytical uses for Čerenkov radiation in determining particle energies or in β detection.

6.3.5. Beta Backscatter

It was noted earlier that β particles undergo interactions that result in large deflections from their original paths. Some number of these interactions will result in the β particles being scattered directly back along their original path of entry. This process is called **backscatter**.

The degree to which β backscatter occurs is related primarily to the atomic number and thickness of the scattering material. Backscattering increases with absorber thickness up to about one-fifth of the extrapolated range of the β particle. After that, the amount of backscattering remains essentially constant. Backscatter processes are not strongly energy dependent, so there is only a slight increase in the amount of backscatter for higher-energy betas. Because the angles of emission and energies of the scattered betas are related to the identity of the absorbing material, backscatter phenomena can be used for analytical purposes.

6.3.6. Positron Interactions

High-energy positrons initially interact with matter through both elastic and inelastic scattering events with the atoms or molecules. In these interactions they lose most of their energy through the formation of ion pairs, as is the case for negatrons. Annihilation (described in Section 2.2.2) occurs when the positron has lost most of its initial kinetic energy.

Prior to annihilation, the positron may undergo a reaction that forms a positronium "atom" (see Section 2.2.2). Ore (1949) proposed an equation that gives the minimum positron energy at which positronium formation is still feasible in a gas:

$$E_{\min} = V - 6.8\ \text{eV} \tag{6.17}$$

where V is the ionization energy of the gas molecule in units of eV. Ionization is the favored interaction when the positron has an energy that is above the ionization energy of the gas. Therefore, only a small energy window, or **Ore gap,** exists where formation of positronium is favored. Positronium may interact with other atoms or molecules in the absorber to cause chemical reactions (e.g., oxidation, reduction, exchange reactions). Eventually, a 2γ-annihilation event which yields two 0.511-MeV photons will occur. The annihilation photons are much more penetrating than the original positrons and can cause further interactions in an absorber. Hence, ion pairs will be produced at depths in the absorber much beyond the original positron's maximum range.

6.4. GAMMA-RAY INTERACTIONS

Because γ rays are electromagnetic radiation, they have no charge and no rest mass. The interactions of γ rays are, therefore, quite different from those experienced by charged particles. Gamma rays are not subject to Coulomb interactions, and so they do not lose energy continuously as do the charged particles. Generally, the γ ray will experience only one, or perhaps a few, catastrophic interactions with the electrons or nuclei of atoms in the absorbing material. In these interactions, the γ ray either disappears completely or has its energy significantly altered. Ionization does take place, but most events are induced by secondary electrons and not by the primary γ ray itself. Gamma rays do not have discrete ranges. Instead, the intensity of a beam of γ rays is continuously reduced as it passes through matter in accordance with an exponential absorption law.

Three major modes of interaction of gamma rays with matter will be discussed: the photoelectric effect (PE), Compton scattering (CS), and pair production (PP). In this section, the terms detector and absorber are used interchangeably. Principles of detectors are discussed fully in Chapter 8.

6.4.1. Photoelectric Effect

In the **photoelectric effect,** a γ ray interacts with an atom in a process that results in the ejection of an electron from the atom and the complete disappearance of the γ ray (Fig. 6.8). The electron receives all the energy of the γ ray, minus its atomic binding energy:

$$E_{\mathrm{e,PE}} = E_{\gamma} - \mathrm{BE}_{\mathrm{e}} \tag{6.18}$$

The ejected electron can then induce secondary ionization events that may be detected.

The probability of occurrence of the photoelectric effect is directly related to the Z of the absorber and inversely related to the energy of the γ ray:

$$\mathrm{Probability}_{(\mathrm{PE})} = \frac{k(Z_{\mathrm{abs}})^5}{E_{\gamma}^{7/2}} \tag{6.19}$$

Thus, the photoelectric effect occurs most often for gammas of lower energy (<1 MeV) in high-Z absorbers.

Emitted γ rays are often used for analytical purposes, so the interpretation of the features of a γ-ray spectrum as recorded with a multichannel pulse-height analyzer (see Chapter 8) is an important skill. A knowledge of interaction mechanisms is essential in this interpretation. In Fig. 6.9, a plot of the number of γ rays against γ-ray energy is shown for an idealized case where the only interaction occurring is the photoelectric effect, and there is 100% detection of the emitted photoelectrons. Only one peak appears, because the γ rays lose all of their energy in the photoelectric interaction, and the photoelectrons generated will yield a voltage pulse in the detector

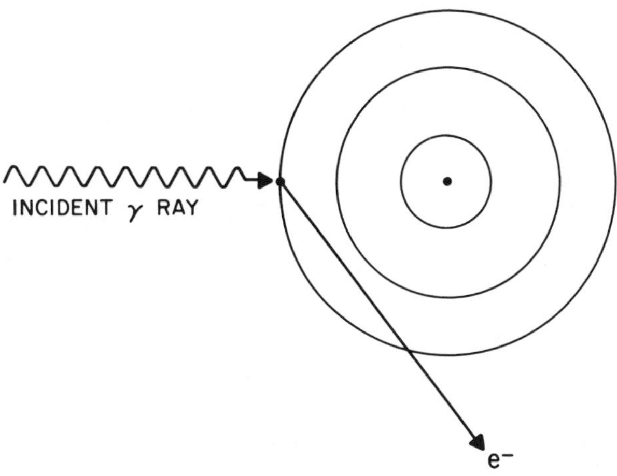

Figure 6.8. A representation of a γ-ray interaction with matter by the photoelectric effect.

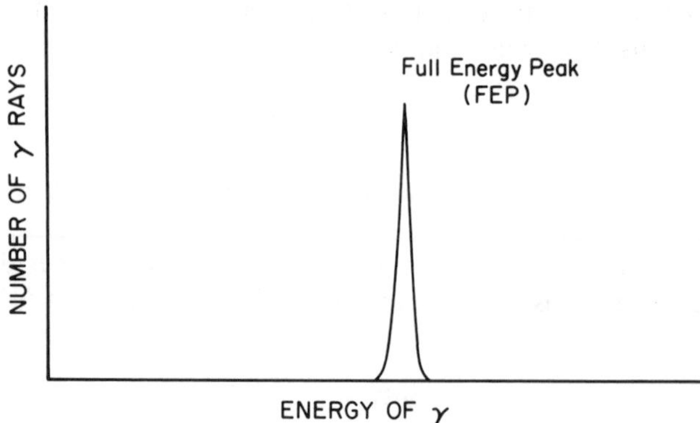

Figure 6.9. The idealized γ-ray energy spectrum that would result from γ rays interacting with the detector solely by the photoelectric effect. The event is recorded in the full-energy peak (FEP), which corresponds to an energy very slightly less than than the energy of the source γ-ray transition.

that has a discrete pulse height. This peak in the γ-ray spectrum is called a **full energy peak (FEP).** Its location in the energy spectrum is slightly lower than that which would correspond to the energy of the incident γ ray. This difference is equivalent to the atomic binding energy of the PE electron. An ideal peak would be a single straight line, but a real peak has a finite **peak width,** due to instrumental factors and the uncertainty principle.

6.4.2. Compton Scattering

In the photoelectric effect, all the γ-ray energy is lost in a single interaction with an atomic electron. A γ ray may also interact with an atomic electron in such a way that it loses only part of its energy. In this process, an electron is scattered away from the atom and gains the energy lost by the γ ray. The γ ray, now with less energy, is deflected from its original path (Fig. 6.10). This process is called **Compton scattering (CS),** after Arthur Compton, who first described it. The CS electron that is ejected from the atom will cause secondary ionization events. It is quite possible that the scattered γ ray will escape from the detector. The electron will probably not escape, because its range is much less than that of the gamma. If the γ ray escapes, only that portion of the original energy that is lost to the CS electron in the scattering event is detected. Therefore, the event will be recorded by a multichannel pulse-height analyzer at an energy less than that of the full-energy peak. If the scattered γ ray does not escape from the detector, but instead undergoes further interactions (e.g., a photoelectric event, or more Compton scattering) it may eventually lose all its energy in the detector. In

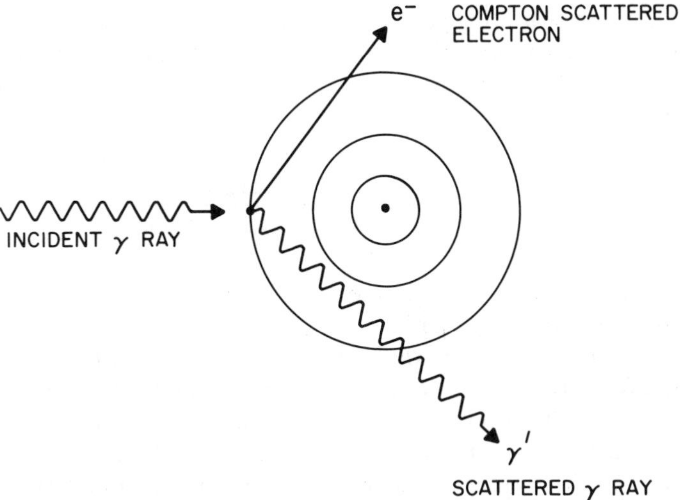

Figure 6.10. A representation of a gamma-ray interaction with matter by Compton scattering.

this case, the event would still be seen by the detector as a single event, and it would therefore be recorded in the full-energy peak.

The probability of Compton scatter is directly related to the number of electrons in the atoms of the absorber material (and thus to Z), and inversely related to the γ energy:

$$\text{Probability}_{(\text{CS})} = \frac{kZ_{\text{abs}}}{E_\gamma} \tag{6.20}$$

Compton scatter is most likely to occur for γ rays in the 0.6- to 4.0-MeV energy range, in high-Z absorbers.

The Compton interactions will result in the production of a continuum of scattered γ-ray energies down to a minimum value. The minimum energy of a Compton scattered γ ray , $E_{\gamma',\text{minimum}}$ would be the energy remaining after it was scattered at 180° from its original direction. This minimum energy can be shown to be equal to:

$$E_{\gamma',\text{minimum}}(180^\circ \text{ scatter}) = \frac{0.511E_{\text{incident } \gamma}}{0.511 + 2E_{\text{incident } \gamma}} \tag{6.21}$$

As the energy of the incident gamma becomes large, this minimum approaches a value of 0.25 MeV.

For Compton scattering generally, it can be shown from conservation of momentum considerations that the energy of the Compton-scattered γ ray, $E_{\gamma'}$, can be calculated for different scattering angles by use of the following equation:

$$E_{\gamma'} = \frac{0.511 E_{\text{incident } \gamma}}{0.511 + (E_{\text{incident } \gamma})(1 - \cos\theta)} \tag{6.22}$$

where θ is the angle that the scattered γ ray makes with the forward projection of the incident γ ray's path.

Figures 6.11*a* and *b* show the effects of Compton scatter on the γ-ray spectrum. If the full energy of the γ ray is lost in the detector , a full-energy peak appears in the spectrum. However, some of the scattered γ rays will escape the detector, carrying away some energy. The amount of energy taken away by these escaped gammas ranges from the minimum 180° backscatter energy to nearly the energy of the incident gamma. These events are recorded in the spectrum at the energies dissipated in the detector by the corresponding Compton scattered electrons. This range of energies is called the **Compton continuum.** There is a gap between the full-energy peak and the Compton continuum. This is because any scattered γ that escapes the detector will take with it *at least* the minimum energy for a 180° backscattered γ ray (Eq. 6.21). This gap approaches 0.25 MeV for high-energy γ rays. Therefore, in the ideal spectrum for Compton scattering of high-energy γ rays from a source emitting γ rays of a single energy, there are no events recorded with energies between the full-energy peak and the energy given by (FEP − 0.25 MeV).

A more realistic gamma spectrum is shown in Fig. 6.12. The full-energy peak still appears, as does the Compton continuum. Note that, in reality, the number of events recorded between the FEP and the Compton continuum is not zero, although there is still a noticeable valley between the two. Events recorded in this interval of the spectrum are in part due to multiple scattering events in the detector and combinations of the interaction effects of more than one incident γ ray that are not time resolved by the detector. The high point on the Compton continuum just before the valley is called the **Compton edge** or **Compton shoulder.** In real γ-ray measurements, the γ rays can interact not only with the detector itself, but also with shielding materials around the absorber. Some of these Compton-scattered γ rays from the shielding can then enter the detector and be recorded. These externally backscattered γ rays will have the minimum energy calculated by Eq. 6.21, and will produce a peak in the spectrum at the minimum backscatter energy. Because some externally scattered γ rays are scattered

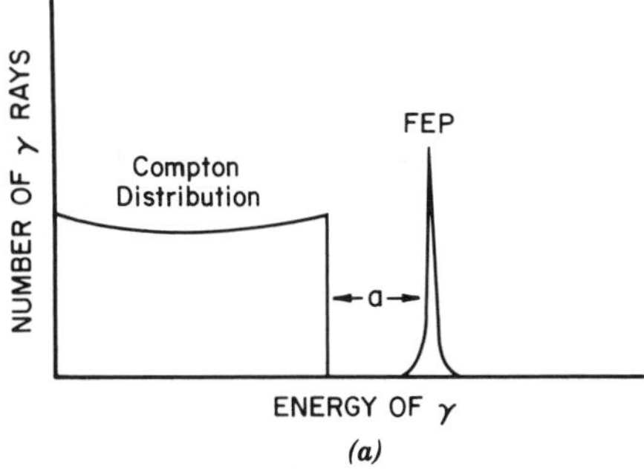

(a)

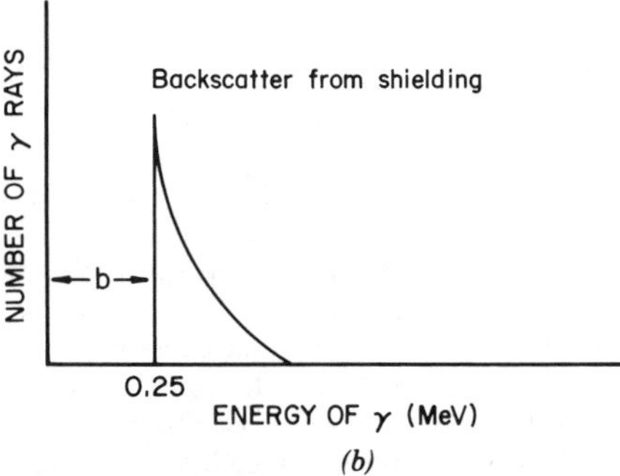

(b)

Figure 6.11. (*a*) An idealized spectrum that would result from source γ rays initially interacting with the detector solely by Compton scattering. Scattering events external to the detector are not considered. If the Compton scattered photon further interacts in the detector by the photoelectric effect, the event will be again be recorded in the FEP, even though the initial interaction was a Compton scattering interaction. (*b*) An idealized spectrum of the photons that are backscattered into the detector by Compton interactions external to the detector. These externally scattered photons fall in a continuum from some minimum energy for a 180° scattered photon, up to the FEP. The experimental γ-ray spectrum produced by the detector would include contributions from both internal and external Compton scattering events.

at angles less than 180°, the backscatter continuum also exists at higher energies, as seen in Fig. 6.11*b*.

6.4.3. Pair Production

The third major mode of interaction of γ rays with matter is less common than the first two, and occurs only for high-energy γ rays. Near the nucleus of an absorber atom, the γ ray is transformed to matter in the form of a

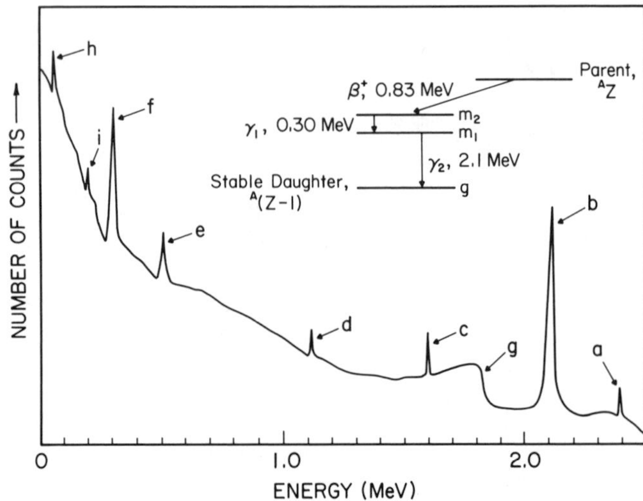

Figure 6.12. A representation of the experimental γ-ray spectrum that would be obtained from a radioactive source that decays by positron emission followed by emission of two γ rays of different energies. The major features of the spectrum are (*a*) 2.4-MeV γ-ray sum peak, (*b*) 2.1-MeV γ-ray FEP, (*c*) 1st escape peak from 2.1-MeV γ ray, (*d*) 2nd escape peak from the 2.1-MeV γ ray, (*e*) 0.511 MeV annihilation radiation peak from positron interactions external to the γ-ray detector, (*f*) 0.30-MeV γ-ray FEP, (*i*) 180° backscatter peak from external scattering of the 2.1-MeV γ ray, (*g*) Compton edge for the 2.1-MeV γ ray (the Compton edge for the 0.30-MeV γ ray is submerged in the backscatter radiation), (*h*) x rays from fluorescence interactions with external lead shielding.

negatron–positron pair (Fig. 6.13). This phenomenon is called **pair production.** Because an electron has a rest mass equivalent to 0.511 MeV of energy, a minimum γ-ray energy of 1.02 MeV is required for this transformation to occur. If the incident γ ray has an energy above 1.02 MeV, the residual energy is given to the e^-–e^+ pair as kinetic energy. In practice, pair production occurs with highest probability for γ rays with energies of 4 MeV and above.

The likelihood of pair production is related to the γ-ray energy and the Z of the absorber. The equation below applies to γ rays with energies above 4 MeV:

$$\text{Probability}_{(\text{PP})} = k(\log E_\gamma)(Z_{\text{abs}})^2 \tag{6.23}$$

The positron created in the event will eventually slow down in the absorber and undergo annihilation, producing two 0.511-MeV photons. The negatron will interact as described in Section 6.3.

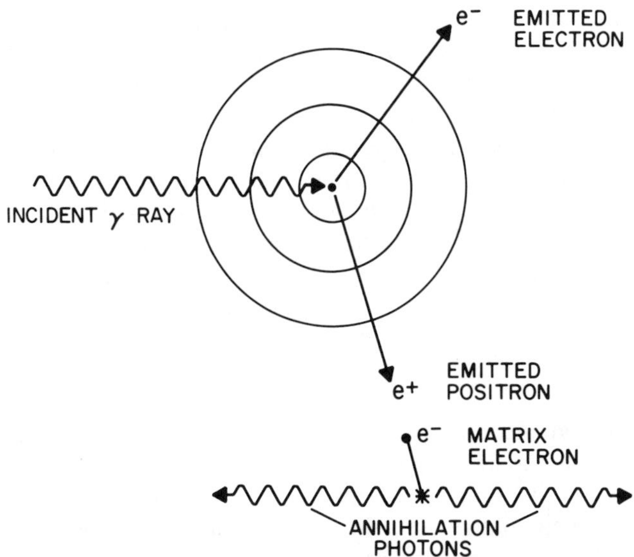

Figure 6.13. A representation of γ-ray interactions with matter by pair production. The positron from the initial γ-ray interaction will encounter a matrix electron, perhaps exist a short time as a positronium atom, then undergo an annihilation reaction to produce two 0.511-MeV photons.

Pair production results in several unique features in the γ-ray spectrum, as shown in Fig. 6.12. A full-energy peak will appear for those events in which all the energy of the electron–positron pair plus the two annihilation photons is dissipated in the detector. The positron interactions can, however, have other outcomes. One of the annihilation photons may escape the absorber, taking with it 0.511 MeV of energy. This would result in an event being recorded in a peak that is 0.511 MeV below the FEP. This is called the **first escape peak.** If both annihilation photons escape, they take with them 1.02 MeV of energy. A peak would then appear in the spectrum at an energy that is 1.02 MeV below that for the FEP. This spectral feature is called the **second escape peak.** In addition, γ rays may interact via pair production events in the shielding surrounding the detector. In this case, it is likely that one of the 0.511-MeV annihilation photons will reenter the detector, resulting in a spectral peak at 0.511 MeV. In all γ-ray spectra, x rays will be recorded at the low-energy end of the spectrum. These may originate from lead shielding that is commonly used to screen out background radiation around the detector. Prominent lead x rays occur in the region from 74 to 88 keV.

The small peak at the highest energy in Fig. 6.12 is called a **sum peak.** It is possible that two γ rays from the radioactive source could enter the detector simultaneously and lose all their energy. The resulting spectral feature would be a peak at an energy equal to the sum of the individual incident γ-ray energies. A sum peak may be distinguished from other peaks in the spectrum by noting how the counting rate varies with source distance from the detector. Normally, radiation is expected to decrease in intensity according to the **inverse square law.** This law is simply a reflection of the fact that the increase in the area of a sphere (over which the radiation is spread) is proportional to the radius squared:

$$\text{Surface area of sphere} = 4\pi r^2 \tag{6.24}$$

Hence, if the sample–detector distance is increased by a factor of two, the counting rate for a single γ ray will drop to $1/(2^2) = 1/4$ of the original value. Because the sum peak depends on the probability of two events being detected simultaneously, the intensity of the sum peak will drop to $1/(2^2) \times 1/(2^2) = 1/16$ of its original intensity for the same factor of two increase in distance.

Figure 6.14 summarizes the roles of photoelectric events, Compton scattering, and pair production in γ-ray interactions as functions of the incident γ-ray energy and the atomic number of the absorber.

6.4.4. Mathematics of Gamma-Ray Absorption

The three major processes by which γ rays interact with matter can all contribute to the absorption of γ rays from a given source. The overall probability of γ-ray interaction is the sum of all the individual probabilities. Observations of the changes in incident intensity as the γ rays pass through an absorber show that the absorption is exponential. For a given γ-ray energy, the following relation holds:

$$I_d = I_0 e^{-\mu d} \tag{6.25}$$

where I_d = intensity at a given thickness, d, of absorber

I_0 = incident intensity

d = thickness of absorber (mg/cm^2)

μ = mass absorption coefficient (cm^2/mg)

The **mass absorption coefficient** is relatively independent of temperature and pressure.

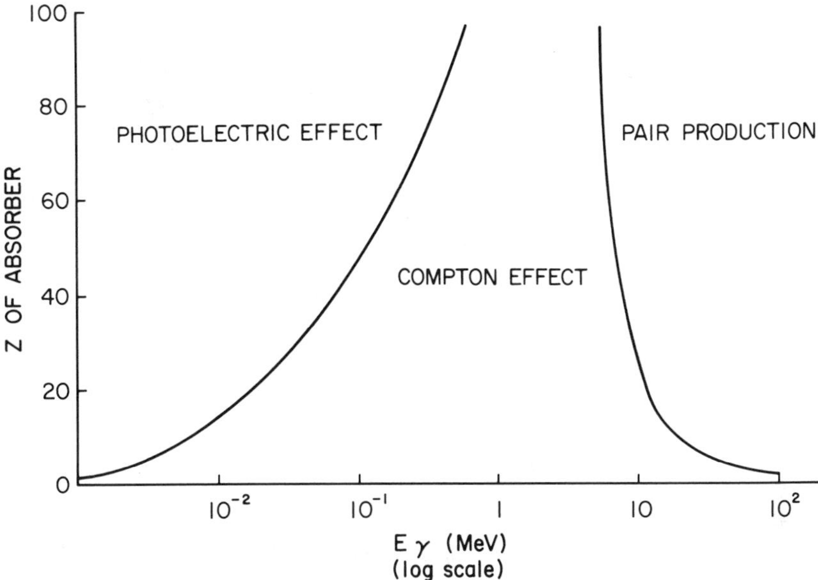

Figure 6.14. Regions where each of the principal modes of γ-ray interaction predominate for differing values of atomic number and γ-ray energy. (Adapted from R.D. Evans, *The Atomic Nucleus*. New York: McGraw-Hill, 1955.)

The thickness of absorber that reduces the γ-ray intensity to half of its original value is called the **half-thickness ($d_{1/2}$).** The relationship between half-thickness and the mass absorption coefficient is

$$d_{1/2} = \frac{0.693}{\mu} \tag{6.26}$$

Plots illustrating the variation of $d_{1/2}$ with logarithms of incident γ-ray energy are shown for Al and Cu in Fig. 6.15. Plots such as these are important in shielding considerations. The most effective absorbers have the smallest $d_{1/2}$ values.

There are places on these curves where sharp decreases in $d_{1/2}$ (and corresponding increases in absorber effectiveness) occur. These are labeled as **K edges** on the plot. These features correspond to the energy required to remove a K-shell electron from an atom. An absorber will be most effective in the absorption of γ rays or x rays that have energies that are just above the K-edge energy. Because Cu has a greater nuclear charge than Al, the K edge is at a higher energy for Cu.

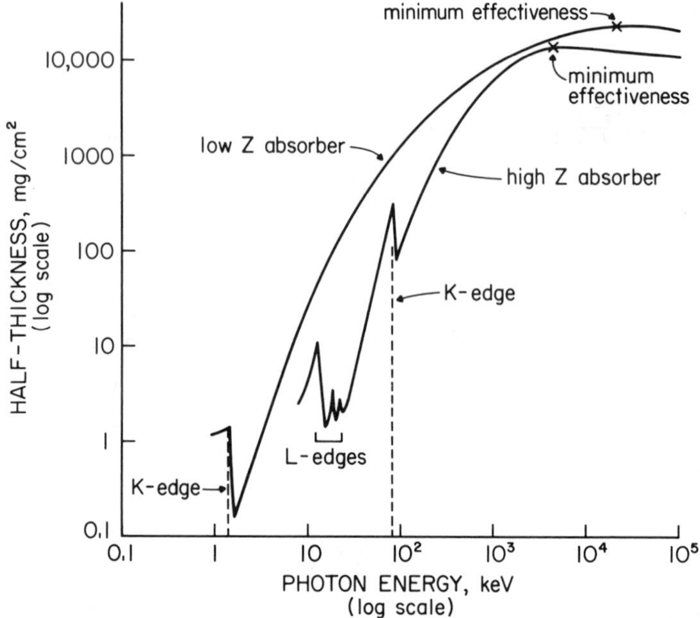

Figure 6.15. Variation of absorber half-thickness (mg/cm^2) with photon energy for both a low-Z and a high-Z absorber. The K edges correspond to energies required to eject a K-shell electron from the absorber atoms. Photons with energies just above the K edges for a given absorber are most effectively absorbed. Each absorber exhibits minimum effectiveness at a given γ-ray energy. The photon energy corresponding to this point of minimum effectiveness is at lower energies for high-Z absorbers.

After the removal of a K-shell electron from the atom, characteristic x rays are emitted as the vacancy is filled. These x rays will have a lower energy than the photon that originally ejected the electron. Thus, an element is not a good absorber of its own x rays, because it does not offer the reaction path opened at the K-edge energy. The best absorber for x rays emitted by an element would be another element that has a smaller Z, and therefore a lower K-edge energy. Often, an element with a Z value that is two units less than that for the x-ray emitter gives the best absorption. Table 6.1 shows the energies of the K edges and the K_α x rays for several elements. Notice that Zn, with $Z = 30$, emits a characteristic x ray at 8.6 keV. Nickel, with $Z = 28$, has a K edge (a point of maximum absorption) at 8.3 keV. Thus, Ni would be a good absorber of the x rays from Zn.

High-energy x rays generated by interactions with the shielding around a detector can often interfere with determinations of low-energy γ rays emitted by a source, or merely increase the gross counting rate of the detector,

Table 6.1. Selected K-Edge and K_α X-Ray Energies

Element	K Edge (keV)	K_α X Ray (keV)
Al	1.56	1.49
Fe	7.1	6.4
Co	7.7	6.9
Ni	8.3	7.5
Cu	9.0	8.0
Zn	9.7	8.6
Ga	10.4	9.2
Sn	29.	25.
Pb	88.	75.
U	116.	98.

Source: Adapted from C. M. Lederer and V. S. Shirley, eds., *Table of Isotopes,* 7th ed. New York: Wiley, 1978.

leading to excessive detector "dead time," that is, time when the detector cannot respond to an incoming event. A **graded shield** can be used to eliminate some of these problems. A graded shield consists of several layers of shielding material, with successively smaller atomic numbers. In principle, each layer will serve to absorb some of the x rays emitted by the previous layer. For example, lead, with $Z = 82$, is usually used as a shielding material for a γ-ray detector. If the Pb shield is first lined with Sn, $Z = 50$, the x rays generated by the Pb would be effectively absorbed. Inside the Sn layer, a sheet of Cu, with $Z = 30$, would effectively absorb Sn x rays. Finally, an inner layer of Al, with $Z = 13$, is used to effectively absorb Cu x rays. Each element absorbs some of the x rays generated in the next outermost one with higher Z. A sheet of a plastic (low average Z), such as Plexiglas, can be placed inside the final metal layer. The graded shield can also lessen the formation of bremsstrahlung radiation (due to interactions of any high-energy β particles emitted by the source), because bremsstrahlung production is dependent on Z_{abs} (see Section 6.3.3).

6.5. NEUTRON INTERACTIONS

Neutron interactions with matter are quite different from those of the charged particles or electromagnetic radiation. The uncharged neutrons interact almost exclusively with nuclei rather than with the atomic electrons. These nuclear interactions include elastic and inelastic scattering, and direct interactions. The capture of a neutron by a nucleus results in the subsequent emission of a charged particle or γ ray. This is used as the basis for detecting neutrons, as is discussed in Chapter 8. Neutron ranges in matter are longer than those of the charged particles with comparable or even greater energies.

The type of interaction that a neutron will experience is very much dependent on its energy. Low-energy neutrons (with a mean energy of about 0.04 eV) are called slow or thermal neutrons. They undergo both elastic scattering and direct interactions. The direct interactions experienced by thermal neutrons are most often of the radiative capture type. In this kind of reaction, the neutron is captured by a nucleus, resulting in an excited state that de-excites by emission of a γ ray. These (n, γ) reactions are commonly employed in determinations by neutron activation analysis.

In the discussion of excitation functions for neutrons (Section 4.3.2), it was noted that the likelihood of a direct interaction between a neutron and a nucleus in the absorber is inversely related to the neutron energy, and thus to its velocity (the $1/v$ law). Thus, as the neutrons increase in energy from thermal to **epithermal** to **resonance** to **fast**, the probability of radiative neutron capture generally decreases (see Fig. 4.3). Energies associated with these different categories of neutrons are given in Table 6.2. In addition, as neutron energy increases, the interactions that do occur are more likely to result in the emission of particles, as in (n, p), (n, 2n), and (n, α) reactions. The likelihood of scattering reactions remains high, and inelastic scattering reactions are especially prevalent for higher-energy neutrons. These scattering processes result in the final reduction of fast neutron energies to energies equivalent to those of gas molecules at standard temperature and pressure (STP). The process of neutron energy reduction to thermal energies is known as **neutron thermalization.**

An understanding of the process of neutron thermalization is important for both shielding and nuclear reactor neutron moderating considerations. The process of neutron energy loss via inelastic scattering can be described by the following equation:

$$\bar{n} = \frac{\ln (E_0/E)}{1 + [(A-1)^2/2A] \ln[(A-1)/(A+1)]} \tag{6.27}$$

where $\bar{n}$ = average number of collisions for energy reduction

E_0 = initial energy of the neutron

E = final energy of the neutron

A = mass number of the absorber

It is important to note the inverse dependence of the thermalization process on A. An absorber that is mostly H ($A = 1$) is not an effective absorber

Table 6.2. Categories of Neutrons

Reactor thermal neutrons	Most probable energy ≈ 0.025 eV Average energy ≈ 0.04 eV
Reactor epithermal neutrons	$0.1 < E_n < 1$ eV
Resonance region neutrons	1 eV $< E_n <$ 1 keV
Reactor fast neutrons	$0.5 < E_n < 2+$ MeV
Neutron generator fast neutrons	14.7 MeV

Note: The definition of energy ranges is not rigorous.

of charged particles or γ rays, but is a good neutron absorber. For example, calculations of the number of collisions, $\bar{n}$, using Eq. 6.27 for $A = 1$, E_o = 4 MeV, $E = 0.04$ eV gives an average number of collisions of 18. Doing the same calculation for lead, $A = 208$, shows that the number of collisions required for thermalization is much higher: $\bar{n} = 1922$. Commonly used low-A shielding materials for neutrons include water, paraffin, graphite, and concrete.

In fission reactors, neutrons produced by fission events (energies of about 2.5 MeV) are reduced to thermal energies by collisions with low-mass nuclei such as ^1H, or ^2H, usually in the form of light water (H_2O) or heavy water (D_2O). These materials are called **moderators**. The lower-energy neutrons have a higher probability of continuing the fission of ^{235}U in a chain-reaction process. Although light water is an effective moderator, it also has a significant neutron capture cross section (0.3 b for ^1H). Thus, heavy water is a better (but much more expensive) moderator because ^2H has a neutron capture cross section of only 0.5 mb and a moderating capacity only slightly inferior to that for ^1H.

6.6. PHYSICAL EFFECTS OF RADIATION ON MATTER

Interactions of radiation with matter result in changes in both the radiation and the matter it encounters. These changes may be useful in some cases and harmful in others. A useful aspect of radiation interaction with matter is that this provides the basis for the detection of radiation. Detecting devices are further discussed in Chapter 8.

The occurrence of a nuclear reaction (e.g., with neutrons or charged particles) may impart significant kinetic energies to the products of the reaction. The energies are greater than the bond energies holding the atoms in a molecule, so a nuclear reaction can result in molecular bond breaking that alters the chemical nature of a material. These highly reactive products include atoms that may be stripped of up to 10 orbital electrons. These species are commonly called **hot atoms,** even though they bear a charge. Hot atoms have many practical uses. They are discussed further in Chapter 11.

Radiation effects on solid inorganic materials may involve changes in either the location or identity of atoms at the lattice positions. The removal of an atom from its position can result in changes in the solid's conductivity, electrical resistance, tensile strength, and other properties. Semiconductors, for example, may have their conductivity altered due to the introduction of lattice defects by radiation. Less desirable physical effects of radiation could be exemplified by the increasing brittleness of structural steel exposed to radiation in a nuclear reactor.

Other radiation effects on solid inorganic materials may involve electron interactions. Radiation may induce or catalyze oxidation or reduction processes, with resulting changes in the color of the material. The color of gemstones may sometimes be altered by placing them into a nuclear fission reactor. Alternatively, excited electrons may become trapped in higher-energy states in the crystal. The trapped energy may be released by heating the solid. The process forms the basis for thermoluminescence dosimetry (TLD), which is discussed further in Chapter 8.

Organic compounds are affected differently by radiation. The excitation induced by the radiation can produce free radicals, which are extremely reactive. The radicals react with a variety of other surrounding molecules. For example, the degree of cross-linking in a polymer may be greatly increased by irradiation during, or even following, the polymerization process. This, of course, has effects on the physical properties of the polymer. The irradiation of food to enhance its preservation, or sterilize it, is another example of a useful effect of radiation on organic matter. These and other physical and chemical effects of radiation on matter are discussed in Chapter 11.

TERMS TO KNOW

Absorber half-thickness
Beta backscatter
Bragg curve
Bremsstrahlung radiation
Čerenkov radiation
Charged particle, extrapolated range
Charged particle, mean range
Compton edge
Compton scattering
Epithermal neutrons
Escape peaks
Fast neutrons
Feather method
Feather range
Full-energy peak
Glendenin equations
Graded shield
Hot atoms
Inverse square law
Ion pairs
K edges
Kinetic energy transfer

Linear range
Mass absorption coefficient
Mass thickness units
Neutron moderator
Neutron thermalization
$1/v$ law
Ore gap
Pair production
Photoelectric effect
Primary ionization
Radiation interactions, inorganic, organic
Radiative processes
Relative stopping power
Resonance neutrons
Secondary ionization
Specific ionization
Stopping power
Sum peak
Thermal neutrons

READING LIST

Crouthamel, C. E., *Applied Gamma-Ray Spectrometry.* Oxford: Pergamon, 1960. [gamma-ray spectrometry]

Evans, R. D., *The Atomic Nucleus.* New York: Krieger, 1955. [basic principles]

Leo, W. R., *Techniques for Nuclear and Particle Experiments.* Berlin: Springer, 1987. [interaction processes]

Siegbahn, K., *Alpha-, Beta-, and Gamma-Ray Spectrometry.* Amsterdam: North-Holland, 1965. [analysis of gamma-ray spectra]

Urbain, W. M., *Food Irradiation.* Orlando, FL: Academic, 1986. [irradiation of foods]

Ziegler, J. F., *The Stopping and Ranges of Ions in Matter.* New York: Pergamon, 1977. [heavy ion ranges]

EXERCISES

1. Using the simple relationships given for relating the ranges of α particles to factors such as Z_{absorber}, E_{α}, and A_{absorber}, calculate the range of 35.0-MeV α particles in an alloy consisting of 10.0% by weight of Be in Al. The alloy has a density of 2.32 g/cm^3. Assume the absorption is due to ionization and scattering. Nuclear reactions do not contribute significantly to the absorption. Express your answer in both mass thickness units (mg/cm^2) and micrometers.

2. What is the ratio of the relativistic mass to the rest mass of a particle traveling at exactly 99% of the speed of light in a vacuum?

3. If the half-thickness for a given energy γ ray is 250 mg/cm^2 in aluminum metal, what thickness of aluminum metal, in centimeters, would be required to reduce the intensity of a beam of these γ rays to 10^{-2} of its initial intensity? The density of aluminum metal is 2.70 g/cm^3.

4. Calculate the energy of the backscatter peak *and* the first escape peak in the γ-ray spectrum for a radionuclide undergoing decay by negatron emission and emission of a 2.13-MeV γ ray.

5. Very pure carbon (graphite) was originally used as a moderator in nuclear reactors. If neutrons with an initial energy of 14.2 MeV undergo 10 collisions in graphite, what is their average energy after the 10 collisions?

6. Which of the primary modes of interaction for γ rays in matter is most affected by the γ-ray energy?

7. If the distance of a γ-ray emitting source from a detector is tripled, the observed counting rate for one peak in the γ-ray spectrum drops from 3.76×10^4 to approximately 4.17×10^3 cpm. Would this peak most likely be a FEP for a single γ ray, or a sum peak?

8. Which element would be the optimum absorber for the x rays emitted from a molybdenum target in a commercial x-ray tube?

9. Both heavy water (D_2O) and light water (H_2O) are used as moderators for neutrons in nuclear reactors. Calculate the number of collisions required to reduce the energy of a neutron from 2 MeV to thermal energy (0.025 eV) for each moderator. (Assume that oxygen plays a minor role in the thermalization. Therefore, mass numbers of $A = 2$ and $A = 1$ can be used for calculations of $\bar{n}$ for heavy and light water, respectively.)

10. If the range of a heavy charged particle in dry air at standard temperature and pressure is known, what other data must be obtained to *calculate* the range of the same particle in the metal titanium?

11. How many 7.0-MeV α particles would have to interact with an absorber to deposit exactly 1.0 J of energy in the absorber? (Assume that 100% of each α particle's energy is deposited in the absorber.)

12. Calculate the number of ion pairs that would be formed in the interaction of one 5.0-MeV α particle with air. Assume that the formation of one ion pair requires 35 eV, and that all of the α energy goes into the formation of ion pairs.

13. Sketch the γ spectrum that would result from a sample emitting two γ rays, one at 1.390 MeV and one at 2.754 MeV. Include the following features, if appropriate: full-energy peaks, any escape peaks, backscatter peaks, annihiliation peaks, and Compton shoulders. Give the energy of each peak.

CHAPTER

7

HEALTH PHYSICS

Health physics is the area of environmental/occupational health that is concerned with the effects of radiation on biological systems, the process of monitoring exposure to radiation, and the means for protecting people who must work with, or in close proximity to, radiation and radioactive materials. The breadth and depth of health physics can hardly be treated adequately in an entire book, let alone in one chapter. Therefore, the topics included in this chapter are meant to give an overview of health physics concepts that all users of radioactive materials, and even the general public, should be familiar with. This includes an explanation of the ways in which radiation affects living organisms, and the units used to measure these effects. Common sources of radiation exposure are discussed, and an attempt is made to give the reader some perspective on the real risks associated with various levels of radiation exposure. Finally, a summary of the legal limits on radiation exposure for both workers and the public is given. More technical aspects of health physics are not covered in this chapter, including detailed calculations of dose and dose rate, the procedures for obtaining licensing for a laboratory, decontamination activities, personnel or facility monitoring, shielding considerations, and the details of regulations governing radiation exposure. The reader is referred to health physics texts and monographs for the latter. Instrumentation for radiation detection is discussed in detail in Chapter 8.

By its nature, health physics is an interdisciplinary field. The health physicist must be familiar with the processes of radioactive decay, understand how radiation interacts with matter, especially in biological systems, and be able to use a wide variety of nuclear instrumentation. A knowledge of radiation dosimetry, the intricacies of dose rate calculations, and the governmental regulations dealing with radiation protection is also essential. Thus, this chapter is one in which many of the principles of radioactivity learned in previous chapters are applied to a particular study.

7.1. RADIATION QUANTITIES AND UNITS

In health physics work, or in related areas where the *effects* of radiation on a sample or subject are of importance, a simple knowledge of the activity level of a radioactive source does not convey sufficient information to assess health risk. Activity was defined in Chapter 5 (Eq. 5.1) as the number of decay events that occur in a given period of time. However, information about the amount of radiation actually received by the absorbing medium and about the potential for adverse effects in the medium are required. Therefore, different units that can describe these effects of radiation are needed.

There is a distinction between radiation exposure and radiation dose. **Exposure** (*X*) is the term used to refer to the total electrical charge (the ionization) produced in a given mass (or volume) of air:

$$X = \frac{dQ}{dm} \tag{7.1}$$

where Q is the charge and m the mass. In SI units, exposure is measured in coulombs/kilogram (C/kg). **Dose** refers to the quantity of energy that is actually put into a medium by the incoming radiation. In SI units, dose is measured in joules/kilogram (J/kg). The term dose is often used with some other modifying term that describes it more fully, for example, effective dose and lethal dose.

7.1.1. Measurement of Exposure

The unit of radiation exposure is the **röntgen (R).** The röntgen was originally defined as that quantity of x or γ radiation that would produce 1 esu of electrical charge of either sign in 0.001293 g of dry air (a volume of 1.00 cm^3 at STP). This can be shown to be approximately equivalent to 1.61×10^{12} ion pairs per gram of dry air, or the release of about 84 ergs of energy per gram of dry air at STP. The equivalent radiation exposure in water or tissue is about 93 ergs per gram of water. The röntgen is now exactly defined as 2.58×10^{-4} C/kg. The number of röntgens produced by a radioactive source is easily measured using air ionization chambers (see Section 8.2.1). However, because the röntgen is a unit of radiation exposure, not dose, it does not provide exact information about the amount of radiation that is actually absorbed by a medium, or about the effects of the radiation on the medium.

7.1.2. Measurement of Dose

The unit traditionally used to measure radiation dose is the **rad**. One rad is defined as that dose of any type of radiation that will deposit 0.01 J of energy in one kilogram of absorbing material:

$$1 \text{ rad} = 10^{-2} \text{ J/kg} \tag{7.2}$$

The newer SI unit of absorbed dose is the **gray (Gy),** defined as that amount of radiation that will deposit one joule of energy in one kilogram of absorbing material:

$$1 \text{ Gy} = 1 \text{ J/kg} \tag{7.3}$$

Hence, the relationship between gray and rad is that 1 Gy equals 100 rad. It is important to remember that the gray and the rad are units of dose, in contrast to the röntgen, which is a measure of exposure or field strength of the radiation. There is no simple relationship between the röntgen and the rad or the gray.

Example. Calculate the amount of energy, in MeV/kg, that is deposited into one kilogram of a material by a one gray radiation dose.

$$1 \text{ Gy} = 1.00 \text{ J/kg} \times 1 \text{ MeV}/1.602 \times 10^{-13} \text{ J} = 6.24 \times 10^{12} \text{ MeV/kg}$$

7.1.3. Measurement of Dose Equivalent

While the rad is more useful than the röntgen in assessing the effects of radiation, it still does not give all the information needed. This is because the physiological effects resulting from radiation absorption will vary greatly depending on which type of radiation deposits the energy. The severity of the molecular damage caused by radiation is directly related to the degree of ionization produced by the radiation in the tissue. Specific energy loss or stopping power (Section 6.2.2) is one way to measure the energy lost by radiation as it passes through an absorbing material. A related quantity used in health physics is the **linear energy transfer (LET),** defined as the rate at which energy is transferred to a given region of matter:

$$\text{LET} = \frac{dE_{\text{abs}}}{dx} \tag{7.4}$$

The units most often used by health physicists to measure LET are keV/μm. The higher the LET of the radiation, the greater the damage it can potentially do in biological tissue.

As discussed in Chapter 6, the specific ionization and thus the LET values of the three main types of radiation are quite different. Alpha particles show the largest specific ionization, β particles have a lower specific ionization than alphas, and γ radiation has a still lower specific ionization for particles and radiations of similar energy. Thus, over a given range, α particles will produce much more biological damage than the other two types of radiation. Therefore, alphas are the most biologically hazardous if they are ingested and incorporated into internal tissue. However, because α particles are not very penetrating, it is easy to shield against them. Alpha particles do not even penetrate the skin, so they are not very hazardous with respect to external exposure. At the other extreme, γ rays do the least biological damage over a given range, but are much more penetrating than α or β particles, and thus require much more protective shielding. The other types of radiation that must be considered besides α, β, and γ radiations include energetic charged particles from cosmic rays and particle accelerators, and both thermal and fast neutrons. Positively charged particles of all masses interact with matter like α particles. They are not very penetrating, but have high LET values, so they are very damaging if acting on internal organs regulating physiological functions. Neutrons are not charged, so they have a high penetrating power. They do not cause ionization directly by interaction with atomic electron clouds, but they will induce nuclear reactions, or undergo elastic scattering processes, which will produce charged particles. Therefore, neutrons are also very biologically hazardous.

The point of the above discussion of LET values and biological effects of radiation is that *identical doses of the various types of radiation may not have identical biological effects*. Therefore, simple knowledge of the absorbed dose will be of limited usefulness when trying to assess the potential for adverse biological effects due to different radiations. To compensate for this, other quantities which assign numerical values to the relative amounts of damage done by radiation have been defined. One of these is called **relative biological effectiveness (RBE),** and is defined as the ratio of the amount of 200-keV x or γ rays needed to produce a specific biological effect to the amount of any other radiation needed to produce the same effect. The RBE is a very specific quantity used in precise radiobiological studies and so has limited usefulness in assessing general biological effects. Therefore, another factor, called the **quality factor (*Q*),** has been defined. Q is related to the LET, as shown in Table 7.1. Note that the higher the LET, the higher is the value of Q. Table 7.2 gives the current value of Q for several common radiations. Recently, there have been recommendations to raise the Q value for neutrons and protons. These proposed new values are shown in parentheses in Table 7.2.

The concepts of quality factor and dose can be used to define the **rem**, which is the traditional unit of **dose equivalent (*H*).** This is the unit used by health physicists in discussing radiation effects. Dose equivalent (or effective dose) is the absorbed dose multiplied by the quality factor. For the rem

$$\text{Dose equivalent in rem} = \text{Dose in rads} \times Q$$

or

$$H_{\text{rem}} = \text{rads} \times Q \tag{7.5}$$

Table 7.1. Relationship Between LET and Q

LET(keV/μm)	Q
< 3.5	1
3.5–7	1–2
7–23	2–5
23–53	5–10
53–175	10–20

Source: Adapted from G. D. Kerr, *Health Phys.* **55**(2):241 (1988).

Table 7.2. Values of Q for Some Common Radiations

Radiation	Q
X and γ rays	1
Beta particles	1
Neutrons, thermal	2 (5)[a]
Neutrons, fast	10 (20)[a]
Protons	10 (20)[a]
Alpha particles	20

Source: Adapted from G. D. Kerr, *Health Phys.* **55**(2):241 (1988).
[a]Numbers in parentheses are new values proposed by NCRP.

The newer SI unit of dose equivalent is the **sievert (Sv),** defined as

$$\text{Dose equivalent in sieverts} = \text{Dose in grays} \times Q \tag{7.6}$$

The relationship between the Sv and the rem is that 1 Sv = 100 rem.

Example. Calculate the dose equivalent, in mrem and in mSv, for 10 mrad of gamma, alpha, and thermal neutron irradiation.

The values of the quality factor for gammas, α particles, and thermal neutrons can be found in Table 7.2. They are 1, 20, and 2, respectively. Equation 7.5 can be used to calculate the dose in rem, and rem coverted to Sv using the relationship above:

$$\gamma: \quad 10 \text{ mrad} \times 1(Q) = 10 \text{ mrem} \times (1 \text{ mSv}/100 \text{ mrem}) = 0.1 \text{ mSv}$$
$$\text{n}: \quad 10 \text{ mrad} \times 2(Q) = 20 \text{ mrem} \times (1 \text{ mSv}/100 \text{ mrem}) = 0.2 \text{ mSv}$$
$$\alpha: \quad 10 \text{ mrad} \times 20(Q) = 200 \text{ mrem} \times (1 \text{ mSv}/100 \text{ mrem}) = 2 \text{ mSv}$$

The above example illustrates that the same amount of absorbed dose does not necessarily result in the same amount of biological damage. In the example, the α radiation would produce more harmful effects than the γ or neutron radiation, even though the dose (in mrads) of all three is the same.

There are two other terms relating to dose equivalent that may be encountered in the literature. One is the **effective dose equivalent.** It refers to the dose equivalent weighted for the differing susceptibilities of specific types of tissue to radiation. The second term comes into play because often doses are expressed in terms of an entire population, rather than just one person. The effective dose equivalent to a whole group of people is called the **collective effective dose equivalent.**

7.1.4. Simple Calculations of Dose and Exposure

Exact calculations of exposure and dose can be quite complicated, and will not be discussed in this chapter. However, there are some simple rules of thumb that can be used to get rough estimates of exposure, or dose. In some cases, these estimates are quite close to the values that would be obtained using more exact methods. In other cases, they are not, so these simple rules must be used with caution.

The user of radioactivity will most commonly encounter β and γ radiation, so rules for these two radiations are the ones given here.

Gamma Rays

The exposure rate from γ rays is directly related to the energy and number of gammas emitted per decay, and inversely related to the square of the distance between the source and the object. An equation that relates exposure rate to these quantities is

$$\text{Exposure (mR/h)} = 6\,AEn/d^2 \qquad (7.7)$$

where A = activity (mCi)

E = energy of the γ ray (MeV)

n = number of γ rays of energy E emitted per decay

d = distance to source (ft)

Distance may be expressed in cm if the coefficient in Eq. 7.7 is changed from 6 to 5.6×10^3. The value of n is needed if a single γ ray of energy E is not emitted with each disintegration. For example, the value of n for ^{54}Mn would be 1, because a single γ ray of 0.835 MeV energy is emitted in the decay of each ^{54}Mn nucleus. However, ^{97}Ru decays by emitting two principal γ rays. One, with energy 0.22 MeV, is emitted in 89% of all decays, while the other, with energy 0.32 MeV, is emitted in approximately 10% of all decays. Thus, the values of n would be 0.89 and 0.10, respectively, for the two γ rays emitted by ^{97}Ru. The total exposure can be calculated by summing the two independent calculations. Values of n for other nuclides can be found in various tables of nuclear data. For γ rays, the exposure rate may be taken to be approximately equal to the dose rate (in mrad/h) and to the dose equivalent rate (in mrem/h), because Q equals 1 for γ rays.

Example. ^{99m}Tc emits a γ ray with an energy of 0.14 MeV. Calculate the exposure rate and state the dose rate for a person standing one foot away from a 100-mCi source of ^{99m}Tc.

Equation 7.7 is used to calculate the exposure:

$$\text{Exposure rate, mR/h} = 6AEn/d^2 = 6(100\text{ mCi})(0.14\text{ MeV})(1)/1^2$$
$$= 84\text{ mR/h}$$

For γ rays, the exposure rate is approximately equal to the dose rate, so

$$84\text{ mR/h} = 84\text{ mrad/h}$$

Example. ^{59}Fe emits two γ rays at 1.10 and 1.29 MeV, with relative intensities of 56% and 44%, respectively. Calculate the exposure rate, and the equivalent dose rate, for a person 2 feet away from a 5 mCi ^{59}Fe source.

Equation 7.7 is also used for this problem, but this time there are two gamma rays to account for:

$$\text{Exposure rate, mR/h} = 6(5\text{ mCi})[(1.10\text{ MeV})(0.56) + (1.29\text{ MeV})(0.44)]/2^2$$
$$= 8.9\text{ mR/h}$$

Again, exposure is approximately equal to equivalent dose, so

$$8.9\text{ mR/h} = 8.9\text{ mrem/h}$$

If the exposure or dose rate is known, along with the total time exposed, the total dose can easily be calculated.

Example. A laboratory technician works for 10 min 1.8 ft away from a ^{99m}Tc source with a 10-mCi activity. The ^{99m}Tc emits a 0.14-MeV γ ray. Calculate the total dose received by the technician in that time.

$$\text{Exposure rate, mR/h} = 6\,AEn/d^2 = 6(10\text{ mCi})(0.14\text{ MeV})(1)/(1.8)^2$$
$$= 2.6\text{ mR/h}$$

and

$$2.6\text{ mR/h} = 2.6\text{ mrem/h}$$

for γ rays. So

$$(2.6\text{ mrem/h})(1\text{ h/60 min}) = 0.043\text{ mrem/min}$$

and

$$(0.043\text{ mrem/min})(10\text{ min}) = 0.43\text{ mrem total dose to technician}$$

Beta Sources

Beta particles are less penetrating than γ rays of similar energies, and the β particles will experience significant attenuation as they traverse the air between source and object. For very low-energy β particles, like the 0.0186-MeV β from ^{3}H, dose rates are small even a few centimeters away from the source. For higher-energy β particles, an estimate of the upper limit of dose can be made using the following equation:

$$\text{Dose rate, mrad/h} = 338\,000\,A/d^2 \tag{7.8}$$

where A = activity of the source (mCi)

d = distance from the source (cm)

Example. Calculate the beta dose rate 1 ft away from a 1-mCi source of ^{32}P.

Equation 7.8 can be used to estimate the dose:

$$1\text{ ft} = 30.5\text{ cm}$$

$$\text{Dose rate, mrad/h} = 338\,000\,(1\text{ mCi})/(30.5\text{ cm})^2 = 363\text{ mrad/h}$$

7.2. BIOLOGICAL EFFECTS OF RADIATION

The dangers of radiation to human health were recognized during the early years of research into the phenomenon, but little was done to recommend safe handling practices. Many researchers suffered radiation burns and eventually died from radiation-induced cancers. Eve Curie writes in her biography of her mother about a gathering of friends held to celebrate the granting of Mme. Curie's doctorate. A ZnS-coated tube containing a strong radium solution was held up for all to admire. The solution glowed beautifully in the dark, but in the light from the radium solution, the radiation burns on Pierre Curie's hands could be seen. P. Curie died prematurely in a street accident, but M. Curie's death in 1934 from a bone marrow disease called aplastic anemia may well have been due to her long years of exposure to radiation.

Radiation interacts with biological material in the same ways that it interacts with other types of matter, as discussed in Section 6.1. Ionization and atomic excitation are the most important effects. In living organisms, where proper physiological functioning is often acutely sensitive to correct chemical structure, the alterations induced in molecules by radiation almost always are deleterious. Radiation effects on biological systems can be put into two broad classes: stochastic and nonstochastic.

7.2.1. Stochastic Effects

The word **stochastic** refers to events that occur by chance, that is, according to the laws of probability. The appearance of a stochastic effect is, therefore, something of a hit-or-miss affair. The relationship between smoking and lung cancer is an example of a stochastic effect. Some persons who have never smoked at all will develop lung cancer, while other people who have been heavy smokers for many years never develop it. The two most important stochastic effects of radiation exposure are cancer and genetic defects in offspring.

Stochastic effects are generally related to long-term, low-level exposures to radiation, and are usually the outcome of radiation damage done initially to only a few cells. It is extremely difficult to determine accurately the effects of low levels of radiation (10–20 rem, or less) because there are no immediately observable effects. Because of the probabilistic nature of stochastic effects, and the fact that the effect may appear decades after the initiating event, it is virtually impossible to link a specific radiation exposure to a given stochastic effect in a single person.

Figure 7.1 illustrates some of the characteristics of stochastic effects. This figure is a **dose–response** curve for stochastic effects. The graph

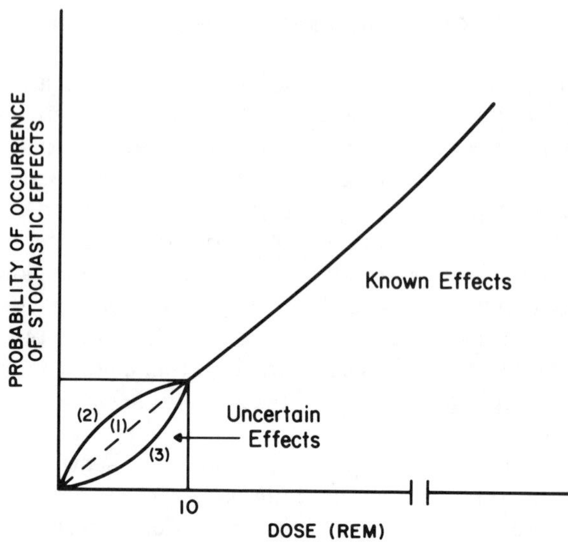

Figure 7.1 Dose–response curve for stochastic effects.

shows the probability of occurrence of a stochastic effect versus the dose of the radiation. Notice that the curve shows no **threshold level,** or minimum radiation dose at which no damage will ever occur. In theory, zero radiation exposure would result in zero effects, but this can never be attained in reality. The area outside the box in the lower left corner represents the levels at which radiation effects are directly observable. The area inside the box is the area of difficulty. Below about 10–20 rem, there are no immediately observable responses to radiation exposure. To predict effects at low levels, the approach that is taken is to extrapolate the known effects of radiation at higher doses down to lower doses. Most health physicists assume that there is a linear relationship (line 1) between dose and effect at these lower levels, rather than an increased risk (line 2) or a decreased risk (line 3), and that there is no threshold. This is referred to as the *linear dose–effect extrapolation with no threshold hypothesis*. The implication of this assumption can be illustrated with a simple example. If one person who is exposed to 300 rems of radiation dies of cancer, the hypothesis above would lead to the conclusion that a dose of 1 rem given to 300 people would cause one of those people to die of cancer. Figure 7.1 also shows that there is an increasing likelihood of a stochastic effect with increasing dose. However, the severity of response is independent of the dose.

Whenever the stochastic effects of radiation are discussed, large populations of subjects must be considered, because prediction of effects in a single individual is nearly impossible. Estimates of the occurrence of radiation-induced stochastic effects in humans have been made by studying groups of people who have been exposed to larger than normal doses of radiation. These groups include the survivors of the atomic bombs dropped on Japan at the end of World War II, survivors of certain nuclear weapons tests, workers in plants where radium-containing paints were used for luminous dials, and persons exposed to large doses of radiation for therapeutic reasons. The Japanese group, in particular, has been regularly studied by several groups, among them the National Academy of Science committee on the Biological Effects of Ionizing Radiation (BEIR), which published its most recent study (BEIR V) in 1990.

The initial effects of radiation occur at the molecular level. The ionization and excitation of the molecules in an organism may alter the chemical nature of the molecules. The biological consequences of these ionizations depend on the identity and number of molecules that are affected. Radiation damage to protein molecules, which are the most important structural and physiological molecules in a living system, could cause immediate adverse effects in the organisms' structure and/or function. It is more likely, however, that the change is one that would not be immediately observable, but would lead to the development of some abnormality in the future. A discussion of the chemical mechanisms by which radiation damages matter is beyond the scope of this chapter, but one important mechanism worth mentioning here is the production of **free radicals.** These are highly reactive species that undergo many reactions with biomolecules. The production of free radicals usually results from radiation interactions with the most abundant molecules in a cell, which are those of water. The radiation-induced dissociation of water can lead to the subsequent formation of hydrogen peroxide, a strong oxidizing agent that can interact adversely with many other molecules.

The observable stochastic effects, as mentioned above, are cancer and hereditary effects. Cancer is a **somatic** effect, meaning that it is a disorder affecting the nonreproductive cells of the organism. Somatic effects will affect only one organism, not its offspring. Leukemia is the most likely type of cancer to be induced, because of the greater susceptibility of blood system components to radiation. Cancers of the breast, thyroid, and lung are also common among radiation-induced cancers. Other types of cancer are less likely to result from radiation exposure. Studies of large groups exposed to radiation indicate that exposure of a population of 10 000 people to 1 rem of radiation would result in three excess cancer deaths. This figure will vary with rate of exposure and the specific type of cancer under consideration.

The molecules that transmit hereditary information can also be damaged by radiation. If damage is done to the genetic material of the organism, mutations may result that would cause alterations that are passed on to the next generation. These are called **genetic** or **hereditary** effects. Extensive studies of the Japanese atomic bomb survivors so far have not shown the presence of significant genetic effects in humans. Therefore, most predictions of the chances of genetic effects are based on animal experiments.

7.2.2. Nonstochastic Effects

Nonstochastic effects are those that do not occur by chance, but rather can be directly related to a particular causative agent. For example, alcohol consumption in humans results in nonstochastic effects. Drinking a small amount of alcohol will not produce any observable effects. However, after a certain level of alcohol is consumed, individuals will exhibit certain systemic effects, which will worsen in direct proportion with the amount of alcohol consumed. A blood–alcohol level of 0.4% is almost certainly fatal. This example illustrates two significant ways in which nonstochastic effects differ from stochastic effects. One is the existence of a threshold level below which no observable changes occur and above which noticeable changes are sure to occur. The second is that there is a direct relationship between the size of the dose and the severity of the effects after the threshold level is reached. Figure 7.2 shows a dose–response curve for a nonstochastic effect, which illustrates the presence of a threshold level and the linear relationship between dose and effect. A useful index that may be

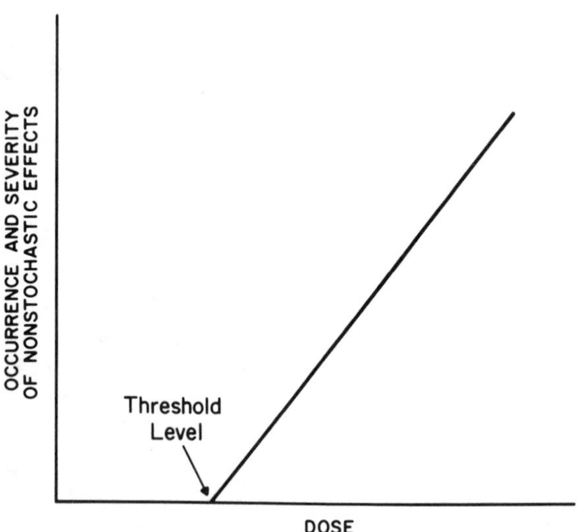

Figure 7.2. Dose–response curve for nonstochastic effects.

obtained from a dose–response curve is the dose to which 50% of the population responds with a specific biological response. If that response is death, this dose is called the **lethal dose** for 50% of the population, or the **LD_{50}**. The time needed for 50% of the population to die is usually specified also. For example, an LD_{50}/30-day dose would be one that would result in death in 50% of the population within 30 days.

The nonstochastic effects of radiation in humans are generally related to large radiation doses received over a short period of time. These effects are usually seen on whole organs or body systems, rather than in only a few cells. In contrast to stochastic effects, nonstochastic effects occur promptly after an acute, high radiation exposure, and a definite link between the cause and the effects can be established. The severity of the effects depends on the total dose, the time over which it was received, the body part irradiated, whether the dose was external or internal, and the age of the person. Young children are much more susceptible to the harmful effects of radiation than older adults are. The most important nonstochastic effects include skin changes, alterations in the blood, gastrointestinal problems, and central nervous system changes.

The first observable changes that take place in a person who has received a dose of radiation are blood alterations. The threshold for these effects for β or γ radiation ($Q = 1$) is around 250–500 mGy (25–50 rem). With a dose of this size, the number of white blood cells is seen to decrease within a few hours after exposure. Red blood cells and platelets decrease at a slower rate, perhaps over days or weeks depending on the dose. Recovery to normal conditions can take weeks or months. The rate and degree of cell loss and time needed for recovery are directly related to the dose received. Doses in this range also induce temporary sterility in both men and women, but fatalities would not be expected to occur.

A dose of 2 Gy (200 rem) will result in damage to the bone marrow, the part of the body responsible for production of blood cells. The bone marrow damage is not complete and is reversible. A dose of this magnitude will also result in gastrointestinal symptoms, including nausea and vomiting, and general malaise and fatigue. Loss of hair is almost always seen. Some deaths may occur.

Doses of 4–6 Gy (400–600 rem) result in complete, although reversible, loss of bone marrow function. More severe levels of the blood cell and gastrointestinal symptoms listed above also occur, but with medical treatment survival is still possible. The LD_{50}/30 for humans is around the 4- to 5-Gy (400- to 500-rem) level.

Above 7 Gy (700 rem) the bone marrow is irreversibly damaged, and survival is very unlikely. At 10 Gy (1000 rem) the desquamation (sloughing off) of the intestinal epithelia leads to severe diarrhea, nausea, and vomiting soon after exposure. Death is highly probable within 1–2 weeks. At 20 Gy (2000 rem) and higher, the central nervous system is damaged. Unconsciousness follows within minutes after exposure, and death occurs within hours or a few days.

7.3. SOURCES OF RADIATION EXPOSURE

The radiation to which we are exposed in everyday life comes from both natural and man-made sources. Figure 7.3 summarizes the most important radiation sources as percentages of the total dose received. It is readily apparent from this figure that the most plentiful source of radiation to which people are exposed is from natural background, which includes cosmic radiation and the natural radioactive materials in the earth, atmosphere, and our own bodies. The average amount of this natural radiation is about 2.4 mSv (240 mrem) per year. Location is important, and the range for natural exposure is 1.6–6.0 mSv per year.

7.3.1. Natural Sources of Radiation

Cosmic rays (Section 13.1.1) originate in outer space and consist mainly of high-energy protons. When these particles encounter the earth's atmosphere, they induce nuclear reactions that produce showers of other subatomic particles, especially electrons, muons, neutrinos, neutrons, and photons. At sea level, a person will receive an annual dose of around 0.36 Sv (36 mrem) from cosmic radiation. The cosmic ray dose increases at higher elevations and at higher latitudes. For example, a person flying in a jet at an altitude of 6 miles (10 km) receives an additional 5 μSv (0.5 mrem) every hour. Similarly, people who live in cities at high altitudes will receive a cosmic ray dose of more than 36 mrem per year.

The earth contains many radioactive elements, with the largest contributors to dose being ^{40}K, ^{87}Rb, ^{232}Th, ^{235}U, ^{238}U, and the radioactive daughters in the U and Th natural decay chains. The dose received averages from 0.3 to 0.7 mSv (30 to 70 mrem) annually, but varies greatly depending on location. There are several "hot spots" that have been identified in Brazil, India, Iran, and some other countries. Many of these are areas where thorium-rich sands are high in abundance. Annual doses in these areas can be up to 250 mSv (25000 mrem) per year (near the city of Pocos de Caldas, Brazil). No significant adverse effects have been observed in persons living in or near these high-radiation background areas.

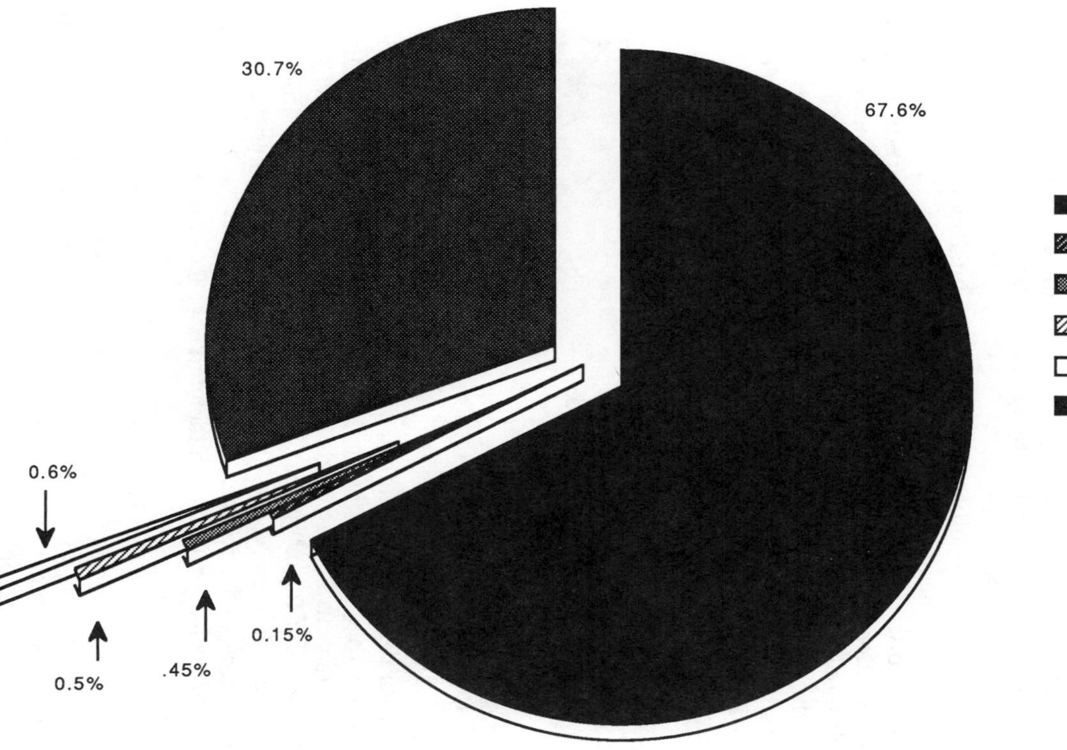

Figure 7.3. Sources of radiation exposure (percentage of total yearly dose).

People are also constantly irradiated by the radioactive atoms within their own bodies and in the food and water they ingest. Potassium is an element that is essential to life, and the ^{40}K isotope is radioactive, with a very long half-life. In a person of 70 kg mass, the ^{40}K contributes an activity of around 10^5 dpm, resulting in an annual dose of about 18 mrem. There are also significant amounts of ^{14}C and ^{3}H present in the body, but these contribute little to the dose because they emit low-energy β radiation. The air we breathe contains radioactive gases, especially the inert gas radon. The food and water we ingest contains ^{40}K, ^{226}Ra, and a variety of radioactive U and Th decay products. All of these sources contribute an additional annual dose of about 40 mrem.

One of the most significant sources of natural radiation is radon gas. The problem of indoor radon gas contamination was first noted in 1984, when it was discovered that one of the employees of a nuclear power plant set off radiation detection alarms when he came *into* the plant. Subsequent investigations led to the finding that the worker's home had high levels (10^5 Bq/m^3) of the radioactive gas radon. For comparison, the average indoor Rn level is only about 50 Bq/m^3 (1.4 pCi/L), and the average outdoor level is 5–10 Bq/m^3.

Radon gas (^{222}Rn) is a natural decay product of ^{238}U. Its decay chain is

$$^{222}\mathrm{Rn} \xrightarrow[3.82\ \mathrm{d}]{\alpha} {}^{218}\mathrm{Po} \xrightarrow[3.10\ \mathrm{min}]{\alpha} {}^{214}\mathrm{Pb} \xrightarrow[27\ \mathrm{min}]{\beta^-} {}^{214}\mathrm{Bi} \xrightarrow[19.9\ \mathrm{min}]{\beta^-} {}^{214}\mathrm{Po} \xrightarrow[163.7\ \mu\mathrm{s}]{\alpha} {}^{210}\mathrm{Pb} \xrightarrow[22.3\ \mathrm{y}]{\beta^-}$$

$$^{210}\mathrm{Bi} \xrightarrow[5.01\ \mathrm{d}]{\beta^-} {}^{210}\mathrm{Po} \xrightarrow[138.4\ \mathrm{d}]{\alpha} {}^{206}\mathrm{Pb}$$

$$^{210}\mathrm{Bi} \xrightarrow[3 \times 10^6\ \mathrm{y}]{\alpha} {}^{206}\mathrm{Tl} \xrightarrow[4.20\ \mathrm{min}]{\beta^-} {}^{206}\mathrm{Pb}$$

Because Rn is gaseous, it is inhaled with the air we breathe into the lungs. If the Rn is trapped there, it emits α radiation and gives rise to the long series of radioactive decay products listed above. The typical levels of Rn in a home ($\sim$50 Bq/m^3) would result in an annual dose of around 2 or 3 mSv (200–300 mrem/year). This dose equivalent is greater than that received from all other sources, and this is why radon has become a great concern. Some homes have been found to have Rn levels well above the average value.

The amount of Rn that is found in any individual home is a function of the amount of Rn generated in the soil underneath the home, the rate at which the Rn enters, and the rate at which air flows through the house (the ventilation rate). In most cases, if excessive Rn levels are found, there are steps that can be taken to prevent the Rn gas from entering the house, or to increase the ventilation rate through it, so that amounts are lowered to acceptable levels.

7.3.2. Man-made Sources of Radiation

The remainder of the radiation we receive comes from man-made sources. As shown in Fig. 7.3, this amounts to only about one-third of the amount received from natural sources (around 80 mrem/year). By far the largest amount of radiation exposure from man-made sources is from medical procedures (see Section 10.5). A small number of these procedures will result in extremely high doses, in cases where radiation is being used therapeutically to treat disease. However, the exposure that most people receive comes from x-ray procedures, both medical and dental. The dose received from an x-ray procedure is variable, depending on the part of the body examined and the type of x-ray procedure performed. A typical chest x-ray procedure would give a dose of around 50 mrem.

The dose received from all other man-made sources is much smaller than that from natural sources or from medical procedures. The dose due to fallout from nuclear weapons testing has decreased significantly in the years following the nuclear test ban treaty. The main nuclides of interest from fallout are ^{14}C, ^{95}Zr, ^{90}Sr, and ^{137}Cs. Presently, fallout contributes less than 1 mrem to the annual dose. The routine operation of nuclear power plants contributes very little to the annual dose. The only radionuclide that is normally emitted from a nuclear power plant is the gaseous fission product ^{85}Kr. The dose to persons in the vicinity of a plant is, at most, a few percent of the dose received from natural sources (a few mrem). Probably, most people fear the potentially high doses that could be received in an accident more than they fear the low-level dose due to routine operation. However, even in the 1979 Three-Mile Island accident, which was the worst in U.S. history, the maximum dose to the surrounding population was only about 100 mrem, due mainly to the release of ^{133}Xe. The accident that occurred at Chernobyl, in the Soviet Union in 1986, was the most severe commercial nuclear power accident that has ever occurred. Thirty-one people died within a short time from a combination of explosion, burns, and radiation exposure as a result of this accident. Surrounding areas, exposed to doses of several rems, had to be permanently evacuated.

Radiation is a natural phenomenon to which living things have been exposed since the beginnings of life. The bulk of the dose (>95%) that is received now by people is either from natural sources that cannot be avoided or comes from voluntary medical procedures. In everyday life, the public is not exposed to untoward amounts of radiation from any man-made sources.

7.4. RADIATION PROTECTION AND CONTROL

As the understanding of the effects of radiation on organisms improved, so did efforts to control unnecessary exposure. The process of setting limits on the amounts of radiation to which people may be exposed at work or in the normal environment is fraught with difficulties and is still an inexact science. The reasons for this include the difficulty of accurately assessing the effects of low-level radiation (discussed in Section 7.2.1) and of defining what is meant by an "acceptable risk."

7.4.1. Risk

Risk can be defined as the chance that a given activity will produce illness, injury, or death in an individual. All life activities involve a certain degree of risk. People accept these risks because they judge that the benefits to be gained from the activity outweigh the risk involved. The difficulty in developing radiation protection standards lies in deciding in the first place what constitutes an "acceptable risk." This judgment is highly variable from one person or group to another. The risks encountered in a sport like skydiving are unacceptably high to some people, but obviously are not too high for others. The risks associated with consumption of alcohol and smoking cigarettes are well documented, but people continue to smoke and drink because, for them, the benefits outweigh the risks. Generally, risks are more acceptable when people feel they have some measure of control over them and when they have the option of whether or not to take the risk. Besides these factors, the better understood a risk is, the more likely it is to be acceptable. It may be that the risks associated with radiation are not well accepted because they are not understood and because people feel that the risks are forced on them. Some life-shortening risks are shown in Table 7.3.

There is no question that excessive radiation exposure is detrimental to living organisms, but the actual risks from the radiation we receive are much less than the level of risk that is perceived by the general public. Exposure to radiation, at the levels commonly encountered in everyday life or even occupationally, poses a very small risk in reality, but it is perceived as a large risk. This is illustrated by a 1979 study in which three groups of Americans were asked to rate the risk of 30 activities. The rankings of the first 20 of these activities are shown in Fig. 7.4. Nuclear power was rated as a high risk by all three groups, but in actuality it results in far fewer deaths than activities, such as smoking, that were rated lower by the three groups. This general problem of the public perception of high risk for nuclear activities and the concomitant lack of appreciation for the positive contributions

Table 7.3. Estimated Loss of Life Expectancy from Health Risks and Occupational Hazards

Source	Average Life Expectancy Lost (days)
Health Risks	
Smoking 20 cigarettes a day	2370
Overweight by 20%	985
Accidents of all types	435
Alcohol consumption (U.S. average)	130
Drowning	41
Natural background radiation	8
Catastrophes (earthquake, etc.)	3.5
Industrial Occupational Hazards	
Mining and quarrying	328
Construction	302
Agriculture	277
Transportation and utilities	164
Service industries	47
All industry (average)	74
Radiation Related	
Medical diagnostic x rays (average)	6
1 rem occupational radiation dose (single dose)	1
1 rem/y dose for 30 years	30

Source: Adapted from B. L. Cohen and I. S. Lee, *Health Phys.* **36**:707 (1979); and World Health Organization, *Health Implications of Nuclear Power Production.* Geneva: WHO, 1975.
Note: The nuclear industry high-dose occupational average is 0.65 rem/y. The maximum population dose associated with the Three Mile Island accident is ≈0.1 rem.

of nuclear science has been discussed in a recent book by H. Wagner and L. Ketchum.

In setting radiation protection standards, a commonly used measure of acceptable risk is that the activity will cause only 1 death in 100 000 people per year in the general population. For industrial or work settings, the level of acceptable risk is 1 death in 10 000 workers per year. Activities that result in this fatality rate are generally considered "safe."

7.4.2. Regulatory Bodies and Objectives of Radiation Standards

The first real step in radiation protection came in the late 1920s when the International Conference on Radiology began work aimed at the definition

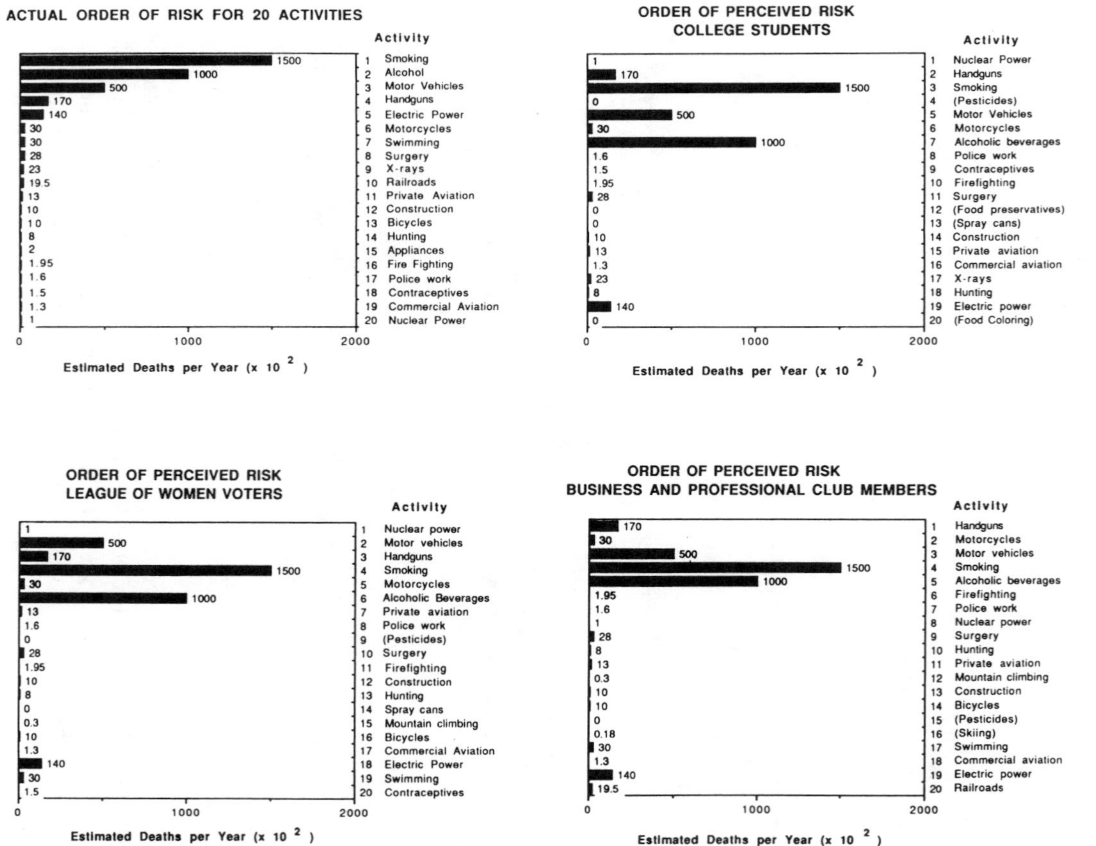

Figure 7.4. Perceived risk versus actual risk for three groups of the American population. (Data from W. K. Sinclair, *Radiology* **138**:1 (1981).)

of methods used for measuring radiation. Some years later, the **International Commission on Radiological Protection (ICRP)** grew out of this. Its purpose is to provide recommendations for safe levels of radiation exposure. In the United States, a similar group, called the **National Council on Radiation Protection and Measurements (NCRP),** was formed. These two agencies, along with the **International Atomic Energy Agency (IAEA)** and the **International Radiation Protection Association (IRPA),** are the professional organizations that provide recommendations on radiation protection to regulatory agencies and promote communication among members of the international radiation protection community. These groups do not write the regulations specifying legal standards for radiation exposure. That task falls to governmental regulatory organizations like the **United States Nuclear Regulatory Commission (NRC)** and the **Environmental Protection Agency (EPA).** The NRC and EPA use the technical information provided by the NCRP and ICRP to formulate their regulations.

The objectives of radiation protection regulations are twofold: (1) to prevent nonstochastic effects, and (2) to limit stochastic effects to an acceptable level. Figure 7.2 illustrated the fact that nonstochastic effects have a threshold level. If the threshold level is known, then limiting dose to an amount below threshold should protect people against the occurrence of nonstochastic effects. The radiation doses required to induce observable nonstochastic effects are really quite large, in relation to exposures that ordinarily occur in daily life, and such doses occur very infrequently. There is greater need for radiation protection standards that deal with routine exposures, rather than with the accidental high-level doses that occur only rarely. So, most regulatory requirements are aimed at reducing stochastic effects to an acceptable limit. Remember that the stochastic effects, because of their probabilistic nature, can never be completely eliminated.

In setting limits for radiation exposure, the following objectives that are stated in the 1977 ICRP report are considered:

1. No practice shall be adopted unless its introduction produces a net positive benefit.
2. All exposures shall be kept **as low as reasonably achieveable (ALARA),** economic and social factors being taken into account.
3. The dose equivalent to individuals shall not exceed the limits recommended for the appropriate circumstances by the Commission.

The quantity that is really of interest in radiation protection studies is the risk to the individual, over a lifetime, of experiencing a severe negative effect (e.g., cancer) from radiation exposure. It is very difficult to state this risk in readily measureable terms. Because the risk is assumed to be related to the dose of radiation received, radiation protection standards are stated in terms of allowable doses for a given period of time.

7.4.3. Recommended Levels for Maximum Exposure

The regulations regarding radiation exposure are very complex. In the United States alone, 23 legally enforceable standards for radiation protection put forth by four different federal agencies can be identified. These standards are for different populations, different sources, and different parts of the body. A detailed explanation of all these is not appropriate here, so only the regulations from the NRC's document 10CFR20 will be discussed.

There are three groups for which the NRC has set radiation standards: the general public, those occupationally exposed, and those exposed for medical purposes. For the last of these three groups, no specific limits have been set. The dose should be the lowest possible dose that can be used that still has the desired beneficial effects for the patient.

There are different sets of standards for occupationally exposed groups and the general public. Limitations are set for total yearly exposure, lifetime exposure, and for specific body parts and organs.

For the general public, the NRC and the ICRP currently recommend a limit of 5 mSv (500 mrem) for an individual per year. This dose excludes that received from natural sources, but proposed revisions in the regulations would change this to include all radiation sources. The average dose to the population should actually be much less than the 5-mSv limit. The NRC standard for the maximum *average* exposure over a lifetime is 1.7 mSv (170 mrem) per year. A modification currently proposed by the ICRP and NCRP would continue to limit individuals to a maximum of 5 mSv in any given year, and also require that the yearly dose averaged over a lifetime be only 1 mSv.

For those who are occupationally exposed, the ICRP specifies a limit of 50 mSv (5 rem) per year for an individual. The accumulated dose is not to exceed $5(N - 18)$, where N = age in years. Individuals under 18 should receive no measured occupational dose.

7.4.4. Health Physics Considerations for the Radioanalytical Laboratory

The radioanalytical chemist using tracers or doing activation analysis is ordinarily working with no more than millicurie levels of activity, most often β- and γ-ray emitters. Rarely are α emitters used as tracers, and only in a few cases have they been used in neutron activation analysis (e.g., determination of Bi by counting ^{210}Po, the α-emitting daughter of β-active ^{210}Bi).

Occasionally, higher activities of short-lived radionuclides are encountered in the use of fast "rabbit" sample transfer systems at reactors. In the latter case, sample handling is usually minimal, since the sample must be counted quickly, although high hand doses may be experienced.

Hence, for most radioanalytical studies, a simple glovebox with lead brick shielding and an independent, filtered exhaust system can be used (Fig. 7.5). Often the most dangerous radioactive samples to handle in the laboratory are dry, finely divided powders. Without a closed, properly ventilated work area, these powders can easily spread and contaminate an entire laboratory. The normal radioanalytical chemistry laboratory will be equipped with area radiation monitors (typically sealed-tube G-M, or proportional counters) to alert the chemist to the escape of radioactive material into the room, and a variety of portable, battery-powered survey meters to localize contaminated areas (Fig. 7.6). The portable survey meters should include a unit with a thin counter window to enable detection of moderate energy β particles. A liquid-scintillation counter should also be available for periodic laboratory surveys. Bench tops, hoods, and floors can be wiped with filter papers moistened with an organic solvent and the papers placed in a scintillation vial containing an appropriate scintillation

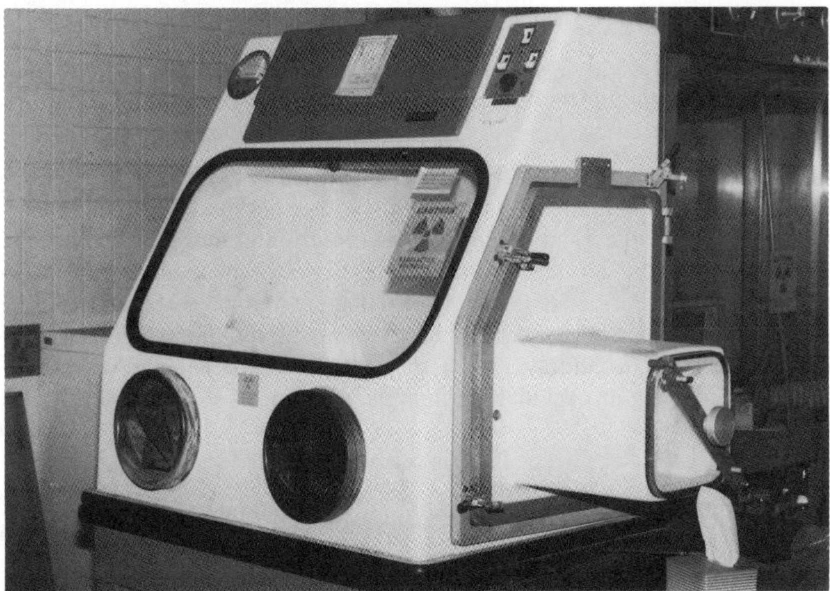

Figure 7.5. A simple glovebox used for handling radioactive samples. Shielding is provided by lead bricks placed inside the box. The box is vented through a filter to an independent stack at the top of the building.

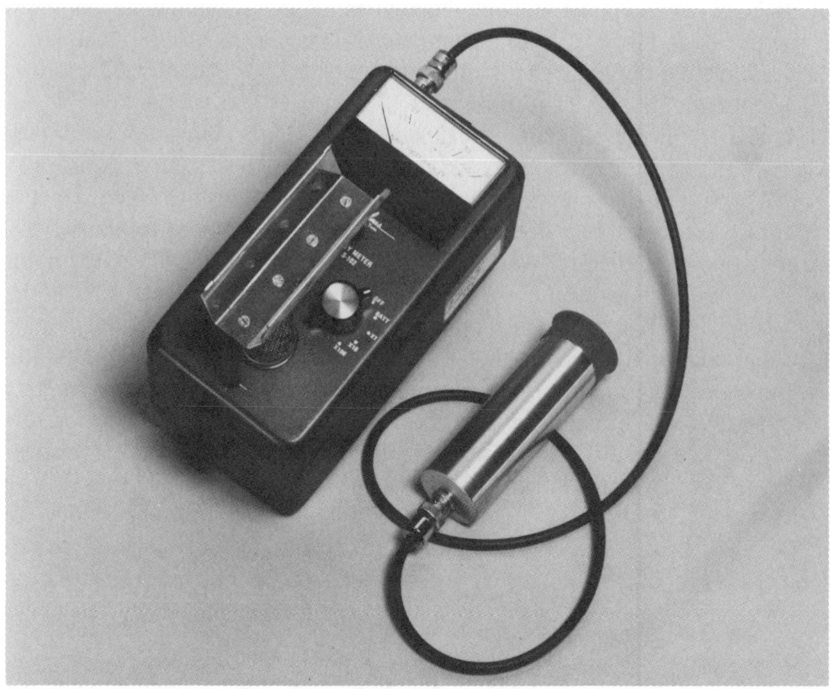

Figure 7.6. A battery-operated portable survey meter for laboratory monitoring. (Courtesy of Tennelec/The Nucleus, Oak Ridge, TN.)

"cocktail" to measure low-energy β and α emitters. A simple but highly effective way of reducing personnel radiation dosage is to carefully plan each step of the radioanalytical procedure in advance, and move away from the site containing the active material during periods when direct manipulation of the sample is not required. Naturally, gloves and protective clothing are required when working with radioactive materials. Film badges or thermoluminescence detectors (TLDs) to measure accumulated dose are required when using any regulated amounts of radioactive materials.

TERMS TO KNOW

ALARA
Collective effective dose equivalent
Dose
Dose equivalent (H)
Dose response curve
Effective dose equivalent
EPA
Exposure
Free radicals
Genetic effects
Gray
Health physics

IAEA	NRC	Risk
ICRP	Quality factor (Q)	Röntgen
Lethal dose (LD_{50})	Rad	Sievert (Sv)
Linear energy transfer (LET)	Radon	Somatic effects
NCRP	Relative biological effectiveness (RBE)	Stochastic effects
Nonstochastic effects	Rem	Threshold level

READING LIST

Cember, H., *Introduction to Health Physics.* New York: Pergamon, 1988. [general]

Hahn, F. F., R. O. McClellan, B. B. Boecker, and B. A. Muggenburg, *Health Phys.* **55(2)**:303 (1988). [nonstochastic effects of radiation]

Howard, N., and S. Atilla, *Dun's Rev.* **114**:48 (1979). [risk perception]

International Atomic Energy Agency, *Radiation: A Fact of Life.* Vienna IAEA, 1979. [general]

Kerr, G. D., *Health Phys.* **55(2)**:241 (1988). [general, quality factors]

Kocher, D. C., K. F. Eckerman, and R.W. Leggett, *Health Phys.* **55(2)**:339 (1988). [relationship between radiation standards and limitation of lifetime risk]

Nero, A., *Phys. Today,* 32–39 (April 1989). [radon in homes]

Shapiro, J., *Radiation Protection,* 2d ed., Cambridge, MA: Harvard University Press, 1981. [general]

Sinclair, W. K., *Radiology* **138**:1 (1981). [radiation risk assessment]

Sinclair, W. K., *Health Phys.* **55(2)**:149 (1988). [trends in radiation protection, NCRP]

Turner, J. E., *Atoms, Radiation and Radiation Protection.* New York: Pergamon, 1986. [general]

United Nations Environment Programme, *Radiation Doses, Effects, Risks.* Nairobi, Kenya: United Nations Publication, 1985. [risk assessment]

United Nations Scientific Committee on the Effects of Atomic Radiation, *Sources, Effects, and Risks of Ionizing Radiation.* New York: United Nations Publication, 1988. [radiation sources and risks]

Wagner, H. N., and L. E. Ketchum, *Living with Radiation, the Risk, the Promise.* Baltimore, MD: The Johns Hopkins University Press, 1989. [politics, public policy, and risks]

EXERCISES

1. Calculate the dose equivalent in humans, in both mrem and mSv, for a 55-mrad dose of α-particle radiation.

2. What is the activity, in Ci, of a γ-ray source that emits a single 1.67-MeV γ ray per disintegration, if the dose received at a distance of 1.0 m from the source is 337 mrad/h?

3. Calculate the number of ion pairs produced if 1 rad is deposited in an absorber. Assume that all of the radiation produces ion pairs and that it requires 35 eV to form one ion pair.

4. 1.0 μCi of radon gas is released into a room that has dimensions of 12 ft $\times$ 12 ft $\times$ 8 ft. Calculate the concentration of Rn, in pCi/L and in Bq/m^3. (Use 1 ft = 30.48 cm.)

5. A ^{60}Co source that had an initial activity of 5.0 Ci has been in a laboratory for 15 years. Calculate the exposure rate 3.0 ft from the source. Assume there is no shielding around the source and remember to account for decay of the source. The value of n is 1 for both of the ^{60}Co γ rays.

6. Calculate the dose equivalent, in mrem and mSv, for a pharmaceutical worker who works at a distance of 0.75 m from a 2.5-mCi, high-energy, β-emitting source for a period of 45 min.

7. From the original definition of the röntgen (1 esu/0.001293 g air), calculate its value in C/kg.

8. If you were setting up a radioanalytical laboratory at your institution, determine what licensing requirements would be required to conduct your research with β- and γ-ray emitters at the millicurie level.

9. Heavy-metal α-particle emitters pose special health problems. Discuss the reasons for this observation.

10. Discuss laboratory shielding requirements for a ^{252}Cf isotopic neutron source. Would lead bricks be effective in shielding the source? Explain your answer.

CHAPTER

8

RADIOCHEMISTRY INSTRUMENTATION

Living organisms do not possess any natural sensors that can directly detect radiation, so instruments that respond to the passage of radiation with some observable signal must be used. For radiation to be detected, it must interact with matter, in one or more of the ways described in Chapter 6. Of all the kinds of interactions, it is the ionization process that forms the basis for most detector systems.

No single type of detection scheme could be equally useful for all types of radiation, because of differences in the ways each interacts with matter. In addition, radiation levels and energies vary drastically. Activities can range from a few counts per hour up to 10^{13} cps, or more. The energies of common radiations can be as low as a fraction of an eV (a slow neutron) up to the GeV level (cosmic rays). Therefore, a variety of detection schemes appropriate for different types, energies, and activities have been developed.

In this chapter, detector systems are divided into two major groups: electronic systems and all others. **Electronic detectors,** which rely on the electrical signals generated when radiation passes through matter, are the most common. Other detectors do not require the direct measurement of an electrical pulse or current. Among the latter are **photographic plates, chemical reaction dosimeters, calorimetric detectors, cloud and bubble chambers,** and **thermoluminescence detectors (TLDs).** Some neutron counters are considered separately. The basic operating principles for each of these detectors are discussed and some typical applications are described.

8.1. DEFINITIONS OF OPERATING CHARACTERISTICS

In the sections that follow, three parameters are mentioned often in characterizing the various types of detectors: the efficiency, the resolution, and the dead time, or resolving time, of the detector.

The **efficiency (ϵ)** of a detector refers to the number of radiations actually detected out of the number emitted by the source. There are two ways to define efficiency, one based on the number of radiations emitted by the source, and one based on the number of radiations that strike the detector.

The former is called the **absolute efficiency,** and the latter the **intrinsic efficiency:**

$$\epsilon_{\text{abs}} = \frac{\text{Number of pulses recorded}}{\text{Number of radiations emitted by the source}} \tag{8.1}$$

$$\epsilon_{\text{int}} = \frac{\text{Number of pulses recorded}}{\text{Number of radiations striking the detector}} \tag{8.2}$$

In most situations, it is desirable to have as high an efficiency as possible. For nonpenetrating radiations, such as heavy charged particles, nearly 100% efficiency can be achieved with appropriate detectors. For the more penetrating uncharged radiations, efficiencies will typically be much lower.

Energy resolution (*R*) refers to the ability of the detector to discriminate between two radiations of different energies. Not all detectors are capable of giving energy information, so for those devices this parameter is not relevant. Resolution is defined with reference to a plot of the number of radiations detected against the radiation energy, as shown in Fig. 8.1:

$$R = \frac{\Delta E}{E_0} \tag{8.3}$$

where E_0 is the energy corresponding to the centroid of the peak, and ΔE refers to the width of the peak halfway between the baseline and the top of the peak. ΔE is also called the **full width at half-maximum,** or the **FWHM.** The resolution is often expressed as a percentage (e.g., $R = 0.10$, or 10%).

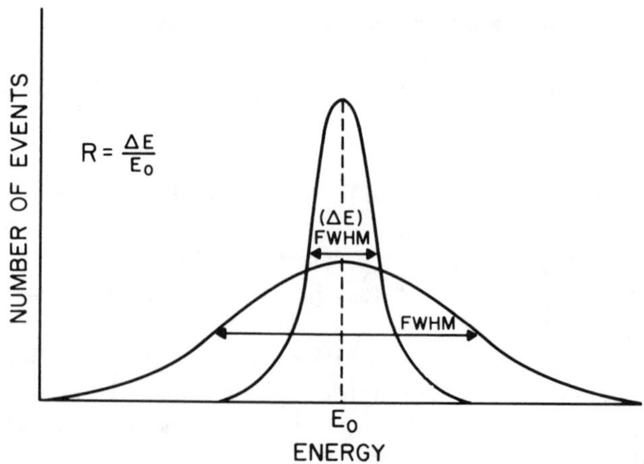

Figure 8.1. Calculation of the energy resolution of a pulse-type detector.

The smaller the value, the better the detector will be at separating two radiations of similar energy. Resolution is never perfect because of electronic noise and the statistical nature of the interactions of radiation with matter. Resolution varies greatly for different types of detectors.

Detector **dead time (τ)** refers to the amount of time needed before a detector can recover from one incoming radiation and respond to the next. The dead time is affected by the nature of the detector and by the associated electronics. Very fast detector systems may be able to respond to a second event within nanoseconds.

8.2. GAS-FILLED DETECTORS

All gas-filled detectors consist of a volume of gas surrounded by a housing that may either be sealed or designed to permit a continuous flow of the counting gas. A voltage is applied across electrodes within the gas volume, creating an electric field. As radiation passes through the gas-filled chamber, it ionizes the gas to form ion pairs consisting of an electron and positive ion. The ions are attracted to the electrodes, and this generates the electrical signal that indicates the passage of radiation. The electric signal that is measured may either be a current, a voltage pulse, or the total accumulated charge. The choice will depend on the way the instrument is configured. The energy lost by the incoming radiation in the formation of one ion pair (called the **W value**) in a gas is about 30–35 eV. The actual value varies by a few eV depending on the type of gas as well as the identity and energy of the incoming radiation.

Three kinds of gas-filled detectors will be discussed: ionization chambers, proportional counters, and Geiger–Müller counters. These differ primarily in the strength of the electric field applied across the electrodes. These variations in voltage lead to differences in the ionization processes taking place in the chamber. A graph illustrating the relationship of the output signal (ion pairs collected) of a gas-filled counter to the applied voltage is shown in Fig. 8.2. Curves for sources undergoing decay via single α- or β-particle transitions are illustrated. In region I, called the **recombination region,** the voltage is too low for all the ion pairs to be collected. Many of the ions simply recombine to form neutral atoms or gas molecules again. This region is not useful for detection. In region II, the voltage is high enough so that virtually all the ion pairs that are formed by a single ionizing event are collected. However, the ions are not accelerated enough to cause secondary ionization events. Therefore, the amplitude of the signal is related to the energy of the incoming radiation and will remain constant even with minor variations in detector voltage. This is the region in which

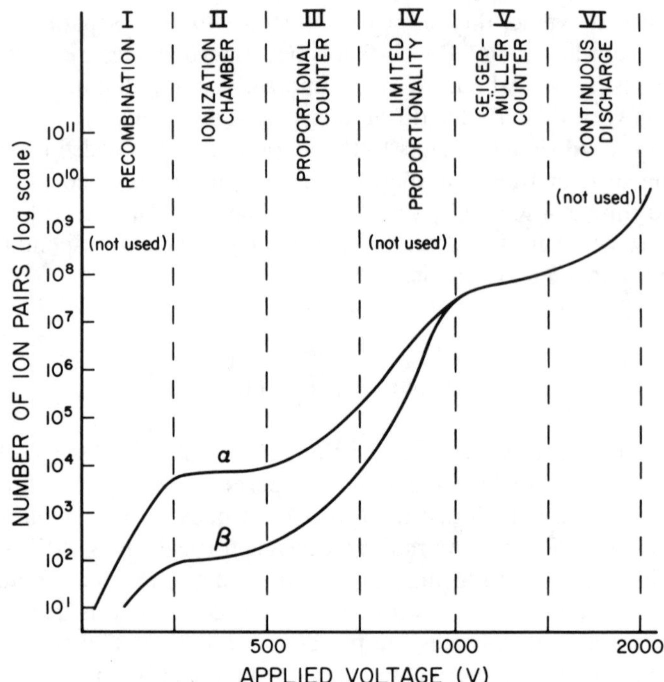

Figure 8.2. Operational regions for gas-filled counters.

the ionization chamber is operated. At higher voltages, in region III, the electrons formed in the original ionization event are accelerated toward the electrodes with enough kinetic energy to induce limited secondary ionization events as they migrate. The amplitude of the signal is not constant with varying detector voltage, but is still linearly related to the energy of the incoming radiation. This is the region of operation for the proportional counter. At still higher voltages, in region IV, the linear relationship between the energy of the incoming radiation and the output signal begins to deteriorate. This **region of limited proportionality** is not used for detection. In region V, the detector voltage has been raised so high that a single ionizing event results in the electrical discharge of the entire tube. The Geiger–Müller (G-M) counter is operated in this region. Because only a single amplitude signal is produced by any radiation interacting with the detector, G-M counters do not permit particle energy discrimination. At very high voltages (region VI), continuous discharge occurs in the gas. This region is not useful for radiation detection.

8.2.1. Ionization Chambers

The **ionization chamber** is one of the oldest and simplest radiation detection devices. A simplified diagram of an ionization chamber is shown in Fig. 8.3. The chamber, typically a few centimeters in diameter, is filled with a gas at pressures ranging from 0.1–10 atm. An electric field is applied across the electrodes. As radiation passes through the gas, ion pairs form. If no electric field were present, these pairs would simply recombine. However, if a large enough electric field is applied across the electrical plates, the ions will drift toward the electrodes. The electrical signal measured can be either a current, a voltage pulse, or the accumulation of the total amount of charge, depending on the type of electrical circuit used.

Current Mode

The most common way of operating an ionization chamber is in the current mode; that is, the electrical signal from the chamber is in the form of a current measurement. When operated in this mode, the chamber is called a **dc ion chamber,** or a mean-level ion chamber. The current is due to the movement of the ion pairs in response to the applied electric field, and the magnitude of the current is related to the rate of formation of ion pairs. Figure 8.4 shows a plot of the current from an ion chamber versus the applied voltage. At low voltages, only a portion of the ion pairs is collected (see recombination region in Fig. 8.2). Eventually a voltage is reached at which virtually all the ion pairs are being collected, and the resulting

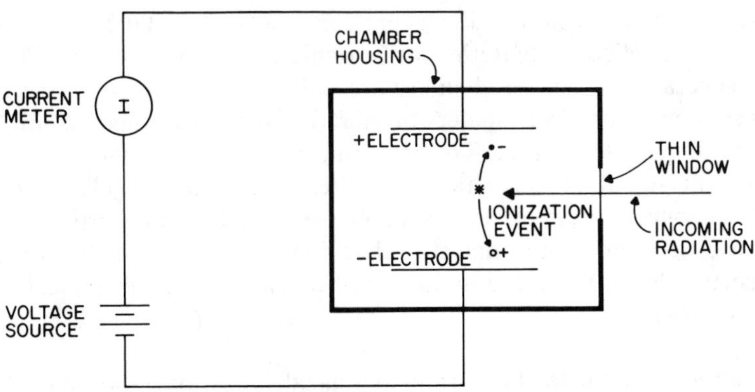

Figure 8.3. Simplified diagram of a dc ionization chamber. The radioactive sample (gas or solid) can also be placed inside the chamber.

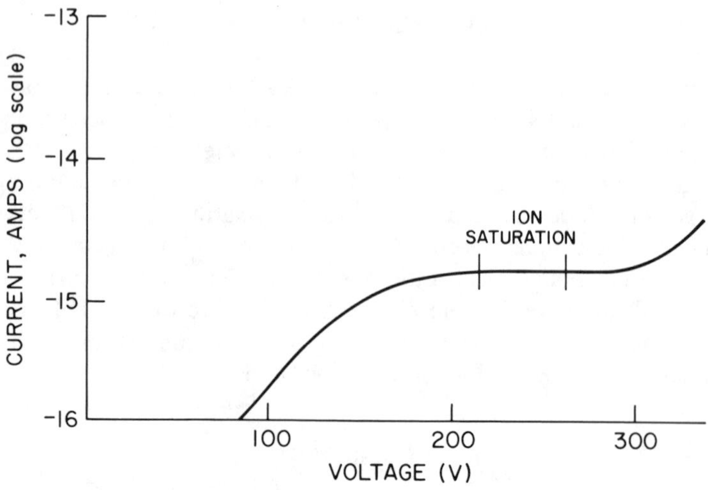

Figure 8.4. Illustration of the saturation region for a dc ionization chamber. The counter is normally operated at a bias voltage one-third to one-half way up this region of constant response.

current is stable for a constant activity source. This is called the **saturation region,** and represents the proper operating voltage for the dc ion chamber.

The output signal from an ionization chamber relies on the collection of both the positive and negative ions. The negative charges are free electrons or any negative ions formed with the electrons. Because only ions produced by primary ionization events are collected, almost any kind of gas can be used as the fill gas in an ionization chamber. Even air is often used, in which case small gas leaks in the chamber are not troublesome. Applied voltages up to a few hundred volts are commonly used. These low voltages are easily supplied by batteries for portable radiation monitors. The currents generated in an ion chamber are quite small. Therefore, electronic devices such as dc electrometers or vibrating reed electrometers are often used to amplify the signal. The **vibrating reed electrometer** converts the dc current signal to ac, which simplifies amplification. The amplitude of the ac voltage in a circuit of suitable design will be proportional to the ionization current in the detector. The following example illustrates the low current levels produced in an ionization chamber under typical counting conditions.

Example. Calculate the current produced in an ionization chamber when an alpha source with an activity of 10^3 dps and particle energy of 5.0 MeV is detected. Assume 100% collection of charge, and a

W value of 35.2 eV. The charge on an electron is 1.6×10^{-19} coulomb (C), and one ampere (A) = 1 C/s.

$$\begin{aligned} 5.0 \text{ MeV}/\alpha &= (5.0 \times 10^6 \text{ eV}/\alpha)/(35.2 \text{ eV/ion pair}) \\ &= 1.4 \times 10^5 \text{ ion pairs}/\alpha \end{aligned}$$

For an activity of 10^3 dps, and 100% collection efficiency, the total number of ion pairs produced is

$$\begin{aligned} (10^3\ \alpha/\text{s})(1.4 \times 10^5 \text{ ion pairs}/\alpha) &= 1.4 \times 10^8 \text{ ion pairs/s} \\ &= 1.4 \times 10^8 \text{ electrons/s} \end{aligned}$$

The total current produced would be

$$\begin{aligned} (1.4 \times 10^8 \text{ electrons/s})(1.6 \times 10^{-19} \text{ C/electron})(1 \text{ A/C s}) \\ = 2.2 \times 10^{-11} \text{ A} \end{aligned}$$

Pulse Mode

The operation of an ionization chamber in **pulse mode** is based on the measurement of a *voltage* pulse produced by individual interactions of charged particles (commonly α particles and fission fragments) with the chamber gas. This is in contrast to the dc ion chamber, where the *current* flow produced by the cumulative interactions of many radiations entering the chamber is measured. An ion chamber configured to operate in the pulse mode is diagrammed in Fig. 8.5. The measured signal is the voltage across

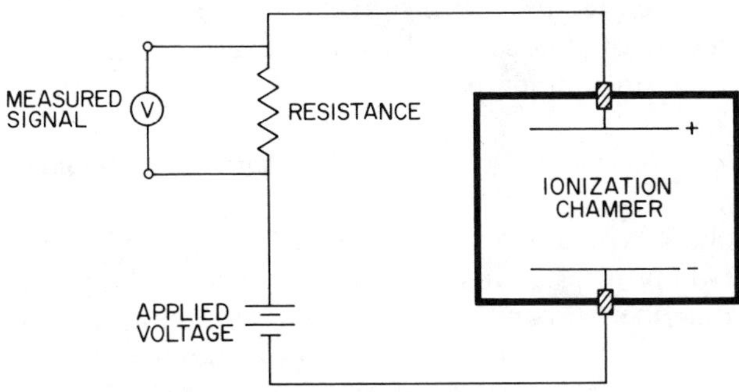

Figure 8.5. Diagram of an ionization chamber system configured to operate in the pulse mode.

the resistor in the circuit. If no radiation enters the chamber, all of the applied voltage appears across the electrodes in the ionization chamber, and the voltage across the resistor is zero. However, when radiation enters the chamber, the ion pairs that are formed collect at the electrodes and reduce the voltage across the electrodes. This results in the appearance of a voltage difference across the resistor that is equal to the voltage drop in the chamber. This voltage difference is the measured quantity. As with the dc ion chamber, the size of the voltage change is quite small, so amplification is required.

Pulse mode ionization chambers may be operated in one of two ways: collection of both the positive ions and the electrons, or collection of electrons only. In the first case, the time needed to collect the positive ions is long (on the millisecond time scale) due to the lower mobility of the heavy positive ions. Therefore, when both positive and negative ion collection is used, only low activities can be measured without experiencing pileup of signals. The advantage of this mode is that the amplitude of the output pulse is strictly proportional to the energy of the ionizing radiation, regardless of the location of the ionizing event in the chamber. Collection of only the electrons results in a faster response time, so higher activity levels can be handled. However, the amplitude of the resulting pulse now is no longer precisely related to the incident particle energy, and is sensitive to the position of the interaction in the chamber. These problems may be overcome through use of a more complex design of ionization chambers, such as the Frisch-gridded ionization chamber.

In a **Frisch-gridded ionization chamber,** a gridded electrode of intermediate potential is placed between the two main chamber electrodes. Due to the gaps in the grid, this electrode is nearly transparent to electron flow. The motion of the electrons after passing through the grid on the way to the anode is no longer influenced by the location and motion of the positive ions formed on the other side of the grid. Therefore, the pulse amplitude measured at the anode is relatively independent of the position of formation of the original ion pair.

Electrostatic or Charge Integration Ionization Chambers

This is the oldest type of ionization chamber. These devices are similar to **electroscopes** in principle. Pairs of lightweight metal foils or fibers (alternatively, a foil or fiber suspended from a rigid conductor) are given a static electric charge, resulting in their physical separation because of the repulsion of like charges. The passage of radiation creates ion pairs which migrate to the foils and gradually discharge them. This causes the foils to return to their original positions. The deflection of the foils over time is measured and is related to the amount of radiation that has passed through the chamber. The **pocket dosimeter** (or pen dosimeter) that is often used in health physics applications for measurement of γ-ray doses is an example of the electrostatic chamber.

Ionization chambers still have many applications in radiochemistry. Their advantages include comparative simplicity in construction and a relatively low cost. These systems can operate with almost any kind of gas in the chamber, making them very convenient for measuring samples of radioactive gases. They can be configured to handle both high and low activity levels, and can give information about the energy of the incoming radiations. The electronics are simple and stable, so they do not require frequent recalibration. They have good detection efficiencies for α and β particles, but poor efficiencies for more penetrating radiation, such as γ rays. A disadvantage of ionization chambers is that the output signal is quite small, so external amplification is usually necessary.

Dc and charge integration ionization chambers are used often as survey instruments or dosimeters, respectively, in health physics. Because of their high efficiency for α particles and ability to give energy information, pulse ionization chambers are quite useful for measuring precise energies of α particles and other heavy charged particles. The dc ion chamber is especially useful for measuring α (Rn measurements) and weak β (^{14}C) activities of gases. For example, ^{14}C dating measurements can be done inexpensively with a simple dc ionization chamber by incorporating the ^{14}C into CO_2 gas, and running the gas into the chamber. In this case, the counting efficiency for ^{14}C is nearly 100%.

8.2.2. Proportional Counters

A **proportional counter** is a gas-filled detector that is operated at a higher voltage than the ionization chamber (see region III in Fig. 8.2). As a consequence of the higher electric field, the ion pairs created by the incident radiation are accelerated to a greater velocity as they drift toward the electrodes. The increased amount of kinetic energy means that the collisions between the electrons and other gas molecules along the path are energetic enough to induce secondary ionization with the release of more free electrons. Therefore, an internal multiplication of the original signal occurs. This multiplication process, called the **Townsend avalanche** or Townsend cascade, is illustrated in Fig. 8.6. If conditions are proper, the amplification can be kept linearly proportional to the original number of ion pairs produced by the incident radiation. Therefore, energy information is retained.

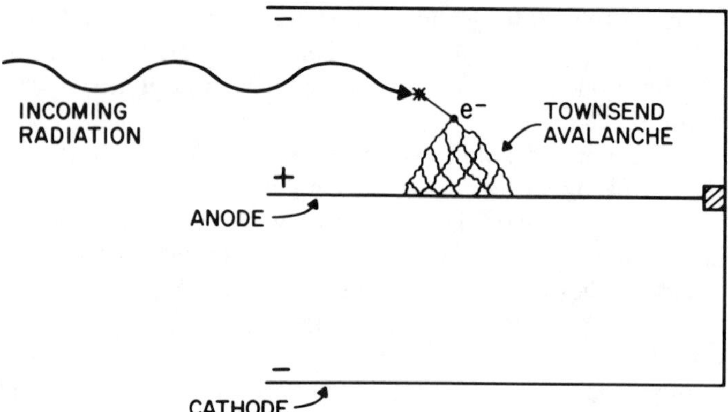

Figure 8.6. Illustration of the Townsend avalanche in a gas-filled radiation detector. (Adapted from G. F. Knoll, *Radiation Detection and Measurement.* New York: Wiley, 1979.)

A diagram showing two common configurations for proportional counters is given in Fig. 8.7. Part *a* shows a sealed, cylindrical tube counter. The tube itself serves as the cathode, and a fine wire passing through the center of the tube acts as the anode. The sample must be placed outside the tube, so a thin window is required to allow passage of the incident radiation. A different configuration, which allows for continuous gas flow and internal sample placement, is shown in Fig. 8.7*b*. This arrangement is better for radiations that have little penetrating power (**soft radiations**) and would be partly absorbed in passage through a detector window. A photograph of a gas-flow proportional counting system is shown in Fig. 8.8. Proportional counters are nearly always operated in the pulse mode.

While nearly any gas will work in an ionization chamber, this is not true for proportional counters. In proportional counters, the signal results from the collection of free electrons, and the formation of negative ions is undesirable. Therefore, the fill gas should not contain any components with a strong electron-attracting tendency that would form anions. The gases that best fill this requirement are the noble gases, mainly He, Ar, and Xe. A second requirement for optimum operation of the proportional counter is that secondary Townsend avalanches not occur with any great frequency. These secondary avalanches can be induced by ultraviolet photons produced by atomic de-excitation processes in the filler gas. To prevent this, another component called a **quenching agent** is added to the filler gas. This component is a molecular species that will receive energy from the excited noble gas ions and then de-excite mainly by dissociation or other nonradiative modes of de-excitation. With proper selection of filler-gas and quenching agent, the Townsend avalanche in the proportional counter

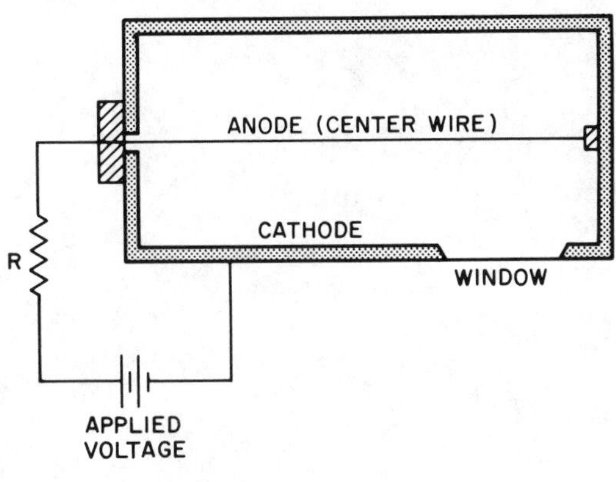

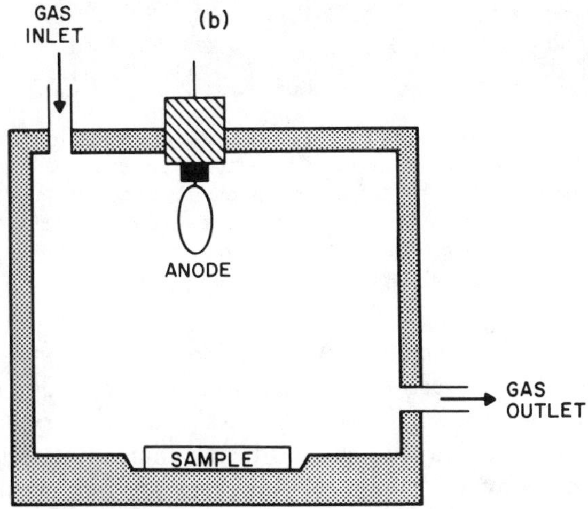

Figure 8.7. Typical proportional detector configurations: (*a*) a sealed cylindrical detector, (*b*) a gas-flow system with a loop electrode.

remains localized and does not spread throughout the tube. A commonly used proportional counter gas mixture is called **P-10 gas.** It consists of 90% Ar, with 10% CH_4 acting as the quenching agent.

Proportional counters have the advantage of producing a large output signal, so less external amplification is required than for ionization chambers. The internal amplification factor may be as much as 10^3 to 10^4. This simplifies circuitry and eliminates some sources of electronic noise. Because only the voltage pulse resulting from electron collection is utilized,

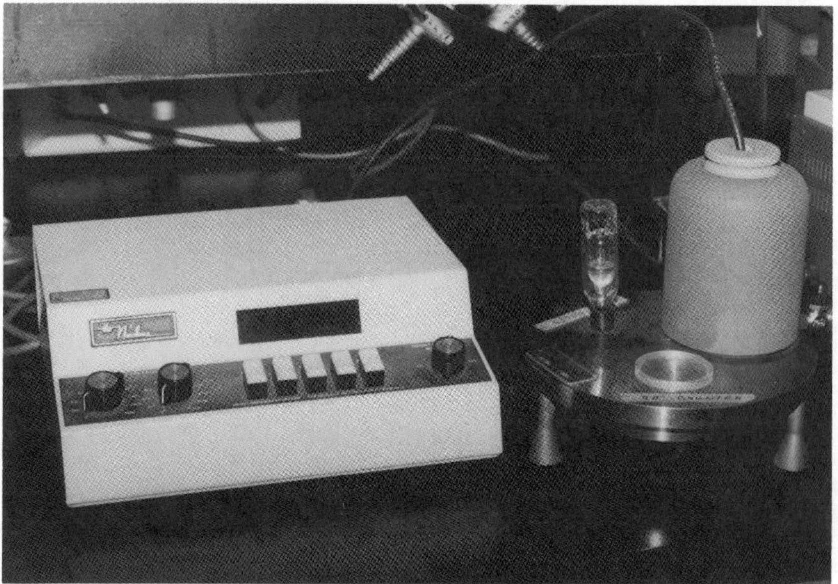

Figure 8.8. A simple gas-flow, 2π geometry, internal sample, proportional or G-M counting system with bias supply and scaler. The sample is placed in a shallow dish in the sample port at the front and then rotated to a gas-purge position, and finally to a counting position under the detector. In this way, the sample may be placed inside the detector without actually opening the detector to the atmosphere. An independent amplifier and discriminator is required for operation as a proportional counter.

and not that of slow-moving positive ions, proportional counters have a fast response time. Dead times as low as a few tenths of a microsecond can be obtained in some applications. They also have a very high efficiency for particles with short ranges, such as heavy charged particles or low-energy β particles. However, they have poorer energy resolution than ionization chambers because of the voltage dependence of the internal secondary ionization process. Proportional counters are commonly used for the detection of low-energy x rays, neutrons (with a BF_3 fill gas), and mixed α and β sources. They are rarely used for γ-ray detection, because of their low efficiency for high-energy, penetrating radiations.

Proportional counters are often used to distinguish α and β activities in a mixed source. This can be accomplished because of differences in the interactions of these two radiations in the proportional counter. For a given voltage, the internal counter multiplication factor is constant. However, the output pulse will be directly related to the number of ion pairs formed in the primary interaction of the incident radiation. Alpha particles will produce a larger output pulse than β particles of the same energy due to their higher specific ionization (more ion pairs/mm of path, see Section 6.2.2), and the fact that they can be completely stopped within the detector volume. Beta particles of moderate energy are more penetrating and may leave the active detector volume before losing all their energy. With a counting system employing a **lower-level discriminator (LLD)** (see Section 8.5), or gate, pulses due to α particles can be detected at a lower applied voltage (as a result of their pulse heights exceeding the LLD) than can β particles that have comparable energies.

The counting characteristics of a proportional counter exposed to a mixed α and β source are illustrated in Fig. 8.9. It is important to note that this figure plots counting rate versus voltage, not number of ion pairs versus voltage as in Fig. 8.2. For very low detector voltages, as in the lowest voltage region of this figure, only high-energy background radiations or sporadic electrical noise would generate voltage pulses sufficient to exceed the LLD setting. This region is not used. In the α plateau region, the voltage is sufficient so that α particles produce pulses large enough to pass the discriminator, but β particles do not. Therefore, with a detector voltage set in this region, only α particles are detected. The third region represents a detector voltage region where all α particles are counted, but only some of the β particles. The number of β particles counted will depend strongly on the detector voltage, because electrons emitted in β decay have a distribution of energies up to $E_{\max}$, not a single energy. This region is not useful for counting. In the $\alpha + \beta$ plateau region, the voltage is high enough so that all α and nearly all the β particles produce pulses large enough to exceed the LLD and be registered.

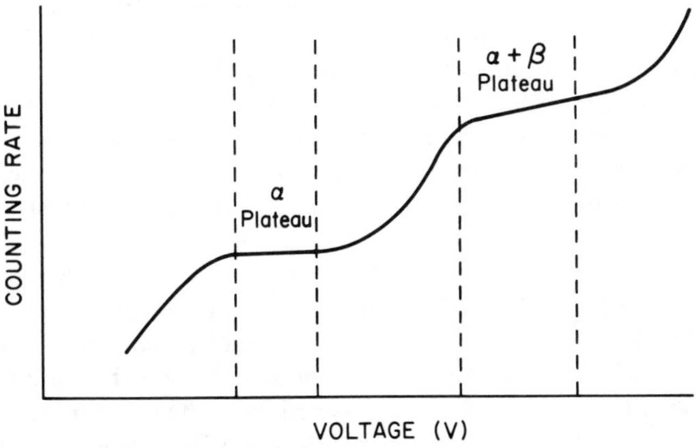

Figure 8.9. A plot illustrating the two plateaus obtained in a counting rate versus bias voltage plot for a proportional counter and a source that is emitting both α and β particles.

The β plateau in the operating region of the proportional counter is not perfectly flat, as it was for a pulse ion chamber or the proportional counter α plateau (see Fig. 8.9). This is because β particles are not monoenergetic and higher detector voltages will cause a small number of lower-energy β particles in the spectrum to pass the LLD. Figure 2.5 shows that β particles with energies that are very low with respect to the most probable energy in the β spectrum are also low in abundance. Hence, small changes in the detector voltage will have only a small effect on the counting rate in the β plateau region. The β plateau is reached after pulses from β particles with energies somewhat below the most probable energy in the spectrum pass the LLD.

Counting efficiencies for α and β particles may vary. This is because some β particles, which are more penetrating than α particles, may escape from the active volume of the counter while depositing an insufficient amount of energy to be recorded. The difference in count rate between region IV and region II in Fig. 8.9 can give a measure of the β activity of the mixed sample.

8.2.3. Geiger–Müller Counters

The third type of gas-filled detector is the **Geiger–Müller (G-M) counter,** often referred to simply as a Geiger counter. These detectors are the most commonly used counting devices when gross activity level alone is of primary interest. In the G-M counter system, the voltage applied to the chamber is higher than for either ionization chambers or proportional counters (see Fig. 8.2). As a result, the electrons formed when radiation passes through the filler gas are accelerated strongly toward the anode and attain high kinetic energies. The electrons will induce secondary ionization, similar to what took place in proportional counters. However, in the G-M tube, the greater energy and larger number of collisions result in the excitation of many gas molecules, some of which will de-excite by emission of energetic photons. These photons can themselves trigger other cascades elsewhere in the tube. The result is that a propagated Townsend avalanche spreads around the entire anode of the detector. Figure 8.10 illustrates this phenomenon.

When a detector is operating in the G-M voltage region, the interaction of any radiation, regardless of type or energy, leads to the discharge of the entire tube. The discharge eventually stops due to the development of a positive ion sheath around the anode. The sheath forms because the more massive positive ions cannot migrate as rapidly as the electrons. In contrast to the other gas-filled counters, the G-M counting rate reaches a single plateau with increasing voltage, regardless of particle type or energy. This

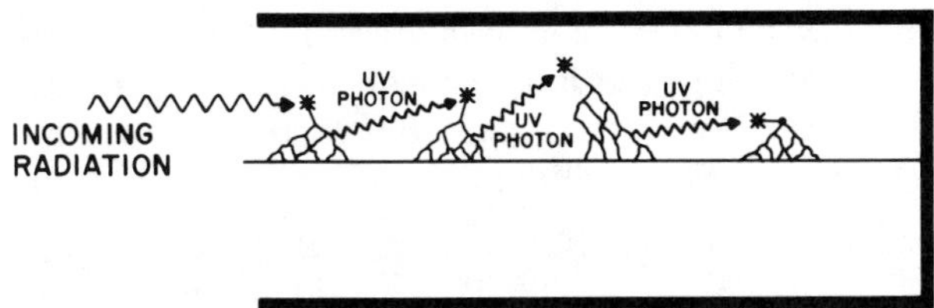

Figure 8.10. The propagated Townsend avalanche in a G-M detector (Adapted from G. F. Knoll, *Radiation Detection and Measurement.* New York: Wiley, 1979.)

is similar to the α plateau in pulse ionization chambers and proportional counters (see Fig. 8.9). If the voltage applied to the detector is increased significantly, the insulating capacity of the fill gas is exceeded and the detector enters the region of continuous discharge (region VI, Fig. 8.2). In this region the detector is not directly responsive to sample activity. The operating voltage for all pulse gas-filled counter systems is set at one-third of the way through the plateau in the counting rate versus applied detector voltage curve.

The G-M counter is similar to the proportional counter in construction and in requirements for fill gas. Sealed tubes are also readily available. Because the output signal from the G-M tube relies on electron production, the fill gas must not contain any high concentrations of strongly electron-attracting species. The noble gases, especially He and Ar, are most used.

The problem of repetitive pulses occurring in the tube due to secondary electron release by the positive ions as they reach the cathode is more severe for the G-M tube than for proportional counters. Quenching agents to suppress this unwanted multiple firing of the counter are essential. In gas-flow detectors, these agents are usually organic molecules, such as ethyl alcohol, isobutane, or ethyl formate. In sealed tubes, halogen gases, mostly small amounts of Br_2 or Cl_2, are used. The excess energy released in the neutralization of the positive ions at the cathode is dissipated by dissociation of the quench gas molecules or excitation of vibrational or rotational modes of freedom in these molecules. Thus, subsequent emission of secondary electrons at the cathode is suppressed. Dissociation results in a gradual depletion of organic quench-gas molecules in a sealed tube over time. Because halogen gases can recombine after dissociation, they retain their usefulness as quenching agents much longer and are preferred for use in sealed G-M tubes. **Q-gas** is a gas mixture commonly used for gas-flow G-M tubes. It contains 97% He and 3% of an organic quenching agent such as isobutane.

The advantages of G-M counters include simplicity of design, durability, insensitivity to small changes in applied voltage (in the plateau region), and low cost. Because the entire tube discharges in response to the interaction of radiation, the signal size is large enough so little external amplification is needed. This simplifies design and construction. However, their disadvantages include the lack of ability to provide information about the energy or identity of incoming radiation, and a relatively long dead time (100–500 μs) required to clear the many positive ions formed. The latter limits their usefulness at high counting rates unless corrections are made. Geiger–Müller tubes are frequently used in survey meters both in the laboratory and under field conditions, because of their low cost and sensitivity to most common types of radiation.

8.3. SCINTILLATION DETECTORS

Scintillation detectors, like ionization chambers, have a long history in the study of radioactivity. By the early 1900s, workers had observed that allowing α particles to impinge on certain materials would cause them to emit a flash of light. (The word scintillation comes from the Greek word for spark). Rutherford used a scintillation detector, in the form of a screen coated with zinc sulfide (ZnS), to detect the α particles used in the experiments that established the nuclear model of the atom. In those days, a microscope was used to watch for and count the light flashes. Radiation detection by this means was extremely tedious, and it provided the incentive for Rutherford's co-worker, Geiger, to begin working on an electronic method (the G-M counter) for counting α radiations. Scintillation detectors operate on a much more sophisticated basis now than the visual counting of light flashes. However, the basis for scintillation detection is still the radiation-induced emission of light by the detecting material, although the light is no longer visually observed.

Although the details of the process will differ among the various types of scintillation detector systems, there are some similiarities in their basic operating principles. The incoming radiation interacts with the scintillating material in the usual ways, producing ionization and atomic or molecular excitation (see Chapter 6). However, scintillation detectors are not based on direct measurement of electrical signals from the ion pairs that are produced in the radiation-absorbing medium. Rather, the excited atoms or molecules undergo de-excitation by emission of a photon of light. This light passes through the scintillating material, then through a **light pipe** to a **photomultiplier tube (PMT),** which converts the light to an electrical signal (see Section 8.3.3). The signal is then processed electronically for display or storage.

From this simple description, some of the properties desirable in a good scintillating material may be deduced. First, it must have the appropriate molecular or crystalline structure to react to the passage of radiation with the emission of light, and it should be able to do this efficiently. The light production process is called **luminescence**. The light must pass through the material to be detected, so the scintillating material must be optically transparent to the wavelength of light generated. Long delays between the time of radiation interaction and the actual emission of light (**phosphorescence**) are undesirable for high counting rates, so the light emission should take place quickly (**fluorescence**) after interaction. The scintillating material should have the appropriate physical properties to allow it to be formulated into crystals and easily machined to enable coupling with other system components. Ideally, the scintillator should be chemically stable and not hygroscopic, so that handling is easier. However, hygroscopic crystals are commonly used and must be maintained in air-tight containers.

Scintillation materials fall into two major groups: inorganic and organic.

8.3.1. Inorganic Scintillation Detectors

The most widely used inorganic scintillation detectors are the alkali halide crystals (e.g., NaI). These crystals can function as radiation detectors because of the electron transitions that take place between energy levels in the crystal in response to the passage of radiation.

In crystalline solids, the individual atomic energy levels form groups of very closely spaced orbitals called **bands**. In the crystal, the electrons can occupy two possible energy levels, as shown in Fig. 8.11. The **valence band** is a lower-energy band containing the electrons bound to the crystal lattice. These electrons are not free to move through the crystal. If only these

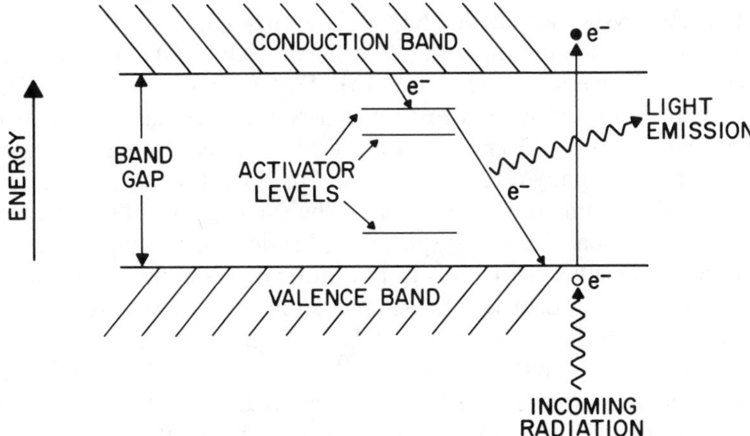

Figure 8.11. Energy bands in a scintillation crystal.

bands are occupied, the material is an electrical **insulator**. The **conduction band** is at a higher energy level, and contains electrons that are free to move within the crystal. If this level is populated with electrons, the material is an electrical **conductor**. Between these two energy levels is the **band gap,** which contains forbidden energy levels. Absorption of energy (from incoming radiation, for example) can excite an electron into the conduction band. As the excited electrons return to the valence band, they may lose energy by emission of photons.

The scintillation process described above is not very efficient for a crystal of pure NaI for two reasons. One is that the band gap is so large that electron transitions are not highly probable. Second, even if electron transitions did occur, the emitted photons would be of such a high energy that they would not be near the visible region of the spectrum where amplification via a PMT is most efficient. To overcome these problems, an impurity called an **activator** is added to the crystal. This creates new energy levels within the band gap (see Fig. 8.11). Electron transitions to these new levels (**activator levels**) do not require as much energy, so they are more likely to occur. In addition, when de-excitation takes place, the emitted light has a lower energy and is more likely to be in the near ultraviolet or visible region of the electromagnetic spectrum. For NaI scintillation crystals, Tl is a good activator. This system is commonly designated as a **NaI(Tl) scintillation crystal.**

The properties of NaI(Tl) scintillation detectors make them best suited for the detection of γ rays, x rays, and sometimes high-energy electrons. These detectors are very widely used, and are likely to be found in almost any laboratory doing radioanalytical work. The characteristics that make them useful for γ spectroscopy include good efficiency, high light output, and a linear energy response. The good efficiency for γ rays arises from the fact that the crystals are moderately high-density solids (in contrast to gas-filled counters) and contain a high-Z element, iodine, which enhances the photoelectric process (see Section 6.4.1). The efficiency advantage is further enhanced by the large size to which NaI(Tl) crystals can be grown. Cylindrical sizes of 12.7 cm in diameter and 12.7 cm in length are common. These are usually referred to in the U.S. literature as 5-by-5's, in reference to their dimensions in inches. The NaI(Tl) crystals can also be formed into unusual shapes for special applications. One shape that is very often used is a **well crystal.** This is a cylindrical crystal into which a cylindrical well has been drilled. The sample is put into the well for counting. For radiations whose energy is not too large, high counting efficiencies can be achieved for this configuration. The light yield from the NaI(Tl) detector is better than from any of the other scintillators, and the relationship between the energy of the incoming radiation and the light output is linear.

The biggest disadvantage of NaI(Tl) crystals for γ detection is their poor energy resolution. The best resolution would be about 6%, as measured for the 662-keV γ ray emitted by ^{137}Cs. The reasons for the poor resolution are intrinsic to the scintillation process itself, so improving technology will not be able to improve the resolution. The time needed for the scintillation to decay is around 230 ns. This is a long time, in comparison with many other scintillation materials. Therefore, NaI(Tl) detectors are not ideal for high-count rate applications or for experiments requiring very fast timing responses. NaI crystals are very hygroscopic and so must always be covered with sealed metal containers. A photograph of a NaI(Tl) scintillation detector with its associated PMT and voltage divider box is shown in Fig. 8.12.

The γ spectrum in Fig. 8.13 illustrates the response of a NaI(Tl) detector to a source consisting of ^{137}Cs and ^{60}Co. A spectrum obtained with a semiconductor detector is also shown. The latter will be discussed in Section 8.4. The broad peaks in the NaI(Tl) spectrum reflect the poor energy resolution of this detector, while the greater areas under the peaks illustrate the superior efficiency of the NaI(Tl) detector over a semiconductor detector of similar size. As will be noted later, high-efficiency semiconductor detectors have recently become available, but they are much more expensive than equivalent-efficiency NaI(Tl) detectors.

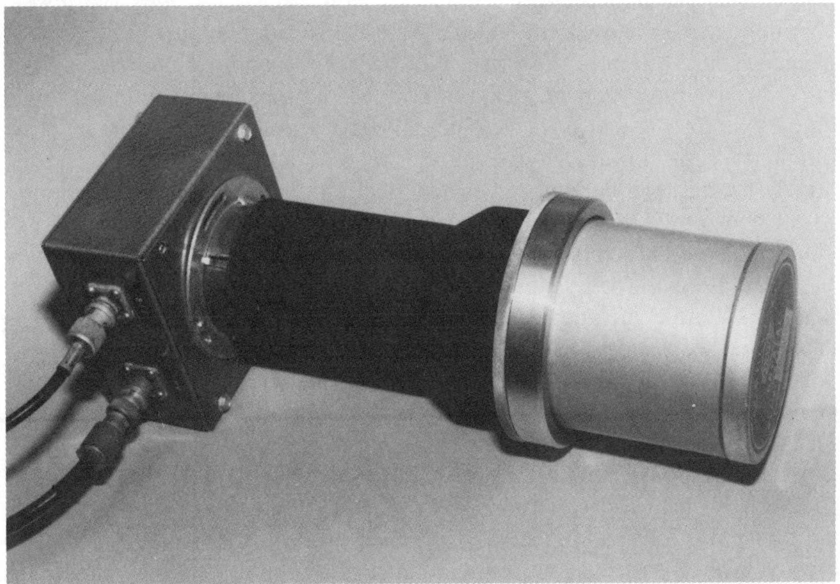

Figure 8.12. A 7.6-cm-long by 7.6-cm-diameter NaI(Tl) scintillation crystal coupled to a PMT and a voltage divider box. Cables provide high voltage for the PMT and signal output.

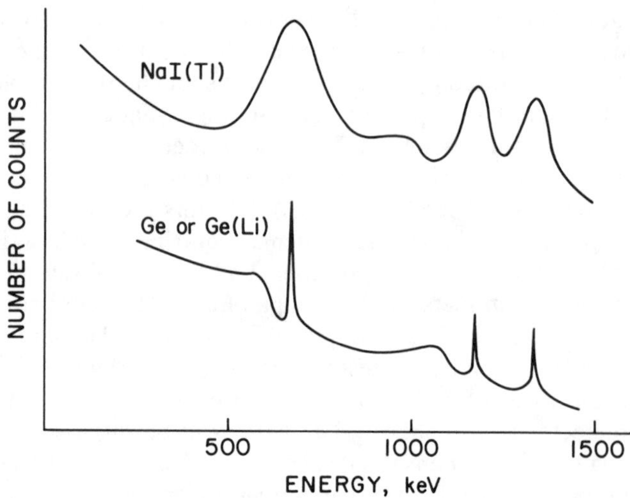

Figure 8.13. Comparison of γ-ray spectra for a NaI(Tl) scintillation detector and a semiconductor detector using a mixed ^{60}Co and ^{137}Cs source.

Other alkali halides used as detectors include CsI(Tl), LiI(Eu), and CsF. CsI(Tl) has the highest γ-ray efficiency (photons produced per MeV of energy deposited) of any scintillation crystal. This is due in part to the high-Z element composition of the crystal. It also has the unusual property of having a variable pulse decay time for different radiations. For example, it would be easy to distinguish positive charged-particle interactions from electron interactions by looking at the pulse shape. The LiI(Eu) crystal is used largely for neutron detection, because of the high capture cross section of Li for neutrons. CsF has the advantage of a very fast decay time (a few ns). None of these detectors is as widely used as the NaI(Tl) detector.

There are a few other inorganic materials that are used as scintillation detectors. These include glasses that contain B and Li (used for neutron detection, see Section 8.7), ZnS-coated screens (used for heavy charged-particle detection), CaF_2 (useful under hostile environmental conditions, because it is not hygroscopic and resists fracture), and some noble gases. A recently developed inorganic scintillation detector that is gaining in popularity is **bismuth germanate** ($Bi_4Ge_3O_{12}$), abbreviated as **BGO**. Because of the high-Z elements it contains, its efficiency for γ rays is very high. Its light output is lower than NaI(Tl), however, and it is more expensive. BGO crystals are often used in **Compton-suppression** spectrometer systems. In these systems a central detector is surrounded by an array of BGO crystals that detect Compton scattered γ rays in coincidence with signals in the central detector. Operating in the anticoincidence mode, the spectrometer does not record events occurring in coincidence. The result is suppression of the Compton distribution in the γ-ray spectrum obtained with the central detector. The use of expensive BGO crystals in this application is justified by the crystal's high efficiency for γ-ray detection.

8.3.2. Organic Scintillators

Most of the organic molecules that are used as scintillation detectors have a π-electron system. Figure 8.14 shows the structure of anthracene and stilbene, two crystalline organic scintillators. For these molecules, light emission occurs as a result of electron transitions within the individual molecules, rather than between the energy levels that exist due to the crystal lattice. Because crystal structure is not important, organic scintillators (also called scintillants) may be either liquid or solid. They are used primarily for α- or β-particle counting rather than for γ counting, because most organic molecules are composed of only low-Z elements.

Anthracene has a good light output for an organic scintillator. Stilbene, like CsI(Tl), exhibits different decay times for different types of radiation, so it can be used to identify the type of particle inducing the scintillations. Another solid inorganic material that has been used as a detector is a polymethyl methacrylate plastic that has had an organic scintillator dissolved in it before polymerization. Plastic has the obvious advantage of being able to be made into many different shapes. None of the solid organic scintillators is used as frequently as the solid inorganic scintillators or the liquid scintillators described below.

Liquid-scintillation detection systems are used extensively in biological, medical, biochemical, and organic chemical studies. The reason is that this

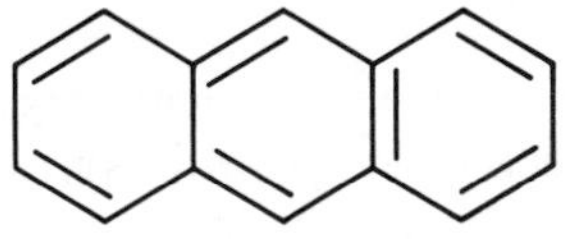

ANTHRACENE

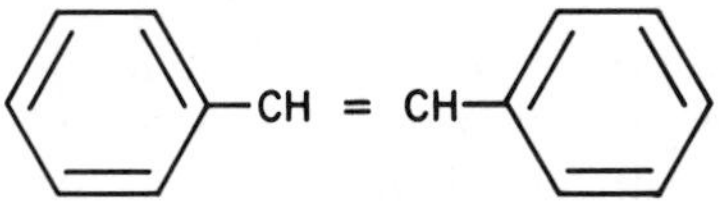

CH = CH

STILBENE

(cis- and trans-configurations exist)

Figure 8.14. Chemical structures for two common solid organic scintillators.

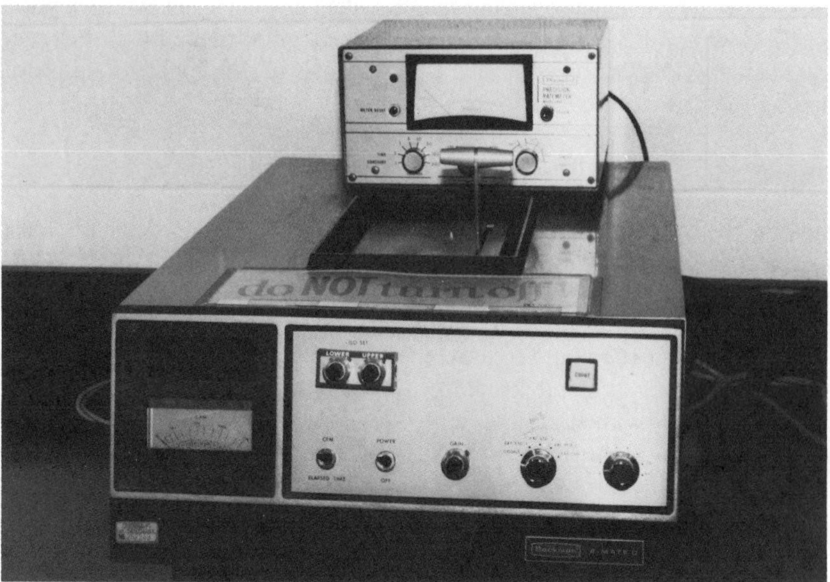

Figure 8.15. A simple liquid-scintillation counting system with power supply. The sample is introduced through a port on the top of the unit which is light-tight when closed. The unit contains an SCA which permits adjustment of both the ULD and the LLD.

method is especially well-suited for the detection of the weak β emitters ^{3}H and ^{14}C, and these are the tracers most used for organic molecules. A simple liquid-scintillation counting system with its power supply is illustrated in Fig. 8.15. The liquid-scintillation mixture (called the **scintillation cocktail**), consists of the scintillating solute, a solvent, and the dissolved sample. Three common scintillating solutes are 2,5-diphenyloxazol (PPO), *p*-terphenyl, and tetraphenylbutadiene. Toluene and *p*-xylene are commonly used solvents. The sample containing the radioactive element (e.g., ^{3}H or ^{14}C) must be dissolved in the solvent. Aqueous samples will not readily dissolve in organic solvents, so another component (e.g., dioxane) may be added to the cocktail to aid in the dissolution process. The cocktail is placed into a small vial, about 6 cm tall and 3 cm wide, made of glass or transluscent plastic. The radiation emitted by the sample is most likely to interact with the solvent molecules first, because there are many more of them than of the solute. The solvent then transfers energy to the scintillator, which de-excites by emission of a photon. This photon must be at a wavelength that the photomultiplier tube can effectively respond to. If it is not, another component, called a **wavelength shifter,** can be added to the liquid scintillation cocktail. The wavelength shifter will efficiently absorb the higher-energy light emitted by the scintillant, and re-emit it at a lower energy that is more suitable for the phototube. POPOP (1,4-bis-[2-(5-phenyloxazolyl)]benzene) is a common wavelength shifter.

8.3.3. The Photomultiplier Tube

The radiation-induced photon emitted by a scintillation detector is only the first step in the detection process. These very weak light signals must be converted into a useful electric signal. The photomultiplier tube (PMT) accomplishes this task.

Figure 8.16 shows the basic structure of a PMT coupled to a scintillation crystal. The PMT is housed in an opaque container, because external light must be completely excluded for the PMT to respond only to the scintillation photons. The scintillation process occurs in the detector and the photon passes first through a light pipe that serves as an optical coupler between the detector and the PMT. At the end of the PMT is a layer of a photosensitive material called the **photocathode**. The photon from the radiation interaction strikes this photocathode and causes the emission of electrons called **photoelectrons**. The photoelectrons are attracted toward the anode. Before reaching the anode, however, they encounter a series of **dynodes** whose function is to increase the number of electrons that ultimately reach the anode. When the photoelectrons strike the first dynode, they induce the emission of more electrons. These electrons are accelerated

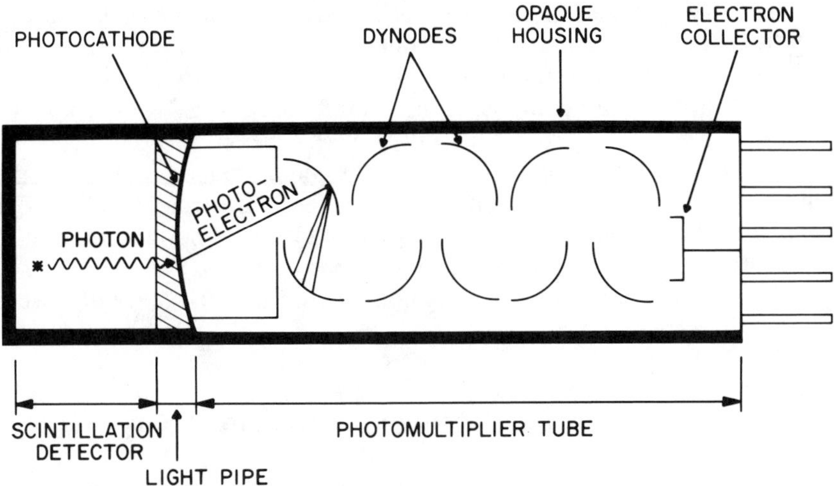

Figure 8.16. Schematic diagram of a solid-crystal scintillator coupled through a light pipe to a PMT.

to the next dynode, where each causes still more electrons to be emitted. This **electron multiplication** process continues for several (8–12) dynodes, with the result being a large shower of electrons collected at the anode for each single photoelectron originally emitted from the photocathode. The **gain** (number of electrons out per photoelectron in) can range from 10^7 to 10^{10} for a typical PMT. The pulse produced, however, is still proportional to the number of original photoelectrons, which is controlled by the intensity of the light produced in the interaction. The light intensity, in turn, is proportional to the energy of the radiation that caused the scintillation. Thus, energy information can be obtained. Photomultiplier tubes are used with all of the scintillation detectors discussed in Section 8.3.

There are sources of **noise** in circuits using a PMT that cause a background counting rate not related to sample activity. It is possible for photoelectrons to be emitted spontaneously from the photocathode due to thermal energy, in a process called **thermionic emission.** Because each background signal from this source is the result of the thermionic emission of a single electron, the amplitude of the signal produced is very small, even after electron multiplication. A discriminator may be used to eliminate this low-amplitude noise. Reductions in thermionic emission may be achieved by cooling the PMT and by varying the photosensitive material used on the photocathode. An electronic solution to the problem is to use a coincidence unit. In this arrangement, two PMTs are arranged at 180° to each other around the sample. A true scintillation event will trigger pulses in both PMTs that will arrive simultaneously at the coincidence unit, and will be recorded as a true event. Thermionic emission in only one PMT will not likely be matched by a simultaneous event in the other, so this false signal will be excluded by the coincidence circuit, which records only coincident events. Another source of PMT noise is the production of photons within the tube by the decay of naturally occurring ^{40}K or U and Th natural decay series radionuclides in the glass envelope or other construction materials of the PMT. These radiations could produce electrons at the photocathode that would be multiplied together with true photoelectrons related to the sample activity. This source of electrical noise is referred to as the **dark current.** These are also low-intensity events. Light leaks in the housing will, of course, also result in noise. However, this source of noise is easily eliminated.

8.4. SOLID-STATE SEMICONDUCTOR DETECTORS

Gamma-ray spectrometry was advanced greatly by the development of solid-state semiconductor detectors in the 1960s and 1970s. These detectors offer a combination of good efficiency and high-energy resolution for γ rays that is not matched by the gas-filled or scintillation detectors.

8.4.1. Theory of Semiconductor Detectors

The semiconductor detectors are usually made of Ge or Si. A band structure exists in the semiconductor crystal, similar to that described earlier for the solid scintillator. However, the band gap is much smaller for semiconductors than for the NaI(Tl) detector. For Si, the gap is about 1.1 eV, and for Ge, about 0.66 eV. The passage of radiation may inject enough energy into the system to raise an electron from the valence band to the conduction band, thus creating an **electron–hole pair,** analagous to the ion pair created in the gas-filled counters. The hole (which is really the absence of an electron) and the electron can both migrate through the crystal in response to an electric field, and produce an electrical signal that marks the passage of the radiation.

Completely pure semiconductor crystals, that is, those composed *only* of atoms of Si or Ge, can be described in theory, but cannot be made in practice. They are called **intrinsic** semiconductors. Real-world semiconductor materials contain various impurities that have a great effect on their electrical properties. Sometimes the impurities are added purposely, in a controlled manner, to effect some desirable change in the electrical characteristics of the semiconductor. Therefore, when discussing the use of semiconductor materials for radiation detection, the nature of the impurities and the effects they have must be understood.

The types of impurities that are important in semiconductor crystals have either one more or one less valence electron than the four normally possessed by neutral Ge or Si atoms. As an example, consider the addition of P to a Ge crystal, as shown in Fig. 8.17. Because P has one more valence electron than Ge, there will be an excess of electrons present. A semiconductor of this type is called an **n-type semiconductor.** The presence of the P also creates a new energy level in the band gap, called a **donor level,** that is very close to the conduction band (see Fig. 8.18). This means that electrons are promoted very easily into the conduction band, and the electrical properties of the semiconductor are controlled by these excess electrons in the conductor band. The electrons in an n-type semiconductor are referred to as the **majority carriers.** On the other hand, if one of the Ge atoms is replaced with a B atom, an excess of holes in the semiconductor will result (Fig. 8.17). This forms a **p-type semiconductor,** in which the electrical properties are dominated by the presence of excess holes. The presence of the B creates an **acceptor energy level** in the band gap that is very close to the valence band (Fig. 8.18). Electrons from the valence band may have

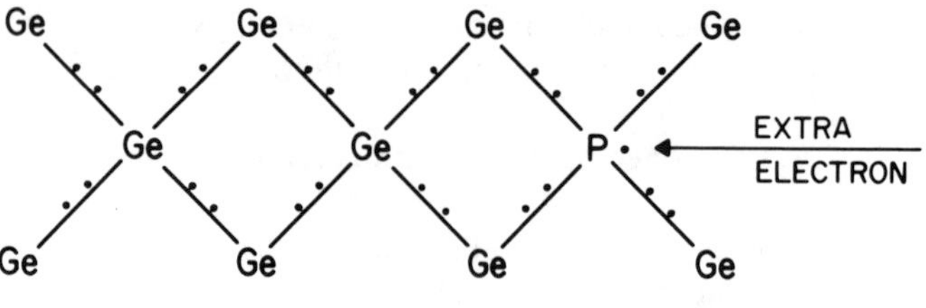

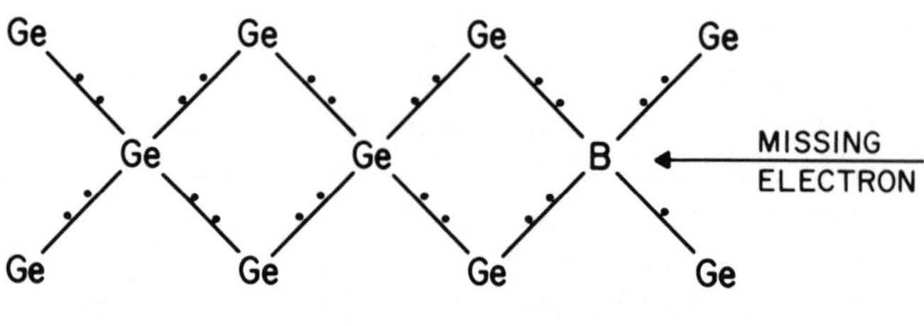

Figure 8.17. Semiconductors with n-type and p-type impurities.

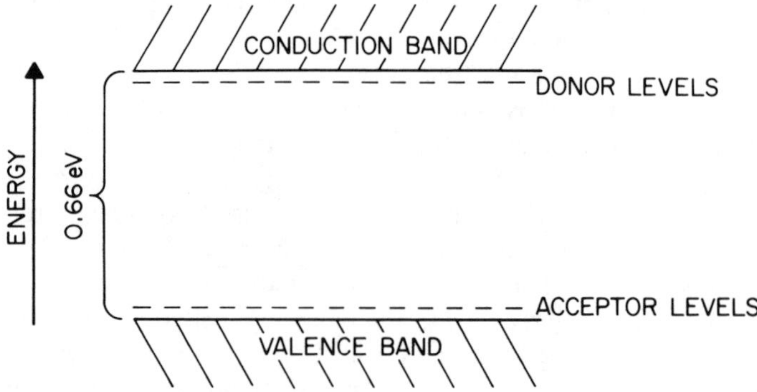

Figure 8.18. Donor and acceptor energy levels in the band gap of a semiconductor crystal.

enough thermal energy to be promoted to the acceptor level, leaving behind the excess holes. Thus, the presence of the impurities increases the number of charge carriers, either electrons or holes, that are available.

To use a semiconductor as a radiation detector, the first approach might be to apply an electrical field to a block of n- or p-type semiconductor material and collect the charge carriers created when ionizing radiation passes

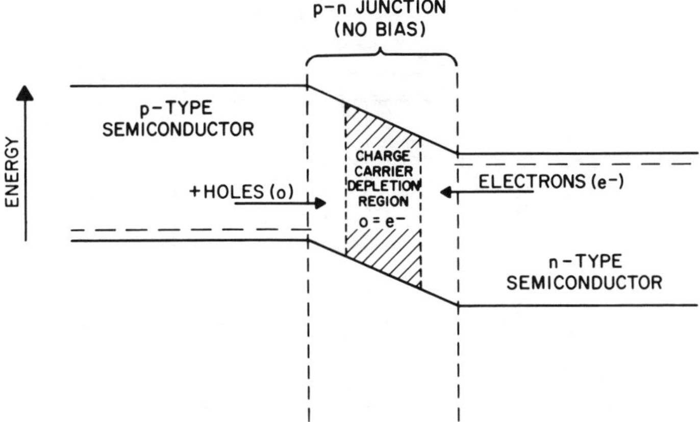

Figure 8.19. Generation of a charge-carrier depletion region in a semiconductor p–n junction with no applied bias.

through the crystal. However, for several reasons, this is not a practical way to use a semiconductor for radiation detection. The designs that have been found to be most useful for radiation detectors are based on the effects that occur when an n-type and a p-type semiconductor are brought into contact, forming a **p–n junction.**

A diagram of a p–n junction is shown in Fig. 8.19. If the two types of semiconductor can be brought into intimate contact, movement of carriers in the immediate vicinity of the p–n junction will occur. Electrons will migrate toward the p zone, and holes toward the n zone. This will create an area where there are neither excess holes nor excess electrons, and thus a depletion of charge carriers. This is called a **depletion region,** and it strongly resists the passage of electric current. If radiation strikes the depletion region, charge carriers are created and are attracted toward the n or p side. Therefore, a depletion zone can serve as a useful radiation detector.

A p–n junction created this way, without application of any external voltage, does not work very well as a detector. The depletion zone is very small, so most types of radiation would pass right through it undetected. In addition, the difference in electrical potential that is naturally created between the p and n sides is not large enough to collect all the charge carriers that form in the depletion zone. Therefore, a more useful detector results if a **reverse bias** is applied to the p–n junction, as diagrammed in Fig. 8.20. Application of a positive voltage to the n side results in the movement of the majority carriers (electrons) away from the depletion zone, while the negative voltage applied to the p side causes the holes to move away from the depletion zone. Thus, the application of a reverse bias intensifies the

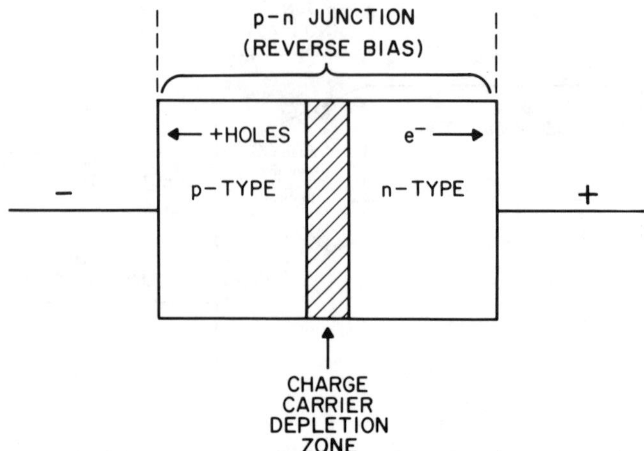

Figure 8.20. Generation of a large charge-carrier depletion zone in a semiconductor p–n junction by applying reverse bias.

situation that developed spontaneously in the p–n junction. As the electrons and holes both move farther away from the junction, the size of the depletion region increases and its resistance to the passage of electricity increases even more. All of the semiconductor detectors are based on the properties of these reverse-biased p–n junctions. It is not practical to create a p–n junction by physically joining the two types of semiconductor, because the spaces left between the p and n types are too large. Therefore, the p–n junction must be created within a single crystal. The following sections describe three kinds of semiconductor detectors that differ primarily in the way in which the p–n junction is formed.

8.4.2. Surface Barrier Detectors

A diagram of a **surface barrier detector** is shown in Fig. 8.21. The base material for these detectors is n-type silicon. The top surface is chemically oxidized to produce a p-type region on the surface. A thin layer of gold is then applied, to provide an electrical contact. Application of a potential of tens of volts creates a small depletion zone with a maximum depth of 2 or 3 mm. There is only a very thin region above the depletion zone that is not responsive to radiation, so these detectors have a small **dead zone.**

Because of their small size and dead zone, the surface barrier detectors are best used for measuring α particles or other heavy charged particles with short ranges. They are not very expensive and have good energy resolution. However, they are not very durable and are not applicable to detection of other types of radiation.

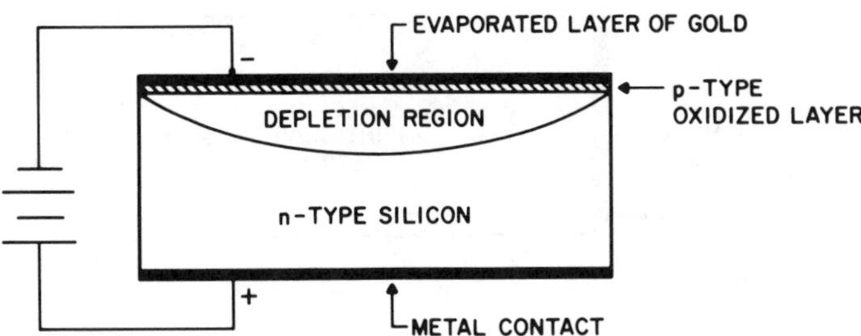

Figure 8.21. Schematic diagram of a surface barrier semiconductor charged-particle detector.

8.4.3. Lithium-Drifted Semiconductor Detectors

It is not possible to use a reverse-bias approach to create depletion zones that are large enough to detect γ rays efficiently. The applied voltage can only be increased to a certain value before the system breaks down. Therefore, an entirely different approach, called **ion drifting,** is used. The principle is that ions opposite in charge to the majority carriers of the semiconductor are made to enter the semiconductor. The numbers of these ions must be sufficient to just equal the numbers of majority carriers. This process is called **compensation**. Compensated semiconductor material is similar to intrinsic material in that there is no excess of either carrier. Therefore, the compensated region, like a depletion zone, can act as a radiation detector.

When Ge and Si are manufactured, both tend to end up as p-type materials. To create a compensated region in a piece of Ge or Si, atoms that will donate electrons and act as n-type material must be added. Lithium is the atom used for this purpose because of its small size and high mobility.

To make the compensation region, lithium metal is coated onto one end of a piece of p-type Ge or Si, and allowed to diffuse into the crystal at elevated temperatures over a period of days or weeks. The electrons donated by the lithium will exactly compensate for the acceptor impurities already present in the Ge or Si, and a depletion zone some 10–15 mm long can be created. Although it is not strictly correct to do so, the depletion zone is sometimes called the intrinsic region. This process is illustrated in Fig. 8.22. Once the lithium-drifting process is completed, the crystal is cooled to prevent the lithium from drifting back out again. The lithium-drifted germanium detector is written as Ge(Li) and is pronounced like the word *jelly.* Similarly, the silicon detectors are written Si(Li) and are pronounced like the word *silly.*

The Ge(Li) and Si(Li) detectors are quite different in their applications. Germanium is more desirable than Si for γ-ray spectrometry applications,

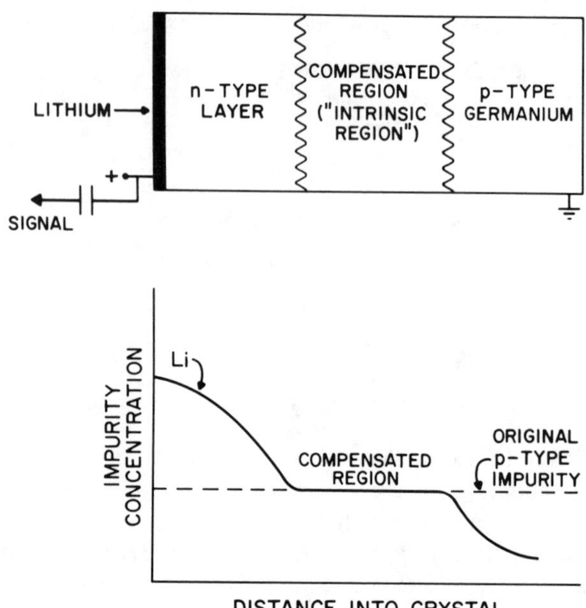

Figure 8.22. Illustration of lithium drifting to form a charge-carrier compensated region in a germanium semiconductor crystal. The lithium concentration as a function of distance is also shown. Note that the lithium concentration in the compensated region is constant.

because its higher Z value makes γ-ray interactions more efficient. Silicon, on the other hand, is more efficient for detection of lower-energy γ rays, x rays, and α and β particles. The difference in the band gap for Ge and Si results in differences in the way the crystals must be used and stored. The smaller band gap in Ge results in more spontaneous electron excitations into the conduction band, and Ge has more noise problems than does Si. This, coupled with the higher mobility of Li in Ge than in Si, means that a Ge(Li) detector must be kept constantly cooled to liquid nitrogen temperatures (77 K). Even one warming cycle is usually enough to destroy the lithium drifting and ruin a Ge(Li) detector. The Si(Li) detectors, on the other hand, are cooled to liquid nitrogen temperatures only when they are in use. They may be stored at room temperature. Because of the need for liquid nitrogen cooling during operation, both Si(Li) and Ge(Li) detectors are attached to large cryostats (see Fig. 8.23), which makes them bulky and difficult to maneuver in some instances.

The semiconductor γ-ray detectors are usually characterized by their resolution, efficiency, and peak-to-Compton ratios. The resolution is given as the value of the FWHM (in keV) for a given γ-ray full-energy peak.

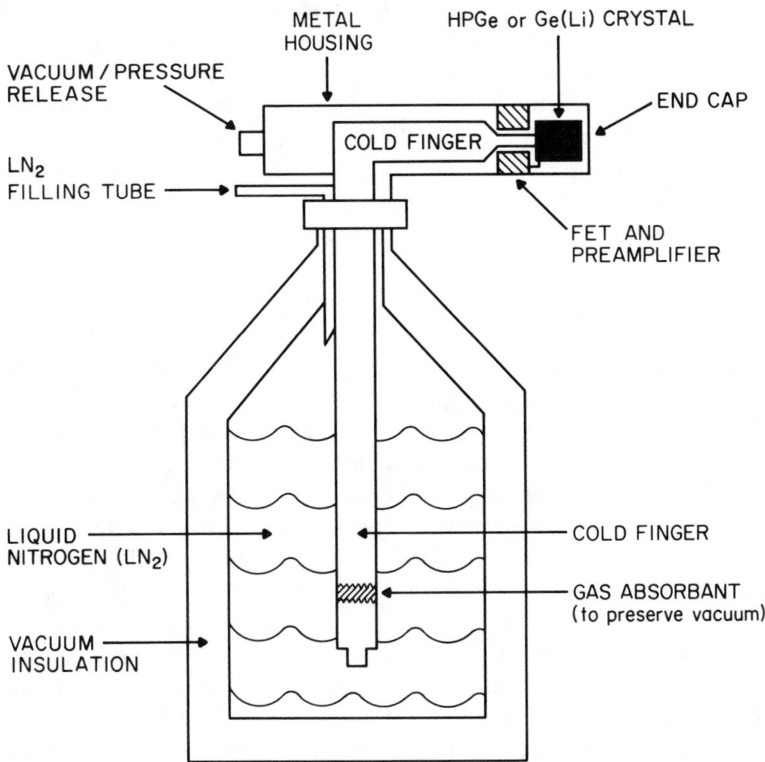

Figure 8.23. Schematic diagram of a HPGe or Ge(Li) detector system with liquid nitrogen Dewar.

Resolutions down to ≈1.5 keV at the 1332 keV ^{60}Co peak can be achieved with the best detectors. The efficiency of a semiconductor detector is usually given as a **relative efficiency,** that is, the efficiency compared to a standard detector. The standard is taken to be a cylindrical 3-in.-diameter by 3-in.-long NaI(Tl) detector with a point source of activity at a distance of 25 cm from the detector. For Ge(Li) detectors, relative efficiencies up to ≈15% with resolutions better than 2.0 keV at ^{60}Co are available for a cost of less than $10 000. The **peak-to-Compton ratio** refers to the ratio of the number of counts in the 1332 keV ^{60}Co peak compared to the number of counts in a selected region of the Compton continuum. Values up to 50 are common for this parameter, with the highest values being preferred.

The advantages of using drifted semiconductor detectors include their good energy resolution and their reasonable efficiency for γ-ray or high-energy particulate radiation. Lithium-drifted silicon detectors can be made with thin windows for α detection. All semiconductor detectors have low dead times and are relatively insensitive to magnetic fields. Their disadvantages include the necessity for liquid nitrogen cooling, the small size of the output signal, which entails the use of more electronic components for amplification, and their greater cost compared to solid-crystal scintillation detectors for γ-ray spectrometry.

8.4.4. Intrinsic Germanium Detectors

In recent years it has become possible to produce Ge in sufficient purity (less than 10^{10} impurity atoms per cm^3) so it approaches the properties of the theoretical true intrinsic (pure) semiconductor material. It can therefore be used as a detector without any lithium drifting. These Ge detectors without any Li are called **high-purity germanium (HPGe),** or **intrinsic germanium** (correctly *near*-intrinsic) detectors. Their great advantage is that they can be stored at room temperature, because there is no Li to drift out of the crystal. They still must be cooled to liquid nitrogen temperatures for operation, because the problem of thermal excitation of electrons still exists. Intrinsic Ge detectors are similar to Ge(Li) detectors in their efficiency and resolution characteristics, but are in some cases even better. Some detectors have been reported to have *relative* efficiencies greater than 100%, but costs can approach \$1000 for each 1% relative efficiency achieved. Other than being able to be stored at room temperature when not in use, intrinsic Ge detectors share most of the advantages and disadvantages of Ge(Li) detectors. They have largely replaced the Ge(Li) detector in the current market.

An intrinsic Ge detector with its large cryostat and shield installed in a laboratory configuration is shown in Fig. 8.24. A portable instrinsic Ge detector with a small cryostat that is useful for field work is shown in Fig. 8.25.

8.5. OTHER COMPONENTS OF ELECTRONIC DETECTOR SYSTEMS

Electronic detector systems based on voltage pulse measurement require many other components besides the detector. A block diagram of some of these is shown in Fig. 8.26. The upper dashed box after the amplifier shows the components for a single channel analyzer, while the bottom dashed box is for a multichannel pulse-height analyzer.

All of the detectors discussed in previous sections require the application of a voltage, often called the **detector bias.** Therefore, a high-voltage

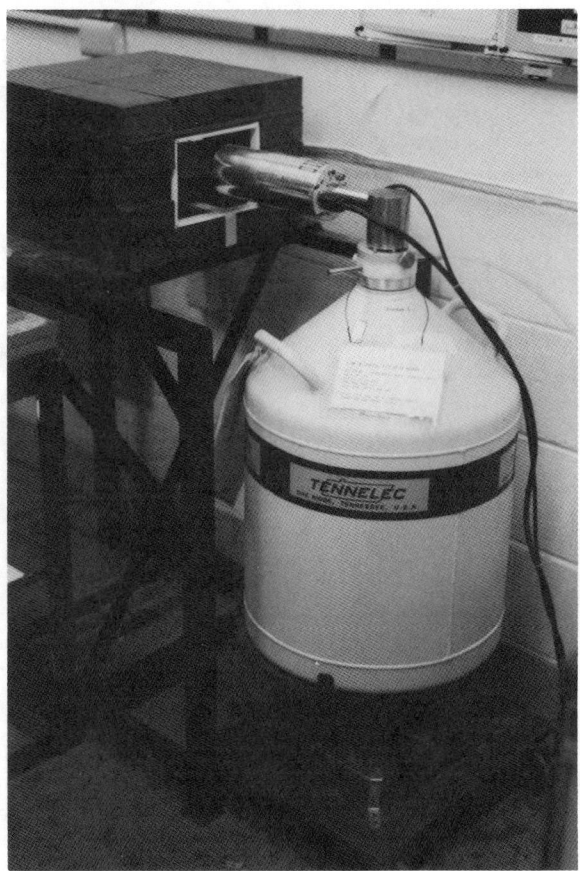

Figure 8.24. A HPGe detector system with Dewar, preamplifier, and counting shield in a laboratory configuration for γ spectrometry. A graded shield (see Section 6.4.4) is used to minimize interferences from lead x rays generated in the outer shield by γ rays from the source and the generation of bremmstrahlung radiation by β particles from the source interacting with the surface of the shield cavity.

supply, or **detector bias supply,** is an essential component of the counting system. The high-voltage supply should be able to deliver a range of stable dc voltages to meet the requirements of the different detectors.

For detectors that innately generate small amplitude pulses, a **preamplifier** is essential. This device is located very close to the detector to avoid loss of signal and interferences. It serves to maximize the signal-to-noise ratio from the detector and to provide preliminary amplification and pulse-shaping of the small detector signal. The **amplifier** accepts the pulse from

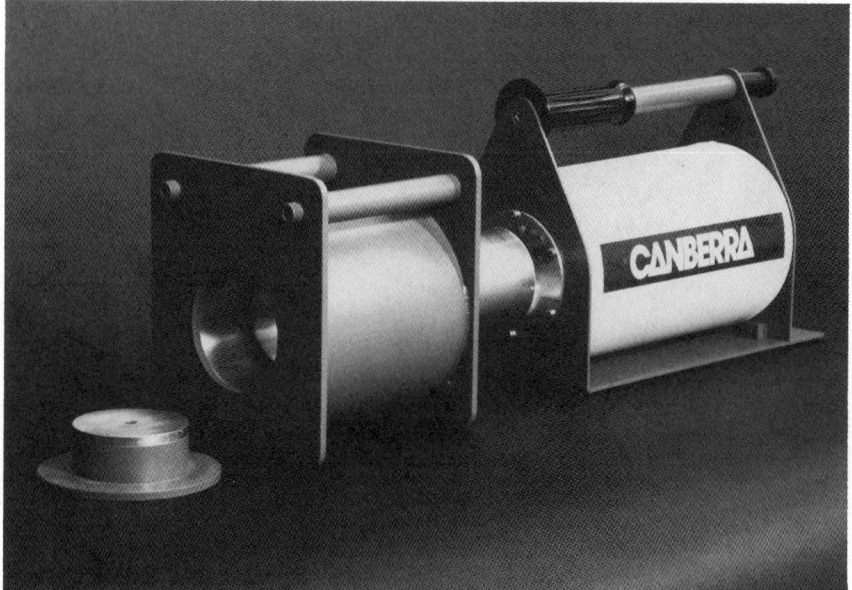

Figure 8.25. A portable HPGe detector with a small liquid nitrogen cryostat that is suitable for field work. (Courtesy of Canberra Detector Products Group, Meriden, CT.)

the preamplifier and provides further amplification. It also further shapes the pulse for improved processing of the signal by the remaining components of the signal processing and data storage system. The height (or amplitude) of the pulse that comes from the amplifier is related to the energy of the radiation producing the pulse.

If radiations of more than one energy are striking the detector, pulses of different amplitudes will be produced. If radiation of only one particular energy is needed, the pulses corresponding to that energy can be selected with a **discriminator**. This is an electronic circuit that can be set to accept pulses above, or between, certain preset amplitudes. A lower level discriminator (LLD) sets the lower amplitude limit, while an **upper level discriminator (ULD)** sets the upper limit. This is illustrated in Fig. 8.27. This figure shows five pulses of differing amplitudes. Pulses 2 and 5 are the only ones that will be accepted by the system, because they fall within the limits set by the LLD and ULD. Pulses 1 and 4 are too low in amplitude, and pulse 3 is too large in amplitude to be accepted. A simple discrimination device that allows both the LLD and the ULD to be preset is called a **single-channel analyzer (SCA).** The SCA can be connected to a **scaler** (a

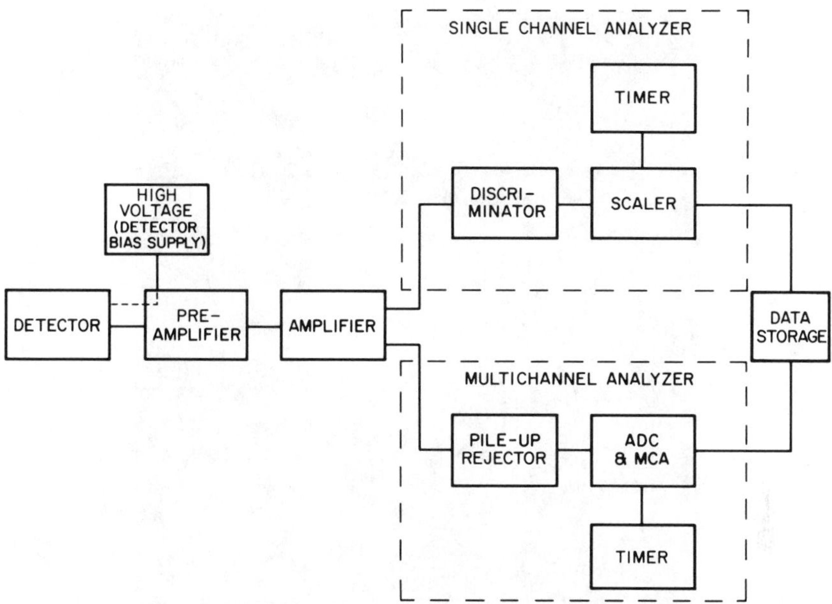

Figure 8.26. Schematic diagram of an electronic counting system configured for pulse analysis. The amplifier output may be routed to a single-channel analyzer (SCA) and scaler to count a single energy radiation (*top*), or to an analog-to-digital converter (ADC) and a multichannel analyzer/computer (MCA) to obtain an energy spectrum (*bottom*).

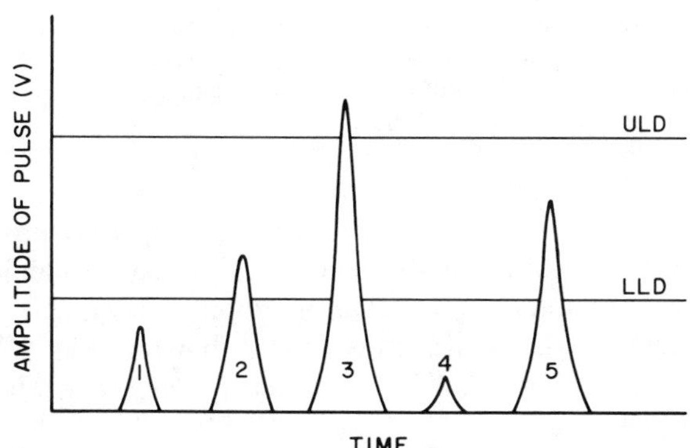

Figure 8.27. Utilization of a single-channel analyzer (SCA) for energy discrimination.

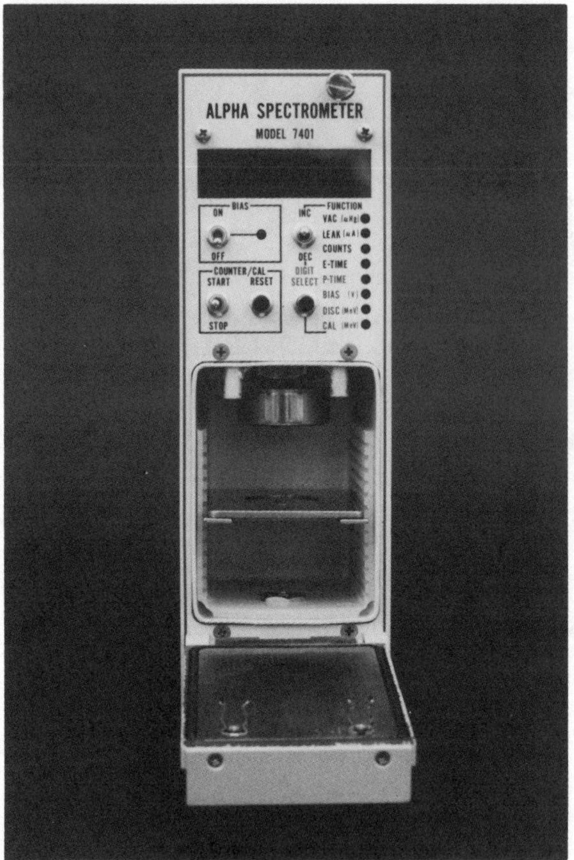

Figure 8.28. A self-contained α-particle spectrometry module in a NIM configuration. The unit contains a vacuum chamber for the sample and surface-barrier detector, bias supply, preamp/amplifier, pulser, discriminator, counter, and digital display. (Courtesy of Canberra Detector Products Group, Meriden, CT.)

device that counts signals received, also called simply a **counter**), so that the number of pulses that pass through the SCA may be recorded. A self-contained α-spectrometer module that incorporates a vacuum chamber for the α source and a surface barrier detector is shown in Fig. 8.28. The unit includes a bias supply, preamplifier/amplifier, pulser, discriminator, scaler, and digital display.

In many spectrometry applications, it is desirable to collect data from all the radiations of differing energies that are emitted by one source. It is then desirable to sort the pulses according to amplitude (and thus according to energy) as they come from the detector, and to record the number of each different amplitude pulse that is received. This task is accomplished by an **analog-to-digital converter (ADC).** The ADC takes analog signals from the amplifier and converts them into digital data for storage in a **multichannel analyzer (MCA).** Each incoming pulse is sorted into an appropriate **channel** (a storage location in a computer memory), which corresponds to a specific incremental range of incident particle energies, and the number of events in each channel is tallied. The information processed by an MCA can be displayed on a computer terminal. The horizontal axis of the display gives the channel number, or equivalent energy, of the radiation, and the vertical axis is the number of pulses or "counts" received for each energy increment. The appearance of a γ-ray spectrum on an MCA was discussed in detail in Chapter 6. Modern spectroscopy systems are almost always computer controlled and the spectra from the MCA are recorded in the computer for future processing with appropriate software. MCA boards are now available that plug directly into today's powerful microcomputers (PCs) and convert them into efficient and relatively inexpensive nuclear spectrometry systems.

All the components of the detector system are linked to one another by appropriate **cables**. Cables used for high dc voltages, low dc voltages, and signals differ from one another and must be properly selected and matched with the circuits they connect. The detector can be placed at some distance from most of the remainder of the counting system, if required by the experiment. The preamplifier, however, is normally attached directly to the detector housing for reasons stated earlier. The high-voltage supply, the amplifier, and the SCA or the ADC are separate modules that fit into special housings. The most popular type of module housing also provides the low-voltage power for these components and is called a **NIM bin. NIM** (Nuclear Instrument Module) and **CAMAC** (Computer Automated Measurement and Control) are international systems of nuclear electronics that allow for maximum compatibility of components. This is achieved through common power supply requirements, connector configurations, pulse shape requirements, and the physical dimensions of modules. The CAMAC system is oriented toward large mainframe computer operations and is much more costly than the more widely used NIM systems. Multichannel analyzers can be stand-alone minicomputer-driven units with a variety of sophisticated data storage capabilities, or simple boards in PCs. A photograph of a typical stand-alone MCA system is shown in Fig. 8.29. In some facilities ADC outputs are fed directly into large mainframe computer systems for processing.

This is by no means an exhaustive list of electronic components that are used in radiation detection systems. Reference should be made to the

Figure 8.29. A multichannel analyzer/computer system (MCA) and associated NIM bin modules in operation as a γ-ray spectrometer. A spectrum is visible on the MCA display screen.

several excellent books in the field of nuclear instrumentation for more detailed information.

8.6. NONELECTRONIC DETECTION SYSTEMS

The bulk of the work in radiochemistry applications is done with one of the electronic systems described in Sections 8.2 through 8.4. The other detection systems discussed briefly here are useful in more specialized situations.

8.6.1. Photographic Plates

Photographic plates were the first type of radiation detector, as used by Becquerel in 1895. Incoming radiation acts on the photographic emulsion just as light does and sensitizes the silver halide grains. Development of the emulsion produces dark areas where radiation interactions have occurred.

In **autoradiography**, a section of material containing radionuclides is placed on a photographic emulsion. The darkened areas that appear after development of the film will produce an image of the object. Autoradiography has been widely used in biological studies to show how a radiotracer is distributed in an organism. Film badges used in health protection to monitor radiation exposure are another way photographic detectors are used. These badges can be quite versatile in measuring different radiations by varying types of filters placed on either side of the film. For example, boron-doped filters or emulsions can be used to detect neutrons by exchanging the neutral, nonionizing neutrons for ionizing α particles via the $^{10}B(n, \alpha)^{7}Li$ reaction.

Photographic emulsions can also be useful to detect tracks of single nuclear particles. For this purpose, a special **nuclear emulsion** is used. It is thicker than the usual photographic emulsions and contains larger amounts of silver halides. As a particle passes through the emulsion, it will leave a track of activated silver halide grains which can be made visible as silver deposits after photographic development. The density (darkness) of the track gives information about the charge on the particle, and the length of the track can provide particle energy information. The direction of deflection of the track, if the plate had been placed in a magnetic or electrical field during data accumulation, can provide the sign of the charge on the particle. This method is still used for cosmic-ray studies.

8.6.2. Chemical Detectors

Chemical detectors are usually used as "disaster dosimeters," where extremely high levels (kCi or more) of γ-ray emitters are present. These detectors rely on the production of free radicals by interactions of radiations with a chemical system. The free radicals induce chemical reactions that may be monitored using conventional chemical techniques. An example is the ferrous sulfate dosimeter. The reactions are

$$\gamma + H_2O \longrightarrow OH^{\cdot} + H^{\cdot}$$

$$H^{\cdot} + H^{\cdot} \longrightarrow H_2(g)$$

$$OH^{\cdot} + Fe^{2+} \longrightarrow Fe^{3+} + OH^{-}$$

$$Fe^{3+} + SCN^{-} \longrightarrow Fe(SCN)^{2+}$$

The $Fe(SCN)^{2+}$ is a red complex that may be detected spectrophotometrically. The system may be calibrated using known radiation doses. To enhance stability, the solutions should be kept in oxygen-free, closed containers.

8.6.3. Calorimetric Detectors

This is a detector system used for very high levels (kCi) of α emitters, or radionuclides undergoing spontaneous fission. The source is placed in a sensitive calorimeter in which all of the energy of the radiations will be completely dissipated. The energy deposited by the emitted α particles or fission fragments eventually shows up as an increase in the temperature of the system. The change in the temperature of the calorimeter with time is used to indicate the activity.

8.6.4. Cloud and Bubble Chambers

The cloud chamber was first used by C.T.R. Wilson in the early 1900s to detect nuclear particles. A diagram of a cloud chamber is shown in Fig. 8.30. A volatile material such as ethanol is added to the chamber, which is then cooled with dry ice. This results in the formation of a supersaturated vapor. As radiation passes through the chamber, ion pairs form in the gas. The supersaturated vapor condenses out as droplets along the path of ion formation, thus marking the particle track. The process is similar to the formation of rain droplets on dust particles in the atmosphere. Photographic and visual observations can be made of the track. As with the photographic plate, track density is related to the particle charge, and path length to particle energy. In Chapter 1, it was noted that Anderson discovered the positron using a cloud chamber. Cloud chambers can be easily set up for classroom demonstrations, but they are no longer used for research, having been supplanted by the bubble chamber.

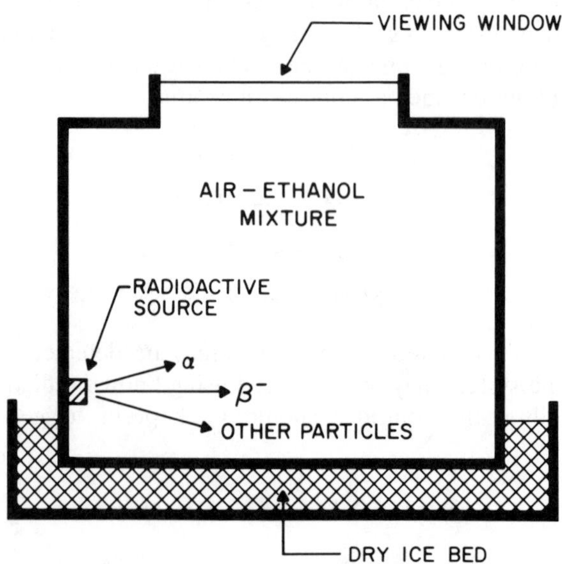

Figure 8.30. Schematic diagram of a Wilson cloud chamber.

The operating principle of the bubble chamber is similar to that of the cloud chamber, except that the bubble chamber contains a superheated liquid (i.e., one that has been heated slightly past its boiling point, but is not yet boiling). The passage of radiation through the liquid forms ion pairs, and bubbles form along the path of the ions to mark the track. The process is similar to the effect obtained by dropping a grain of salt into pure water heated slowly to just above its boiling point. Both cloud and bubble chambers are used primarily by nuclear physicists interested in studies of cosmic rays or subatomic particles. They do not have analytical applications.

8.6.5. Thermoluminescence Detectors (TLDs)

The exposure of certain inorganic crystals to radiation results in excitation of electrons from the valence to the conduction band. This process is familiar from the discussion of scintillation detectors in Section 8.3. Crystals used in thermoluminescence dosimetry, however, have the property of "trapping" the electrons, so light emission does not occur immediately. If the crystals are heated at some time after their radiation exposure, sufficient energy is provided to release the electrons from the traps, and de-excitation occurs with emission of photons. The intensity of the light emitted is proportional to the dose of radiation received by the crystal.

The type of material chosen for a TLD depends on what kind of radiation is to be detected and the time before detection is to occur. TLD crystals can respond to x rays, γ particles, β particles, and protons over a wide range of exposure intensities. Lithium fluoride is a favorite choice. It is reasonably sensitive, yet also stable enough so it does not re-emit energy too easily. Its effective atomic number is also close to that of biological tissue (8.1 vs. 7.4 for biological tissue). A $CaSO_4$:Mn (where the Mn is an activating impurity) crystal is very radiation sensitive, but does not hold the trapped electrons very well, so **fading** is a problem for this material. On the other hand, CaF_2:Mn does not fade, but is not very sensitive to radiation. Selection of the appropriate crystal is dictated by the needs of the situation.

8.7. SPECIAL NEUTRON DETECTORS

Neutrons will not be detected efficiently by any of the basic detectors discussed in preceding sections, because neutrons interact with matter via scattering and nuclear reactions, rather than by atomic ionization or excitation (see Section 6.5). However, the products of neutron-induced reactions are often charged particles or γ rays, which *can* be detected with the instruments we have discussed. Therefore, modifications of the detector systems are needed for neutron detection.

8.7.1. Thermal Neutron Detectors

Thermal neutrons do not undergo significant scattering reactions, so their nuclear reactions that form charged particles are the basis for their detection.

The thermal neutron interaction most often used for its detection is the $^{10}B(n, \alpha)^{7}Li$ reaction. It is actually the α particles formed in the reaction that are detected. Ionization chambers and proportional counters may be adapted for neutron detection by using $^{10}BF_3$ as the fill gas in the chamber, or by lining the chamber with a material containing ^{10}B. The use of a boron-containing gas is preferred, because it is difficult to make the solid lining material thin enough for the α particles to escape after they are produced. The $^{3}He(n, p)^{3}H$ reaction can also be used in ionization chambers and proportional counters. ^{3}He is used as the gas, and the proton is the particle detected. ^{3}He gas is expensive, so this approach is not as widely used as the ^{10}B reaction. Another possible reaction appropriate for ionization chambers and proportional counters is the neutron-induced fission of ^{235}U. The chamber can be lined with this (or another) fissile material, and the fission products detected.

Another useful reaction with a high cross section for thermal neutron interaction is the $^{6}Li(n, \alpha)^{3}H$ reaction. This is not used in ionization chambers or proportional counters because no suitable gas exists, but it is possible to use this reaction conveniently with scintillation detectors. Lithium iodide crystals, analagous to NaI crystals, can serve as both the neutron detector and the scintillating medium. This method gives good neutron detection efficiency because the detecting medium is a moderately high-density solid, rather than a gas. Lithium compounds may also be dispersed in other scintillating media, such as ZnS(Ag). Light produced by the secondary α particles interacting with the ZnS scintillator is measured by use of a PMT.

The radiative capture reactions (n, γ) are rarely used for simple neutron detection. The $^{113}Cd(n, \gamma)^{114}Cd$ and $^{115}In(n, \gamma)^{116}In$ reactions both have quite high cross sections, and can be used to determine neutron fluxes from a nuclear reactor, or other neutron sources, by methods discussed in Sections 4.3.1 and 4.3.2.

8.7.2. Fast Neutron Detectors

Fast neutrons undergo nuclear reactions to produce charged particles, as do the slow neutrons. However, for fast neutrons, scattering is also an important means of interaction and can be used for detection.

The detection schemes for slow neutrons discussed in the preceding section can all be used for fast neutron detection if a suitable way to moderate the fast neutrons can be found. The thermalization process involves the collisions of the fast neutrons with moderating material and this requires time. Thus, this method is not appropriate where fast response is needed, nor is it good when information regarding neutron energy is important.

Scintillation detection is often used to detect the recoil protons produced by fast neutron collisions with hydrogen. Many of the scintillation materials are organic, and therefore contain a large amount of hydrogen. Inorganic scintillators like ZnS can act as neutron detectors when they are embedded in a hydrogenous medium, such as a plastic. This type of fast neutron detector is called a **Hornyak button.**

All of these neutron detectors simply detect the presence of neutrons, but do not give much energy information beyond "thermal" and "fast" classifications. Determination of neutron energies requires more sophisticated techniques.

TERMS TO KNOW

Absolute detector efficiency
Activators
ADC
Amplifier
Autoradiography
Band gap
BF_3 neutron counter
BGO detector
Bubble chambers
Calorimetric detectors
Chemical detectors
Chemical dosimeter
Chemical reaction detectors
Cloud chambers
Compensation
Conduction band in scintillators
CsI(Tl) detector
Dark current
Dc ion chamber
Dead time
Depletion region
Detector bias
Detector energy resolution
Dynodes
Electron–hole pair
Electron multiplication
Electronic detectors
Electronic discriminator
Fluorescence
FWHM
G-M counters
Gain
Gas-filled counters
Hornyak button
Intrinsic detector efficiency
Intrinsic Ge detector (HPGe)
Ionization chambers
Light pipe
Li(Eu) detector
Lithium-drifted detector
Lower level discriminator
Majority carriers
Multichannel analyzer (MCA)
N-type semiconductor
NaI(Tl) detector
Nuclear emulsion
p–n junction
P-type semiconductor
Peak-to-Compton ratio
Phosphorescence
Photocathode
Photoelectrons
Photographic plates
Photomultiplier tube
Preamplifier
Proportional counters
Quenching agent
Recombination region
Region of limited proportionality
Relative efficiency

Reverse bias
Saturation region
Scintillation cocktail
Scintillation counters
Scintillators
Semiconductor detectors
Single-channel analyzer (SCA)
Surface barrier detector
Thermionic emission
Thermoluminescence detectors
Townsend avalanche
Upper level discriminator
Valence band in scintillators
Vibrating reed electrometer
W value
Wavelength shifter
Well crystal

READING LIST

Bertolini, G., and A. Coche, eds., *Semiconductor Detectors.* New York: Wiley, 1968. [semiconductor detectors]

Birks, J. B., *The Theory and Practice of Scintillation Counting.* New York: Macmillan, 1964. [scintillation counting]

Cooper, P. N., *Introduction to Nuclear Radiation Detectors.* Cambridge: Cambridge University Press, 1986. [a brief introduction to commonly used detectors]

Eichholz, G. G., and J.W. Poston, *Principles of Nuclear Radiation Detection.* Ann Arbor, Ann Arbor Science, 1980. [general, detectors]

Fernow, R. C., *Introduction to Experimental Particle Physics.* Cambridge: Cambridge University Press, 1986. [general, detectors]

Fox, B.W., *Techniques of Sample Preparation for Liquid Scintillation Counting.* Amsterdam: North-Holland, 1976. [liquid-scintillation counting]

Horrocks, D. L., *Applications of Liquid Scintillation Counting.* New York: Academic, 1974. [liquid scintillation counting]

International Atomic Energy Agency (IAEA), *Neutron Detectors.* STI/PUB/21/18, IAEA, Vienna, 1966. [neutron detectors]

Knoll, G. F., Some recent developments in charged-particle and gamma-ray detectors. *Nucl. Instrum. Methods Phys. Res.* **B24/25**:1021–1027 (1987). [new instrumentation]

Knoll, G. F., *Radiation Detection and Measurement,* 2nd ed. New York: Wiley, 1989. [all aspects of radiation detection]

Krugers, J., ed., *Instrumentation in Applied Nuclear Chemistry.* New York: Plenum. 1973. [theory and applications]

Miller, G. D., *Radioactivity and Radiation Detection.* New York: Gordon & Breach Science, 1972. [general, detectors]

Nicholson, P.W., *Nuclear Electronics.* New York: Wiley-Interscience, 1974. [general, detectors]

Noujaim, A. A., C. Ediss, and L. I. Wiebe, *Liquid Scintillation Science and Technology.* New York:Academic, 1971. [liquid-scintillation counting]

Wilkinson, D. H., *Ionization Chambers and Counters.* Cambridge: Cambridge University Press, 1950. [ionization chambers]

Note: The journal *Nuclear Instruments and Methods in Physics Research* publishes papers on many of the new advances in radiation detectors.

EXERCISES

1. A dc ion chamber is used to measure the activity of an air sample containing radon (Rn) gas. ^{222}Rn emits a 5.49-MeV α particle and has a half-life of 3.82 d. Assuming 100% counting efficiency and a W value of 34.5 eV, calculate the number of moles of ^{222}Rn in the chamber, if the ionization current is 4.10×10^{-12} A.

2. Below is a list of parameters of interest in γ-ray spectrometry with solid-crystal scintillation counters. Identify which are important in determining the absolute efficiency (ϵ_{abs}), the intrinsic efficiency (ϵ_{int}), and the energy resolution (R) of the scintillation counting system.

(a) density of the detector

(b) average atomic number of the scintillator

(c) geometry of the source–detector system

(d) gain of the PMT

(e) type of solid scintillator used

3. Which of the detectors discussed in this chapter would be best for the following situations:

(a) a high detection efficiency for γ rays is required

(b) detection of α particles with good energy resolution

(c) detection of low-energy β particles with good efficiency from gas samples

(d) inexpensive detection of a thermal neutron counting rate

4. Describe the advantages and disadvantages of proportional counters compared to Geiger–Müller (G-M) counters.

5. List an important application for each of the following types of detectors and state why it would be selected for that application.

(a) ferrous sulfate chemical dosimeter

(b) a liquid-scintillation counter

(c) an ionization chamber operating in the pulse mode

(d) an intrinsic (HPGe) detector

(e) a proportional counter with a thin interior layer of enriched ^{235}U

6. Describe the advantage of Cl_2 gas over an organic compound when used as a quenching agent in a G-M counter.

7. By reference to literature sources, describe methods that have been developed to detect neutrons and simultaneously measure their energies.

8. Thermoluminescence has been used as a dating method as well as for radiation dosimeters. By reference to the literature, describe the principles of thermoluminescence dating.

9. You wish to do an aerial survey to detect uranium deposits in the western United States and need to discriminate between the γ-ray signals produced by ^{40}K and those produced by members of the U/Th decay series. Which detection system discussed in this chapter would you select to install in your survey aircraft? Why?

10. Many of the counting systems described in this chapter will exhibit significant dead times with high-activity samples. Suggest methods that could be used to determine the dead time of a detector system and one or more methods to correct experimental counting rates for dead-time losses. Use of the literature listed in the Reading List is suggested.

11. In the use of a liquid-scintillation counter, the efficiency for counting ^{14}C is greater than for ^{3}H. How is this explained?

12. Can a liquid-scintillation counter be used to detect γ rays? Explain your answer.

13. Plot a typical curve of count rate against bias voltage for a G-M counter. Explain the features of the curve.

14. An aliquot of a solution containing a pure β-particle emitter is counted in a liquid-scintillation counter, while an equal aliquot is evaporated and counted with a thin window G-M counter. The observed background-corrected counting rates were 5835 ± 70 cpm and 32 030 ± 215 cpm. Which detector is associated with each counting rate and why do they differ?

15. How would you measure the flux density (neutrons cm^{-2} s^{-1}) at a position close to a ^{252}Cf isotopic neutron source?

16. Why should solid-state semiconductor detectors exhibit better energy resolution than gas-filled ionization chambers?

17. Recently an ad for an HPGe detector boasted of a *relative efficiency* for the detector greater than 100%. Can this claim be true?

18. **(a)** Calculate the resolution for a 1115.5-keV Zn peak that has a FWHM of 2.01 keV.

(b) Calculate the resolution for the same Zn peak that has a FWHM of 200 keV.

(c) Which of these peaks would have appeared on a spectrum taken with an HPGe detector and which with an NaI(Tl) detector?

19. If As were used as a dopant (impurity) for a semiconductor material, what type of semiconductor (p or n) would result?

20. Given the following components of a nuclear spectroscopy counting system, make a block diagram that shows their correct relative positions in the system.

high-voltage supply
ADC
computer storage of spectra
PMT
MCA
NaI(Tl) detector
amplifier

CHAPTER

9

NUCLEAR ACTIVATION ANALYSIS

9.1. PRINCIPLES OF ACTIVATION ANALYSIS

Activation analysis techniques as described in this chapter include those methods that use nuclear reactions to form product radionuclides or excited states of the target nucleus as the basis for the analytical determination. Figure 9.1 shows a general scheme for the activation process. An element is bombarded with either neutrons, charged particles, or photons. This induces a nuclear reaction, and an excited intermediate is formed. The intermediate may de-excite by emitting *prompt* γ *rays,* so named because they are typically emitted within less than 10^{-14} s after formation of the excited intermediate. The radioactive product then decays via α, β, γ, or delayed neutron emission processes. The γ rays emitted in this latter decay process are known as **delayed γ rays.** The emitted radiations or particles are monitored to obtain both qualitative and quantitative analytical information. If an incoming charged particle is used merely to induce x-ray emission via interactions with the electron shells of the target element, this nonnuclear excitation is the basis for an analytical method known as **particle-induced x-ray emission analysis (PIXE).** PIXE, although it is not a nuclear technique, does use equipment and procedures common to radiochemistry and is considered separately in Chapter 10.

An overview of the many forms of activation analysis and some of their advantages and disadvantages is given in the next section. A brief description of the sources of activating particles is also given. In succeeding sections, each technique is discussed in more detail.

9.1.1. Overview of Nuclear Activation Analysis Methods

Neutron Activation Analysis

Neutron activation analysis (NAA) is the most common form of activation analysis. In NAA, neutrons are the activating particle. Neutrons induce different kinds of nuclear reactions, depending on their energy. For most applications, it is the delayed γ radiation from the radioactive product that

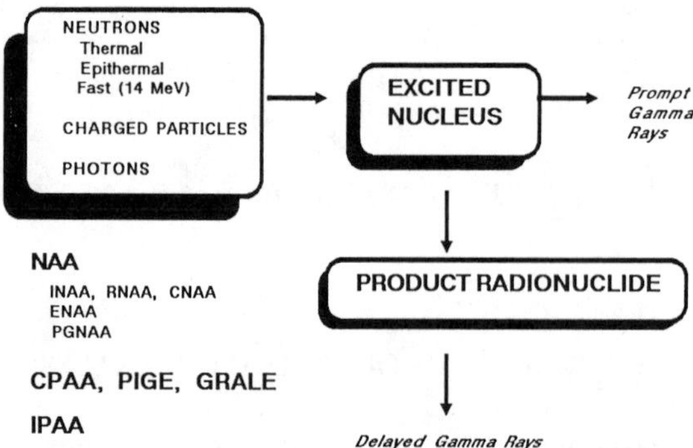

Figure 9.1. Representation of the general scheme for commonly used methods of nuclear activation analysis.

is detected after activation. A different form of NAA called **prompt-gamma neutron activation analysis (PGNAA)** is also used, though less frequently. In this latter procedure, it is the prompt γ rays emitted by the excited intermediate nucleus that are monitored.

Neutron activation techniques are usually divided into three categories based on whether any chemical separations are required in the procedure, and if so, whether they are done before or after the irradiation. If no chemical treatment is done, the process is called **instrumental neutron activation analysis (INAA).** If chemical separations are done after irradiation to remove interferences or to concentrate the radionuclide of interest, the technique is called **radiochemical neutron activation analysis (RNAA).** If preirradiation chemical separations are employed, the procedure is called **chemical neutron activation analysis (CNAA),** or if specific molecular components are determined, this technique may be called **molecular activation analysis.**

Another way to group NAA methods is according to the energy of the incoming neutrons. Thermal neutrons are the most commonly used activating particle or radiation. It has been noted in earlier chapters that thermal neutrons are also called slow neutrons, because their velocites are low ($\approx$2200 m/s) and their mean energy is only about 0.04 eV (the most probable energy is about 0.025 eV). Activation analysis employing these thermal neutrons is called **thermal neutron activation analysis (TNAA).** Neutrons with slightly higher energies (0.1–1 eV) are called **epithermal neutrons,** and serve as the activating particles for **epithermal neutron activation analysis (ENAA). Resonance neutrons** in the 1–eV to 1–keV energy region are often grouped with epithermal neutrons in discussions of ENAA. Any neutrons with energies greater than 0.5 MeV are called **fast neutrons** and are used in a technique called **fast neutron activation analysis (FNAA).** A common and particularly useful form of FNAA, called **14-MeV INAA,** is based on reactions with 14-MeV neutrons that are produced by small accelerators known as neutron generators.

Charged-Particle Activation Analysis

In **charged-particle activation analysis (CPAA),** a variety of charged particles is used to activate the analyte. The particles usually used include protons, deuterons, and α particles. The activating particles have energies in the MeV range, and delayed γ rays emitted by the product radionuclide are usually measured. CPAA is not used as widely as NAA, partly because there are fewer convenient sources of charged particles.

One variation of CPAA detects the prompt γ rays emitted by the excited intermediate in the activation process. This approach is called **particle-induced gamma-ray emission analysis (PIGE),** or sometimes **particle-induced prompt photon spectroscopy (PIPPS).** Because it is often used for the determination of low-Z elements such as Li, F, H, O, and Na, it may also be called **gamma-ray analysis of light elements (GRALE).**

Instrumental Photon Activation Analysis

In **instrumental photon activation analysis (IPAA),** energetic photons are used to activate the sample. The bremsstrahlung radiation resulting from the acceleration of electrons in a linear accelerator or a synchrotron light source is used as the photon source. This method is less widely used than either NAA or CPAA.

9.1.2. Advantages and Disadvantages of Nuclear Activation Methods

All of the techniques summarized above have unique strengths and weaknesses, which will be noted when each is discussed. However, some general observations apply to most of the nuclear activation methods.

Nuclear activation methods can provide *very high sensitivities* for many elements. Thermal INAA can determine two-thirds of the elements in the periodic table at 10^{-6} g/g, or below, as shown in Fig. 9.2. FNAA is less sensitive than thermal NAA, but can still determine O, N, and other light elements at concentrations down to 0.01%. Analyses for O and N are very easy to perform by FNAA, while they are rather lengthy and tedious determinations by more conventional methods. CPAA is even more sensitive than FNAA for the light elements. Concentrations of 10^{-6} g/g can be determined for many elements.

ACTIVATION ANALYSIS SENSITIVITIES

Sensitivities are expressed as the micrograms of the element that must be in the sample to be detected and determined by the Activation Analysis Service. THE SENSITIVITIES OF MANY ELEMENTS CAN BE INCREASED UP TO 100 - FOLD. Sensitivities are for interference-free conditions.* For more information, contact Lawrence E. Kovar.

1 H Hydrogen NA 1.00797																	2 He Helium NA 4.0026
3 Li Lithium NA 6.939	4 Be Beryllium NA 9.0122											5 B Boron NA 10.811	6 C Carbon NA 12.01115	7 N Nitrogen NA 14.0067	8 O Oxygen NA 15.9994	9 F Fluorine 0.4 18.9984	10 Ne Neon 2. 20.183
11 Na Sodium 0.004 22.9877	12 Mg Magnesium 0.5 24.305											13 Al Aluminum 0.004 26.98154	14 Si Silicon I.fs 28.0855	15 P Phosphorus NA 30.97376	16 S Sulfur NA 32.064	17 Cl Chlorine 0.05 35.453	18 Ar Argon 0.002 39.948
19 K Potassium 0.2 39.0983	20 Ca Calcium 4. 40.08	21 Sc Scandium 0.001 44.9559	22 Ti Titanium 0.1 47.90	23 V Vanadium 0.002 50.942	24 Cr Chromium 0.3 51.996	25 Mn Manganese 0.0001 54.9380	26 Fe Iron 2.fs 55.847	27 Co Cobalt 0.01 58.9332	28 Ni Nickel 0.7 58.71	29 Cu Copper 0.002 63.546	30 Zn Zinc 0.1 65.38	31 Ga Gallium 0.002 69.72	32 Ge Germanium 0.1 72.59	33 As Arsenic 0.005 74.9216	34 Se Selenium 0.01 78.96	35 Br Bromine 0.003 79.904	36 Kr Krypton 0.01 83.80
37 Rb Rubidium 0.02 85.467	38 Sr Strontium 0.005 87.62	39 Y Yttrium 0.4 88.9059	40 Zr Zirconium 0.8 91.22	41 Nb Niobium 3. 92.9064	42 Mo Molybdenum 0.1 95.94	43 Tc Technetium NA 98.9062	44 Ru Ruthenium 0.04 101.07	45 Rh Rhodium 0.005 102.9055	46 Pd Palladium 0.03 106.4	47 Ag Silver 0.004 107.868	48 Cd Cadmium 0.005 112.41	49 In Indium 0.00006 114.82	50 Sn Tin 0.03 118.69	51 Sb Antimony 0.007 121.75	52 Te Tellurium 0.03 127.60	53 I Iodine 0.002 126.9045	54 Xe Xenon 0.1 131.30
55 Cs Cesium 0.001 132.9054	56 Ba Barium 0.02 137.33	57 Ⓛ La Lanthanum 0.005 138.9055	72 Hf Hafnium 0.0006 178.49	73 Ta Tantalum 0.1 180.948	74 W Tungsten-Wolfram 0.004 183.85	75 Re Rhenium 0.0008 186.2	76 Os Osmium 1. 190.2	77 Ir Iridium 0.0003 192.22	78 Pt Platinum 0.1 195.09	79 Au Gold 0.0005 196.9665	80 Hg Mercury 0.003 200.59	81 Tl Thallium NA 204.37	82 Pb Lead NA 207.19	83 Bi Bismuth NA 208.9808	84 Po Polonium NA (209)	85 At Astatine NA (210)	86 Rn Radon NA (222)
87 Fr Francium NA (223)	88 Ra Radium NA 226.0254	89 Ⓐ Ac Actinium NA (227)	104 (Rf) (Rutherfordium) NA (261)	105 (Ha) (Hahnium) NA (262)	106 (263)												

* Interference-free implies that only the element of interest will become radioactive in the sample. If interferences exist, radiochemistry may be necessary for optimum sensitivity.

Ⓛ	58 Ce Cerium 0.2 140.12	59 Pr Praseodymium 0.03 140.9077	60 Nd Neodymium 0.03 144.24	61 Pm Promethium NA (145)	62 Sm Samarium 0.001 150.4	63 Eu Europium 0.0001 151.96	64 Gd Gadolinium 0.007 157.25	65 Tb Terbium 0.03 158.9254	66 Dy Dysprosium 0.00003 162.50	67 Ho Holmium 0.003 164.9304	68 Er Erbium 0.002 167.26	69 Tm Thulium 0.2 164.0342	70 Yb Ytterbium 0.02 173.04	71 Lu Lutetium 0.0003 174.97
Ⓐ	90 Th Thorium 0.2 232.0381	91 Pa Protactinium NA 231.0359	92 U Uranium 0.003 238.029	93 Np Neptunium NA 237.0482	94 Pu Plutonium NA (244)	95 Am Americium NA (243)	96 Cm Curium NA (247)	97 Bk Berkelium NA (247)	98 Cf Californium NA (251)	99 Es Einsteinium NA (254)	100 Fm Fermium NA (257)	101 Md Mendelevium NA (258)	102 No Nobelium NA (259)	103 (Lr) (Lawrencium) NA (257)

KEY

ATOMIC NUMBER → 33 As ← SYMBOL
Arsenic
0.005 ← SENSITIVITY IN MICROGRAMS (INTERFERENCE FREE)
74.9216 ↑ ATOMIC WEIGHT (CARBON - 12)

COUNTING BY GAMMA–RAY SPECTROMETRY, UNLESS OTHERWISE STATED

fs – FAST NEUTRONS, FISSION SPECTRUM

p – REACTOR PULSE

NA– ANALYSIS NOT NORMALLY PERFORMED

Figure 9.2. A periodic table of the elements with sensitivities for NAA in terms of micrograms of the element determined by γ-ray spectrometry under interference-free conditions. A thermal neutron flux density of at least 10^{13} n cm^{-2} s^{-1} is assumed. (Courtesy of Lawrence Kovar, General Activation Analysis Inc., San Diego, CA.)

A major advantage of nuclear activation methods for complex matrices is that the techniques can be made highly selective. There are *many adjustable experimental parameters* that can be exploited so that maximum sensitivity for the desired element is achieved. The type, energy, and flux of the irradiating particles or radiations may be varied. This allows for selective activation of certain elements, because not all elements are identically activated. The time of irradiation can be varied, resulting in enhanced degrees of activation of elements with either short or long half-life indicator radionuclides. The differences in half-lives among the activated species can also be exploited as a way to discriminate among elements in the counting process. If two interfering species were activated simultaneously and the desired species had a longer half-life than the interference, the interferent activity could simply be allowed to decay before counting the sample for the desired analyte indicator radionuclide.

The nuclear analytical methods are capable of *simultaneous multielement analysis.* Irradiation of a sample will induce activity in many elements. Use of a multichannel analyzer to record all the γ rays emitted by the sample permits the analyst to obtain information on many different elements. In recent years trace analysis (i.e., determination of substances at levels of 10^{-6} g/g, or less) has assumed increasing importance, due to medical and environmental concerns. In trace analytical work, contamination is an ever-present problem. Nuclear analytical methods often require much less sample manipulation than do other methods, so they are *relatively free of reagent and laboratory contamination.* Even in RNAA, where chemical processing is done after irradiation, contamination is not a factor, because any analyte inadvertently added during processing will not be active, and so it will not add to the measured sample activity.

Because nuclear reactions are the basis for activation methods, the information obtained by nuclear analysis is generally *independent of matrix or chemical form.* This may be either an advantage or a disadvantage, depending on the analytical information needed for a particular problem.

In some cases, activation analysis may be an essentially *nondestructive method.* That is, the sample may not be visually or measureably chemically altered. This is obviously a valuable characteristic for situations where extremely valuable samples must be analyzed, such as ancient coins or paintings. However, lengthy irradiations of some materials, such as biological tissue, can destroy the chemical nature and physical appearance of the sample.

Many nuclear activation methods are capable of *rapid analysis,* particularly those based on use of short-lived indicator radionuclides. An FNAA analysis for O, for example, can easily be accomplished in a minute, using a 14-MeV neutron generator with a rapid sample transfer system. Longer-lived indicator radionuclides often require a longer analysis time, due to the long delay times prior to counting (sometimes days or weeks) required to permit short-lived interfering radionuclides to decay.

Although the expenses associated with activation analysis methods can be high, much information is obtained in each analysis. In TNAA, many samples are irradiated with neutrons simultaneously in a single reactor irradiation unit, and many elements are determined simultaneously in each sample. Thus, there is a *low unit cost* for the data obtained in the large irradiation unit. A common TNAA irradiation vessel may hold as many as 70 samples and standards, and 20–30 elements are often determined simultaneously in geological and biological samples. Use of activation methods for only a few elements in a small number of samples may be cost prohibitive, unless alternative techniques are not applicable. CPAA, FNAA, and IPAA do not ordinarily have the advantage of simultaneous multisample irradiation, but are capable of multielement determinations in single samples.

Nuclear activation methods also have disadvantages compared with other analytical techniques. The *lack of information on chemical form* was mentioned above. There are times when the chemical form of the element (**speciation**) is a critical piece of information, and activation analysis cannot provide this directly. Some options for speciation do exist with combined nuclear and conventional techniques, as will be discussed in Section 9.6.1.

Start-up costs for a nuclear instrumentation laboratory are high, but are not greatly different from those of most sophisticated modern analytical instrumentation facilities. A detection system that would include a HPGe detector, associated electronics, and a PC-based analyzer could be purchased for approximately the same cost as a modern computerized atomic absorption spectrometer. It is not absolutely necessary to have a source of activating particles on site, because irradiations can be done off site at a number of facilities that provide activation services for a fee. The irradiated samples can then be returned to the home facility for processing and counting.

Perhaps a more formidable start-up problem than the initial instrumentation cost is the necessity for establishing a laboratory that meets legal requirements for *radiological safety.* The activity levels from most activated samples are not excessive, so exotic "hot cells" and massive shielding are usually not necessary. However, laboratories handling radioactive samples must meet certain federal or state standards and have provisions for the disposal of radioactive waste.

Finally, a single nuclear activation technique is *not applicable to all elements.* Lead and Sn are examples of two elements that cannot be sensitively determined with TNAA, ENAA, or FNAA, although they can be determined by CPAA.

9.1.3. Sources of Activating Particles or Radiations

A wide range of devices is used to produce the particles or photons needed for activation analysis. Some are sophisticated and extremely expensive, while others are rather simple and modest in cost. Table 9.1 lists a variety of commonly used neutron sources, and Table 9.2 summarizes types of charged-particle sources.

Nuclear Reactors

A nuclear reactor produces neutrons as a by-product of uranium fission processes occurring within the reactor core (see Chapter 14 for a discussion

Table 9.1. Common Sources of Neutrons for NAA

Reaction	Half-life	Average Neutron Energy (MeV)	Neutron Yield ($n\ s^{-1}\ Ci^{-1}$, unless otherwise stated)
I. Photonuclear sources			
^{88}Y with ^{9}Be	106.6 d	0.16	1×10^5
^{124}Sb with ^{9}Be	60.2 d	0.02	1.9×10^5
II. Alpha emitter (α, n) sources			
^{239}Pu with ^{9}Be	2.4×10^4 y	3–5	$\approx 10^7$
^{226}Ra with ^{9}Be	1600 y	3.6	1.1×10^7
^{241}Am with ^{9}Be	433 y	3–5	2.2×10^6
III. Spontaneous fission sources			
^{252}Cf	2.64 y	2.3	$2.3 \times 10^{12}\ n\ s^{-1}\ g^{-1}$
IV. Cockroft–Walton accelerators			
$^{3}H(d, n)^{4}He$	—	14.7	$10^8 - 10^{11}$ n/s
V. Cyclotron			
10 μA of 30-MeV deuterons on Be	—	Broad distribution	2×10^{11} n/s
VI. Nuclear reactor			
Induced fission	—	Broad distribution	10^{12}–$10^{15}\ n\ cm^{-2}\ s^{-1}$ typical research reactors

Source: Adapted in part from A. N. Garg and R. J. Banta, *J. Radioanal. Nucl. Chem.* **98:**167 (1986).

Table 9.2. Common Sources of Charged Particles for CPAA

Accelerator Type	Particle Energies	Advantages/Disadvantages
Van de Graaff electrostatic generator	2- to 7-MeV protons typical for single-stage machines; 2× higher for He^{2+} ions; tandem generators used to reach higher energies and to accelerate heavy ions	Precisely controlled energies, with an energy spread of ≤0.1%, relatively inexpensive, easy to operate, limited energy range, extracted beams typically ≤0.1 mA
Cyclotron	A few MeV to several hundred MeV	Higher energies available, expensive and more complex to operate, extracted beams ≈0.1 mA
Neutron reactions $^{6}Li(n,t)^{4}He$	4.8-MeV tritons	Simple, no accelerator needed, must mix Li salt with sample, utilizes reactor neutron flux

of reactor principles). The large reactors used to produce electrical power are not used by researchers as neutron sources. Instead, smaller reactors, called **research reactors,** have been designed to provide the neutrons needed for physics research and applications such as nuclear activation analysis. Research reactors are often located at universities and government laboratories, but some industrial irradiation facilities are also available to the analyst. Table 9.3 provides a partial list of North American reactor facilities offering routine neutron activation analysis services in 1989.

Nuclear reactors emit neutrons that have a wide range of energies. The fission process itself results in the emission of fast neutrons (mean energy ≈2.5 MeV). Moderation of these fast neutrons produces thermal (mean energy ≈0.04 eV, most probable energy ≈0.025 eV), epithermal (0.1–1 eV), and resonance (1 eV to 1 keV) neutrons at the available NAA irradiation positions. A reactor characteristic that is often of interest to an activation analyst is the ratio of the number of thermal neutrons to the number of combined epithermal and resonance neutrons. This ratio is often referred to as the **cadmium ratio (Au),** because Au foils are irradiated with and without a cover of Cd metal foil (0.7 to 1 mm thick). The ratio is determined by dividing the ^{198}Au activity produced in a bare Au foil by the activity produced in the same (or equivalent) Au foil irradiated under the same conditions, but covered with Cd. Cd has a high cross section for neutrons with energies below ≈0.55 eV, and the covered foil is thus irradiated predominately with the higher-energy epithermal and resonance

Table 9.3. Listing of North American Neutron Activation Services Available to Outside Users

Activation Laboratories Ltd. P.O. Box 1420 Brantford, Ontario, Canada N3T 5T6	Becquerel Labs, Inc. Buffalo Materials Research Center SUNY/Buffalo-Rotary Road Buffalo, NY 14214
Dalhousie University SLOWPOKE-2 Reactor Facility Life Sciences Building Halifax, NS B3H 4J1, Canada	General Atomics TRIGA Reactors Facility P.O. Box 85608 San Diego, CA 92138-5608
General Activation Analysis, Inc 11575 Sorrento Valley Road San Diego, CA 92121-9990	NEA, Inc. 10950 S.W. 5th Street, Suite 380 Beaverton, OR 97005
North Carolina State University Nuclear Services Laboratory Raleigh, NC 27675	Oregon State University Radiation Center Corvallis, OR 97331
Pennsylvania State University Breazeale Nuclear Reactor University Park, PA 16802	Texas A&M University Center for Chemical Characterization College Station, TX 77843 (reactor and 14 MeV)
University of Illinois Department of Nuclear Engineering 103 South Goodwin Urbana, IL 61801	University of Kentucky Department of Chemistry Radioanalytical Services Lexington, KY 40506-0055 (14 MeV INAA only)
University of Maryland Department of Nuclear Engineering College Park, MD 20742	University of Michigan Nuclear Reactor Laboratory 2301 Bonisteel Blvd. Ann Arbor, MI 48109-2100
University of Missouri Research Reactor Facility Columbia, MO 65211	University of Missouri Nuclear Reactor Rolla, MO 65401
Washington State University Nuclear Radiation Center Pullman, WA 99164	

Note: Reactor irradiation facilities, unless otherwise specified. List derived from a survey by the ASTM Task Force on Nuclear Methods of Chemical Analysis in 1989. The list may not be complete, but is representative of the facilities available. Government laboratories are not included.

neutrons. The latter are collectively known as **epicadmium neutrons,** but may sometimes just be grouped together as part of the "epithermal neutron flux." To prevent self-absorption effects, the Au foil must be very thin (<0.5 mg/cm^2), or be in the form of a dilute alloy with a low cross section metal. Metals other than Au (e.g., Co, In) may be used to characterize different portions of the epithermal–resonance neutron energy band. From an analytical standpoint, the (n, γ) reaction induced by thermal neutrons is the most important in NAA and produces a product radionuclide that is an isotope of the analyte element. This product radionuclide is usually the indicator radionuclide that is counted in the NAA determination. A high thermal-to-epicadmium neutron ratio is often desirable to eliminate interference reactions that may be induced by the higher-energy neutrons on elements other than those of interest, and resonance absorption effects in larger samples.

The largest research reactors can produce thermal neutron flux densities up to approximately 10^{15} n cm^{-2} s^{-1} and are most often located at national laboratories. Smaller reactors, like the TRIGA® and SLOWPOKE, produce useful flux densities on the order of 10^{11}–10^{13} n cm^{-2} s^{-1} and are more often found in university and industrial research facilities.

The epicadmium neutron flux density is typically only about one-tenth that of the thermal neutron flux in irradiation positions used for NAA. The fast neutron flux (>0.5 MeV) is even smaller. The actual values are dependent on the particular reactor (core design, type of moderator), and also on the location within a given reactor. Both reactor epithermal and thermal neutrons are used analytically, so a knowledge of the cadmium ratio is important. The fast neutrons from a reactor are not often used analytically, but can cause interference reactions.

Isotopic Neutron Sources

Isotopic neutron sources are quite simple in design and operation compared to the other sources of neutrons. They rely on spontaneous nuclear processes for neutron production. The isotopic sources include α-emitter sources, photon sources, and ^{252}Cf sources.

The **α-emitter sources** are the oldest isotopic neutron sources. In his work to discover the neutron, Chadwick placed radium, an α emitter, in close proximity to ^{9}Be. The nuclear reaction that occurs is ^{9}Be$(\alpha, n)^{12}$C. This (α, n) reaction is the basis of the neutron emission in many isotopic neutron sources. Modern α-emitting sources use Am or Pu in place of Ra, but the principle is the same as in Chadwick's source. The α-emitter sources produce neutrons of varying energies, mostly at higher energies ($\approx$3–4 MeV) that must be moderated for use in NAA. The neutron flux varies with the type and amounts of the α emitter and target, but ranges from 10^5 to 10^7 n s^{-1} Ci^{-1}. The low fluxes dictate that the analytical sensitivity, when they are used for NAA, is also low. The α-emitter sources are the simplest and least expensive of the neutron sources. They require relatively little shielding, and their neutron yield per curie is better than the photon sources. They are good for instructional purposes or for applications in which great analytical sensitivity is not needed.

Photon sources of neutrons rely on the $^9Be(\gamma, n)2\text{-}^4He$ reaction. The neutrons produced in this reaction are of lower energy than those from the α-emitter sources, with most being around 26 keV. The γ ray must have an energy of at least 1.67 MeV to induce this nuclear reaction. ^{124}Sb, which emits a γ ray at 1.69 MeV, is the most widely used photon emitter for neutron sources. The neutron flux produced by these sources is rather low, only about 10^5 n s^{-1} Ci^{-1}. The photon sources are not widely used because the flux is low, the half-life of ^{124}Sb is only 60.2 days, and it is necessary to construct shielding sufficient to guard against the high-energy γ rays, in contrast to the more easily shielded α-emitter sources.

It has already been noted that the fission process produces neutrons, and that the neutron-induced uranium fission taking place in a nuclear reactor provides the best source of neutrons. The **spontaneous fission of ^{252}Cf** provides an alternative isotopic fission neutron source. A range of neutron energies is produced in the spontaneous fission decay of ^{252}Cf, with most being in the 1- to 3-MeV range. The neutron yield depends on the size and age of the ^{252}Cf source. For one gram of a freshly produced ^{252}Cf source, a flux of 2.3×10^{12} n/s is calculated. The half-life of ^{252}Cf is 2.64 years, with approximately 3% of the decays occurring by spontaneous fission and the remainder by α decay. The moderately short half-life is a disadvantage, because the neutron output diminishes appreciably over times equivalent to the useful lives of many other instruments. The cost of purchasing ^{252}Cf is a significant outlay ($\approx$\$25 000 for a fabricated 1-mg source and shield), but once in place the ^{252}Cf neutron source requires no maintenance and produces an analytically useful flux. A simple small ^{252}Cf source installation is shown in Fig. 9.3.

Accelerator Sources

Accelerators are instruments that receive ions from an ion source and then accelerate these ions to high kinetic energies. Accelerator principles are discussed in Chapter 14. The energetic ions produced in an accelerator can be used directly as activating particles for charged particle activation analysis, or they can induce nuclear reactions on a target, which will, in turn, produce the activating particles.

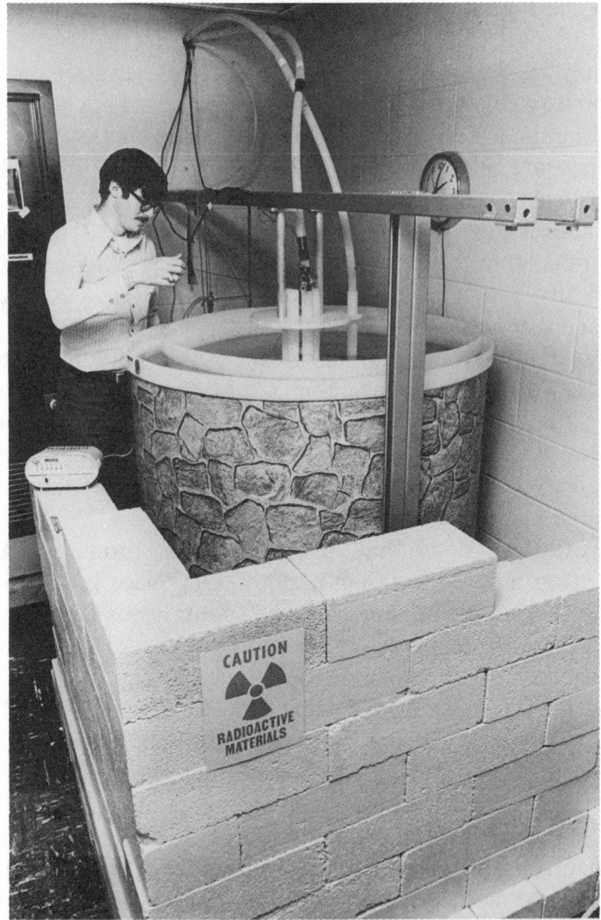

Figure 9.3. An irradiation facility for a small ^{252}Cf isotopic neutron source. Two concentric heavy-walled polyethylene tanks contain water for moderation of the fission neutrons and shielding. The outside of the largest tank is also covered with cadmium sheet and a decorative cover. The Cf source is suspended in the center of the interior tank and several rabbit pneumatic sample transfer tubes connect the irradiation facility to the counting room.

The **neutron generator** is a small accelerating device that produces energetic neutrons as a result of a nuclear reaction. The design most used is the **Cockroft–Walton accelerator,** developed in 1932 by J. D. Cockroft and E.T. S. Walton. Photographs of a neutron generator and neutron tube are shown in Fig. 9.4*a*, *b*. In generators with pumped vacuum systems deuterium ions (deuterons) are accelerated through a high electrical potential gap

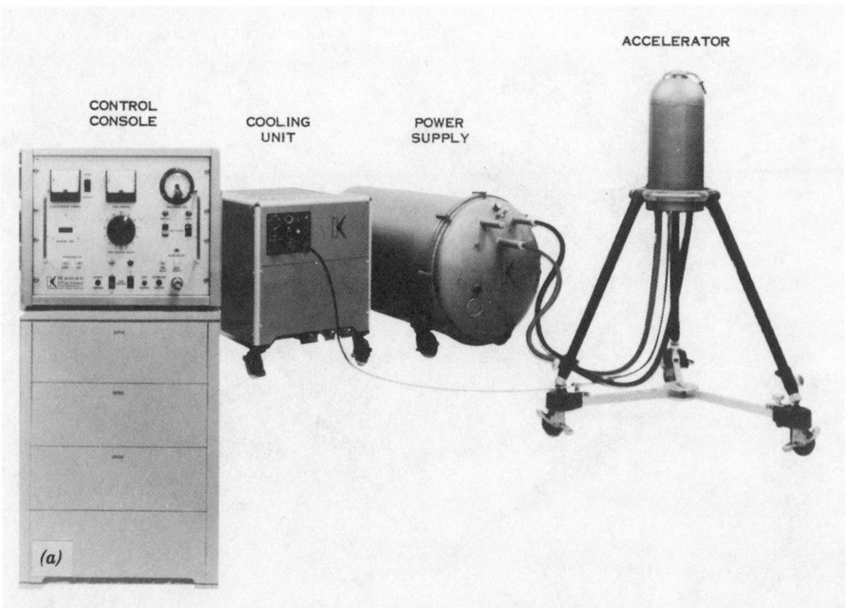

Figure 9.4. (*a*) A modern sealed-tube 14-MeV neutron generator system designed by Kaman Instrumentation Corp., now a product line of MF Physics Corp., Colorado Springs, CO. (Courtesy of M. F. Frey, MF Physics Corp., Colorado Springs, CO.)

toward a tritium-containing target. The target commonly consists of titanium tritide on a copper backing. Alternatively, in modern sealed neutron tubes mixed beams of deuterium and tritium ions may be accelerated into an initially blank metal target. Both deuterium and tritium will become embedded in the target and interact with the incident mixed beam to produce neutrons. The nuclear reaction that occurs is a fusion reaction:

$$^{2}\mathrm{H}(^{3}\mathrm{H}, \mathrm{n})^{4}\mathrm{He}$$

The neutrons produced in the reaction above have energies of approximately 14.7 MeV. A typical research neutron generator produces a total yield of 2–3 $\times$ 10^{11} n/s, but the useful flux density is only about 10^9 n cm^{-2} s^{-1} at the sample irradiation position. Neutron generators are rather compact units, and can be designed to fit into boreholes used for oil and mineral prospecting. Laboratory 14-MeV neutron irradiation facilities can be installed for $50–100 000, utilizing existing buildings and below-ground shielding. Many universities and industrial laboratories have 14-MeV neutron generators. **Cyclotrons** and **linear accelerators** can be used to produce either neutrons or charged particles. Typically protons, deuterons, tritons, alphas, and ^{3}He

Figure 9.4. (*b*) The accelerator head contains the small sealed tube (pictured here) only about 8 in. in length, in which deuterium and tritium ions are accelerated through a potential of 175–190 kV to effect the fusion reaction that produces 14-MeV neutrons. (Courtesy of M. F. Frey, MF Physics Corp., Colorado Springs, CO.)

ions are accelerated. These particles may be used directly for CPAA. Alternatively, they may be allowed to strike a target to induce nuclear reactions that will produce neutrons. For example, the reaction of deuterons on a Be target produces a good flux of fast neutrons.

Electron linear accelerators are devices specifically designed to accelerate electrons. The electrons themselves are not used much analytically, although a few applications exist. More often, the bremsstrahlung radiation emitted by the accelerated electrons is used for instrumental photon activation analysis. The primary analytical use of **synchrotrons**, like the linear electron accelerators, is as a photon source for IPAA. The **Van de Graaff accelerator** is used analytically to generate 1- to 5-MeV protons or other charged particles for PIXE and PIGE. The operation of all these devices is discussed in Chapter 14.

Most of the accelerating devices mentioned above, except for the Cockroft–Walton neutron generator, are large and expensive devices that were constructed primarily for nuclear physics research. Due to limited time availability on the larger accelerators, they are less often used for analytical chemistry than nuclear reactors and isotopic neutron sources.

9.1.4. Interferences in Activation Analysis

When properly performed, activation analysis can be among the most precise and accurate of all analytical methods for the determination of trace elements. However, there are potential interferences and problems that an analyst should be aware of in the practice of activation analysis. These include interference reactions (primary, secondary, and second order), overlapping γ-ray photopeaks in the spectrum, absorption of activating particles entering or radiations being emitted by the sample, pulse pileup for "hot" samples, and associated conventional instrumental and analytical errors (e.g., electronic instability, weighing errors, inhomogeneous standards).

Primary Interference Reactions

A given radionuclide can often be produced in more than one way. If the indicator radionuclide being used for analysis is produced by a reaction on an element other than the analyte, then a **primary interference reaction** occurs. An example of this would be the determination of Si by FNAA. The desired analytical reaction is $^{28}Si(n,p)^{28}Al$. Another reaction that also produces the ^{28}Al indicator radionuclide with fast neutrons is $^{31}P(n,\alpha)^{28}Al$. Therefore, if a significant proportion of P were present in the Si-containing sample, the measured ^{28}Al activity would be greater than that due only to the presence of ^{28}Si, and the results would be in error.

It was mentioned in Section 9.1.3 that fast neutrons from a reactor can cause interfering reactions, and that it is sometimes important to know their flux. A specific example of a primary interference reaction where the ratio of fast to thermal neutrons is of importance is in the determination of Cr using the $^{50}Cr(n,\gamma)^{51}Cr$ reaction. The desired analytical reaction occurs with thermal neutrons. A competing reaction that takes place with fast neutrons is $^{54}Fe(n,\alpha)^{51}Cr$. Determination of Cr in a matrix high in Fe, using a reactor with a significant fast neutron flux, will give erroneous results. Such a situation occurs for the determination of Cr in blood, which contains much higher Fe than Cr levels.

In some cases, corrections can be made for primary interference reactions. In other cases, the presence of the interfering reaction may preclude the possibility of performing the analysis.

Secondary Interference Reactions

These interferences occur less often than the primary ones. In a **secondary interference reaction,** particles produced in the sample matrix generate a reaction that produces the same indicator nuclide as that produced by the analytical reaction. For example, the determination of N by 14-MeV FNAA uses the (n, 2n) reaction on ^{14}N to produce ^{13}N. If the sample matrix is high in H, the fast neutrons could induce a significant proton flux through scattering reactions. The protons could then induce a (p, n) reaction on any C present in the sample: $^{13}C(p, n)^{13}N$. The product is the same (^{13}N) as in the desired analytical reaction. In most cases, this type of interference is not a serious problem, because the flux of secondary particles is quite small.

Second-Order Additive Interference

A **second-order additive interference** is a rather rare situation where a nuclear reaction on some matrix element produces a measureable amount of the stable trace nuclide that is being determined by the analytical nuclear reaction. For example, phosphorus may be determined via the radiative capture reaction on ^{31}P: $^{31}P(n, \gamma)^{32}P$. The following reaction on ^{30}Si could be an interfering reaction: $^{30}Si(n, \gamma)^{31}Si$. The ^{31}Si formed here will decay by β emission to form ^{31}P, which is the target of the original analytical reaction. Only in very unusual cases would the amount of stable target nuclide formed during the irradiation be a significant problem.

Gamma-Ray Spectral Interferences

It is possible that two radionuclides will emit γ rays of the same, or nearly the same, energy. In this case, a **gamma-ray spectral interference** occurs. An example of this would be the 846.8-keV γ ray emitted by ^{56}Mn, half-life of 2.58 h, and the 843.8-keV γ ray emitted by ^{27}Mg, half-life 9.45 min. These interfering γ rays can be used as examples in discussing ways in which spectral interferences may be overcome. The simplest way to overcome overlapping peaks is to have a very good resolution detector coupled with a good peak analysis program. This would provide an instrumental solution to the problem, because the system should be able to resolve these two peaks and extract good data from each. The advent of the high-resolution HPGe detectors and continually improved software for peak analysis makes this a likely possibility for the Mn–Mg example.

Another solution to the spectral interference problem utilizes the difference in half-lives between the two mutual interferents. The sample containing the 9.45-min ^{27}Mg radionuclide could be allowed to decay for 10 half-lives, thus reducing its activity to 0.1% of the original value. After 90 min there would still be significant ^{56}Mn activity in the sample that could be measured, because its half-life is 2.58 h. The ^{56}Mn activity could be measured directly only after decay of the ^{27}Mg. Initial measurement of the gross counting rate, followed by subtraction of the decay-corrected ^{56}Mn counting rate would also provide a way to measure the ^{27}Mg. Adjustment of irradiation time might also be used to enhance measurement of one of the radionuclides. For example, a short irradiation would discriminate against the longer-lived interferent.

Sometimes two γ-rays overlap each other so closely that no currently available detector is able to separate the two peaks, and the half-lives are not sufficiently different to allow discrimination by adjusting the decay period or irradiation time. When this occurs, it may be that the analysis is simply not possible, and another indicator nuclide, γ-ray peak, or analytical method must be found. In some cases, corrections may be made for this kind of interference. This is the situation for the ^{203}Hg and ^{75}Se indicator radionuclides, which emit γ rays of 279.2 and 279.5 keV, respectively. These two peaks cannot be resolved and both half-lives are relatively long, precluding discrimination by decay time. However, the ^{75}Se also emits a γ-ray with a **clean photopeak** at 264.7 keV. A clean peak is one that experiences no interferences from other γ rays produced in INAA. The relative intensities of the clean and interference peaks for ^{75}Se can be measured for a pure source. This value is then used to calculate what proportion of the $\approx$279-keV composite peak must be due to ^{75}Se, and the number of counts subtracted from the total to get the counts due to ^{203}Hg.

Still another way to correct for spectral interferences is the use of coincidence counting techniques. This is appropriate in cases where a cascade of γ rays is emitted in the decay of the indicator radionuclide formed by the activation reaction.

Other Interferences

Two interferences that can occur during the irradiation process itself are due to **self-shadowing** (or **self-shielding**) and **resonance capture.** Thermal neutron self-shadowing takes place when one part of a sample shields another part from the incident neutron flux. Matrices that contain Cd and B are examples of this. Isotopic cross sections for neutron capture are 20 600 b for ^{113}Cd and 3838 b for ^{10}B. The result is that different nuclear production rates occur in different parts of the sample, and this is undesirable. A simple approach to solution of this difficulty is to keep sample size small, so that the flux of activating particles penetrates the entire sample and remains nearly constant throughout. The resonance capture problem occurs for certain elements that have extremely large neutron capture cross sections in the epithermal neutron energy range. These large cross sections are referred to as resonance absorptions and were discussed in Section 4.3.2. If a sample or comparator standard contains an analyte element with high resonance capture cross-section peaks in its excitation function (e.g., Au at 4.9 eV), small differences in the neutron energy spectrum with depth in the sample due to matrix scattering can bias the sample-to-comparator indicator radionuclide production rates. Again, small samples and diluted comparator standards can be used to minimize the problem.

A related problem can occur during the counting process due to **self-shadowing of γ rays.** If the samples and standards used in the comparator method have different densities or bulk matrix compositions, the emitted γ rays may not be attenuated equally by the samples and standards. This can lead to self-shadowing errors that may produce results that are either too high or too low, depending on whether the self-shadowing is greater in the sample or the comparator standard. To avoid this problem, comparator standards that have matrices similar to those of the samples are often chosen, and sizes of standards and samples are made nearly equal.

When using the comparator method for activation analysis (see Section 9.2.2), the assumption is made that the samples and standards are treated in the same way before irradiation, that they have similar placement in the reactor during irradiation, and that they are counted in identical locations relative to the detector after irradiation. Nonequivalent placement during either irradiation or counting can cause errors due to **geometry considerations,** because the assumption of identical treatment for samples and standards is no longer true. For example, the neutron flux densities in a reactor can vary significantly (>10%) over a distance of only a few centimeters. Thus, if a sample were placed close to the reactor core and the standard happened to be located on the other side of the irradiation vessel 2–4 cm away, they could experience significantly different fluxes. Similarly, the efficiency with which the detector detects emitted radiation is highly dependent on the location of the active material. Ideally, the source is a *point source* located several centimeters away from the detector. Significant errors can occur for samples and standards that have large, irregular shapes and are placed very close to the detector for counting. Where possible, sample and comparator samples should have the same size and shape, or if not, be placed at a greater distance from the detector so they more closely resemble point sources.

Instrumental Errors

All components of the analyzer system can cause errors for the careless analyst. Serious detector malfunctions are normally observable from the distortions they will cause in peak shapes. Changes in the electronic components may cause gain shifts or timing differences. Attempting to count highly active samples can lead to problems with pulse pileup and peak distortion. The peak integration procedures used by various software programs should be verified. For detailed discussions of these problems, the references at the end of the chapter should be consulted.

9.2. NEUTRON ACTIVATION ANALYSIS (NAA)

Of all the activation analysis methods, neutron activation is the most widely used. In the 1930s, Georg Hevesy and his student Hilde Levi were investigating the effects of neutron irradiation on the rare-earth elements. They observed that the rare earths reacted quite differently to the irradiation. Most striking was dysprosium, which became highly radioactive after neutron irradiation. Hevesy recognized that this method could be used for qualitative detection of the rare earths, and he and Levi reported on this new method in 1936. Their identification was based on discrimination by half-life rather than by energy of the emitted radiation. Figure 9.5*a* shows Hilde Levi in the ^{14}C laboratory at the University of Copenhagen in about 1951 and Fig. 9.5*b* shows her at the Modern Trends in Activation Analysis

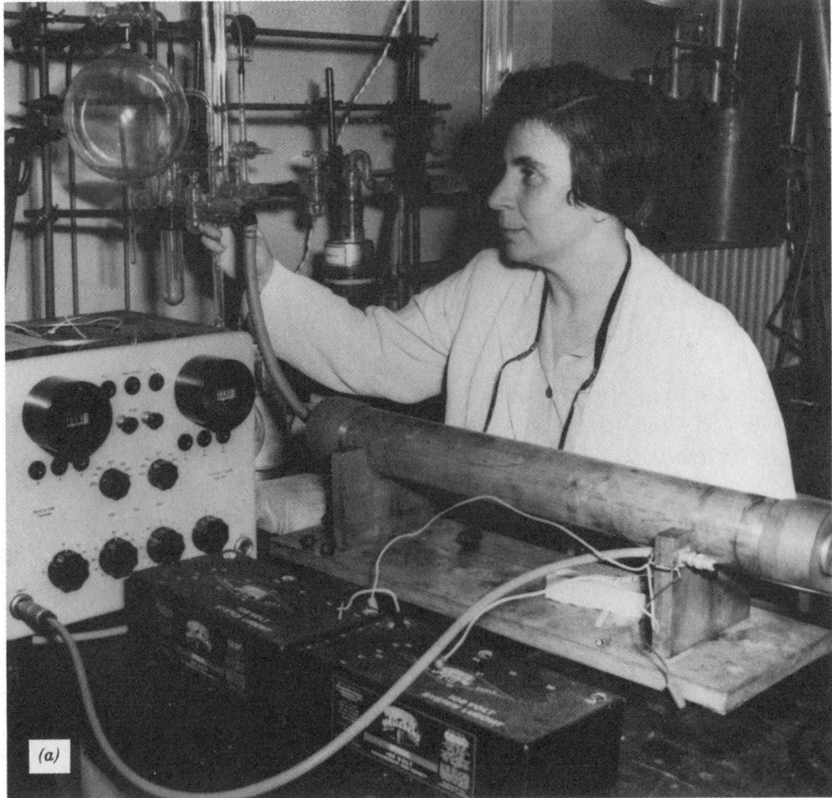

Figure 9.5. (*a*) Hilde Levi in the ^{14}C laboratory at the University of Copenhagen about 1951. Levi and Georg Hevesy are credited with the first use of neutron activation analysis in 1936. (Courtesy of Hilde Levi, through the assistance of K. Heydorn, Risø National Laboratory, Roskilde, Denmark.)

Figure 9.5. (*b*) Hilde Levi at the celebration of the 50th anniversary of NAA, 7th International Conference Modern Trends in Activation Analysis, in Copenhagen in 1986. (Courtesy of K. Heydorn.)

Conference in Copenhagen in 1986 (50th anniversary of NAA). Figure 9.6 shows Georg Hevesy early in his career. The counting equipment used by Hevesy and Levi in their first activation analysis experiments in shown in Fig. 9.7.

The technique of neutron activation analysis was not used much after its discovery, because there were few neutron sources available. However, as reactor technology developed in the 1950s, this situation changed and NAA grew very rapidly in the 1950s and 1960s. It is now over 50 years old and is considered to be a mature technique. Although neutron activation analysis is not as widely practiced as are other elemental analysis techniques such as atomic absorption or inductively coupled plasma emission, it still retains a solid niche in analytical chemistry because it can yield very accurate and precise results for trace and ultratrace elemental determinations and is relatively free of matrix effects and laboratory contamination problems.

9.2.1. Overview of NAA Procedures

The actual performance of a neutron activation analysis procedure begins with obtaining a representative sample. The process of sampling is a critical part of any analytical scheme and much has been written about it.

Figure 9.6. Georg Hevesy. (Courtesy of the American Institute of Physics, Niels Bohr Library.)

Because NAA is mostly used for trace analysis, avoidance of contamination in sampling is paramount. Use of a laminar flow hood, or a Class 100 clean room (room with a filtered air flow in which >99.97% of the particles greater than 0.3 μm in size are removed) is recommended. Once the sample has been obtained, very little pretreatment is needed before activation. The sample, whose mass may vary from milligrams to several grams, is weighed into an irradiation vessel appropriate for the type of irradiation to be done. This is likely to be a polyethylene vial if the irradiation is short (20 min or less) and at a low or moderate flux density ($<10^{13}$ n cm^{-2} s^{-1}), or a high-purity quartz glass vial if the irradiation is to be long (hours or days) and at moderate to high flux densities. A photograph of several types of

Figure 9.7. Gas-filled counter and electronics used by Hevesy and Levi for their early experiments in activation analysis. A large battery rack for power is located below the table.

quartz glass vials, plastic "rabbits" for rapid transfer systems, and the cans that are placed into the reactor are shown in Fig. 9.8*a* and *b*.

For a long irradiation, many (30–100) samples and standards are bundled and sent to a nuclear reactor, where they are placed into an appropriate irradiation can. This can is then lowered into the reactor and left there for the required time, after which it is removed and allowed to **cool** (i.e., the short-lived radioactivity levels are allowed to abate somewhat). The samples and standards are then returned to the analyst for processing and counting. If INAA is being done, the irradiated vials are washed to remove surface contamination and then counted directly. If RNAA is to be performed, the vials must be broken open, the contents removed, and the separations

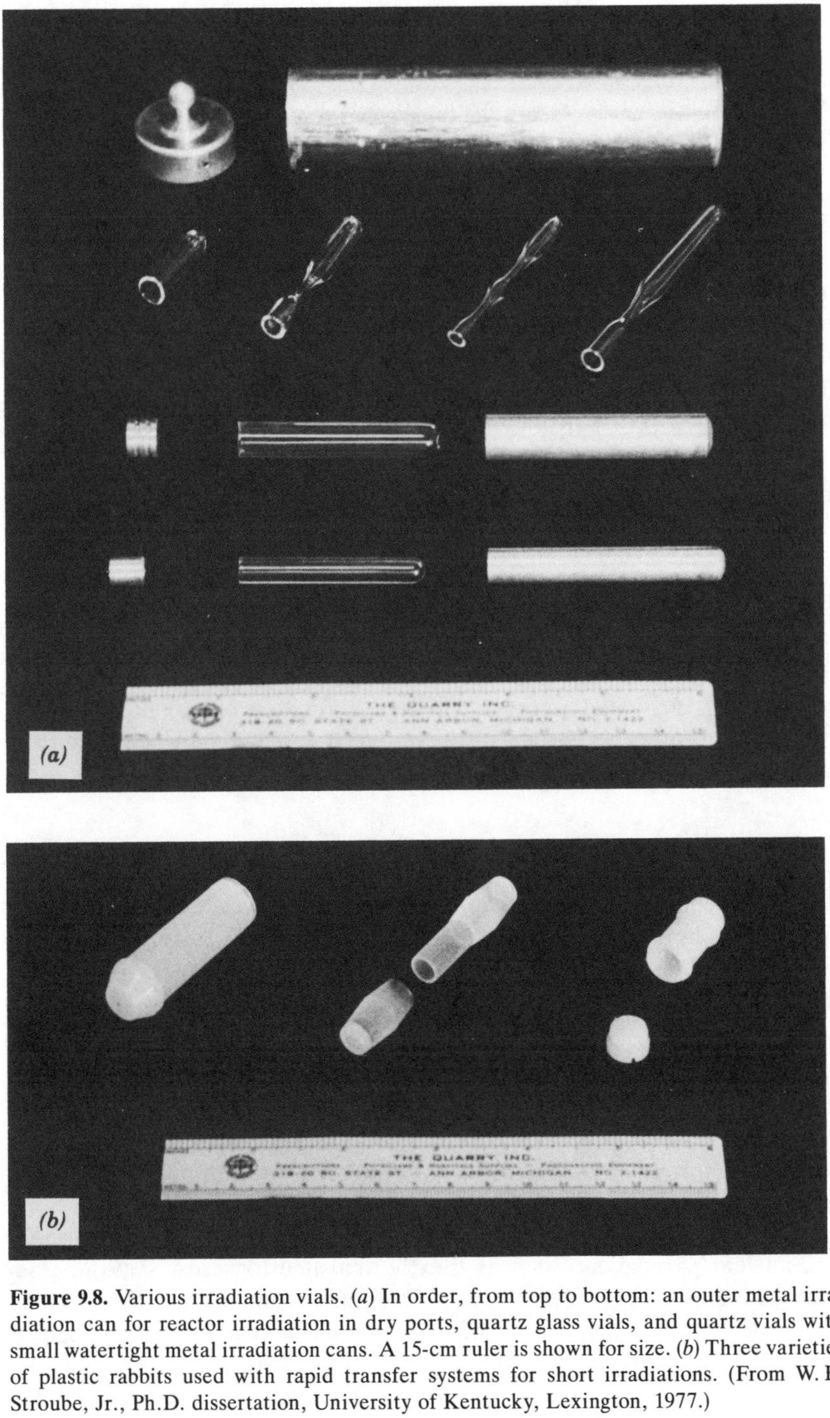

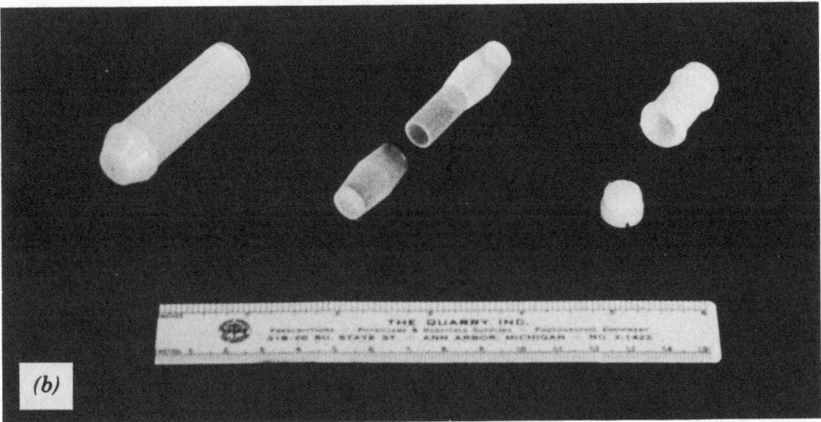

Figure 9.8. Various irradiation vials. (*a*) In order, from top to bottom: an outer metal irradiation can for reactor irradiation in dry ports, quartz glass vials, and quartz vials with small watertight metal irradiation cans. A 15-cm ruler is shown for size. (*b*) Three varieties of plastic rabbits used with rapid transfer systems for short irradiations. (From W. B. Stroube, Jr., Ph.D. dissertation, University of Kentucky, Lexington, 1977.)

procedure performed before counting. It should be noted that the laboratory does not need to be at the reactor site for the analysis of long-lived nuclides. The samples can be shipped to and from the reactor using commercial shipping operations.

To perform the analysis of short-lived nuclides, the analyst must go to the reactor or other neutron source. The polyethylene vial containing the sample is put into a larger vial commonly called a **rabbit**. This rabbit is made to travel between the laboratory and the reactor through a system of transfer tubes. Rabbits are ordinarily sent to the reactor one-at-a-time using the transfer system. This is in contrast to the irradiations producing long-lived nuclides, where many samples and standards are irradiated simultaneously. Upon return from the reactor, samples are removed from the rabbit, transferred to nonirradiated vials, and counted. Counting is usually done in a low-background area located outside the reactor containment.

The detectors used for counting the activated samples are either NaI(Tl), Ge(Li), or HPGe detectors. The latter two are most likely for multielemental analyses performed with INAA, because the large number of peaks in the γ-ray spectra will require the best resolution possible. The NaI(Tl) detector may be satisfactory, or even preferable, for RNAA, because most of the interfering activities will have been separated out. The γ-ray spectra from the multichannel analyzer can be recorded on a computer and processed later using appropriate software that is commercially available from a number of companies.

9.2.2. Calculations of NAA

The basic equation used for NAA calculations was developed in Section 5.7. When the term $e^{-\lambda T}$ is added to account for the decay of the radionuclides after the irradiation period, the equation becomes

$$A = n\phi\sigma(1 - e^{-\lambda t})(e^{-\lambda T}) \tag{9.1}$$

where T is the time of decay, and the meaning of the other terms is the same as that given in Section 5.7. In theory, it is possible to perform **absolute activation analysis** and use Eq. 9.1 to calculate the number of target atoms directly. In practice, this is not usually done. The reason is that there may be significant uncertainties in the values available in the literature for the nuclear parameters, such as the thermal neutron cross section, resonance integral, and decay constants. Even more important is the fact that the neutron energy spectrum and flux density at the irradiation position are not well known or are variable during irradiation periods. Also, for Eq. 9.1 to be used directly it is necessary that the *absolute activity* of the irradiated sample be measured. Therefore, a **comparator method** is normally used. In this latter approach, a standard containing a known amount of the element to be determined is irradiated along with the samples. It is assumed that the neutron flux, cross sections, irradiation times, and all other variables associated with counting are constant for both the standard and the sample. We can then write the equation used for calculations in NAA using the comparator method:

$$\frac{R_{\text{std}}}{R_{\text{sam}}} = \frac{W_{\text{std}}(e^{-\lambda T})_{\text{std}}}{W_{\text{sam}}(e^{-\lambda T})_{\text{sam}}} \tag{9.2}$$

where R = counting rates of standard (std) and sample (sam)

W = mass of the element

T = decay time

Example. A standard containing 10.5 μg of Al and an unknown sample weighing 240 mg were irradiated simultaneously in a nuclear reactor. The standard was counted 3.50 min after the end of the irradiation, and the ^{28}Al counting rate was determined to be 5.37×10^3 cpm. The sample was counted 5.00 min after the end of the irradiation and had an ^{28}Al counting rate of 1.37×10^3 cpm. Assume the counting times are short with respect to the ^{28}Al half-life. Calculate the concentration of Al in the sample, in μg/g.

Equation 9.2 is used to solve this problem. The half life of ^{28}Al is 2.25 min:

$$\frac{5.37 \times 10^3}{1.37 \times 10^3} = \frac{10.5\ \mu\text{g}(e^{-(0.693/2.25\ \text{min})(3.50\ \text{min})})}{x\ \mu\text{g}(e^{-(0.693/2.25\ \text{min})(5.00\ \text{min})})}$$

Solving for x gives 4.25 μg Al in the sample. The concentration of Al in the sample would be

$$4.25\ \mu\text{g Al}/0.240\ \text{g sample} = 17.7\ \mu\text{g/g}$$

The best type of standard material to use is one prepared from very pure, stoichiometrically well-defined compounds, either in solid or solution. When many elements are to be determined in each sample, preparation of individual standards becomes tedious and impractical, because the standards would occupy an excessive amount of space in the irradiation can. Thus, either multielemental standards or the use of a single comparator for all elements becomes necessary.

The **standard reference materials (SRM)** that are prepared by the National Institute of Standards and Technology (NIST) and the U.S. Geological Survey (USGS) in the United States, the International Atomic Energy Agency (IAEA) in Vienna, or other sources are often used as primary standards for NAA. This approach of using SRMs as primary standards is not recommended by these agencies, because uncertainties may be large even for some of the certified elements and homogeneity may not have been established for the small sample sizes normally used in NAA. These materials are also very difficult and expensive to prepare. Nevertheless, they are frequently used as primary standards in laboratories where small numbers of analyses are done on matrices that are similar to those available as SRMs.

The possibility of using a single element as a comparator for multielement NAA is attractive, and several schemes have been devised to accomplish this. The one that has become most widely used is called the $\boldsymbol{k_0}$ **method.** The equation used to calculate concentrations using this method is

$$\rho = \frac{A_{sp}}{A_{sp}^*} \cdot \frac{1}{k_0} \cdot \left(\frac{f + Q_0^*(\alpha)}{f + Q_0(\alpha)}\right) \cdot \frac{\varepsilon^*}{\varepsilon} \tag{9.3}$$

where ρ = concentration, in μg/g

A_{sp} = specific activity of comparator element (*), or sample

f = the thermal-to-epithermal flux ratio

Q_0 = the ratio of the resonance integral to the thermal neutron cross section

ε = the detector efficiency

k_0 = the k_0 factor

α = the deviation of the epithermal flux from ideality

The k_0 factor is an experimentally determined value that, in essence, contains the values for the nuclear constants needed in the activation analysis calculation, such as the cross section and isotopic abundance. The k_0 factors for many nuclides have been determined and published in a joint project involving the Activation Analysis Laboratory of the Central Research Institute for Physics in Budapest and the Institute for Nuclear Sciences in Ghent using gold as a comparator element.

Lower limits of detection in INAA can be calculated for a given type of sample matrix using an Advance Prediction Computer Program (APCP), such as that developed by V. P. Guinn at the University of California,

Irvine. The method takes into account cumulative Compton continuum effects and allows advance estimates of the best analytical photopeaks, measurement precisions, lower limits of detection, maximum allowable sample size, and optimum irradiation, decay, and counting parameters for a given sample matrix.

9.2.3. Thermal Neutron Activation Analysis (TNAA)

Thermal neutrons are the activating particles for TNAA. With rare exception, thermal neutrons induce radiative capture (n, γ) reactions in the target nuclei. The delayed γ rays from the sample are most often detected. In the thermal neutron region, cross sections follow the $1/v$ law, as described earlier in Section 4.3.2. These cross sections are generally higher than those for other types of activation reactions. Neutrons have no charge and thus do not experience electrostatic repulsion as they approach the nucleus. At low kinetic energies, the neutrons have longer deBroglie wavelengths and low velocities ($\approx 2.2 \times 10^5$ cm/s), and are more likely to interact with target nuclei (see Section 4.3.2). Therefore, if high fluxes of thermal neutrons are available, TNAA offers potentially higher sensitivity for a greater number of elements than any of the other activation methods.

Applications of Thermal NAA

Thermal NAA is used to determine elemental concentrations at trace and ultratrace levels in a wide variety of sample types. The earliest applications were in the areas of **geochemistry** and **cosmochemistry**. Because TNAA is a good method for the determination of the rare-earth elements (REE), it was used to determine those elements in terrestrial rocks and also in meteorites. These data, especially those from meteorites, were used by cosmochemists in formulating theories about the nuclear processes occurring in stars (see Chapter 13). Many of the trace element compositions of the lunar rock samples brought back to earth by the Apollo missions were determined by TNAA. A γ-ray spectrum from a USGS standard rock after a 20-s thermal neutron irradiation and a 20-min decay period is shown in Fig. 9.9. The spectrum is dominated by γ rays from short-lived radionuclides, such as ^{24}Na. Figure 9.10 shows the spectrum from a natural terrestrial glass irradiated with thermal neutrons for 4 h and allowed to decay for 27 days prior to counting. Many more photopeaks are seen than in Fig. 9.9, because short-lived radionuclide photopeaks with their associated Compton distributions have decayed. In Fig. 9.11, a spectrum for a glassy sample collected from the lunar soil is shown. This sample was irradiated for 8 h and

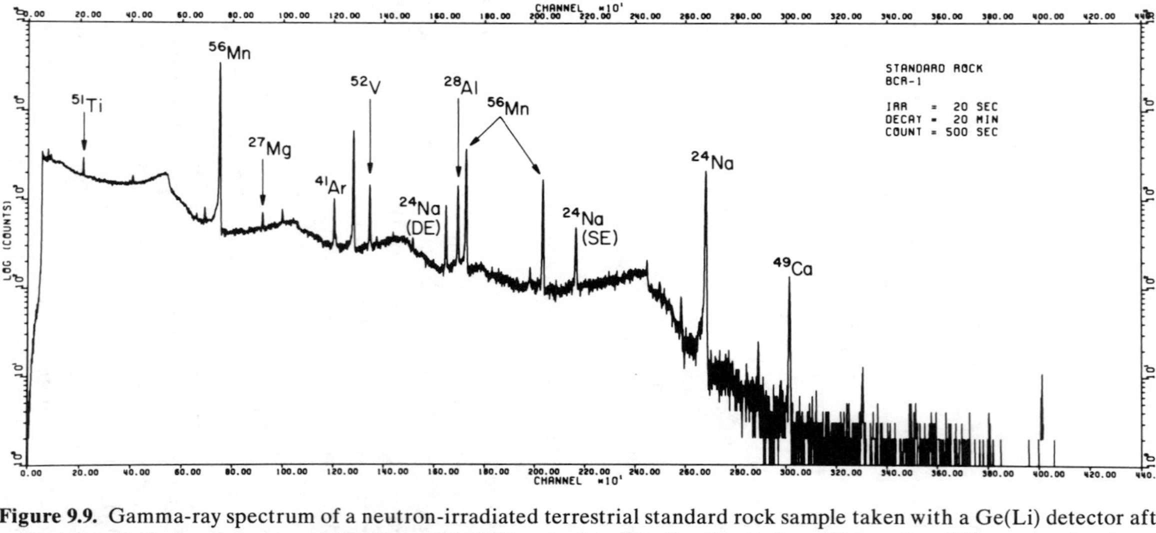

Figure 9.9. Gamma-ray spectrum of a neutron-irradiated terrestrial standard rock sample taken with a Ge(Li) detector after a short irradiation and decay period. (From W. B. Stroube, Jr., Ph. D. dissertation, University of Kentucky, Lexington, 1977.)

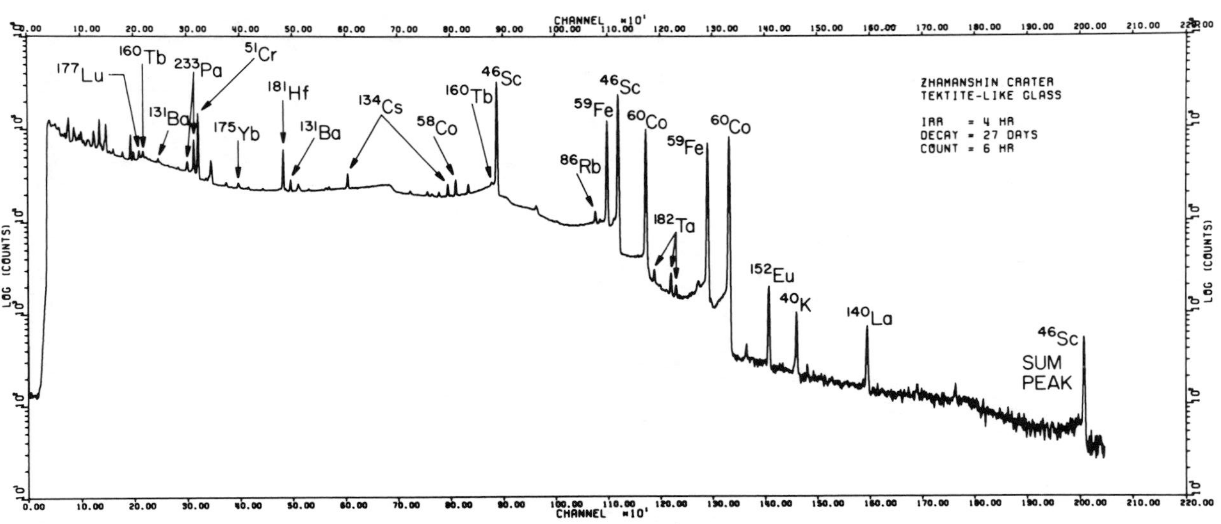

Figure 9.10. Gamma-ray spectrum of a natural terrestrial glass sample after a 27 day decay period. (From W. B. Stroube, Jr., Ph.D. dissertation, University of Kentucky, Lexington, 1977.)

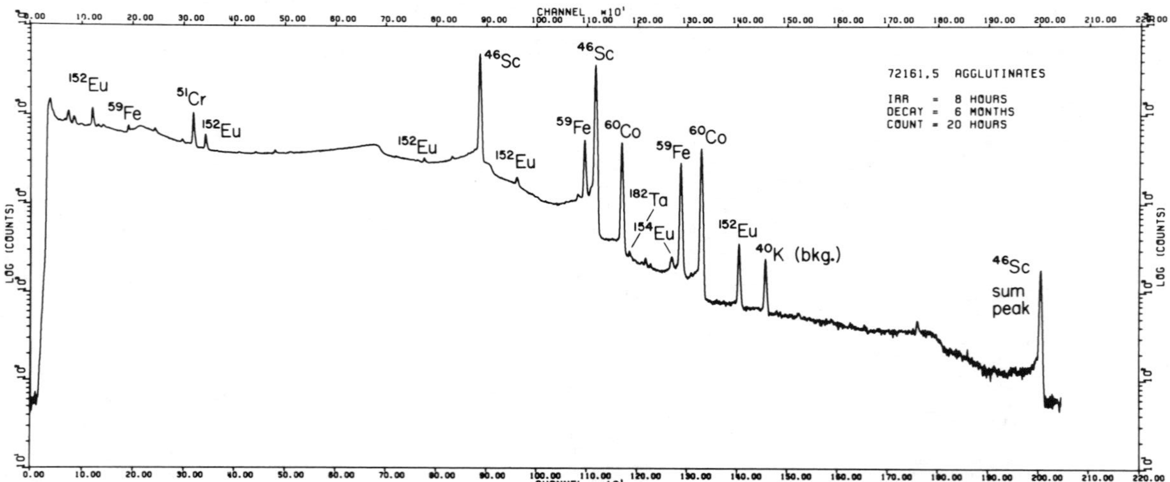

Figure 9.11. Gamma-ray spectrum of a sample from the lunar soil after a 6-month decay period.

was allowed to decay for 6 months. After this long decay period, the number of peaks in the spectrum again decreases, because only a few very long-lived radionuclides remain.

In recent years, TNAA has been increasingly used for trace element analyses of **biological materials.** The sensitivity and selectivity of TNAA make it an excellent technique for analyses of low levels of elements in the complex biological matrix. Figure 9.12 shows a γ-ray spectrum of neutron-irradiated human brain after a decay period of 12 days. Although most of the ^{24}Na that dominates the spectrum shortly after the end of the irradiation has decayed away, irradiated human tissue samples may exhibit spectra that are dominated by γ rays from intermediate half-life radionuclides, such as ^{82}Br ($t_{1/2}$ = 1.47 days) and the bremsstrahlung radiation from the

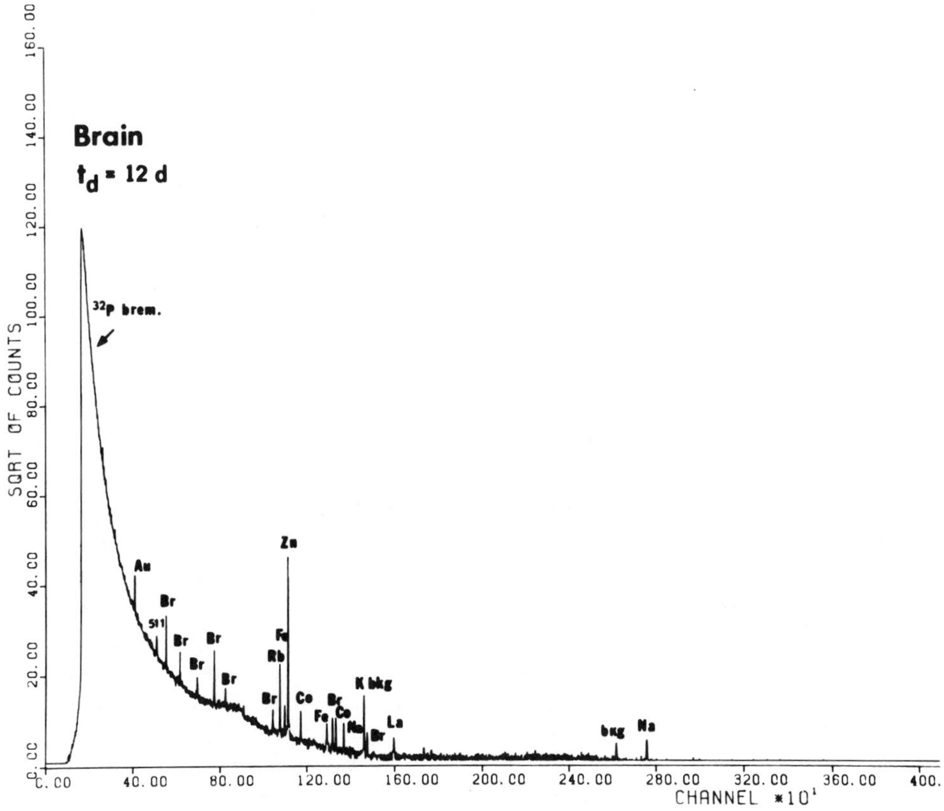

Figure 9.12. Gamma-ray spectrum of neutron-irradiated human brain after a decay period of 12 days.

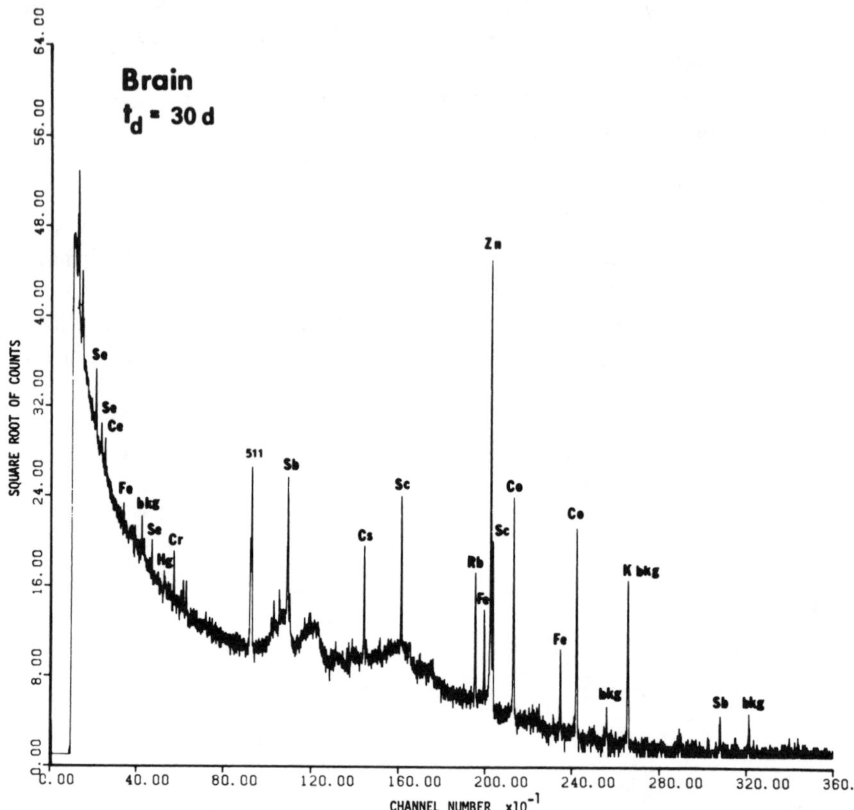

Figure 9.13. Gamma-ray spectrum of neutron-irradiated human brain after a 30-day decay period.

β decay of ^{32}P ($t_{1/2}$ = 14.28 days). Figure 9.13 shows the spectrum obtained for human brain tissue after a decay period of 30 days. Although the bremsstrahlung contribution to the spectrum is still high, many additional analytical photopeaks may now be seen. All human body tissues have been analyzed by NAA, along with many kinds of plant and animal materials. Trace element levels in both normal and diseased states have been studied. There is much interest currently in nutrition and food composition, and trace element characterization of food is often done using TNAA. In addition to the analysis of in vitro samples (i.e., samples removed from an organism), TNAA is also used for in vivo activation analysis. In the latter procedure, a living patient's entire body, or a particular part of the body, is irradiated with neutrons from an isotopic neutron source. Information on major body elements, such as N, Ca, O, Na, and H, along with some toxic trace elements like Cd, can be obtained with this technique.

Concerns about the quality of our environment have led to a great increase in the number and types of analyses performed on **environmental samples.** Thermal NAA is particularly useful in the assessment of metal contamination, such as the elements Hg, Cd, As, Cu, and Sb in sewage and mining runoff waters. Thermal NAA has been used on air samples, including particulates, water, fuels such as coal and petroleum, the fly ash from coal combustion, and industrial and municipal wastes. The determination of U using the technique of delayed-neutron counting (measuring neutrons emitted by the delayed-neutron-emitter fission products of uranium) is also important in areas where uranium may be a natural or artificial environmental contaminant.

The analysis of **high-purity materials** for trace contamination is a difficult analytical task. The class of high-purity materials that is most often subjected to TNAA is the semiconductor material used in electronic components. Very small amounts of certain elements, such as B, can drastically alter the properties of the semiconductor. Boron analyses are difficult to perform using most analytical methods, but can be done with high sensitivity by TNAA. Other kinds of high-purity materials that can be analyzed using TNAA include quartz glass, metals, plastics, and ceramics. Although TNAA is usually used as a bulk analysis technique (i.e., to give information only on the content and not on the location of the elements), new applications of TNAA have also made localization possible. This is discussed further in Section 9.6.5.

Thermal NAA has been used in the analysis of **archaeological artifacts.** The multielemental information provided by NAA can be analyzed using statistical techniques such as cluster and factor analysis to give information about the original location of starting materials used in the object's creation, and sometimes indirect information about the time of its creation. **Art objects** can also be analyzed using nondestructive TNAA. Knowledge of the trace element profiles of the pigments used in paints can be used to establish the authenticity of a painting. The field of **forensics** has found many uses for TNAA. Explosive materials, for example, have unique trace element compositions that may aid in identifying the source of the explosive. Bullet leads also have unique trace element compositions that can be used to match slugs from crime scenes with specific batches of cartridges.

Many review articles have been published that discuss the myriad applications of TNAA. Some of these are listed at the end of the chapter.

9.2.4. Epithermal Neutron Activation Analysis (ENAA)

Epithermal neutrons are narrowly defined as those having energies from approximately 0.1 to 1.0 eV. In practice, the lower end of the neutron energy range used in ENAA is closer to 0.5 eV which is the Cd cutoff energy

($\approx$0.55 eV), and the higher end of the range extends through the resonance neutron region to several keV. These epicadmium neutrons, like thermal neutrons, usually induce only (n, γ) reactions. Delayed gammas are detected in most applications.

In the broadly defined epithermal (or more accurately, the epicadmium) energy region, the neutrons no longer follow the $1/v$ law closely. Rather, this region is characterized by the presence of resonances superimposed on the $1/v$ line. This was discussed in Chapter 4, and Figure 4.3 shows an excitation function for an element with very large resonance peaks in the epithermal region just above 1 eV. The total cross section for capture of epithermal neutrons is, therefore, the sum of the $1/v$ probability and that associated with the superimposed resonance peaks. These epithermal region neutron cross sections are called **resonance integrals (*I*).** For those elements with large resonance integrals, ENAA is a very sensitive and selective technique.

The main use of ENAA is to suppress interfering activities from product radionuclides produced in high yields from nonanalyte elements with high thermal neutron cross sections. As an example, geological materials contain significant levels of elements such as Na, Al, K, Sc, Cr, Mn, Fe, and La, all of which are strongly activated in a thermal neutron flux. Irradiation of a geological sample in a reactor will therefore result in such high activities from these elements that other trace element activities are swamped. If a desired low-level analyte has a large resonance integral and interfering elements do not, irradiation of the sample with epithermal neutrons will result in a better lower limit of detection and improved precision in the determination of the desired analyte. Epithermal irradiation is physically accomplished by putting the sample into an irradiation unit constructed of, or lined with, Cd or B. A small boron carbide thermal neutron shield is illustrated in Fig. 9.14. Some reactor facilities have a permanently shielded reactor irradiation position for ENAA. The Cd or B shield filters out most of the thermal neutrons and allows epithermal neutrons to pass into the sample.

The sensitivity enhancement produced by using ENAA may be expressed as an **advantage factor** (or ***F* value**). There are several ways to define the advantage factors. One is to compare the critical detection limits (Currie 1968) for an element produced by using both ENAA and TNAA with otherwise identical irradiation and counting parameters:

$$F = \frac{L_c}{L_{ce}} \tag{9.4}$$

where L_c = thermal (TNAA) critical detection limit

L_{ce} = epithermal (ENAA) critical detection limit

Figure 9.14. A small irradiation container constructed of boron carbide for use in ENAA.

The advantage factor may be further generalized by taking into account that the irradiated sample in ENAA may often be placed much closer to the detector than the "hot" TNAA sample, hence improving the counting geometry and further improving the sensitivity. In addition, higher flux reactor irradiation positions with lower Cd ratios may also be selected with advantage for ENAA and again improve detection limits for a given element. Simple advantage factors (excluding advantages of counting geometry and irradiation position) are usually in the range of 2–5, but some may range as high as 10^3.

ENAA is most used for geological and biological samples to discriminate against the high thermal neutron induced activities from matrix elements such as Na. Elements with high resonance integral/thermal neutron capture cross section ratios (I/σ_{th}) whose determination is usually enhanced by using ENAA include Ag, As, Au, Ba, Br, Cd, Cs, Ga, Gd, In, Mo, Ni, Pd, Pt, Rb, Sb, Se, Sm, Sr, Ta, Tb, Th, Tm, U, and W. Spectra of a neutron-irradiated standard coal sample with and without boron shielding are shown in Fig. 9.15. Elements more readily determined by TNAA or ENAA in this sample are identified.

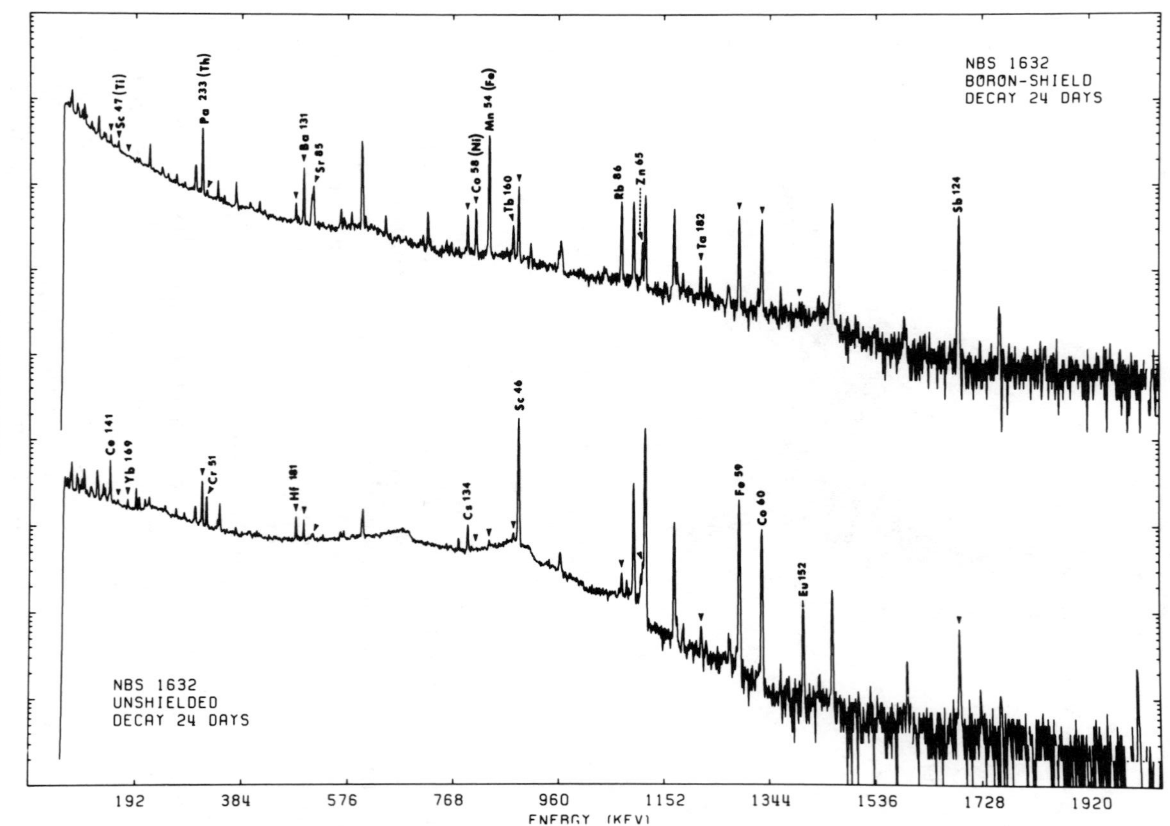

Figure 9.15. Spectra of a neutron-irradiated standard coal sample with and without the use of a boron carbide thermal neutron shield.

9.2.5. Fast Neutron Activation Analysis (FNAA)

The fast neutrons encompass the widest range of energies, from 0.5 MeV up. Neutrons of this energy do not induce as many (n,γ) reactions as thermal neutrons, but are instead responsible for threshold reactions, such as (n,p), (n,α), (n,n'), and $(n,2n)$ reactions. Most FNAA procedures use 14-MeV neutrons produced by a neutron generator. The fast neutron flux in a nuclear fission reactor is occasionally used for analytical reactions with low threshold energies.

As noted earlier, the lower cross sections and smaller fluxes of fast neutrons mean that FNAA is not as sensitive as TNAA. However, FNAA is capable of determining many of the light elements that cannot be done with either thermal or epithermal NAA. Oxygen analyses are often difficult to perform with conventional analytical methods, but quite simple with 14-MeV FNAA. The reaction used is $^{16}O(n,p)^{16}N$. The ^{16}N indicator radionuclide has a half-life of 7.13 s, and emits high-energy gammas (6.13 and 7.12 MeV), which are easy to detect and to distinguish from other NAA product γ rays. Oxygen levels down to 0.01% (or even lower in certain types of large samples) can be determined in many matrices using this approach. The reaction is also used for in vivo NAA determinations for oxygen.

Another element that is easily determined using 14-MeV neutrons is nitrogen. The reaction is $^{14}N(n,2n)^{13}N$. The ^{13}N indicator radionuclide is a positron emitter, and the 511-keV annihilation photons are detected. This reaction is also used in in vivo activation analysis.

Other elements that are routinely determined by FNAA are F, Mg, Al, Si, Cu, Fe, P, and Zn. A summary of the most commonly used 14-MeV FNAA determinations is given in Table 9.4. It should be remembered, however, that FNAA employing a neutron generator is not normally a trace-level determination technique. Conventional detection limits with small (several grams) samples are generally no better than tens or hundreds of micrograms per gram.

9.2.6. Radiochemical Neutron Activation Analysis (RNAA)

Radiochemical NAA includes any type of NAA in which sample irradiation is followed by chemical processing. The chemical treatment is designed to separate single elements or groups of elements so that the ultimate selectivity and sensitivity for the analyte is obtained. The separated fractions are then assayed for the analyte activity. Some elaborate RNAA separation schemes for biological matrices have been completely automated.

Table 9.4. Common 14-MeV FNAA Determinations

Element	Analytical Reaction	Product Half-Life	Gamma Rays (MeV)
Oxygen	$^{16}O(n,p)^{16}N$	7.13 s	6.13, 7.12
Nitrogen	$^{14}N(n,2n)^{13}N$	9.97 min	0.511[a]
Silicon	$^{28}Si(n,p)^{28}Al$	2.25 min	1.78
Phosphorus	$^{31}P(n,\alpha)^{28}Al$	2.25 min	1.78
Copper	$^{63}Cu(n,2n)^{62}Cu$	9.74 min	0.511[a]
Zirconium	$^{90}Zr(n,2n)^{89m}Zr$	4.18 min	0.588
Thorium	Fission	—	Delayed neutrons
Uranium	Fission	—	Delayed neutrons

[a] β^+ annihilation radiation.

Because the actual amounts of radioactive material are extremely small, a chemical "carrier" is added to the solutions. This **carrier**, chemically identical to the analyte, provides a more manageable amount of sample for the analyst. The carrier also lets the analyst determine the efficiency of the separation process from a knowledge of the amount of carrier added at the beginning of the separation (greatly in excess of the amount of analyte present) and the amount present at the completion of the separation. Care must be taken that the carrier equilibrates with the chemical species of the analyte present in the sample and that volatile elements are not lost in the sample dissolution process. Sealed Teflon Parr bombs are commonly used for dissolution of biological samples prior to radiochemical separations.

Many different separation methods have been used in RNAA. The same types of chemical treatments used in conventional analyses are often used. Sample dissolution procedures depend on the type of sample. Many separation schemes use conventional solvent extraction and ion exchange procedures, although unique approaches have been devised for various elements. A good separation of a single radionuclide would mean that NaI(Tl) detectors, or even liquid-scintillation counters, with their high relative detection efficiency could be used for counting. If the sample is not radiochemically pure, Ge(Li) or HPGe detectors, with their good energy resolution, can be used to discriminate against residual interferences.

RNAA is most often used on samples that have complex matrices producing activities from many elements. Geological and biological samples fit into this category, and it is for these types of samples that most RNAA procedures have been developed. In geological samples, the rare-earth elements, noble metals, and transuranium elements are frequent targets of RNAA. In biological samples, the elements As, Cd, Cu, Hg, Mo, Se, and Zn are often determined using an RNAA approach.

9.3. PROMPT-GAMMA NEUTRON ACTIVATION ANALYSIS (PGNAA)

Neutrons of any energy may be used for PGNAA. When a nucleus captures a neutron, it gains energy which results in excited nuclear states. These de-excite very quickly (in less than 10^{-13} s) by **prompt-gamma emission.** These γ rays are detected and used analytically. Because the half-life of these excited states is so short, samples cannot be transferred from the irradiation location to the counting location as is usually done with delayed-γ NAA. Systems for PGNAA must be designed so that irradiation and counting can be done simultaneously. This necessitates the presence of the counting equipment very close to the reactor, which is normally an environment with a very high background of γ rays and neutrons. In most PGNAA procedures, a beam of neutrons is actually extracted from the reactor. A great deal of shielding must be provided to protect the detector from scattered radiation. Long irradiation times are often needed because extracted beams have lower intensities than irradiation positions inside the reactor. Generally, only one sample at a time can be analyzed.

Altogether, the actual practice of PGNAA is more difficult than that for delayed-γ NAA. Despite these experimental difficulties, PGNAA is done because it can give information complementary to that of delayed-γ NAA. It is quite sensitive to the major components of many samples, such as H, Si, S, C, N, and P, elements for which delayed-γ NAA is not as sensitive. A recent development that may alleviate some of the difficulties associated with PGNAA and also enhance the sensitivity of the method is the production of **cold neutrons** at several reactor sites worldwide. Cold neutrons have very low energies, about 0.005 eV. They are produced by placing a cold source, generally a refrigerated tank of liquefied deuterium, inside a nuclear reactor in an area of high neutron flux. The neutrons are cooled in this tank and guide tubes are provided for the neutrons to leave the cold source.

The very low energy of the cold neutrons means that they have very large de Broglie wavelengths, which result in large capture cross sections. The enhanced wave properties of the neutron also lead to the possibility of guiding and even focusing neutron beams using well-known principles of optical wave guides. If the cold neutron beam can be guided to an area farther from the reactor, the background problems associated with PGNAA can be reduced. Both the enhanced cross section and the lowered background result in better sensitivities for many elements of interest. A cold neutron facility is installed at the National Institute for Standards and Technology in Gaithersburg, Maryland.

Applications of PGNAA

PGNAA is used to perform analyses of the same types of samples as in delayed-γ NAA. The prompt γ rays from Ca, N, Cd, H, Cl, and P are often used for in vivo NAA. An industrially important application of PGNAA is the analysis of semiconductor materials for very low levels of B. PGNAA has also been used for **well-logging** procedures. A portable PGNAA system containing a neutron generator, or an isotopic neutron source, can be lowered into the borehole drilled while prospecting for oil or mineral deposits. Analysis of the composition of the surrounding materials is done on the spot by monitoring the prompt γ rays emitted. In a similar vein, PGNAA has been considered as a technique for the **remote analysis** of extraterrestrial bodies and can be used industrially for **on-line analysis.**

9.4. CHARGED-PARTICLE ACTIVATION ANALYSIS (CPAA)

9.4.1. Principles of CPAA

In CPAA, the activating particles are the charged particles generated by accelerators. Protons, deuterons, tritons, ^{3}He, and α particles are the most commonly used particles.

The type of nuclear reaction induced in the sample nuclei depends on the identity and energy of the incoming charged particle. Protons can be easily accelerated and have low Coulomb barriers, so they are a favorite choice for many applications. The (p, n) reaction results most often with protons up to approximately 15 MeV in energy. Higher-energy protons may also induce (p, α), (p, d), or (p, γ) reactions. With deuterons, it is mainly (d, n) and (d, p) reactions that are utilized. Alpha particles and ^{3}He ions have higher Coulomb barriers than the proton and other singly charged particles, so higher initial energies are needed for them to induce nuclear reactions. This, in turn, imparts a greater amount of energy to the target nucleus during reaction, which means that the nucleus can de-excite in a greater variety of ways. Therefore, a greater variety of nuclear reactions can occur and more detection options are available to the analyst.

The elements that can be determined well by CPAA are different from those ordinarily determined with NAA. The light elements, which have low Coulomb barriers and low capture cross sections for neutrons, are easily done using CPAA. Because charged particles may not penetrate the entire sample as neutrons do, CPAA is often used as a thin sample, or surface analysis technique. Ordinarily, only one sample at a time is analyzed by CPAA, so it is not a good technique for performing large numbers of routine analyses. Also, samples are likely to be damaged more by CPAA than by NAA, due to the charged particles' greater specific ionization in matter.

Charged particles interact with matter in ways other than nuclear reactions, and these interactions also have analytical applications. Charged particles may interact with the atomic electrons of the sample atoms rather than with the nucleus, with resultant ejection of an electron from the atom. The internal electron rearrangements that occur subsequent to electron removal produce x rays that are characteristic of the element. As noted previously, this technique is called PIXE. Scattering of charged particles from target nuclei also occurs. Detection of the backscattered particles is the basis for **Rutherford backscattering,** or **RBS**. Analysis of very small areas of a sample can be accomplished by using charged-particle beams that have been finely focused. These **nuclear microprobes,** with spatial resolutions of 1 μm, or better, can be used for PIXE and PIGE analyses of small mineral grains in geological materials and subcellular components in biological tissue. These techniques (PIXE, RBS, and nuclear microprobes) are discussed in Chapter 11.

9.4.2. Calculations of CPAA

The equations used for calculations in NAA are not directly applicable for CPAA for several reasons. One is that the charged particles are not able to penetrate the entire sample as neutrons do, so the number of target atoms may not be the same as the total number of atoms present. As the charged particles pass through the sample, they lose energy, so the cross section for a given reaction is not constant. An activation equation modified for use in CPAA is

$$A = \frac{CNaI}{M} \int_0^R \sigma(x)\, dx \tag{9.5}$$

where A = absolute activity

C = concentration

N = Avogadro's number

M = atomic weight

I = beam intensity (particles/s)

R = range in sample (cm)

x = depth in sample (cm)

σ = cross section (cm^2)

a = isotopic abundance

As with NAA, the fundamental equation is not used directly in practice. A comparator method is used. A working equation for the comparator method is

$$C_x = \frac{C_s A_x I_s R_s}{A_s I_x R_x} \tag{9.6}$$

where C = concentrations of an unknown (x) and standard (s)

A = measured activity

I = beam intensity

R = particle range dependent on matrix (from tabled values)

9.4.3. Applications of CPAA

CPAA has been applied in the same areas as mentioned for NAA. Of these, the analysis of light elements in metals or semiconductors is one of the most important industrially. The importance of B in semiconductor function has already been mentioned. The amounts of boron in metals used for nuclear reactors is also an important quantity to determine, because B has such a large cross section for neutrons. Boron can be determined using the (p, n) reaction on ^{11}B, the more abundant isotope, or the (d, n) or (p, α) reactions on ^{10}B. Carbon can be determined in semiconductors and other metals via the ^{12}C(^{3}He, α)^{11}C or ^{12}C(d, n)^{13}N reactions. The ^{14}N(p, α)^{11}C reaction has been used to determine N in metals. For O, irradiation with ^{3}He to produce ^{18}F is used in metallic materials. Heavier elements can also be determined using CPAA. Table 9.5 summarizes some of the most useful CPAA determinations.

Irradiation of samples with heavier charged particles (A = 4 or higher) is used primarily with detection of the prompt γ rays emitted by the excited nuclei. This specialized form of the CPAA branch of activation analysis is referred to as particle-induced γ-ray emission (PIGE).

9.4.4. Particle-Induced Gamma-Ray Emission (PIGE)

In PIGE (also referred to as PIGME or GRALE) the prompt γ rays emitted following a charged-particle-induced nuclear reaction are detected and used for both qualitative and quantitative analysis. PIGE is thus analagous to PGNAA in neutron activation methods. Protons, deuterons, tritons, α particles, and heavy ions have all been used as activating particles in various applications. PIGE is one of several techniques included under the heading of **nuclear reaction analysis** (NRA) which use ion beams to induce nuclear

Table 9.5. Representative CPAA Determinations

Element	Analytical Reaction	Product Half-Life	Gamma Rays (MeV)
Boron	$^{11}B(p,n)^{11}C$	20.3 min	0.511[a]
	$^{10}B(d,n)^{11}C$	20.3 min	0.511[a]
	$^{10}B(p,\alpha)^{7}Be$	53.28 d	0.478
Carbon	$^{12}C(d,n)^{13}N$	9.97 min	0.511[a]
	$^{12}C(^{3}He,\alpha)^{11}C$	20.3 min	0.511[a]
Nitrogen	$^{14}N(p,\alpha)^{11}C$	20.3 min	0.511[a]
	$^{14}N(p,n)^{14}O$	70.6 s	2.31, 0.511[a]
	$^{14}N(d,n)^{15}O$	122 s	0.511[a]
Oxygen	$^{16}O(p,\alpha)^{13}N$	9.97 min	0.511[a]
	$^{16}O(^{3}He,p)^{18}F$	1.83 h	0.511[a]
Phosphorus	$^{31}P(\alpha,n)^{34m}Cl$	32.2 min	2.13, 1.18, others
Calcium	$^{40}Ca(\alpha,p)^{43}Sc$	3.89 h	0.373, others
Titanium	$^{48}Ti(p,n)^{48}V$	15.98 d	0.984, 1.31, others
Iron	$^{56}Fe(p,n)^{56}Co$	77.3 d	0.847, 1.24, others
Zirconium	$^{90}Zr(p,n)^{90}Nb$	14.6 h	1.13, 2.32, others
Silver	$^{107}Ag(^{3}He,2n)^{108m}In$	57 min	0.633, 0.876, others
Cadmium	$^{111}Cd(p,n)^{111}In$	2.80 d	0.245, 0.171
Tungsten	$^{182}W(p,n)^{182m}Re$	12.7 h	1.12, 1.22, others
Lead	$^{206}Pb(p,n)^{206}Bi$	6.24 d	0.803, 0.881, others

[a] β^+ annihilation radiation.

Note: Many other product radionuclides listed are also β^+ emitters, but the γ rays listed are usually measured, rather than annihilation radiation.

reactions for analytical use. PIGE is the most commonly used of the NRA techniques.

As for PGNAA, the experimental work is complicated by the fact that detection must occur simultaneously with irradiation. Either the sample itself may be located within the beam vacuum system, or the beam can be extracted for irradiation of the sample outside the vacuum. The detectors used are the same as those used for γ-ray detection in neutron activation analysis, including BGO, NaI(Tl), Ge(Li), and HPGe. The choice of detector depends on the particular needs for efficiency and energy resolution in a given application.

PIGE is most useful for the analysis of light elements ($Z < 15$ and $A < 30$). In this, it is complementary to PIXE, and the two techniques are often used together because they require similar experimental apparatus. PIGE does have simultaneous multielement capability for the lighter elements. Variation of the type and energy of the incident particles permits some selectivity in the atoms activated and can also be used to minimize

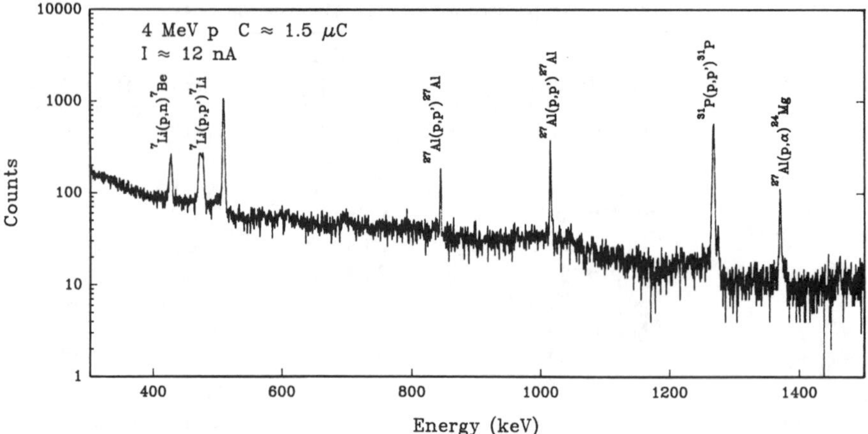

Figure 9.16. Gamma rays from the proton bombardment of a 1-μm-thick (1-x)Li_4SiO_4: $x Li_3PO_4$ film. The PIGE analysis was used to determine the Li and P content of the film. (Courtesy of J. D. Robertson, University of Kentucky, Lexington.)

potential interfering background. Some heavier elements can be determined in matrices where few lighter nuclides are present. The γ-ray yield is not strongly affected by surface roughness, which can be a problem for techniques that detect lower-energy emissions.

PIGE is mainly used to determine light elements at a surface or in thin films. Depth profiling, with a resolution of 1–10 nm for elements such as H, ^{19}F, and ^{27}Al, is possible. The elements Li, Be, F, and Na are frequent targets for PIGE, and can be determined at the μg/g level. Figure 9.16 shows a spectrum from a PIGE analysis used to determine the Li and P content of a thin film. PIGE has been used to characterize ^{15}N profiles in metals. The ability to determine F is especially notable, because analysis for this element is quite difficult by nonnuclear methods. Table 11.1 (in Chapter 11) summarizes some joint applications of PIGE and PIXE.

Energetic neutrons from accelerators may also be used to excite nuclei through inelastic scattering reactions. The prompt gamma rays emitted are used analytically, as in PIGE.

9.5. INSTRUMENTAL PHOTON ACTIVATION ANALYSIS (IPAA)

Photons may be used to induce threshold nuclear reactions in a technique called **instrumental photon activation analysis (IPAA).** The nuclear reaction that actually occurs will depend on the atomic number of the target atom and on the energy of the photons used for irradiation. For photons with energies of 15–20 MeV, the (γ, n) reaction predominates. Other reactions that can be used include (γ, p), $(\gamma, 2n)$, and (γ, α). Although photon energies between 15 and 20 MeV are commonly used in PAA, the lower-energy 2.1-MeV γ rays from ^{124}Sb can be used for the determination of Be by the $^{9}\text{Be}(\gamma, n)2\text{-}^{4}\text{He}$ reaction with prompt neutron detection. The latter reaction is possible because of the low neutron binding energy of the Be. The source of photons used for IPAA is nearly always the bremsstrahlung radiation produced by electron accelerators.

Like CPAA, IPAA yields information complementary to NAA. Analyses for the light elements C, N, O, and F are good examples of useful applications where absolute detection limits of $\leq 0.5\ \mu g$ are possible. The (γ, n) reaction on these elements leaves the product nucleus proton rich, so the analytical nuclide is frequently a positron emitter. This means that radiochemical separations usually must be done, because several elements in a sample will emit 511-keV annihilation photons. Discrimination by half-life may also be possible. Heavier elements can be determined with IPAA too, with the advantage that the product radionuclides emit their own characteristic γ rays, rather than just positron annihilation radiation.

IPAA is similar to NAA in that the photons can completely penetrate most samples. Thus, the procedures and calculations used for IPAA are like those used in NAA, except for the irradiation source. IPAA has not been as widely used as NAA or even CPAA, due in large part to the limited availability of photon sources. The technique has been used to analyze semiconductor materials and metals for trace amounts of C, N, O, and F. The N content of biological materials, which is an indicator of protein content, can readily be determined with IPAA. Lead, an element of great environmental interest, cannot be determined with TNAA, but can be done with IPAA with an absolute detection limit of $\approx 0.5\ \mu g$ in small samples.

9.6. SPECIAL ACTIVATION ANALYSIS TECHNIQUES

9.6.1. Derivative Activation Analysis (DAA)

Although NAA is applicable to a majority of the chemical elements, there are some that cannot be determined by this direct activation technique. It may be that the cross section for the analytical neutron capture reaction is too small, or that of an interfering reaction (direct or spectral interference) is too large. The product radionuclide may emit radiations that are unsuitable for analytical purposes, or the half-life may be either too long (yielding a low specific activity), or too short (allowing insufficient time for transfer and counting). **Derivative activation analysis (DAA)** is a novel composite analytical approach that may be used for elements that are poorly determined using conventional NAA. In DAA, the poorly determined element is chemically exchanged for, or complexed with, an element that is amenable to NAA. After the chemical reaction, the sample is irradiated, and the amount of the element of interest is calculated from the activity of the surrogate element. For DAA to be successfully used, the derivative formed must have a known and constant stoichiometry, and it must be able to be isolated cleanly after its formation.

Examples of elements not ordinarily determined, or determined with poor sensitivity, by conventional reactor INAA include Li, Be, Ni, P, Nb, Rh, Si, Sn, Tl, Pb, and Bi. Determination of several of these elements in biological and environmental samples is of interest because they may be either toxic or essential. PGNAA, FNAA, CPAA, or IPAA can be used sometimes, where the special facilities required are available. However, in a given matrix, the purely instrumental determination of any element can be restricted by reaction or spectral interferences, even if the fundamental nuclear properties of both the stable target nuclide and the product indicator radionuclide are favorable. Published DAA studies have been limited to NAA.

In practice, DAA uses a preirradiation chemical reaction on the sample to initiate the formation of, or an exchange with, a chemical complex that contains a surrogate element, S. As a result, the amount of the element or the chemical species to be determined, X, is now represented by measurement of the amount of the surrogate element, S, that is made part of, or released by the complex species. The surrogate element is selected for its superior properties for nuclear activation analysis and the absence of interference reactions in its determination by NAA after some separation chemistry.

Ideally, element S should have a naturally occurring stable target isotope with a high isotopic abundance, a large thermal neutron cross section, and an irradiation product indicator radionuclide with a suitable half-life for easy handling and counting. The indicator radionuclide produced from irradiation of S should undergo decay with the emission of one or more γ rays that yield photopeaks in interference-free regions of the γ-ray spectrum obtained from the neutron-irradiated, chemically processed sample matrix.

When element S is to be complexed with element X, additional characteristics are desirable. The molar stoichiometric ratio, S/X, should be 1.0 or greater, any excess S from the original complexing medium that has not reacted stoichiometrically with element X should be easy to remove, and the final chemical form containing the surrogate element should be stable and easily isolated.

Three approaches by which the surrogate element, S, may be substituted for the analyte of interest, X, can be considered. In the first, the element of interest in an aqueous phase, X_{aq}, is combined with the surrogate indicator element, S_{aq}. The complex formed, (XS), is extracted into an organic solvent to separate it from excess complexing reagent, S_{aq}, and to effect a concentration of the analyte. The separated complex $(XS)_{org}$ is then irradiated and the induced activity produced from S is used as an indirect measure of X in the original sample. The separation procedure in all DAA approaches must be quantitative or have a reproducible, measured chemical yield.

In the second approach, the element or chemical species of interest, X_{aq}, is exchanged for the surrogate element, S, which has been introduced to the aqueous solution of the analyte as a complex, $(SL)_{org}$, in an organic solvent. L is usually an organic ligand. The surrogate element released from the complex and transferred to the aqueous solution, S_{aq}, is collected, quantified by NAA, and used as a measure of X in the original sample solution. This method has the advantage that the complex, $(SL)_{org}$, is introduced in an organic solvent which is immiscible with the analyte solution. Thus, the excess complexing agent can be separated from the released S simultaneously with the exchange reaction.

The third DAA approach is essentially the reverse of the second approach. The element or species of interest, X_{aq}, is first quantitatively converted to the complex $(XL)_{aq}$ and then isolated by solvent extraction into an organic solvent of volume V_{org}. $(XL)_{org}$ is then exchanged with the surrogate element, S_{aq}, to form $(SL)_{org}$ which is separated and irradiated, and the induced activity from S is counted. This method has the potential for preconcentration of X, if V_{org} is smaller than V_{aq}. An additional advantage is that both the exchange reaction and the separation of the complex from excess reagent can be completed simultaneously.

An example of a DAA procedure is the determination of P. Activation of P does occur, but the resulting radionuclide (^{32}P) emits only high-energy negatrons. Beta-particle counting is not always suitable for quantitative analysis, because there are often many β emitters produced in the irradiation, and energy discrimination is inferior to that in γ-ray spectrometry. Phosphorus is easily complexed to form a vanadium-containing phosphovanadomolybdate moiety of known stoichiometry. Vanadium is an element that is readily amenable to NAA. Irradiation of the complex and subsequent counting of the sample for ^{52}V activity will enable the analyst to determine the amount of P in the sample. The method has been used to determine P in natural waters, brain tissue, and commercial plant foods.

The DAA approach has also been used for the elements Mg (using ^{82}Br activity), Si (using ^{101}Mo–^{101}Tc activity), and Ni (using ^{198}Au activity). It has been used for functional group analysis of coal. The hydroxyl concentrations in coal samples are inferred by forming silyl ether moieties at –OH sites by treatment of the coal with hexamethyldisilazane and measuring the increase in Si concentration in the coal following derivatization by using

14-MeV FNAA. The carbonyl contents of the coals are determined by treating the samples with hydroxylamine to form oximes and measuring the increase in N content of the sample after derivatization by using 14-MeV FNAA.

DAA has the potential for use in other functional group analyses and for elemental speciation in natural systems. The method is still largely unexplored.

9.6.2. Cyclic Activation Analysis

Cyclic activation analysis is based on the concept of enhancing the sensitivity of the activation method for the determination of elements with short-lived indicator radionuclides by use of repetitive short irradiation and counting periods and summing of the γ-ray spectra obtained. Use of the procedure will result in improved counting statistics for the short-lived species due to summing and thereby increase the effective signal-to-noise ratio in the determination of short half-life indicator radiouclides with respect to interfering longer-lived radionuclides.

With use of the short irradiation times, saturation activity levels can be reached for the short-lived species in each irradiation. In contrast, longer-lived species activity levels increase only slowly with successive irradiations, with the use of appropriate decay intervals between irradiations. The total detector response (events recorded) for a short-lived indicator radionuclide determined by cyclic NAA will increase linearly with experiment time (irradiation time + counting time), as shown in Fig. 9.17. For conventional "one-shot" NAA, the saturation factor in the activation equation approaches unity after an irradiation time equal to several half-lives of the shorter-lived species, and increased irradiation time enhances only the activity levels of the longer-lived interference radionuclides. It should be noted that discrimination between a short-lived and a long-lived radionuclide produced in an irradiation is always enhanced by using the shortest possible irradiation and counting periods. However, short irradiations and short counting periods result in poor counting statistics. The use of the cyclic procedure results only in enhanced counting statistics for the shorter-lived radionuclide while keeping signal-to-noise ratios low.

The detector response in cyclic activation analysis is given by the following equation:

$$\mathrm{DR} = \frac{N\phi\sigma k}{\lambda}(1 - e^{-\lambda t})(1 - e^{-\lambda t'})(e^{-\lambda t''})\left[\frac{n}{1 - e^{-\lambda T}} - \frac{(e^{-\lambda T})(1 - e^{-n\lambda T})}{(1 - e^{-\lambda T})^2}\right] \tag{9.7}$$

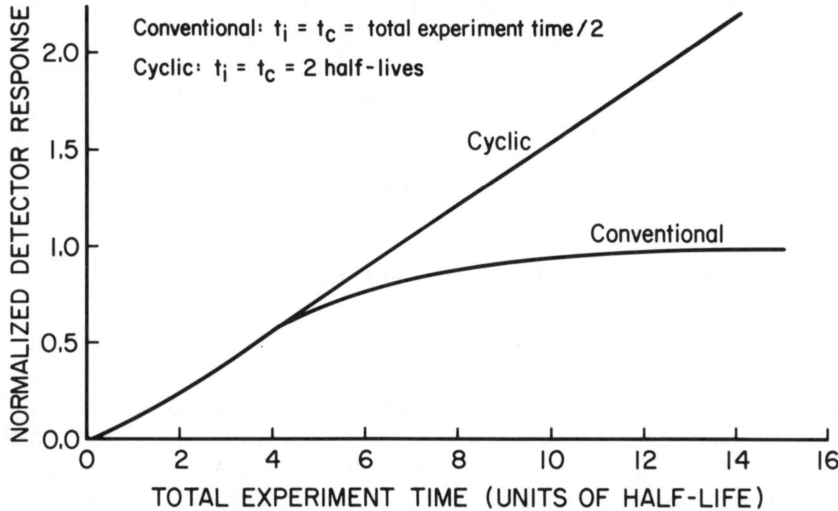

Figure 9.17. Increase in detector response by use of cyclic activation analysis for short-lived indicator radionuclides.

where DR = the detector response in terms of cumulative counts recorded in n counting periods

N = number of stable target nuclei of interest in the sample

ϕ = flux density of incident particles (n cm^{-2} s^{-1} in NAA)

k = factor containing the detector efficiency and branching ratio for the decay path

λ = the decay constant for the indicator radionuclide (s^{-1})

t = irradiation time for each cycle (s)

t' = counting time for each irradiation (s)

t'' = delay time prior to counting for each irradiation (s)

T = the delay period from the end of a previous delay period to the end of the next delay period (cycle time, s)

n = the number of cycles

Optimization of the process depends on selection of the appropriate times in Eq. 9.7. Many short cycles are most desirable for discrimination from long-lived interferences and optimum results may be obtained by setting the counting time equal to the time of irradiation in each cycle.

When neutrons are used as the activating particle, the technique is called **cyclic instrumental neutron activation analysis (CINAA).** The technique has been used to determine Se in foods and short-lived indicator radionuclides in zoological and botanical samples. Both reactor thermal neutrons and 14-MeV neutrons are used with CINAA.

9.6.3. Secondary Particle Activation Analysis

Secondary particle activation analysis is actually a unique approach to charged-particle activation analysis (CPAA) in which a neutron flux is converted to a charged-particle flux. Typically, a Li salt such as LiF is mixed with the sample to be analyzed and the mixture is irradiated with thermal neutrons. The $^{6}Li(n,t)^{4}He$ reaction produces ≈4.8-MeV tritons which induce charged-particle reactions in the sample. Product radionuclides from these reactions are then counted, as in conventional NAA. Sample particle sizes must be kept small so that the tritons will not suffer severe energy degradation. While the concept is interesting, there have been few applications of this technique.

9.6.4. Coincidence Techniques in Activation Analysis

An alternative approach to minimizing spectral interferences in NAA is to use time discrimination as well as energy resolution. Significant improvements in sensitivity and precision can be achieved in special cases through the use of **γ–γ fast coincidence spectrometry.** One of the simplest examples is the use of the 180°-correlated annihilation photons emitted by positron-emitting indicator radionuclides. The coincidence requirement and the defined angular correlation for photon emission results in reduction of spectral interferences from single γ-ray emitting interferences, as well as discrimination from cascade γ-ray emitters whose emissions are isotropic (emitted equally in all directions) or correlated at other angles.

If a simple decay scheme for an indicator radionuclide containing two coincident isotropic γ rays, γ_a and γ_b, in every decay is assumed, the following equation will give the number of coincidence events recorded if the coincidence resolving time is assumed to be zero:

$$N_{ab} = \epsilon_a \epsilon_b A t \tag{9.8}$$

where N_{ab} = the total number of coincidence events recorded

ϵ_a and ϵ_b = the detection efficiencies of γ_a in detector 1 and γ_b in detector 2, respectively

A = indicator radionuclide decays per unit time

t = counting time

However, coincidence resolving times (τ) for real systems are not zero and accidental coincidences may be recorded. The number of accidental coincidences (N^*_{ab}) due to finite coincidence resolving times under the stated conditions is approximated as follows:

$$N^*_{ab} = 2\tau\epsilon_a\epsilon_b A^2 t \tag{9.9}$$

The ratio of accidental to true coincidences, R, is:

$$R = 2\tau A \tag{9.10}$$

Thus, to minimize accidental coincidences, it is necessary to use instrumentation with a short coincidence resolving time and adjust counting geometries to achieve appropriate counting rates to optimize the system.

Instead of using two detectors with energy windows set to accept selectively the two coincident γ rays of interest, it is possible to use multidimensional γ-ray spectrometry and simultaneously record coincidence events at all energies. This is particularly useful when determining several short-lived radionuclides in an irradiated sample where time for counting is limited. Figure 9.18 illustrates the output of a multiparameter coincidence spectrometer, when counting a neutron-irradiated lunar rock sample.

9.6.5. Localization Methods in Activation Analysis

Activation analysis has traditionally been used to determine total elemental levels, without regard to the location of the elements in the sample. However, for many applications, the location of the element is of primary interest. Some of the activation techniques that have been adapted or developed to respond to this analytical need are described in this section.

Neutron Depth Profiling

The familiar analytical problem of the determination of B in semiconductor materials was the impetus for the initial efforts at developing the technique

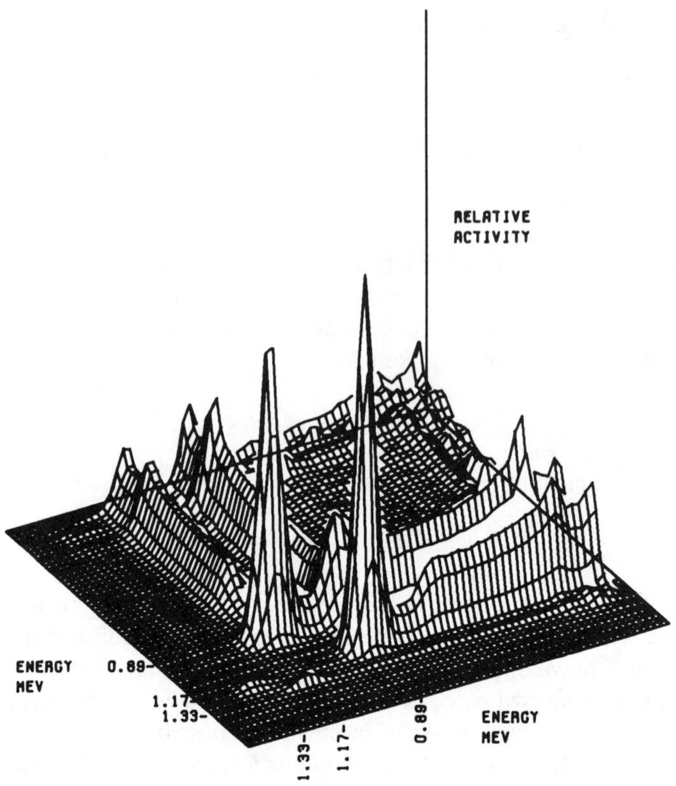

Figure 9.18. Multiparameter coincidence γ-ray spectrum of a neutron-irradiated lunar rock sample.

of **neutron depth profiling (NDP).** In NDP, a beam of thermal neutrons is allowed to impinge on the sample material. For B, and some other light nuclides, the thermal neutrons induce the emission of charged particles, probably protons or α particles, and a recoil nucleus. Each of these particles will have a definite energy, defined by the Q value for the nuclear reaction. As the charged particles pass through the sample, they will lose energy. The particles are detected after they leave the sample, and the difference in the initial energy and measured energy can be related to the depth of the original target nucleus in the sample.

Table 9.6 shows some elements that can be determined using neutron depth profiling. Using thermal neutrons, depth profiles of 1–10 μm are common. The use of cold neutron beams for NDP should further enhance the sensitivity of this method.

Table 9.6. Representative Determinations by Neutron Depth Profiling

Element	Reaction	Energy of Emitted Particles (MeV)		Detection Limit (atoms/cm^2)
He	$^{3}He(n,p)^{3}H$	0.572	0.191	3.1×10^{13}
Li	$^{6}Li(n,\alpha)^{3}H$	2.055	2.727	1.8×10^{14}
Be	$^{7}Be(n,p)^{7}Li$	0.143	0.207	3.5×10^{12}
B	$^{10}B(n,\alpha)^{7}Li$	1.472	0.840	4.3×10^{13}
N	$^{14}N(n,p)^{14}C$	0.584	0.042	9.1×10^{16}
O	$^{17}O(n,\alpha)^{14}C$	1.413	0.404	7.1×10^{17}
Na	$^{23}Na(n,p)^{23}Ne$	2.247	0.103	4.7×10^{12}
S	$^{33}S(n,\alpha)^{30}Si$	3.081	0.411	1.2×10^{18}
Cl	$^{35}Cl(n,p)^{35}S$	0.598	0.017	3.4×10^{17}
K	$^{40}K(n,p)^{40}Ar$	2.231	0.056	3.8×10^{16}
Ni	$^{59}Ni(n,\alpha)^{56}Fe$	4.757	0.340	1.4×10^{16}

Source: Adapted from R. G. Downing et al., *Nucl. Instrum. Methods Phys. Res.* **218:**47 (1983).

Neutron Activation Tomography

Tomographic techniques refer to those in which many individual projections through an unknown structure are mathematically manipulated to construct an image of the unknown. A technique that uses tomographic reconstruction that is probably familiar to most people is computerized axial

Table 9.7. Representative in Vivo Determinations by NAA

Element	Reaction(s)	Gamma Ray Detected (MeV)
Ca	$^{48}Ca(n,\gamma)^{49}Ca$	Delayed: 3.10
		Prompt: many
N	$^{14}N(n,2n)^{13}N$	Delayed: 0.511
	$^{14}N(n,\gamma)^{15}N$	Prompt: 10.8
Cd	$^{113}Cd(n,\gamma)^{114}Cd$	Prompt: 0.559
O	$^{16}O(n,p)^{16}N$	Delayed: 6.1
H	$^{1}H(n,\gamma)^{2}H$	Prompt: 2.223
Na	$^{23}Na(n,\gamma)^{24}Na$	Delayed: 2.75, 1.369
Cl	$^{37}Cl(n,\gamma)^{38}Cl$	Delayed: 2.168
		Prompt: many
P	$^{31}P(n,\alpha)^{28}Al$	Delayed: 1.78
	$^{31}P(n,\gamma)^{32}P$	Prompt: 0.08

Source: Adapted from D. R. Chettle and J. H. Fremlin, *Phys. Med. Biol.* **29:**1011 (1984), and S. H. Cohen, *Textbook of Nuclear Medicine,* Vol. 1, *Basic Science.* New York: Lea & Febiger, 1984.

tomography, or CAT scanning, used for medical imaging. In **neutron activation tomography,** the sample is activated with neutrons, and the γ rays emitted by the sample are used to reconstruct the image. Both two- and three-dimensional images have been formed using neutron activation tomography. If the radionuclides emit positrons, as is the case in positron emission tomography (PET), spatial configurations are more easily inferred due to the angular correlation of the annihilation photons generated. Representative in vivo determinations are listed in Table 9.7.

TERMS TO KNOW

Absolute NAA
Accelerator neutron source
Activating particles
Activation analysis
Adjustable experimental parameters
Advantage factor in ENAA
α-emitter neutron source
Cadmium ratio
Carriers in RNAA
CNAA
Cockroft–Walton neutron generator
Cold neutrons
Comparator NAA
CPAA
Cyclic activation analysis, CINAA
Cyclotron
Delayed γ rays
Derivative activation analysis, DAA
ENAA
Epicadmium neutrons
Fast neutrons
FNAA
γ-γ fast coincidence spectrometry in NAA
γ-ray spectral interference
γ self-shadowing
GRALE
INAA
In vivo NAA
IPAA
k_0 method
Linear accelerator
Molecular activation analysis
NAA
Neutron activation tomography
Neutron depth profiling, NDP
Particle-induced x-ray emission analysis, PIXE
PGNAA
PGNAA well-logging
Photon neutron source
PIGE
PIPPS
Primary interference reaction in activation analysis
Radiologic safety
RBS
Research reactors
Resonance capture
Resonance integrals
RNAA
Second order additive interference
Secondary interference reaction
Secondary particle activation analysis
Self-shadowing, incident particles/radiations
Simultaneous multielement analysis
Speciation
Spontaneous fission neutron source
SRM
Thermal neutrons
TNAA
Van de Graaff accelerator

READING LIST

Alfassi, Z. B., ed., *Activation Analysis,* Vols. I and II. Boca Raton, FL: CRC Press, 1989. [topics in activation analysis, including many special methods (DAA, RNAA, FNAA, IPAA, NDP) and applications for different types of matrices]

Al-Mugrabi, M. A., and N. M. Spyrou, *J. Radioanal. Nuclear Chem.* **110**:67–77 (1987). [cyclic activation analysis]

Becker, D. A., *J. Radioanal. Nuclear Chem.* **113**:5–18 (1987). [primary standards in activation analysis]

Bird, J. R., and J. S. Williams, eds., *Ion Beams for Materials Analysis.* Sydney: Academic, 1989.

Brune, D., B. Forkman, B. Persson, *Nuclear Analytical Chemistry,* Verlag Chemie, Weinheim, Fed. Rep. Ger., 1984. [principles of activation analysis and other nuclear methods of analysis]

Cesareo, R., ed., *Techniques and Instrumentation in Analytical Chemistry*, Vol. 8, *Nuclear Analytical Techniques in Medicine.* Amsterdam: Elsevier, 1988. (applications of NAA and other nuclear methods in medicine]

Csikai, J., *CRC Handbook of Fast Neutron Generators,* Vols. I and II. Boca Raton, FL: CRC Press, 1987. [neutron generator technology and FNAA]

Currie, L. A., *Anal. Chem.* **40**:587 (1968). [detection limits in NAA]

Das, H. A., A. Faanhof, and H. A. van der Sloot, *Developments in Geochemistry,* Vol. 5, *Radioanalysis in Geochemistry* Amsterdam: Elsevier, 1989. [applications of NAA and other nuclear methods in geochemistry]

De Corte, F., A. Simonits, A. De Wispelaere, and J. Hoste, *J. Radioanal. Nuclear Chem.* **113**:145–161 (1987). [k_0 standardization method]

De Soete, D., R. Gijbels, and J. Hoste, *Chemical Analysis, a Series of Monographs on Analytical Chemistry and Its Applications,* Vol. 34, *Neutron Activation Analysis,* P. J. Elving and I. M. Kolthoff, eds. London: Wiley-Interscience, 1972. [principles and applications of NAA for nonspecialists in radiochemistry]

Downing, R. G., J.T. Maki, and R. F. Fleming, *ACS Symposium Series,* No. 295. Washington, DC: American Chemical Society, DC, 1988. [applications of NDP]

Ehmann, W. D., and D. E. Vance, *CRC Critical Reviews in Analytical Chemistry,* Vol. 20(6). Boca Raton, FL: CRC Press, 1989, pp. 405–443. [review of recent advances in activation analysis]

Ehmann, W. D., D. M. McKown, and J.W. Morgan, Coincidence counting applied to the activation analysis of meteorites and rocks. In *Activation Analysis in Geochemistry and Cosmochemistry,* Oslo: Universitetsforlaget, 1971. [coincidence counting techniques in NAA]

Ehmann, W. D., R. C. Young III, D.W. Koppenaal, W. C. Jones, and M. N. Prasad, *J. Radioanal. Nuclear Chem.* **112**:71–87 (1987). [derivative activation analysis]

Erdtmann, G., *Kernchemie in Einzeldarstellungen,* Vol. 6, *Neutron Activation Tables.* Weinheim: Verlag Chemie, 1976. [data on half-lives, cross sections, energies of analytical γ rays, and detection sensitivities for TNAA, ENAA, and FNAA]

Filby, R. H., ed., *Atomic and Nuclear Methods in Fossil Energy Research.* New York: Plenum, 1982. [NAA in fossil fuel research]

Guinn, V. P., *J. Radioanal. Nuclear Chem.* **110**:5–8 (1987). [purposes and capabilities of the Advance Prediction Computer Program (APCP) in INAA]

Heydorn, K., *Neutron Activation Analysis for Clinical Trace Element Research,* Vols. I and II. Boca Raton, FL: CRC Press, 1984. [NAA of biological materials]

Kruger, P., *Principles of Activation Analysis.* New York: Wiley-Interscience, 1971. [a textbook on all types of activation analysis]

Lindstrom, R. M., R. Zeisler, and M. Rossbach, *J. Radioanal. Nuclear Chem.* **112**:321–330 (1987). [use of cold neutrons in NAA]

Mandelkow, E., ed., *Synchrotron Radiation in Chemistry and Biology—II.* New York: Springer, 1988. [synchrotron radiation for IPAA]

Morgan, J.W., and W. D. Ehmann, *Anal. Lett.* **2**:537–545 (1969). [multiparameter coincidence counting in NAA]

Nargolwalla, S. S., and E. P. Przybylowicz, Activation Analysis with Neutron Generators, *Chemical Analysis, A Series of Monographs on Analytical Chemistry,* Vol. 39, P. J. Elving and I. M. Kolthoff, eds. New York: Wiley-Interscience, 1973. [principles and applications of FNAA]

Nuclear activation and radioisotopic methods of analysis. In *Treatise on Analytical Chemistry,* Vol. 14, Part 1, Section K, I. M. Kolthoff and P. J. Elving, eds. New York: Wiley, 1986, Chapters 1–8. [a comprehensive treatment of all aspects of activation analysis]

Segebade, C., H. P. Weise, and G. J. Lutz, *Photon Activation Analysis,* Hawthorne, NY: de Gruyter, 1987. [principles of IPAA]

Tian, W., and W. D. Ehmann, *J. Radioanal. Nuclear Chem.* **84**:89–102 (1984). [advantage factors in ENAA]

Tolgyessy, J., and E. H. Klehr, *Nuclear Environmental Chemical Analysis.* New York: Wiley, 1987. [applications of NAA and other nuclear methods in environmental chemistry]

Vandecasteele, C., *Activation Analysis with Charged Particles,* Chichester, UK: Ellis Horwood, 1988. [principles of CPAA]

Vis, R. D., *The Proton Microprobe: Applications in the Biomedical Field.* Boca Raton, FL: CRC Press, 1986. [capabilities of the proton microprobe]

Yates, S.W., A. J. Filo, C.Y. Cheng, and D. F. Coope, *J. Radioanal.* Chem. **46**:343 (1978). [inelastic neutron scattering analysis]

[*Note:* Many articles on various types of activation analysis appear in the *Journal of Radioanalytical Chemistry*, Elsevier Sequoia S. A., Lausanne/Akadémiai, Budapest.]

EXERCISES

1. A 32.50-mg sample of naturally occurring elemental silicon was irradiated with neutrons in a nuclear reactor at a flux density of 2.75×10^{13} thermal neutrons $cm^{-2}\ s^{-1}$. The activity of ^{31}Si from the $^{30}Si(n, \gamma)^{31}Si$ reaction was determined at a remote site with a β particle detector that has a counting efficiency of 16.5%. Transport of the sample to the remote counting site and chemical processing (RNAA) to isolate a pure Si counting sample required a delay time of 8.00 h prior to counting. The chemical yield of the RNAA procedure was 97.5%. If permitted by the experimental parameters, how long a reactor irradiation would be required to produce the following counting rates at the remote site? (a) 6.00×10^5 cpm (b) 9.78×10^8 cpm
Note: The cross section for the reaction = 107 mb and $t_{1/2}$ of ^{31}Si = 2.62 h. ^{31}Si decays >99.9% of the time to the ground state of ^{31}P via negatron emission.

2. The radionuclide ^{53}Mn is produced in meteorites and lunar rocks by the action of cosmic rays. ^{53}Mn ($t_{1/2} = 3.7 \times 10^6$ y) decays via electron capture directly to the ground state of its daughter. (a) It has been proposed that the amount of ^{53}Mn in meteorites and lunar samples be determined by neutron activation analysis, using the following reaction which has a σ_{th} = 70 b:

$^{53}Mn\ (n, \gamma)^{54}Mn$ ($t_{1/2}$ = 312 d, decays by EC, 0.835-MeV γ ray in 100% of decays)

Why would one elect the NAA approach, rather than just measure the decay of ^{53}Mn directly?
(b) In one meteorite, the activity of ^{53}Mn was determined to be 50.0 disintegrations/min in exactly 1.0 kg of meteorite. Calculate the number of ^{53}Mn atoms there would be in a 1.0-g sample of this meteorite.
(c) The 1-g sample of the meteorite in (b) above was irradiated with thermal neutrons at a flux density of 1.50×10^{14} n $cm^{-2}\ s^{-1}$ for a period of time sufficient to reach 15.0% of the saturation level of the reaction product indicator radionuclide, ^{54}Mn. If the counting efficiency is 20.0%, what is the measured counting rate of ^{54}Mn in the sample at the end of the irradiation period?

3. Thallium is difficult to determine by INAA. Why? Thallium is known to form a complex ion containing iodine which is extractable into an organic solvent. Propose a derivative activation analysis method for the determination of Tl based on the above information.

4. By reference to a chart of the nuclides, discuss possible interferences to the determination of oxygen by FNAA. List both the analytical and the interference reactions. What advantages would CPAA have over FNAA in the determination of oxygen in high-purity materials?

5. Characterization of the bullet lead from the assassination of President Kennedy was done by INAA. List several reasons for selecting INAA for this study, rather than some of the more widely used methods of trace element analysis. Include major reasons for the use of INAA, and also the potential limitations of INAA for this specific study.

6. As discussed in Chapter 4, the cross section (σ) for radiative capture reactions with thermal neutrons may be approximated by the expression

$$\sigma = \pi(R + \lambdabar)^2$$

where R is the radius of the target nucleus and $\lambdabar$ is the de Broglie wavelength of the neutron divided by 2π. Calculate the de Broglie wavelength of a neutron that has a velocity of 2.20×10^5 cm/s. *Given:* Plank's constant $= 6.63 \times 10^{-34}$ J $\cdot$ s and the mass of a neutron $=$ 1.00866 daltons.

7. Thin metal wires or foils are often used as flux monitors in TNAA. If these metals have a high thermal neutron capture cross section, self-shadowing can occur. If a pure indium (In) foil is to be used as a flux monitor, how thin would the foil have to be so that the flux gradient through the foil is only 1.0%? Assume the thermal neutron capture cross section of In metal is 3.2×10^3 b.

8. Several types of neutron sources were discussed in this chapter. Select one of these sources that could be used appropriately in the situations below. Consider such factors as neutron flux, availability of source, cost, and on- or off-site irradiation:
   (a) the determination of μg/g levels of As in a hair sample
   (b) the determination of % levels of V in an experiment performed as part of a radiochemistry laboratory course
   (c) the determination of % levels of N in a large sample of biological tissue

9. Which nuclear activation analysis technique discussed in this chapter would be best used for the following determinations? Consider such points as the amount of the analyte, cross section, isotopic abundances,

identity of indicator nuclide, and possible interferences. Use tables in the chapter and a chart of the nuclides.

(a) Al in a nitrogenous matrix

(b) trace Au in a mineral sample

(c) trace Mn in an air filter

(d) % levels of O in silicate rocks

(e) the trace element pattern of the pigment in a valuable painting

10. An RNAA experiment is done to determine the amount of Sc in a 100-mg sample of Pt metal. Calculate the amount of Sc in the sample in g, % by mass, and μg/g, given the following data: neutron flux = 1.00×10^{13} n cm^{-2} s^{-1}; irradiation and decay times 40.0 h and 20 days, respectively; detector efficiency 40.0%; and a measured activity after the decay of 86 cps. (Find other needed data in a chart of the nuclides, or in nuclear data tables.)

CHAPTER

10

RADIOTRACER METHODS

The use of radionuclides as tracers illustrates better than any other topic in this book the myriad applications of radioactivity. Fully addressing these applications in a single chapter is an impossible task. There are numerous books and articles written about the many uses of radiotracers, and these should be consulted for details or for applications that cannot be mentioned here. The intent of this chapter is to state some basic principles of tracer use and to give an overview of selected applications. The selections are primarily those of analytical, chemical, medical, biological, and industrial applications. The hope is that the reader will gain an appreciation of the tremendous scope of application of radiotracer methodology.

10.1. GENERAL ASPECTS OF RADIOTRACER USE

Georg Hevesy, the co-discoverer of the technique of neutron activation analysis, is also credited with being the originator of the use of radionuclides as tracers. He was awarded the Nobel Prize in 1943 for his work. As early as 1913, he was using radium-D (^{210}Pb) to determine the solubility of lead salts in water. He was also the first person to use radiotracers to follow biological processes, tracing the movement of radionuclides from soil into plants and the movement of food through animal systems. The first production of artificial radionuclides by I. Curie and F. Joliot gave a great impetus to applications of the tracer concept, because now radiotracers could be produced to fit specific needs.

Tracers are materials that are used as markers to follow the course of a chemical reaction or physical process or to show the location of a substance. **Radiotracers** are chemical species that contain a radionuclide, and it is the activity of the radioisotope that is monitored to follow the process under investigation. Radiotracers can be used either qualitatively, as simple markers of a process, or quantitatively, to determine the amounts of nonradioactive species.

10.1.1. Assumptions Made in Tracer Studies

The most important assumption made in the use of radiotracers is that the radioactive material will blend in perfectly with the system under study.

This implies that the radioactivity emitted by the tracer will not adversely affect any components of the system, and that the tracer will behave in a way that is indistinguishable from the nonradioactive materials, except for the emitted radiation. The first assumption is nearly always true for routine tracer work. The tracer is present in such small amounts that its radioactivity does not significantly affect the system. The second assumption holds for most tracer applications, but there are some exceptions.

Because the radiotracer is a different isotope from that of the element to be traced, there will be a mass difference between the active and nonactive species. Isotopes of differing masses do behave differently in chemical reactions, particularly with respect to reaction rate. For example, in a diffusion-controlled process, a more massive isotope will react more slowly than a less massive one. Recall that according to Graham's law the rate of diffusion is inversely proportional to the square root of the molar mass of the substance. Randomizing isotopic distributions leads to entropy changes that can alter values of equilibrium constants in exchange reactions. In addition, electron binding energies in atoms are related to the nuclear mass, and the fundamental vibrational frequency between two dissimilar atoms (A and B) in a molecule is related to the reduced mass of the system ($\mu_{A-B} = m_A m_B/[m_A + m_B]$). The combination of kinetic, entropy, and energy effects when using isotopic tracers is collectively called the **isotope effect.** Corrections for the isotope effect can be quite complex and the reader is referred to the Reading List at the end of this chapter for sources of detailed mathematical treatments of this subject.

For heavy elements, the relative mass difference between the active and nonactive isotopes is often small and isotope effects are not significant. For example, the relative mass difference between the isotopes ^{235}U and ^{238}U is only 1.3%. These two isotopes behave so similarly in chemical and physical processes that they are very difficult to separate by chemical means. In fact, separation of the uranium isotopes by physical means (gaseous diffusion) was a major obstacle to be overcome during the development of the first atomic bomb during World War II. For lighter mass elements, the relative mass difference can be substantial. Hydrogen is an extreme example of this. The three isotopes of hydrogen are ^{1}H, ^{2}H, and ^{3}H. Tritium has a mass three times greater than protium and this large mass difference will result in significant isotope effects for the hydrogen isotopes.

10.1.2. Factors in the Choice of a Radiotracer

There are several factors that should be considered when choosing the proper radiotracer for a study. First, the tracer should be chemically and physically appropriate for the system to be studied under real experimental conditions. For example, in imaging studies of the thyroid gland ^{125}I or ^{131}I would be the most appropriate tracers to use because iodine is known to be essential for thyroid gland function and preferentially accumulates there. However, ^{131}I would not be a very good tracer for bone-imaging studies. For these studies, isotopes like ^{99m}Tc or ^{89}Sr would be more appropriate. Potential chemical and physical interactions of the tracer with the system under study should be considered. Attention must also be given to the possible interactions of the other components of the tracer solution with the system because very few isotopes can be obtained as pure elements. Most are chemically associated with other chemical species and physically dissolved in a solution that has still other chemical components, such as an acid or buffer.

The half-life of the tracer is obviously an important factor to consider. It must be long enough so that there is sufficient activity present throughout the course of the experiment for detection and good counting statistics. The half-life should be at least as long as the duration of the experiment itself. On the other hand, the half-life should not be excessively long. Long half-lives mean low specific activity levels and can result in problems with disposal, storage, long-term environmental contamination in field studies, and excessive radiation dose in patients who retain the tracer. The latter problem is especially important when human subjects are receiving radiotracers, because the goal is always to keep radiation dose to a minimum (ALARA; see Section 7.4.2).

The type of radiation emitted by the tracer must be considered, mainly with respect to penetrating power, ease of measurement, and potential for damage to the sample. For samples that are very thick or dense, γ-emitting tracers would be a necessary choice, because β or α particles cannot penetrate the sample sufficiently to be detected. Beta emitters may be used as tracers for thinner samples, such as films and surfaces, and in solutions to be counted in liquid-scintillation counters. Alpha emitters are the least useful for tracer studies because their penetrating power is so low and because they have the potential to be very damaging to the sample.

Some other practical points that need to be considered when choosing a tracer for a proposed study include the availability and cost of the desired tracer, the accessibility of the appropriate detector systems for the radiation emitted, and the availability of a laboratory that is equipped and licensed for the activity levels to be used.

10.1.3. Production of Radiotracers

A few useful radiotracers, such as ^{3}H, ^{14}C, and the products of U and Th decay, are naturally occurring. However, most radiotracers used in laboratory studies are now produced artificially. Some devices used to produce radionuclides are nuclear reactors and various kinds of accelerators. The operating principles of these devices are discussed in Chapter 14. Nuclear reactors, which subject samples to intense neutron fluxes, create both gamma- and negatron-emitting nuclides via the (n, γ) reaction. Reactors are also sources of fission products, which are negatron emitters. To produce proton-rich nuclides that decay by positron emission, charged-particle reactions induced by particles generated by accelerators must be used.

There are hundreds of radiotracers that are commercially produced by many companies. Some of these tracers are listed in Table 10.1. Unfortunately, there are no suitable ($t_{1/2} > 10$ m) radiotracers for the elements He, Li, B, N, O, or Ne.

10.1.4. Advantages and Disadvantages of Radiotracer Use

For some studies, radiotracers are the only possible type of tracer that can be used, since the naturally occurring element may be monoisotopic, or separated stable isotopes are either unavailable or too expensive. If nonactive tracers are available, however, why might radiotracers still be the preferred method? An important advantage of radiotracers is the sensitivity of detection that can be attained. Amounts of radiotracer that would not be detectable using ordinary chemical analysis methods are easily measured with appropriate radiation detectors. This minimizes the amount of tracer needed and the potential effects of the tracer on the system to be studied. This is especially important in the use of elemental or molecular tracers in biological systems, where toxicity must be considered. Radioactive tracers also have the potential of specificity due to their unique decay properties, and detection equipment is less expensive than the mass spectrometric instrumentation required for measurement of stable isotopic ratios. Another advantage of radiotracers is that the decay process itself is not influenced by ambient conditions of temperature, light level, concentration of reagents, pH, and so on. The chemical nature of many nonradioactive tracers, such as organic dyes used in tracing natural water systems, can be profoundly affected by some, or all, of these conditions.

Localization of a radiotracer is relatively straightforward. In several kinds of tracer studies, the intent is to physically follow the metabolic or environmental path of a species and to determine its ultimate destination. Radiotracers lend themselves well to this purpose, as illustrated by the technique of **autoradiography**. The material that contains the radiotracer is placed on an undeveloped photographic plate. The emitted radiation will expose the film, and development of the plates will produce an image of the radioactive areas of the object. An example would be the use of autoradiography to characterize separations in gel chromatography and related biochemical separations methods.

Table 10.1 Isotopes Commonly Used as Tracers

Isotope	Half-life	Commonly Used Tracer Radiations/Energies (MeV)
^{3}H	12.3 y	β^-/0.0186
^{14}C	5730 y	β^-/0.157
^{24}Na	14.96 h	β^-/1.391
		γ/2.754, 1.369
^{32}P	14.28 d	β^-/1.709
^{36}Cl	3.01×10^5 y	β^-/0.709
^{40}K	1.28×10^9 y	β^-/1.33
		γ/1.461
^{42}K	12.36 h	β^-/3.52
		γ/1.525
^{45}Ca	162.7 d	β^-/0.258
		γ/0.0124
^{51}Cr	27.70 d	γ/0.3201
^{56}Mn	2.578 h	β^-/2.84
		γ/0.8468
^{52}Fe	8.28 h	β^+/0.80
		γ/0.1687
^{59}Fe	44.51 d	β^-/0.466, 0.271
		γ/1.099, 1.292
^{60}Co	5.271 y	β^-/0.318
		γ/1.3325, 1.1732
^{65}Zn	243.8 d	β^+/0.325
		γ/1.1155
^{67}Ga	3.260 d	γ/0.093, 0.1846, 0.3002
^{72}Ga	14.10 h	β^-/0.96, 0.64
		γ/0.8341, 2.2017
^{75}Se	119.78 d	γ/0.265, 0.136
^{82}Br	35.30 h	β^-/0.444
		γ/0.7765
^{90}Sr	29.1 y	β^-/0.546
^{99}Mo–^{99m}Tc	65.95 h − 6.01 h	β^-/1.214–IT
		γ/0.1405, 0.7395–0.1405
^{128}I	25.00 min	β^-/2.13
		γ/0.4429
^{133m}Xe	5.243 d	β^-/0.346
		γ/0.0810
^{137}Cs	30.17 y	β^-/0.514
		γ/0.6617
^{140}Ba	12.75 d	β^-/1.0, 0.48
		γ/0.5373, 0.030
^{192}Ir	73.83 d	β^-/0.672, 0.535
		γ/0.3165, 0.4681
^{198}Au	2.694 d	β^-/0.962
		γ/0.4118
^{204}Tl	3.78 y	β^-/0.7634
^{210}Pb	22.3 y	β^-/0.017, 0.061
		γ/0.0465

There are also disadvantages to the use of radiotracers. One is the necessity to have access to laboratories that are properly equipped and licensed to handle the types and levels of radioactivity that will be used. Personnel must be properly trained in the handling of the radioactive tracers. Active waste must be disposed of in approved ways. A problem mentioned earlier is that there are some elements for which no suitable radiotracers exist. Finally, to meet the requirement that the tracer be chemically similar to the components of interest, chemical synthesis may need to be done to incorporate a tracer atom into a molecule. Synthesizing a complex molecule can be a difficult and time-consuming process that is made more difficult when using radioactive materials.

10.2. ISOTOPE DILUTION ANALYSIS

Isotope dilution analysis (IDA) is a quantitative analytical technique that was first developed by Hevesy and Hobbie in the late 1930s. The basis of IDA is that the specific activity of a mixture of stable and radioactive isotopes is not changed by chemical processing. In **direct isotope dilution analysis (DIDA),** a known amount of a radioactive isotope of the element of interest is added to the sample containing the analyte. Then, a portion of the analyte is isolated in high purity from the sample. This separation step need not be quantitative, but the portion isolated must be pure. The mass and activity of the isolated portion are used to calculate the amount of analyte in the original sample.

10.2.1. Theory and Calculations for DIDA

Consider a sample that contains an unknown mass, m_a, of the analyte. A radioactive isotope of the analyte, whose mass of analyte species (m_1) and activity (A_1) are known, is added to the sample. This isotope is referred to as the **spike**. The spike and the sample are well mixed to facilitate equilibration of the spike with the analyte. An arbitrary, but weighable, portion of the analyte–spike component in the sample with mass m_2 and activity A_2 is isolated in very pure form. Because specific activity is an intrinsic property, the specific activity of the spike–sample mixture will be the same as the specific activity of the isolated portion.

$$\text{SA, spike + analyte mix.} = \text{SA, isolated portion} \tag{10.1}$$

Using the definition of specific activity (SA $= A/m$), we can write

$$\frac{\text{Activity added, spike}}{\text{Mass spike + Mass analyte}} = \frac{\text{Activity separated portion}}{\text{Mass separated portion}} \tag{10.2}$$

Rearranging the above equation to solve for the mass of the analyte in the original sample, we have

$$m_a = \left(\frac{SA_1}{SA_2} - 1\right)m_1 \tag{10.3}$$

where SA_1 is the specific activity of the pure spike and SA_2 is the specific activity of the isolated portion from the spike-analyte mixture.

Example. A solution contains an unknown amount of cobalt. A spike solution, which contains 7.50 mg of cobalt spiked with ^{60}Co and has a measured activity of 340 cpm is added to 10.0 mL of the unknown. After mixing, a portion of the cobalt is isolated as a pure cobalt metal sample by electrodeposition. The isolated cobalt has a mass of 10.3 mg, and a measured activity of 178 cpm. Calculate the mass of cobalt in the original sample solution.

Equation 10.3 is used to solve the problem. The specific activities of the spike and isolated portion are easily calculated:

$$SA_1 = 340 \text{ cpm}/7.50 \text{ mg} = 45.3 \text{ cpm/mg}$$

$$SA_2 = 178 \text{ cpm}/10.3 \text{ mg} = 17.3 \text{ cpm/mg}$$

Substituting these values and the mass of the spike into Eq. 10.3 gives $m_a = (\frac{45.3}{17.3} - 1)$ 7.50 mg = 12.1 mg Co in the 10-mL sample, or 1.21 mg Co/mL sample.

10.2.2. Applications of IDA

Isotope dilution analysis is applied to several quantitative analytical problems that are difficult to deal with using other methods. The fact that the analytical sample does not have to be quantitatively isolated makes IDA especially useful in the situations given below.

1. *A quantitative isolation of the element or compound is difficult or impossible.* An example of this would be the determination of I^- in a mixture of halides. Quantitative separation of the I^- from the other halides would be difficult, but this analysis can be accomplished without quantitative separation using IDA. A good radiotracer for I is ^{128}I, which is formed from the radiative neutron capture reaction on ^{127}I. A spike of ^{128}I tracer, with known mass and specific activity, could be added to the halogen mixture. Then, all of the halogens could be precipitated by addition of excess silver ion. Addition of sulfuric acid and manganese dioxide would oxidize the I^- to I_2. Heating the mixture would sublime some of the I_2 that is present,

and the I_2 vapors could be condensed on a cold surface. The mass and activity of an arbitrary amount of the condensed sample can be readily obtained and the mass of I^- in the original sample calculated using Eq. 10.3. The fact that a significant amount of I_2 is lost in the separation from the other halogens will not affect the results, as long as enough I_2 is collected to weigh and provide adequate counting statistics.

2. *Quantitative isolation is possible, but very time-consuming, and a more rapid analysis is needed.* The amount of Co present in steel can be quickly determined using IDA. A small sample of steel extracted from the special furnace melt is quickly weighed and dissolved in acid. An essentially "carrier-free" spike solution of ^{60}Co of known activity is added to the dissolved steel solution. (**Carrier-free** spike solutions are those that are virtually free of stable isotopes of the same element and, hence, have a high specific activity. Thus, m_1 in Eq. 10.3 is insignificant as compared to m_a.) A small, arbitrary amount of pure cobalt metal is isolated by electrodeposition, and its mass and activity are determined. The amount of cobalt in the original steel sample is simply determined by use of Eq. 10.1, since the spike was carrier-free. This IDA approach is fast and may be used for routine determination of $\leq 0.1\%$ Co in specialty steels.

3. *The analyte is present in very small amounts that might be easily lost through exchange with the glassware or analytical manipulation.* In the calculations of the Rb–Sr dating method (Chapter 12), it is necessary to know the concentration of Sr in small mineral grains where the Sr content is very low. Ordinary silicate rock dissolution and chemical analysis processes could result in loss of the very small amount of Sr in the sample due to adsorption on labware, or formation of insoluble residues. Either a stable (there are four stable isotopes of Sr, only one of which is produced by a naturally occurring decay process) or a radioactive Sr spike can be added in the dissolution process followed by IDA to accurately determine the Sr content of the sample, even if only a fraction of the total Sr is recovered after isolation of a pure Sr fraction. If stable isotopes, rather than radioisotopes, are used, detection must be by mass spectrometry.

4. *It is impossible to actually obtain the entire sample for analysis.* An example of this situation would be the determination of total blood volume or the blood cell volume of a living organism. An IDA procedure to determine total blood cell volume could involve the addition of carrier-free ^{52}Fe ($t_{1/2}$ = 8.28 h), ^{59}Fe ($t_{1/2}$ = 44.51 d), or ^{51}Cr ($t_{1/2}$ = 27.7 d) that would bind with a blood component. Iron is a good element for this study because it binds to the hemoglobin molecules in the red blood cells. The reactor-produced radioisotope of iron, ^{59}Fe, has a long half-life, which would argue against its use at high activity levels in humans. However, the short half-life of cyclotron-produced ^{52}Fe (a positron emitter) is a disadvantage in

some metabolic studies that can extend over several days. After injection of radioisotope spike, some time is allowed for equilibration of the iron or chromium into the hemoglobin and for circulation to distribute the label. After about an hour, a small aliquot of blood (usually 1 mL) can be removed and counted. The red cell volume can be determined in the aliquot and the total blood cell volume calculated from these measurements. Where total blood volume, not just blood cell volume, is desired, the above isotopes, or the relatively short-lived ^{24}Na ($t_{1/2}$ = 14.96 h) can be used as the radiotracer. In this case, calculations are made on the basis of the volume of the blood extracted, not the volume of blood cells.

Example. A spike containing 100 000 cpm of carrier-free ^{24}Na in an appropriate biological solution is injected into the blood stream of an organism. Later, a 1.00-mL aliquot of blood is withdrawn and found to have an activity of 50.0 cpm. Calculate the total blood volume, in mL.

A method of simple proportions can be used to solve this problem:

$$1.00 \text{ mL}/50.0 \text{ cpm} = x \text{ mL}/100\,000 \text{ cpm}$$

$$x = 2000 \text{ mL, total blood volume}$$

An important difference between IDA and some other nuclear methods of analysis is that IDA can be used to quantitatively determine molecular species, as well as elements. The species to be determined must contain an element that has a suitable tracer available. For example, vitamin B_{12} can be determined using an IDA approach. This vitamin contains Co as an essential cofactor. Molecules of vitamin B_{12} containing ^{60}Co as a label can be used as the spike solution.

10.2.3. Variations of IDA

Substoichiometric Isotope Dilution Analysis

Substoichiometric IDA was developed by J. Ruzicka and J. Stary. In this approach, both the spike solution and the unknown solution to which an aliquot of the spike solution has been added are subjected to an extraction process. The extraction is purposely designed to be **substoichiometric**; that is, there is too little extractant to remove all of the analyte from either the spike or the unknown-spike mixture. However, exactly the same amount of extracting agent is added to each solution, so the same mass of the analyte is removed from both and counted. Since the masses of the portions counted are identical, the specific activities (SA_1 and SA_2) in Eq. 10.3 can be replaced with just the measured activities of the two portions. This eliminates the mass-determining step for the samples that are counted. Ideally, the spike added to the unknown should have approximately the same mass of analyte as is expected to be found in the unknown.

A simple example of a substoichiometric IDA procedure is the determination of trace amounts of Ag^+ in a solution. A spike containing ^{110}Ag is added to the unknown solution. Then, both the original spike and the mixture are extracted with equal, substoichiometric amounts of dithiozone in chloroform, which complexes Ag^+. The activity of each extracted portion is determined, and Eq. 10.3 (with activities instead of specific activities) is used to calculate the mass of Ag^+ in the unknown sample.

An important procedure that utilizes the principles of substoichiometric IDA is radioimmunoassay. This medical diagnostic technique is discussed later in this chapter (see Section 10.5.4).

Inverse Isotope Dilution Analysis (IIDA)

In **inverse isotope dilution analysis (IIDA),** the aim is to determine the amount of a radioactive substance in a sample, rather than of a stable substance. In IIDA, the spike is a known amount of a nonactive species, chemically identical to the radioactive analyte to be determined in the unknown. Eq. 10.3 becomes

$$m_{\text{radio}} = \frac{A_{\text{unk}}}{A_{\text{mix}}} m_{\text{mix}} - m_{\text{stable}} \tag{10.4}$$

where m_{radio} is the mass of the radioactive species in the total volume of analyte solution, m_{stable} is the mass of a stable spike species that is added to the analyte solution, m_{mix} is the mass of purified analyte species in an arbitrary aliquot of the mixed solution of spike and analyte, A_{mix} is the activity of the purified analyte in the same aliquot, and A_{unk} is the total activity of the analyte solution computed from the measured activity of an aliquot withdrawn from the original analyte solution and a knowledge, or estimate, of the total analyte solution volume.

This procedure can be used to calculate the mass of a radioactive substance in a highly radioactive sample, such as the contaminated cooling water that might result from a nuclear reactor accident. Determination of the mass of each radioactive species present would allow for calculation of the amounts of cleanup materials (e.g., ion exchange resins) that would be required. The rapidity of the procedure and relative ease with which it can be conducted would minimize radiation exposure to personnel.

Stable Isotope Dilution Analysis

Stable isotope dilution analysis is based on the same principles as ordinary IDA, except that a stable isotope is used as the spike. There is no activity to measure, so isotopic ratios, determined with a mass spectrometer, are used in place of specific activities. This technique is commonly used in geology for isotopic dating.

Example. A solution contains an unknown amount of Sr, which is to be determined using stable IDA. A spike containing 1.00×10^{-9} g of ^{84}Sr is added to the unknown solution. The natural isotopic abundances of ^{88}Sr and ^{84}Sr in the unknown are 82.58 and 0.56%, respectively. The ^{88}Sr/^{84}Sr isotopic ratio is not altered by natural radioactive decay processes. The ^{88}Sr/^{84}Sr ratio in an arbitrary-mass purified Sr portion of the spike–sample mixture is determined by mass spectrometry to be 50.0. Calculate the mass of Sr in the original sample.

Let m stand for the total amount of Sr in the sample. The mass of ^{88}Sr in the unknown solution would be $0.8258m$ and the mass of ^{84}Sr would be $0.0056m$, based on the natural isotopic ratio. After mixing of sample and spike, the ratio in the mixture becomes 50.0. Therefore,

$$\left(\frac{^{88}\text{Sr}}{^{84}\text{Sr}}\right)_{\text{exp}} = \frac{0.8258m \text{ g } ^{88}\text{Sr in sample}}{(10^{-9} \text{ g } ^{84}\text{Sr in spike}) + (0.0056m \text{ g } ^{84}\text{Sr in sample})} = 50.0 \tag{10.5}$$

Solving the above equation for m, using the experimental isotopic ratio of 50.0, gives $m = 91.6 \times 10^{-9}$ g, or 91.6 ng of Sr in the original sample.

10.3. TRACERS IN THE STUDY OF CHEMICAL PROCESSES

For most chemical systems, the assumption that a radioactive isotope behaves in the same way as a nonactive one is true. Therefore, radiotracers offer a convenient way to study many chemical processes and reactions that often could not be examined in any other way.

10.3.1. Equilibrium Processes

Equilibrium processes that can be studied using radiotracers include self-diffusion in solids, liquid–liquid diffusion, isotope exchange reactions, and complex formation.

Diffusion is the movement of matter due to the random motions of atoms in a material. In a pure substance, the process is called **self-diffusion.** Because it is impossible to chemically distinguish among the atoms, the only practical way to study self-diffusion in a pure substance is through the use of radiotracers. The quantity measured in self-diffusion studies is the **diffusion coefficient, *D*.** The diffusion coefficient is a constant that relates the rate of atom transfer (J, atoms crossing a 1-cm^2 boundary per unit time) and the concentration gradient (dc/dx, where c = g/cm^3 of solution and x = distance in cm).

$$J = -D\frac{dc}{dx} \tag{10.6}$$

Equation 10.6 is **Fick's first law of diffusion.**

The process of self-diffusion in solids was first studied with radiotracers in the 1920s. The technique has been used mostly with metals. In a self-diffusion study, the idea is to put a very thin section of an active isotope in intimate contact with a much thicker section of an inactive one. This can be done by physically welding the two pieces of metal together, plating the active metal onto a nonactive strip, or bombarding the surface layer of atoms with nonpenetrating particles to produce a thin surface layer of radioactive atoms. After the two portions are brought into contact, diffusion is allowed to occur for some period of time. After this, a determination must be made of the change in concentration of the active tracer with distance into the inactive portion. The most reliable way to do this is by the **serial sectioning method.** The whole specimen is weighed, and then a very thin section is removed with a precision grinding tool. The activity of the section is measured and its thickness determined from the mass. A plot of the log of the counting rate against the square of the distance moved can be used to calculate the value of the diffusion coefficient.

There are other methods for doing this kind of analysis. One is the **residual activity method,** in which the residual activity in the specimen, rather than the activity of the section removed, is measured. A method that does not require sectioning is the **surface decrease** method. Here, the activity of the surface of the sample is measured over a period of time. Hevesy and Seith (1929) used a variation of this technique in an early study of the self-diffusion of lead. ^{212}Pb was allowed to diffuse into inactive lead halide crystals. ^{212}Pb decays as follows:

$$^{212}\text{Pb} \xrightarrow[10.6\text{ h}]{\beta^-} {}^{212}\text{Bi} \xrightarrow[1.009\text{ h}]{\alpha} {}^{208}\text{Tl} \xrightarrow[3.05\text{ min}]{\beta^-} {}^{208}\text{Pb (stable)}$$

If the ^{212}Pb penetrates very deeply into the inactive lead halide crystal, the α activity will not be detected. Therefore, measurements of the α activity taken over a period of time can be used to determine how deeply the ^{212}Pb atoms have penetrated. The results of studies like this show that Pb in Pb diffusion is about 10 times slower than tin in lead, and 10^5 times slower than gold in lead.

Liquid–liquid diffusion can also be studied using tracers. The general scheme is to put a tracer into a diffusion tube, capillary tube, or container with a porous plate, and allow diffusion to occur. The decrease in activity of the tracer, or increase in activity of the surrounding solution, is monitored as a function of time, and the rate of diffusion can be calculated from these measurements. If only self-diffusion is to be measured, any sources of vibration, thermal gradients, and convection must be eliminated.

A general equation for an **isotopic exchange reaction** is

$$AB + A'C \rightleftharpoons A'B + AC$$

In this exchange reaction, the atoms A and A′ have switched places. This process would not normally be discernible from a chemical point of view, but if A′ were a radioactive isotope of A, it would be possible to follow the course of the reaction and determine the rate of exchange. Knowledge of the rate of exchange is useful for determinations of the reversibility and mechanisms of certain kinds of reactions. It is also important to know the rate of exchange for reactions involved in the Szilard–Chalmers process (see Chapter 11), because this process is only feasible when exchange rates are sufficiently low. Radiotracer exchange studies can be used to determine solubility products and partition coefficients.

Radiotracers are also useful in determining the **formation constant** of complex ions and for the study of reaction rates and mechanisms. They are used to best advantage in the study of slow reactions.

10.3.2. Analytical Applications

Radioactive materials can be used for direct analysis or to test the efficacy of separations procedures.

In a **radiometric titration,** the detection of radioactivity in a separable species is used to indicate the endpoint of a titration. The most successful radiometric titration procedures have been developed for precipitation reactions, where the separation process occurs automatically.

The determination of halides using silver ion can serve as an example of a radiometric titration. Either the titrant or the unknown can contain the radioactive component. Figure 10.1 illustrates the experimental apparatus and the resulting titration curves for the titration of Cl^- with $^{110}Ag^+$ solution. At the beginning of the titration, there is no activity in the titration flask. As $^{110}Ag^+$ is added, it combines with the Cl^- in the sample to form

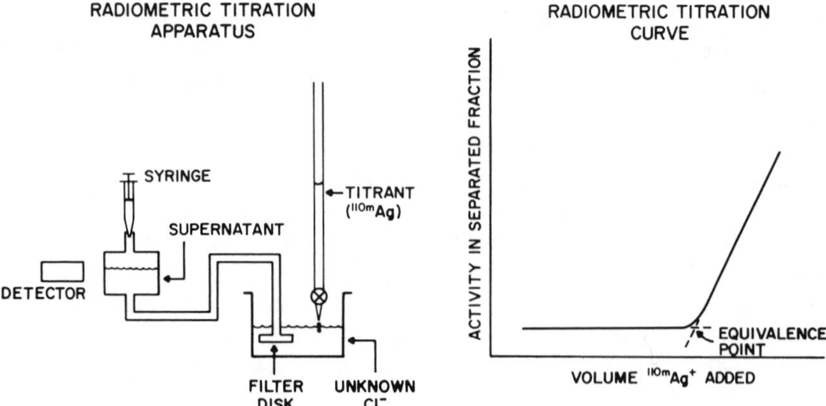

Figure 10.1. Representation of the apparatus and a titration curve for a radiometric titration using a radioactive titrant. The supernatant counting chamber and tubing connected to the filter disk are enlarged here for clarity. Normally small-bore tubing is used and the counting chamber is small with respect to the counter window. The syringe is used to move supernatant into and out of the counting chamber.

insoluble AgCl, which precipitates out of solution. A sample of the supernatent solution would still indicate no activity. Eventually, after the equivalence point is reached, there is no more Cl^- ion to combine with the added $^{110}Ag^+$, so it remains in the solution. Therefore, the solution will show a sudden increase in activity levels, as illustrated in the titration curve shown on the right in Fig. 10.1. The endpoint is taken to be that volume reached by extrapolating the two linear portions of the titration curve.

If the titrant were the inactive species, the titration curve would look like that in Figure 10.2. Here, the activity would slowly decrease as the

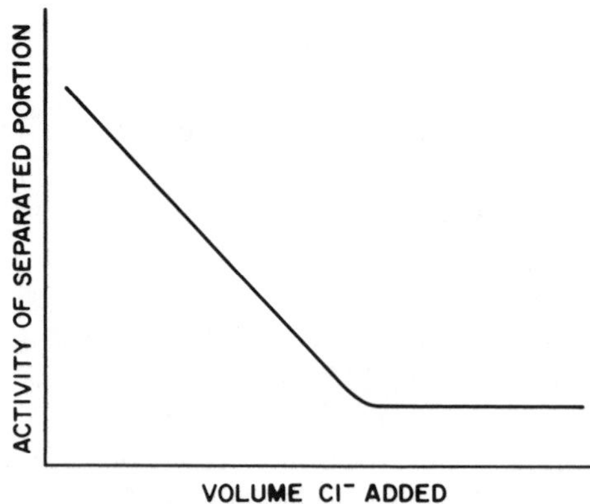

Figure 10.2. Titration curve for a radiometric titration using an inactive titrant.

active species is precipitated out of solution, until a constant low level of only background activity is reached.

The primary advantage of radiometric titrations is their sensitivity. Traditional titration analyses are not useful for trace amounts of substances, because endpoints become very difficult to detect. Also, ambient conditions of temperature, pH, turbidity, and so on, have no effect on the activity and so do not adversely affect endpoint determination. However, radiometric titrations are lengthy and cumbersome due to the need for separation and measurement of activity, so they are not widely used.

Techniques of chemical separation have assumed a greater role in analytical chemistry in recent years. Radiotracers have proved useful both in testing the efficacy of these separations and in locating components that have been separated.

Many variations of gel chromatography are performed for molecular separations. It is simple to locate any of these molecules that are radioactively labeled by doing autoradiography. The gel is placed on the photographic plate, and the labeled molecules form an image. Radiotracers can be used in paper chromatography the same way. The efficacy of solvent extraction or ion exchange separations could be easily verified using radiotracers.

10.3.3. Studies of Reaction Mechanisms

Tracers are useful to elucidate reaction mechanisms for both simple and complex reactions. They have been indispensible, for example, in establishing the mechanisms for the complex series of reactions that occur in biological systems. Pathways for photosynthesis and almost all other metabolic processes were outlined using radioactively labeled molecules as tracers. A classic example of this was the work done by M. Calvin and his associates in elucidating the mechanisms of photosynthesis.

One interesting example is the study of the mechanism for the chromium plating of metals. Plating is normally done from a Cr(VI) solution. A solution containing Cr^{3+} and CrO_4^{2-} ions, with the Cr^{3+} labeled with ^{51}Cr radiotracer, was electroplated. It has already been shown that Cr exchange between Cr^{3+} and CrO_4^{2-} is negligible at room temperature. The Cr metal deposit obtained contained essentially no activity. Hence, the experiment established that Cr^{3+} is not an intermediate in the chromate-plating process.

10.4. OTHER APPLICATIONS OF RADIOTRACERS AND RADIONUCLIDES

This section gives a brief look at a variety of other ways radioactive materials can be put to good use.

One of the earliest industrial uses of a radiotracer was in the monitoring of the effectiveness of various motor oils in the **lubrication** of automobile engines. Piston rings containing ^{60}Co tracer were run for various times with each oil, and the amount of ^{60}Co activity found in the oil was taken as a measure of the wear due to friction in the engine. Oils where the least amount of activity was transferred to the oil were, obviously, the best lubricants.

The use of radiation for **industrial process control** has many advantages. In a production facility, it is often necessary to measure parameters for process streams where the reagents are very hazardous, are under high pressure, or are hard to access. This makes direct sampling or observation impossible, or at least very difficult. Radiation measurements can often be made continuously, through the walls of the containers, so that operators do not actually have to open vessels that contain dangerous reagents or those at high pressure. The radioactive sources used for continual monitoring are usually completely sealed, which reduces potential handling problems. Once in place, the source requires little maintenance. In addition, the radioactivity is not affected by ambient conditions that might affect chemical indicators, such as temperature, pH, and pressure. Only a few of the many kinds of industrial process applications are mentioned below.

Measurements of the **thickness** or **density** of a material may be done using γ-ray absorption measurements. A γ- or x-ray source is placed on one side of the material to be measured, and an appropriate detector is placed on the other side. Variations in the thickness or density of the material will result in different levels of γ-ray absorption. If the density of the material being monitored is constant, thickness is determined from the γ-ray absorption measurement, while if the thickness of the stream is constant, density may be measured. Continual in-line monitoring of many types of materials is easily done in this way. If very thin films are to be measured (e.g., aluminum foils), negatron emitters can be used instead of x-ray or γ-ray emitting sources. In manufacturing processes, the output of the detectors may be fed directly to computers that control the spacing of the rollers that determine foil thickness. These devices are called **beta gauges.**

Determination of the **level** of material in a container is important for control of many processes. Over- or underfilling of a reagent vessel in a process stream could have adverse effects on the products, or even be dangerous. Level gauging is perhaps the most widely used radiotracer technique in industrial process control. A γ-ray source, usually ^{60}Co or ^{137}Cs, is placed on one side of the vessel, and the detector is placed on the other side. The level of the liquid in between these two components affects the signal received by the detector.

Small ^{60}Co sources in the form of ceramic beads or wire needles are often buried in the refractory linings of furnaces in the steel industry.

Gamma-ray detectors outside the furnace monitor the loss of activity as the lining erodes and can indicate the time for lining replacement without an interior inspection of the furnace. Gamma-ray scattering measurements may be used to measure the **thickness of coatings,** or the buildup of deposits inside reaction vessels or pipes.

Moisture measurements are based on the principle of neutron moderation. A probe containing an isotopic energetic neutron source and a slow neutron detector is used. The fast neutrons are slowed primarily by collisions with the hydrogen in water, and the resulting count rate can be related to the moisture content. Obviously, this method will work well only in media that do not already contain a significant fraction of hydrogen.

The measurements of **flowrate** and of **residence time** are two of the most important applications of radiotracers in industrial processes. Both are critical to the proper output and quality of the product. Tracers can be injected into the process stream and monitored to obtain information on these parameters. Tracers are also useful to detect **leaks** in a process.

A group of methods collectively called **radiorelease methods** involve the release of a radiotracer to signal the end of a chemical reaction, the percentage of a reaction that is completed, or a phase change. With the availability of high specific activity (or carrier-free) tracers, these methods may be far more sensitive, rapid, and convenient than conventional methods. An example of the use of a radiorelease method would be the determination of SO_2 concentration in air samples. The sample containing the SO_2 gas could be bubbled through a solution of $K^{128}IO_3$, and this reaction would occur:

$$5\ SO_2 + 2\ K^{128}IO_3 + 4\ H_2O \longrightarrow K_2SO_4 + 4\ H_2SO_4 + {}^{128}I_2$$

After a given volume of air has passed through the solution, the ${}^{128}I_2$ is extracted with chloroform. The activity of the chloroform layer is directly proportional to the amount of SO_2 in the air that passed through the solution. The system could be calibrated using standard amounts of SO_2 in air and the unknown amount of SO_2 read from a calibration curve.

Transitions in solid materials may be studied using **emanation** methods. The release of an entrapped radioactive gas can be used to indicate phase transitions, solid-state deformations, or rates of corrosion. The method involves trapping a chemically inert gas in the solid to be studied. Gases commonly used include ${}^{85}Kr$ and ${}^{222}Rn$. The gases are introduced into the solid in several ways. A solid parent nuclide that decays to give a gaseous product (e.g., ${}^{226}Ra$) may be incorporated into the solid material to be studied. The solid can be irradiated to produce the gaseous products in situ, or ions can be implanted by an accelerator, and then activated. In some cases, the solid can simply be crystallized in an atmosphere of the tracer gas.

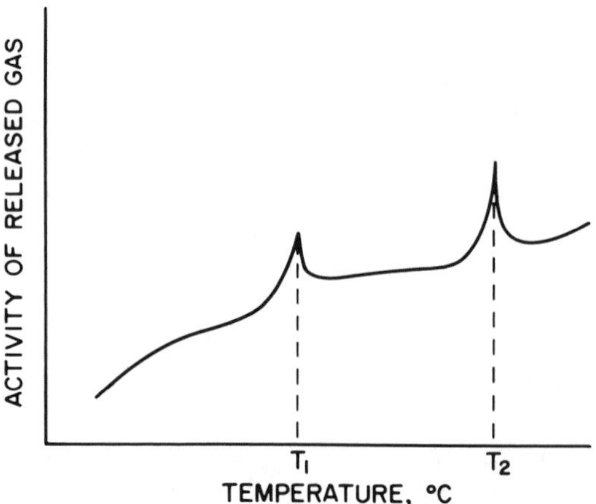

Figure 10.3. A radiorelease method for the determination of the temperatures of phase transitions in a solid. The peaks correspond to the release of a radioactive gas at temperatures associated with phase transformations.

Figure 10.3 shows an example in which a solid that has two phase transitions has been heated. The entrapped tracer gas is released most strongly during the phase transitions, resulting in the peaks on the graph.

Radiographic techniques can be used to determine the structural integrity of an object. It is common to take x rays of welds to check for integrity. For some very large objects (e.g., drive shafts for oceangoing ships), the use of higher-energy γ rays from ^{60}Co sources is preferable because they are much more penetrating. Neutron radiography with ^{252}Cf sources is useful for detection of low-Z materials in an object because low-Z atoms are effective neutron moderators. A neutron radiograph of an automobile would show shadow images of the gasoline in the tank and the plastic parts of the car. In contrast, x-ray radiography of the automobile would show shadow images of the high-Z, metal components of the car.

Radiotracers are also very useful in environmental work. Tracers can be used in several kinds of groundwater studies, including mapping groundwater flow, assessing the movement of contaminants in groundwater, and studying the entire water cycle. Atmospheric flow can also be followed using radiotracers.

Radionuclides are also used in applications other than tracer studies. The radionuclide ^{85}Kr is used as a static eliminator to neutralize the charged dust particles that can be attracted to a phonograph needle. Radionuclides can be used to sterilize male insects. The sterile males may breed with many female insects, leading to nonproductive mating and reduction of the insect population. This technique has been successfully used to control fruit flies. Large doses of radiation from ^{60}Co isotopic sources have been used to sterilize foods used in the field by military forces and kill insects that might get into grains during shipment. A radionuclide source of electrical power has been used in space vehicles and submarines. The heat produced by the β decay of ^{90}Sr or the α decay of a transuranium element such as ^{238}Pu acts on thermocouples to produce thermoelectric power. These long-life power sources require kilocurie levels of the radionuclide, and power output levels are relatively low. Radionuclides have also been used as polymerization catalysts in the enhancement of cross-linking in specialty plastics.

10.5. NUCLEAR MEDICINE AND PHARMACY

The fields of nuclear medicine and nuclear pharmacy have undergone tremendous growth in the past 20 years. Advances in computing capability, detector technology, and development of appropriate radiopharmaceuticals have all contributed to this growth.

Nuclear medicine is that branch of medicine that uses radiation and the nuclear properties of radioactive and stable nuclides for both diagnosis and treatment of disease. This can involve either the direct irradiation of the patient by an external source of radiation, or the administration of radioactively labeled drugs to the patient. Radioactively labeled drugs that are given to the patients are called **radiopharmaceuticals. Nuclear pharmacy,** or radiopharmacology, is a pharmacy specialty that deals with the compounding, testing, and proper administration of radiopharmaceuticals to patients. The field of nuclear pharmacy is recognized as a specialty by the American Pharmaceutical Association.

In the following sections, a brief overview of some important aspects of nuclear medicine will be given. This includes a survey of the nuclear properties of the most used radionuclides and a look at the types of diagnostic and treatment regimes that can be performed with these tracers. The important technique of radioimmunoassay (RIA) is discussed briefly in Section 10.5.4.

10.5.1. General Aspects of Radiopharmaceutical Use

Like other tracer techniques, nuclear medicine procedures are often able to provide important information that could not be obtained in any other way. This is the justification for intentionally exposing the patient to radiation. As noted in Chapter 7, there are no limits specified for medical uses of radiation. In this case, the risk/benefit question is decided by the physician and patient. Nuclear medicine procedures are done either for **diagnostic** or **therapeutic** purposes, and the radiation may come from either external or internal sources. External radiation sources include high-energy electron sources and gamma sources. Internal radiation will come from the administration of radiopharmaceuticals.

A radiopharmaceutical that is administered to a patient should remain in the target organ long enough to do its job properly, but no longer than necessary, so that radiation exposure is minimized. The amount of time that a drug is useful depends on both its radioactive half-life and its biological half-life; that is, the amount of time the drug remains in the body before being deactivated by metabolic processes or excreted entirely from the system. The radioactive half-lives of nuclides commonly used in nuclear medicine are already well known, but the biological behavior and residence time of the molecule to which the tracer is attached may not be so well characterized. It is one of the jobs of the nuclear pharmacist to establish values for these parameters during initial drug testing stages.

The type of radiation emitted by the radionuclide is also an important factor to consider. The use to which the nuclide will be put determines which type of radiation is most appropriate. In diagnosis, a major use of radiopharmaceuticals is for imaging biological structures. For images to be formed, the radiation must be penetrating enough to pass through the subject and into the detector. During its passage, it is desirable for the radiation to undergo minimal interaction with the subject, so that radiation dose is lessened. Therefore, the best radionuclides for many diagnostic procedures are those that emit only γ rays or x rays with no associated particulate radiation. This includes primarily nuclides that undergo decay via electron capture or isomeric transitions.

The characteristics of a radionuclide that would make it useful for imaging are different from those useful for treatment. In radiation treatments, the point is to destroy diseased tissue. Tissue destruction is accomplished through the ionization and free-radical production induced by the radiation. Thus, radiations that have a high specific ionization and short, well-defined ranges are most desirable because these characteristics lead to a great amount of tissue destruction in a small, confined area. The best nuclides for therapeutic purposes are those that emit alphas, low-energy betas, or Auger electrons.

10.5.2. Nuclear Properties of Indicator Nuclides

The bulk of nuclear medicine procedures are carried out using just a few radionuclides. Over 90% of all diagnostic procedures, for example, use either ^{99m}Tc or an iodine isotope as the radioactive label. In this section, the nuclear properties of the most commonly used nuclides are summarized and a few examples of how they are used are given. The elements with good properties for imaging or treatment do not necessarily occur in natural biomolecules, so much research effort has been directed at incorporating those nuclides that are suitable tracers into organic carrier molecules. Radiopharmaceuticals are used in a variety of chemical forms. A few are pure elements, such as ^{133}Xe, or ionic, like $^{131}I^-$. Many others are incorporated into molecules. These range from small inorganic species like $Na_3{}^{32}PO_4$ to large proteins, such as ^{125}I serum albumin. On a still larger scale, cells may be labeled with tracers like ^{51}Cr. Table 10.2 lists some

Table 10.2. Radiopharmaceutical Tracers

Radionuclide	Chemical Form	Use
^{99m}Tc	Sodium pertechnetate	Brain, thyroid, salivary gland, and blood pool imaging; placenta localization
^{99m}Tc	Albumin colloid	Liver, spleen, and bone marrow imaging
^{99m}Tc	Etidrontate (EHDP)	Bone imaging
^{99m}Tc	Pentetate (DTPA)	Brain imaging, renal perfusion, kidney renograms, lung inhalation imaging
^{99m}Tc	Pyrophosphate (PPi)	Bone imaging, infarction avid imaging
^{131}I	Sodium iodide	Thyroid function diagnosis, thyroid imaging
^{125}I	Albumin	Blood and plasma volume determination, turnover studies
^{123}I	Sodium iodide	Thyroid function diagnosis, thyroid imaging
^{201}Tl	Thallous chloride	Myocardial infarction imaging
^{133}Xe	Gas	Pulmonary inhalation imaging, cerebral blood flow studies
^{67}Ga	Gallium citrate	Tumor imaging

Note: Activity levels administered range from 5 μCi ^{125}I for blood volume determinations and turnover studies to 20 mCi for ^{99m}Tc brain imaging.

Source: Adapted from R. J. Kowalsky and J. R. Perry, *Radiopharmaceuticals in Nuclear Medicine Practice,* Norwalk, CT: Appleton & Lange, 1987.

chemical forms and uses for the indicator nuclides discussed in this section.

Technetium

^{99m}Tc is by far the single most used radionuclide for diagnostic nuclear medicine procedures. Its formation and decay are shown in these reactions:

$$^{98}\text{Mo}(n, \gamma)^{99}\text{Mo} \xrightarrow{\beta^-} {}^{99m}\text{Tc} \xrightarrow{\text{IT(142.7-keV)}} {}^{99g}\text{Tc}$$

^{99m}Tc has a number of characteristics that make it well suited for use in medical diagnosis. Its half-life of 6.01 h is sufficiently long so that needed medical information can be obtained, but not so long that the patient suffers needless radiation exposure. The 142.7-keV γ ray emitted by the ^{99m}Tc has an energy high enough to penetrate tissue well and low enough for efficient detection. ^{99m}Tc is easily obtained from a technetium generator (described below) that can be kept on site at the hospital, and its cost is reasonable. Finally, Tc has a versatile chemistry that allows it to be incorporated into a wide variety of molecules, so that labels that seek out specific organs may be prepared.

The **Tc generator** mentioned above serves as an example of what is called an **isotopic radionuclide generator** or, more whimsically, a **cow**. As noted above, ^{99m}Tc is produced by the negatron decay of ^{99}Mo. The ^{99}Mo is obtained either from a nuclear reactor as a fission by-product or by the neutron irradiation of ^{98}Mo, using the $^{98}Mo(n, \gamma)^{99}Mo$ reaction. The ^{99}Mo is purified and chemically treated to produce anionic species, such as MoO_4^{2-}. This anionic form is then loaded onto an alumina (Al_2O_3) column that is kept acidic, so that a positive charge is maintained, and the MoO_4^{2-} binds to the alumina. The prepared column containing the ^{99}Mo (the cow) may then be "milked" for the active Tc daughter. A schematic diagram of a ^{99m}Tc generator is shown in Figure 10.4. A plot of the transient equilibria that exist during two "milkings" is given in Figure 10.5. Decay of the Mo produces the $[^{99m}TcO_4]^{1-}$ (**pertechnetate**) species. The pertechnetate does not bind to the alumina column as strongly as the molybdate anion, so the addition of physiological saline (0.9% NaCl) to the column results in the elution of the ^{99m}Tc as pertechnetate without removing the Mo. Once the pertechnetate is eluted from the cow, there are a variety of commercially available kits that may be used conveniently in the hospital laboratory to transform it into the specific chemical form needed for a particular diagnostic procedure. In its various forms, ^{99m}Tc is used for studies of nearly all of the body's organs and systems. Table 10.2 lists only a few of the many types of molecules into which ^{99m}Tc has been incorporated.

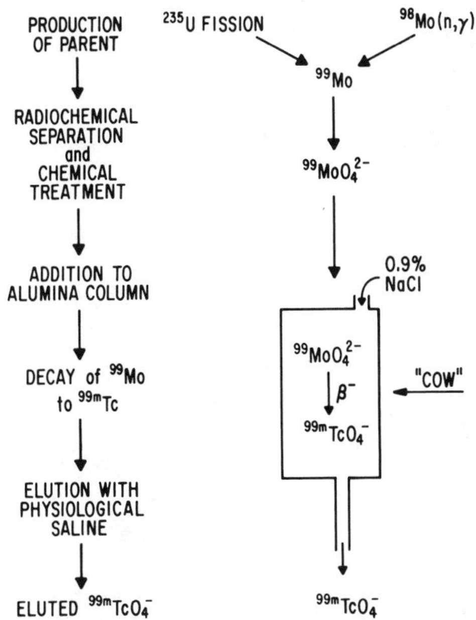

Figure 10.4. A schematic diagram of a ^{99m}Tc generator.

Iodine

There are three iodine isotopes commonly used for diagnostic imaging, ^{131}I, ^{125}I, and ^{123}I. The special affinity of I for the thyroid makes it most useful for imaging and treatment of this gland, but there are other uses for the iodine isotopes. They are usually chemically processed into a sodium iodide solution.

^{131}I is produced by the negatron decay of ^{131}Te. Its formation and decay are shown below:

$$^{130}\mathrm{Te}(n, \gamma)^{131}\mathrm{Te} \xrightarrow{\beta^-} {}^{131}\mathrm{I} \xrightarrow{\beta^-, \gamma(364.5\,\mathrm{keV})} {}^{131}\mathrm{Xe}$$

Like Mo, the ^{131}Te is obtained either as a fission by-product or from the neutron irradiation of ^{130}Te. The half-life of ^{131}I is 8.04 days. Because ^{131}I emits both β^- and γ radiation, it has less ideal characteristics as a diagnostic agent than ^{99m}Tc. In addition, ^{131}I emits several γ rays, most of which have energies that are too high for efficient detection by thin scintillation crystals. The 364.5-keV γ ray is the one usually used for imaging.

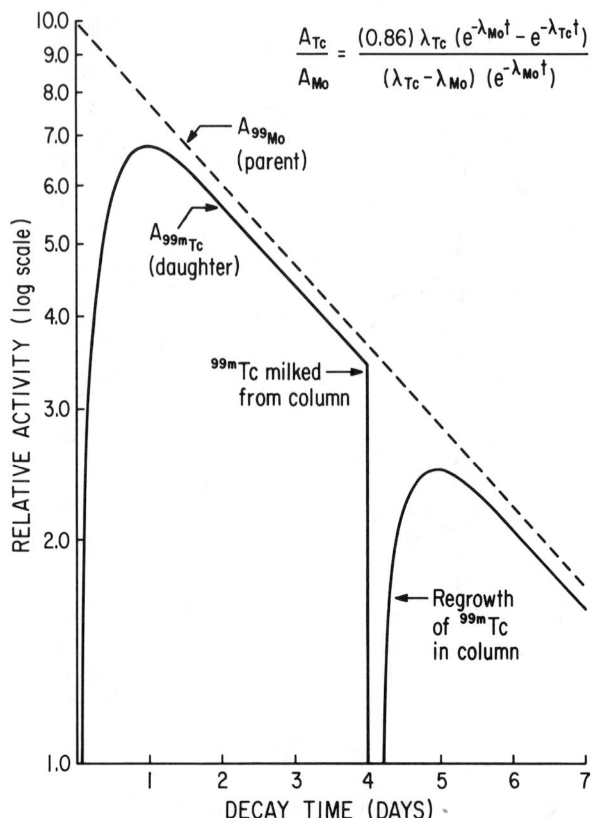

Figure 10.5. The parent–daughter equilibrium in the ^{99}Mo–^{99m}Tc generator. Transient equilibrium is reached, at which time the Tc daughter is eluted from the column. The daughter then grows in again until transient equilibrium is again approached and the column can again be "milked" for Tc. Only ≈86% of the ^{99}Mo decays yield ^{99m}Tc, which accounts for the 0.86 factor in the equilibrium equation and the fact that the ^{99m}Tc daughter activity does not exceed the parent ^{99}Mo activity at equilibrium. In this figure it is assumed that the column is milked for 100% of the ^{99m}Tc present after a decay time of 4.0 days from the preparation of the ^{99}Mo source at $t = 0$. In practice, it is common to elute only about 80% of the ^{99m}Tc from the column.

^{125}I is produced from the electron capture decay of ^{125}Xe. The xenon is produced by neutron irradiation of ^{124}Xe, as shown below:

$$^{124}\text{Xe}(\text{n}, \gamma)^{125}\text{Xe} \xrightarrow{\text{EC}} {}^{125}\text{I} \xrightarrow{\text{EC}} {}^{125}\text{Te}$$

^{125}I itself undergoes electron capture decay to ^{125}Te, with a half-life of 60.1 days. The x rays emitted following electron capture, and some low-energy γ rays (e.g., 35.5 keV), are used for detection. The γ rays do not penetrate tissue very efficiently, so the usefulness of ^{125}I for imaging is less than that of ^{131}I. However, because ^{125}I emits no particulate radiation, it still has some use as an imaging agent.

The proton-rich ^{123}I isotope is produced in a cyclotron by a variety of reactions, including $^{124}Te(p, 2n)^{123}I$ and $^{122}Te(d, n)^{123}I$. ^{123}I decays by electron capture to ^{123}Te, with a half-life of 13.2 h. A useful γ ray at 159 keV is emitted. ^{123}I has good properties for imaging, but is less convenient to use than the other two iodine isotopes because a cyclotron is necessary for its production.

Like technetium, iodine has a rich and varied chemistry, so it can be incorporated into many different molecules.

Thallium

^{201}Tl results from the decay of ^{201}Pb. The lead is produced initially by proton bombardment of thallium metal, in this reaction:

$$^{203}\mathrm{Tl}(p, 3n)^{201}\mathrm{Pb} \xrightarrow{\beta^+} {}^{201}\mathrm{Tl} \xrightarrow{\mathrm{EC}} {}^{201}\mathrm{Hg}$$

^{201}Tl decays by electron capture to stable ^{201}Hg, with a half-life of 72.9 h. The γ rays with energies of 167.4 and 135.3 keV are used for imaging. Because the Tl^+ ion is chemically similar to K^+, ^{201}Tl is often used for heart imaging. Potassium accumulates in a normally functioning heart during exercise.

Xenon

This chemically inert gas is a fission by-product. ^{133}Xe has a half-life of 5.243 days and decays by β^- and γ-ray emission to ^{133}Cs. The presence of the negatron and the low energy of the γ rays (e.g., 81 keV) are disadvantages in the use of this radionuclide. Because ^{133}Xe is gaseous, it can be inhaled, and is used primarily for lung imaging, such as assessment of the regional ventilation of the lungs.

Gallium

Gallium is in the same group of the periodic table as aluminum. The radiotracer ^{67}Ga decays by electron capture to ^{67}Zn with the emission of γ rays (93.3, 184.6, and 300.2 keV). Its half-life is 3.26 days. ^{67}Ga as citrate provides for good uptake in many types of tumors, and is commonly used for screening of soft-tissue tumors. Another isotope of gallium, ^{72}Ga, is a negatron and γ-ray emitter ($\beta^- = 0.96, 0.64$ MeV; $\gamma = 834.1, 2202$ keV) with a half-life of 14.10 h. The latter isotope shows a greater affinity for skeletal tissue than for soft tissue, and has been used in scanning bone tumors. The negatron emission is a disadvantage with respect to patient dose. Gallium is known to be toxic, but the levels used as radiotracers ($\leq 10^{-7}$ mg/kg of human body weight) are very low. The precise reason for the affinity of gallium for tumors is not known. Several days after administration, the highest levels of radiogallium are found in the liver, spleen, kidneys, and in the skeletal structure. ^{67}Ga is cyclotron produced, while ^{72}Ga is produced by radiative neutron capture in nuclear reactor irradiations.

10.5.3. In Vivo Diagnostic Procedures

In vivo diagnostic procedures are those in which a radiopharmaceutical is put into the system of a living patient, either orally, by injection, or by inhalation. The γ rays emitted by the radiopharmaceutical are monitored to provide the desired information. The γ-ray detectors used in medical imaging are often referred to as "gamma cameras." Sodium iodide detectors coupled to photomultiplier tubes are most often used, because high efficiency is of more importance than good energy resolution. In vivo procedures may give either anatomical (structural) or physiological (functional) information.

If a radiopharmaceutical is administered to a patient and then given time to accumulate in the target organ, a static image of the organ or system targeted can be obtained. This is analagous to a still photograph of the organ, and it can give information about the organ's size, shape, and location. Indications as to the location of diseased tissue can also be obtained, because the diseased portions of an organ may differ from the normal tissue in their tendency to take up the radiopharmaceutical. This can lead to the presence of hot spots or cold spots on the image. **Hot spots** refer to areas of increased radiopharmaceutical uptake, and **cold spots** to areas of decreased uptake. An example of reduced uptake might be observed with a ^{201}Tl heart scan. This radionuclide is accumulated by the normal cells of an actively exercising heart. A patient is given the ^{201}Tl and then asked to perform some exercise, such as walking a treadmill. After some time, the heart is scanned, and an image is formed. Areas of the heart that may have been damaged because of insufficient blood flow (called **infarcts**) will not take up the ^{201}Tl, and cold spots will appear in an image. In contrast, images of cancerous bone taken using ^{99m}Tc will show hot spots, where the rapidly metabolizing cancer cells have accumulated more ^{99m}Tc than surrounding normal bone.

Dynamic imaging procedures can also be done. In these, the passage of the radiopharmaceutical into and out of an organ is monitored continuously with a gamma camera. An example is the imaging of the liver and gallbladder using a ^{99m}Tc-labeled molecule. Timed images taken of this system after administration of the ^{99m}Tc can give information about impaired function of these two important organs.

Still another type of diagnostic procedure attempts to assess the function of an organ by measuring the rate of uptake of an administered radiopharmaceutical. No image of the organ is formed; only count rates are measured. An example of this would be the measurement of the rate of uptake of ^{131}I by the thyroid gland. The patient might be given around 5–10 μCi of Na^{131}I orally and the thyroid scanned at 4 and 24 h after administration to determine the portion of the original dose that has been concentrated in the thyroid. A patient with a hyperactive thyroid will show an increased uptake of the ^{131}I relative to a normal person, while a hypothyroid condition will result in decreased uptake.

A unique imaging technique that relies on the use of positron-emitting radionuclides is **positron emission tomography (PET).** Unlike many imaging techniques which give only anatomic information, PET is unique in that it can also give physiological information. This method has experienced rapid growth in recent years, due in large part to the development of computer-aided tomographic techniques and to the synthesis of ^{18}F-fluorodeoxyglucose, which can be used for brain-imaging studies.

Figure 10.6 shows a photograph of a PET facility. In PET scanning, the patient is given a positron-emitting radiopharmaceutical. The positron travels a short distance after it is emitted and then undergoes annihilation, with the emission of two 0.511-MeV annihilation photons oriented at 180° from each other. If two detectors also oriented at 180° detect the photon pair simultaneously, then the possibility exists for localizing the source of the positron emission along the line between the two detectors. The detectors used in PET scanning are either NaI(Tl) or BGO detectors, both of which have good efficiency and fast response times. By moving the patient, moving the paired detectors, or using a large multidetector array, three-dimensional localization of the source of the activity is possible. In contrast to some other types of medical imaging techniques, PET scanning with multiple detectors allows dynamic measurements of metabolic processes and organ function, rather than snapshot images at a single point in time.

The nuclides most used for PET scanning include ^{11}C, ^{13}N, ^{15}O, and ^{18}F. Because they are positron emitters, they are best produced using charged-particle-induced reactions. Small medical cyclotrons have been designed that can make these nuclides on site, and many fast synthetic schemes have been devised to incorporate them into a variety of metabolic substrates, such as fatty acids, proteins, and amino acids. PET has been applied most to studies of brain (Fig. 10.7) and heart function. In brain, the rate of oxygen consumption and blood flow through various regions can be studied.

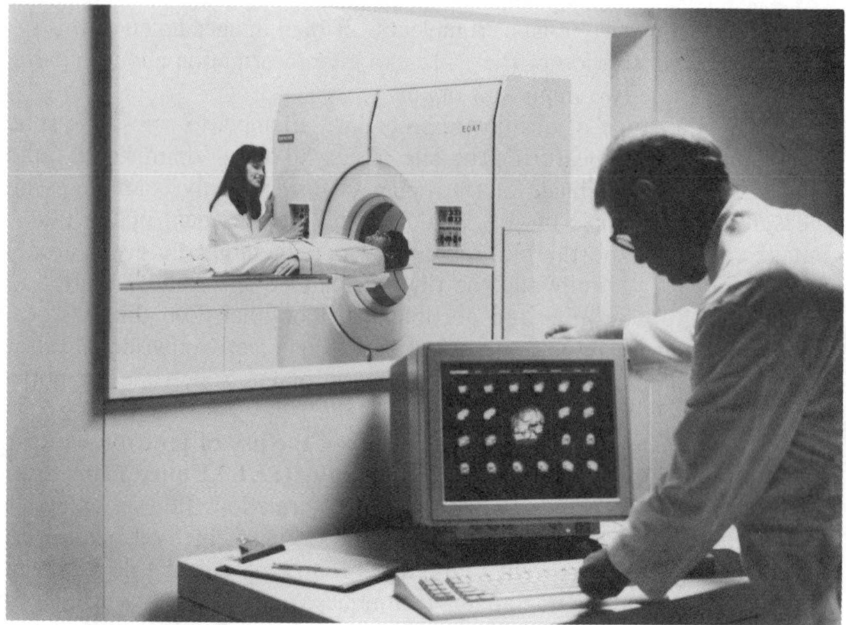

Figure 10.6. A positron emission tomography (PET) facility. The physician in the foreground is reviewing clinical and patient information on the Sun workstation while the technologist in the background positions the patient into the ECAT's gantry aperture for imaging. (Courtesy of Siemens Medical Systems, Inc., Hoffman Estates, IL.)

Dramatic alterations from the normal state can be seen in victims of brain disorders, such as those resulting from stroke or Alzheimer's disease.

10.5.4. In Vitro Diagnostic Testing: Radioimmunoassay

In vitro procedures are those performed on tissue samples removed from a patient. There are a number of these that involve use of radiopharmaceuticals, but the most important is the technique of **radioimmunoassay, RIA.** Radioimmunoassay is a type of substoichiometric IDA, in which labeled and unlabeled analyte compete for limited amounts of a molecule that binds the analyte very specifically. RIA is widely used in medical laboratories worldwide for the determination of hormones, drugs, viruses, and other organic species.

RIA had its beginnings in the 1950s with the investigations of S. Berson and R. Yalow on the metabolism of [^{131}I]insulin in diabetics. Berson and Yalow recognized that diabetics had substances in their blood serum that bound insulin. They observed that labeled and unlabeled insulin competed

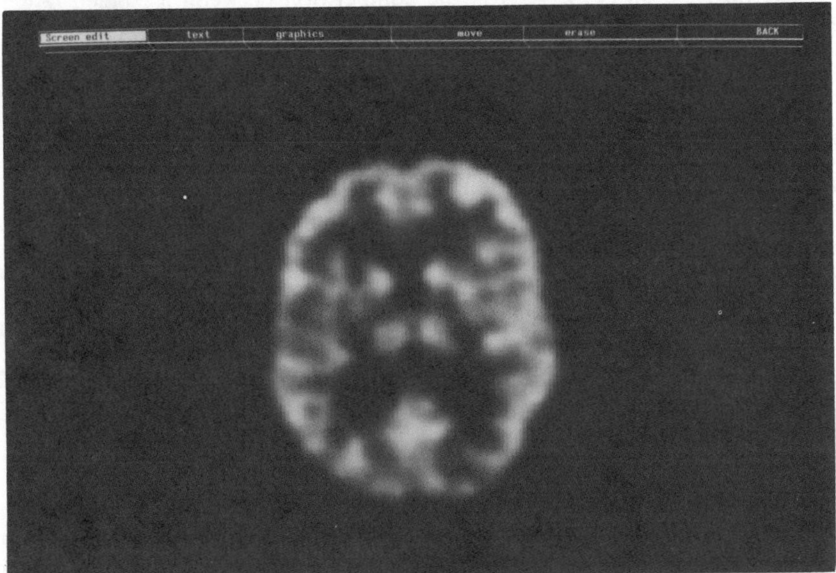

Figure 10.7. PET image of a transverse human brain slice. (Courtesy of Siemens Medical Systems, Inc., Hoffman Estates, IL.)

for this binding substance, and that the amount of unlabeled insulin present affected the amounts of labeled insulin that became bound. They recognized in this the potential for an assay method for insulin. RIA has since developed into a very widely used medical technique, with broad application for the measurement of minute quantities of many important biomolecules. Yalow was awarded the Nobel Prize in medicine in 1977 for the development of this technique. (Berson had died, and Nobel Prizes are not awarded posthumously.)

Radioimmunoassay procedures are based on the reaction between an antigen and an antibody. An **antigen (Ag)** is a molecule that can induce an organism to produce other molecules in response to the antigen's presence. These other molecules are called **antibodies (Ab).** Usually, antigens are molecules that are "foreign" to an organism, and the body's defense system reacts with the formation of antibodies that can bind to the antigens and render them inactive. Antibodies are very specific for a given type of antigen. Antigenic substances are large molecules, with molecular masses of at least 5000 daltons, that have certain chemical species or configurations that the antibody can attach to. Smaller molecules that are not antigenic themselves (**haptens**) can be made antigenic by conjugating them with larger molecules.

The physical basis for RIA is the competition between a labeled (Ag*) and unlabeled (Ag) antigen for a limited amount of antibody (Ab):

$$\begin{matrix} \text{Ag} \\ \text{Ag}^* \end{matrix} + \text{Ab} \longrightarrow \begin{matrix} \text{Ag} \cdot \text{Ab} \\ \text{Ag}^* \cdot \text{Ab} \end{matrix}$$

The unlabeled antigen is the analyte (drug, hormone, etc.). If fixed amounts of the antibody and the labeled antigen are present, the amount of the labeled antigen–antibody complex that forms is inversely related to the amount of unlabeled antigen present. Thus, if the antigen–antibody complex can be separated from the mixture and its activity determined, the amount of analyte (the unlabeled antigen) can be determined. In practice, quantitative determinations of the analyte are performed by use of a calibration curve. The curve is prepared by adding standard amounts of labeled antigen to a fixed amount of antibody and measuring the activity of the separated Ag · Ab complex. A plot is made of activity versus the amount of labeled antigen. The complex activity in the presence of the actual sample is then used to determine the amount of antigen.

The advantages of RIA include its excellent sensitivity (down to 10^{-13} g for some molecules), and its high specificity, due to the special nature of the antigen–antibody reaction. The accuracy and precision attainable using RIA are good. Although RIA procedures are time-consuming and sometimes difficult to develop at first, they can later be standardized so that simple kits for a given procedure are commercially available. These kits are commonly used by personnel in medical laboratories.

10.5.5. Therapeutic Uses of Radiation

The therapeutic uses of radiation and radiopharmaceuticals are more limited than the diagnostic uses. When radiation is used therapeutically, the aim is to destroy a specific portion of a diseased tissue with the radiation. The source of the radiation can be either external or internal.

External radiation sources are now mainly in the form of electron beams or x rays. Many devices can be used to produce these radiations, but small linear accelerators are most used. Electrons with energies of 4 to 15 MeV are used to treat cancers that are near the body surface, such as those of the skin, breast, head, and neck.

When greater penetration is needed, γ radiation from a sealed radionuclide source can be used. ^{60}Co was used extensively for this purpose for many years, but now ^{137}Cs is preferred.

In addition to the external irradiation of an organ, it is also possible to implant a radioactive "needle" or "grain" into the diseased area and thereby irradiate even more specifically the area to be destroyed. Implants of ^{198}Au and ^{125}I are common.

TERMS TO KNOW

Antibody
Antigen
Autoradiography
Beta gauges
Carrier-free tracer
Direct IDA (DIDA)
Emanation analysis method
Fick's first law of diffusion
Formation constants
Hapten
Inverse IDA (IIDA)
Isotope dilution analysis (IDA)
Isotopic exchange reactions
Liquid–liquid diffusion measurements
Moisture gauges
Nuclear medicine
Nuclear pharmacy
Positron emission tomography (PET)
Radiographic methods
Radioimmunoassay (RIA)
Radiometric titrations
Radionuclide generator
Radiopharmaceuticals
Radiorelease methods
Radiotracer
Self-diffusion measurements
Serial sectioning
Spike
Stable isotope dilution analysis
Substoichiometric IDA
Surface decrease methods
Therapeutic methods

READING LIST

Arnikar, H. J., *Isotopes in the Atomic Age.* New York: Wiley, 1989. [isotope effect, applications of tracers, isotopic separations]

Berson, S. A., and R. S. Yalow, *J. Clin. Invest.* **36**:873 (1957). [first RIA work]

Bisker, J., *Clinical Applications of Medical Imaging.* New York: Plenum Medical, 1986. [computerized tomography, many radiographic illustrations and applications]

Bowen, H. J. M., *Chemical Applications of Radioisotopes.* London: Metheun, 1969. [general applications]

Chilton, H. M., and R. L. Witcofski, *Nuclear Pharmacy.* Philadelphia: Lea & Febiger, 1986. [production and applications of radiopharmaceuticals]

Fritzberg, A. R., ed. *Radiopharmaceuticals: Progress and Clinical Perspectives,* Vols. 1 and 2. Boca Raton, FL: CRC Press, 1986. [nuclear pharmacy]

Harbert, J. C., *Nuclear Medicine Therapy.* New York: Thieme Medical, 1987. [nuclear medicine]

Hevesy, G., and W. Seith, *Z. Phys.* **56**:790 (1929). [self-diffusion experiments]

Hladik, W. B., G. B. Saha, and K.T. Study, eds. *Essentials of Nuclear Medical Science.* Baltimore: Williams & Wilkins, 1987. [nuclear medicine]

International Atomic Energy Agency, *Guidebook on Radioisotope Tracers in Industry,* Tech. Report Ser. No. 316, Vienna: IAEA, 1990. [industrial applications]

Kowalsky, R. J., and J. R. Perry, *Radiopharmaceuticals in Nuclear Medicine Practice.* Norwalk, CT: Appleton & Lange, 1987. [general medical applications]

Nuclear activation and radioisotopic methods of analysis. In *Treatise on Analytical Chemistry,* Vol. 14, Part 1, Section K, I. M. Kolthoff and P. J. Elving, eds. New York: Wiley, 1986, Chapters 1–8. [a comprehensive treatment of IDA, radiorelease methods, radiometric titrations, RIA, and general tracer methodology]

Parker, C.W., *Radioimmunoassay of Biologically Active Compounds.* Englewood Cliffs, NJ: Prentice-Hall, 1976. [applications of RIA]

Spencer, R. P., R. H. Seevers, Jr., and A. M. Friedman, *Radionuclides in Therapy.* Boca Raton, FL: CRC Press, 1987. [therapeutic applications]

Starý, J., and J. Ružička, Substoichiometric analytical methods. In *Wilson and Wilson's Comprehensive Analytical Chemistry*, Vol. VII. Amsterdam: Elsevier, 1976. [substoichiometric IDA]

EXERCISES

1. A steel sample is to be analyzed for its cobalt content by DIDA. A 1.37-g sample of the steel was dissolved in acid and 2.00 mL of a ^{60}Co spike solution was added. The spike solution contained 2.00 mg of Co/mL. Separate portions of both the pure spike solution and the spike–sample solution were processed by electrodeposition. A deposit of 10.5 mg of pure Co on the electrode from the pure spike solution was found to have a counting rate of 1.50×10^4 cpm, while a 6.25-mg Co deposit from electroplating the spike–sample solution had a counting rate of 2750 cpm, under the same counting conditions. Neither of the plating operations were quantitative and the plating yields were not known. Calculate the percentage of Co in the steel sample.

2. A solution contains an unknown amount of osmium at the μg/g level, or below. Chemically pure osmium (Os) may be separated from the solution, but the chemical yields are variable and low. The determination is done by direct isotope dilution analysis (DIDA). The isotopic abundances of ^{186}Os and ^{190}Os are 1.58 and 26.4%, respectively. 7.80×10^{-9} g of isotopically pure, stable ^{186}Os is added to the unknown solution and mixed thoroughly, and then a small amount of pure Os is extracted from the mixture. With a mass spectrometer, the ^{190}Os/^{186}Os mass ratio in the extracted sample is found to be 7.40. Calculate the amount of Os originally present in the unknown solution.

3. Compound C is to be determined in some strip mine runoff waters by use of substoichiometric IDA. A spike solution of radioactive tagged compound C contains 55.0 ng of tagged C per milliliter. Exactly 200 μL of the spike solution is added to 15.0 mL of the runoff water sample and is well mixed. By using a complexing agent and a solvent extraction procedure, equivalent mass substoichiometric amounts of C

were extracted from both the pure spike solution and the spike–sample mixture. The equal volume extracts were counted under the same conditions and yielded 2.73×10^3 cpm for the sample–spike extract and 1.72×10^4 cpm for the pure spike extract. Calculate the concentration of C in the runoff water sample in units of g/L.

4. A 3.00-kg batch of a synthetic mixture containing a new antibiotic drug was prepared in a "pilot plant" operation. The crude product was analyzed by IDA. A spike of exactly 8.25 mg of pure radiolabeled antibiotic having an activity of 8600 dpm was added to a 10.0-g sample of the crude product. After thorough mixing, a sample of pure antibiotic was extracted from the spike-sample mixture in low chemical yield. A 50.0-mg portion of this pure fraction was counted and yielded an activity of 230 dpm. Calculate the percent by mass of the antibiotic in the crude product.

5. Discuss how a tracer experiment could be used to determine that the Walden inversion of optically active organic compounds is a one-step concerted process. The Walden inversion involves substitution at a chiral center (e.g., Cl substitution in 2-chlorobutane) with a change in optical configuration. In fact, the substitution and the inversion do take place simultaneously and not sequentially.

6. Calculate the time required for the ^{99m}Tc activity in a Tc isotopic generator to reach 75% of the activity of the parent ^{99}Mo.

CHAPTER

11

ION BEAM ANALYSIS AND CHEMICAL APPLICATIONS OF RADIOACTIVITY

There are numerous other applications of radioactivity in chemistry and related fields that are neither nuclear activation nor tracer techniques. **Ion beam analysis (IBA)** refers to techniques that use ion beams as part of an analytical procedure. One of these, PIGE, was discussed briefly in Chapter 9. Two other ion beam techniques, particle-induced x-ray emission (PIXE) and Rutherford backscattering (RBS), are discussed in this chapter. Another technique, Mössbauer spectroscopy, uses the resonance absorption and emission of γ rays by nuclei to discern chemical information about a system. The field of hot atom chemistry deals with the effects and final disposition of energetic ions produced as a result of nuclear transformations. Finally, in the field of radiation chemistry, researchers use the chemical effects of radiation for both theoretical studies and a variety of practical applications.

11.1. PARTICLE-INDUCED X-RAY EMISSION

There are several analytical methods that use the x rays emitted from excited atoms as the basis for obtaining qualitative and quantitative data about the elemental composition of a sample. These include x-ray fluorescence, particle-induced x-ray emission, x-ray absorption, electron-induced x-ray emission, and other related methods of analysis. The primary difference among most of these methods lies in the way in which the electrons are initially excited. In **particle-induced x-ray emission (PIXE),** charged particles produced by an accelerator, or emitted by the decay of a radionuclide, are used for electron excitation. The most common particle used for this purpose is the proton, so the acronym PIXE is also used to mean *proton*-induced x-ray emission.

PIXE is commonly classified as a nuclear analytical technique, even though the relevant interactions occur with orbital electrons, rather than with the nuclei of the target atoms. Its placement with nuclear analytical techniques arises from the facts that PIXE uses subatomic particles to induce the electron changes and that the detector systems used for x rays are similar to those used for detection of γ rays. In these respects, PIXE has much in common with the nuclear analytical techniques discussed in Chapter 9. PIXE is often omitted from books on instrumental analysis, so it seems both useful and appropriate to include it here.

11.1.1. Overview of the PIXE Process

The basic process occurring in PIXE is the excitation or ionization of orbital electrons by a beam of energetic charged particles. This excitation is followed by the emission of characteristic x rays as the electrons undergo internal rearrangements to return to the atom's ground-state configuration. The emitted x rays are detected, and qualitative and quantitative information is gleaned from them.

PIXE is a relatively new technique, having been developed in the early 1970s. It blossomed quickly as x-ray detectors improved, and as more low-energy accelerators previously used for nuclear physics research became accessible for analytical purposes. One might wonder why an analyst would go to the trouble of using protons to induce x-ray emission when there are other excitation modes that involve much simpler experimental techniques. The answer is that protons, or heavier charged particles, induce much less background bremsstrahlung radiation than do electrons. Therefore, the potential sensitivity of PIXE is much greater. In addition, the protons can be focused on very small areas (micro-PIXE) permitting spatial resolution. Depth profiling is possible by variation of the incident proton energy. These latter applications are discussed later in this chapter.

11.1.2. Projectile Acceleration and Target Preparation

Figure 11.1 shows a block diagram for a PIXE experimental arrangement. The first step in PIXE is the generation of the charged particles used to excite the electrons. In theory, many kinds of charged particles from a variety of accelerators could be used for this purpose. However, proton beams with energies from 1 to 3 MeV have been found to provide good sensitivity for most elements and are by far the most commonly used particles. Van de Graaff accelerators or Cockroft–Walton accelerators are good devices for the generation of proton beams of these energies. (The operation of accelerators is discussed in more detail in Chapter 14.) As mentioned above, one reason for the rapid growth of PIXE has been the increasing availability of single-stage or tandem Van de Graaff accelerators for analytical applications. Beam intensities of a few nanoamperes are ordinarily used. For best results, the proton beam should have parallel trajectories and have a uniform areal intensity when it strikes the target. Collimators are included to

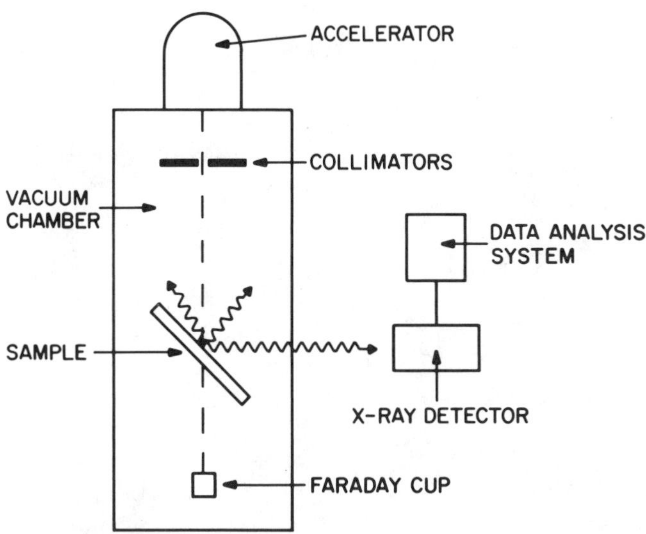

Figure 11.1. Experimental configuration for PIXE analysis.

accomplish this. The collimators should be made of a material such as graphite that will produce minimum x- and γ-ray background as the proton beam impinges upon it. A typical proton beam in a PIXE system will have a diameter of about 1 mm. The beam line and sample chamber are both normally under a vacuum of around 10^{-6} torr. For thin samples, that is, those that the beam passes through virtually undiminished in intensity, beam intensity is measured by a Faraday cup at the end of the beam line. More complicated schemes are needed to measure beam intensities when thick samples are used, because these either stop the particle beam or significantly diminish its intensity.

Because the sample to be analyzed must ordinarily be placed inside a vacuum chamber, sample holders are made to contain several samples at one time. This minimizes the time and effort involved in changing samples, because otherwise it would be necessary to break and reform the vacuum for each sample change. A wide variety of sample holders have been designed. An example of a ladder-type changer is shown in Fig. 11.2. For some samples sensitive to alteration in a vacuum, it is necessary to extract the proton beam from the beam line through a thin window and into a sample chamber which is at 100 torr to 1 atm pressure.

Because of beam size, PIXE is most often used to analyze small samples. The samples must be mounted in the sample chamber in fixed positions relative to the proton beam, so it is desirable for the sample to possess a certain degree of mechanical rigidity. Preparation methods for the target will

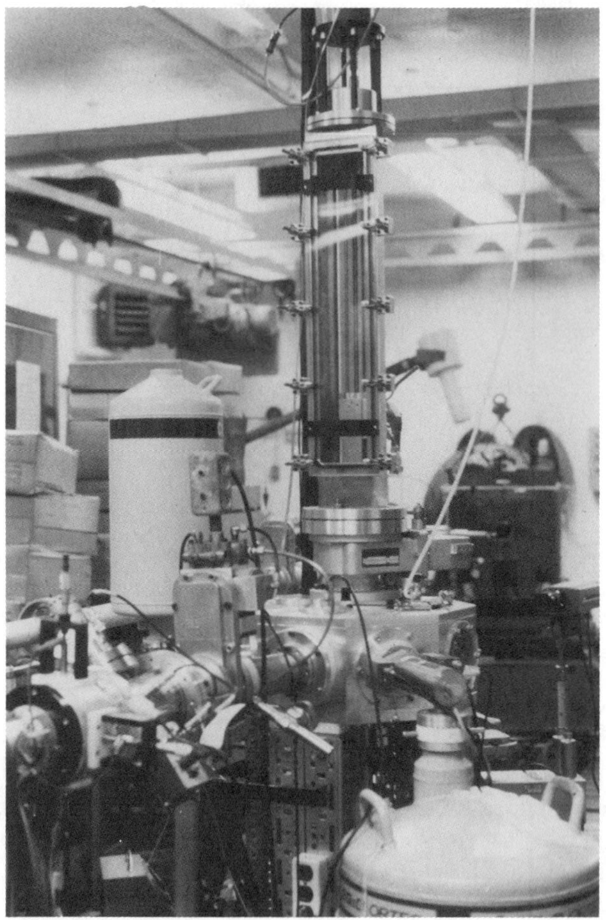

Figure 11.2. A PIXE counting chamber with a vertical "ladder" sample changer connected to a beam line from a single-stage Van de Graaff accelerator. A Si(Li) detector in a large Dewar of liquid nitrogen is positioned to count the x rays emitted. The vertical sample rack is under vacuum and permits sample changing without opening the chamber. (Courtesy of J. D. Robertson, University of Kentucky, Lexington.)

vary with sample type. Materials that have sufficient strength may be mounted directly in the chamber and irradiated. Examples of these types of samples would be hair fibers, small pieces of bone or leaves, or tiny mineral grains.

For samples that cannot be mounted alone, a backing material is needed to provide support. The backing material, like that for the collimators, must be carefully chosen. Minimal background radiation should be produced as a result of irradiation by the proton beam. Other desirable characteristics of the backing material are that it be of high purity, inert to any solvents used in sample mounting, strong enough so that very thin pieces are sufficient to support the sample in the vacuum chamber, thermally and electrically conductive, inexpensive, and easy to manipulate. Thin foils made of carbon, aluminum, or a plastic are commonly used for backing material. Liquid samples may be deposited on the backing and then dried. Solid samples can be embedded into a matrix such as a wax or a plastic, sectioned with a microtome, and then placed on the backing.

11.1.3. Ionization and X-Ray Emission

When the particle beam from the accelerator strikes the target, it interacts in the ways described in Chapter 6 for heavy charged particles. For PIXE, the most important interaction is ionization of electrons through Coulomb interactions. The ionization events of greatest utility are for electrons that are lost from the first or second energy level (K or L atomic shell).

Immediately after removal of an electron from an inner shell, electron rearrangements occur to fill the vacant spot. These processes were described in Chapter 2 in the discussion of electron capture and internal conversion decay. The electron rearrangements result either in Auger electron emission or in x-ray emission. The ratio of vacancies filled by x-ray emission to the number of vacancies created is the fluorescence yield (ω). Figure 2.6 shows the relationship of ω to atomic number. As can be seen from the graph, the emission of Auger electrons is more common for lower-Z elements, and the emission of x rays is more common for higher-Z elements.

The emitted x rays correspond to energy level spacings in the electron orbitals and so are characteristic of the target nucleus. This provides a basis for qualitative identification of elements in a sample. The x rays are designated by using the symbol of the element from which they originated, along with a subscript that indicates what shell the electron came from (K, L, M, etc.). Sometimes, a Greek letter is added to show the origin of the replacement electron. The letter α indicates that the replacement electron came from the next highest shell, β designates a replacement electron from the second higher shell, and so on. For example, the designation $Mn_{K\alpha}$ would refer to the x ray emitted when a K electron was replaced by an electron from the L level (α) in manganese.

11.1.4. Detection and Analysis of X Rays

The development of the Si(Li) detector for x rays greatly enhanced the usefulness of PIXE as an analytical technique. Prior to this, x-ray detection was generally done using a **crystal spectrometer.** In this type of spectrometer, the x rays emitted by the sample were allowed to impinge on a crystal which diffracted the x rays according to their wavelengths (an application of Bragg's law). These devices are known as **wavelength-dispersive spectrometers.** These spectrometers offer outstanding energy resolution, but are quite expensive, are difficult to use, and have a limited range of useful wavelengths. They are still used for research purposes, but are not often found in analytical laboratories. Instead, an **energy-dispersive system,** using a semiconductor Si(Li) detector, is preferred. These are very similar to the detection systems discussed in Chapter 9 for γ-ray spectrometry that use Ge(Li) or HPGe detectors. The x rays strike the Si(Li) detector, where they undergo interactions similar to those experienced by γ rays, and lose their energy. The signals from the detector are fed to a multichannel analyzer, where they are sorted according to energy, and an x-ray spectrum is displayed and recorded with a multichannel pulse-height analyzer. The use of the Si(Li) detector confers the ability to do multielemental analysis with PIXE, which is a great analytical advantage over the wavelength-dispersive systems. The Si(Li) detector has better characteristics for x-ray detection than the Ge(Li) or HPGe detector. It can offer an energy resolution of 0.15 to 0.18 keV at the 5.9-keV Mn K line. For x rays of low energy (4–30 keV), a Si(Li) detector several millimeters thick with a thin window can have nearly a 100% detection efficiency for x rays that are incident on the window.

For PIXE, the detector is placed at an angle to the sample which is adjusted to minimize the amount of background radiation reaching it. Typically, the sample is placed from 10–40 mm away from the detector. The remainder of the analyzer system, including amplifiers, multichannel analyzers, and computer processing, is similar to that described in Chapter 9 for activation analysis techniques.

Figure 11.3 shows a representation of an x-ray spectrum obtained by PIXE analysis of a natural bone sample using a Si(Li) detector. The spectrum consists of x-ray peaks superimposed on a background that is generated primarily from matrix interactions. The majority of analytically useful peaks are found in the region from 1 to 20 keV. K-level x rays are observed for lighter elements, and L x rays are used for the heavier elements. The most significant contributor to the background is the bremsstrahlung radiation produced as the incident projectiles, and the secondary electrons produced by the projectiles, pass through the matrix. A lesser contributor is the γ radiation produced as a result of nuclear reactions taking place in either the target or backing. The importance of minimizing this background by choosing appropriate target and target support materials has already been mentioned.

Most PIXE spectra are complex, with many single and overlapping peaks, so manual peak analysis is not usually attempted. A variety of

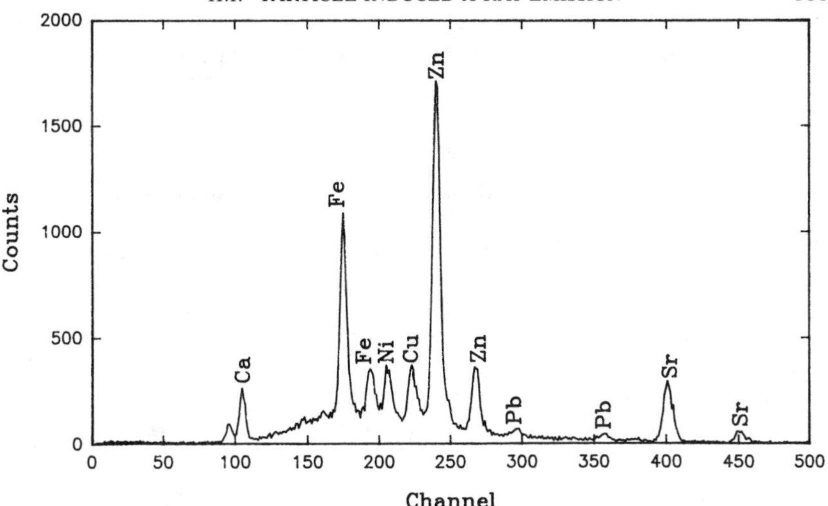

Figure 11.3. PIXE spectrum of a natural bone sample. The spectrum was acquired with an external 3-MeV proton beam in a sample chamber under 1 atm of helium. A polycarbonate filter was placed in front of the x-ray detector to reduce the intensity of the calcium x rays. The labeled peaks are K-level x rays, except for lead which are L-level. (Courtesy of J. D. Robertson, University of Kentucky, Lexington.)

computer programs have been developed for deconvolution and peak analysis of PIXE spectra. As with activation analysis, both *absolute* and *comparator* methods may be used to determine amounts of analyte. For absolute calculation on thin samples, information on the proton ionization cross section, fluorescence yield, detection efficiency, experimental geometry, and beam current must be available. Calculations are more complicated for the case of thick samples. Detailed equations for PIXE calculations can be found in the monograph by Johansson and Campbell (see the Reading List at the end of this chapter).

11.1.5. Applications of PIXE

PIXE is best used as an elemental trace analysis technique on small, thin samples. PIXE has very good sensitivity for elements from $Z = 20$ to $Z = 90$, and analyses have been performed on many different types of samples. The high-sensitivity, multielemental capability, short analysis time, and nondestructive analysis capability afforded by PIXE make it an especially useful technique for analysis of biological and geological samples. For biological samples, which have a low-Z matrix, many trace elements, such as Cu, Fe, and Zn, can be determined at the μg/g level, or below. PIXE methods can also determine the major elements such as Ca, K, S, Cl, and P,

Table 11.1. Examples of Recent PIXE/PIGE Applications

Area	References
Archaeology	
Pottery	Deurden et al., *Nucl. Instrum. Methods Phys. Res.* **B14**(1):50 (1986)
Stone artifacts	Pineda et al., *ibid.* **B35**(3–4):463 (1988)
Art	
Paintings	Tuurnala et al., ibid. **B14**(1):70 (1986)
Biology and medicine	
Spermatozoa, teeth	Hall et al., ibid. **B15**(1–6):629 (1986)
Phospholipid bilayers	Georgallas et al., *J. Chem. Phys.* **86**(12):7218 (1987)
Geology	
Aerosols from volcanos	Quisefit et al., *Nucl. Instrum. Methods Phys. Res.* **B22**(1–3):301 (1987)
Obsidian, desert varnish	Deurden et al., ibid. **B14**(1):50 (1986)
Zeolites	Frey et al., *J. Radioanal. Nucl. Chem.* **120**(2):281 (1988)
Industrial	
Borosilicate glasses	Borbely–Kiss et al., ibid. **92**(2):391 (1985)
Waste waters	Fakhouri et al., ibid. **129**(1):163 (1989)

which are all biologically important. For geological samples, PIXE has good sensitivity for some of the important matrix elements, such as Al, Si, F, O, Mg, Ca, Fe, and K. Other areas where PIXE has found wide application are similar to those where activation analysis is used. They include atmospheric studies, archaeology, art, forensics, and environmental studies. As noted in Section 9.4.4, PIXE is often used in conjunction with PIGE because PIGE has a high sensitivity for low-Z elements ($Z < 15$) while PIXE is most effectively used to determine medium or high-Z elements. Table 11.1 lists some of the applications taken from current literature in which both PIXE and PIGE are used.

11.1.6. PIXE Variations

Microbeam techniques have assumed increasing analytical importance as knowledge of elemental distribution in samples becomes necessary for many applications. In **micro-PIXE,** the proton beam is focused to a much smaller dimension than the 1-mm diameter usually used. A proton beam that gives a spatial resolution for analysis of 1 μm can be achieved. The microbeam techniques are useful in biological and geological studies. They can be used, for example, to determine the spatial distribution of elements in a cell or in a single mineral grain in a rock section.

A few special applications of PIXE require that the analyses be performed under nonvacuum conditions. Examples of this would be the nondestructive analysis of delicate materials, such as art works, archaeological artifacts, or biological specimens. In some cases, the end of the beam line and sample chamber are isolated by a thin foil from the accelerator stage and filled with an appropriate gas, following which analyses are carried out as usual. Another design, referred to as **external beam analysis,** extracts the proton beam through a thin foil in the beam line into the laboratory atmosphere where sample changing and detector positioning is less restricted and the integrity of fragile samples is maintained.

11.2. RUTHERFORD BACKSCATTERING SPECTROMETRY

Scattering reactions were defined in Chapter 4 as those in which a projectile is deflected away from its original path due to Coulomb interactions with target nuclei. Scattering can be either elastic, in which total kinetic energy of the system is conserved, or inelastic, where some kinetic energy goes to excite the target nucleus. **Rutherford scattering** refers to the elastic scattering of charged particles by target nuclei. It was this type of reaction that was used by Rutherford in his early studies of atomic structure which formed the experimental basis for the nuclear model of the atom. Scattering reactions continue to be important tools in nuclear physics studies of nuclear structure. They also have some analytical uses, as exemplified by the technique called **Rutherford backscattering spectrometry (RBS).**

The basic premise of RBS is that the angle, energy, and number of projectiles scattered from a target can be related to the identity, location, and abundance of the atoms in the target. Scattering occurs in all directions from the target atoms, but most often it is the particles scattered at large angles from the target that are detected. This is the reason for the name of *backscattering* spectrometry.

Figure 11.4 shows a block diagram of an RBS system. A beam of monoenergetic ions, usually $^4He^+$ from an accelerator, is aimed at a target. The ions are elastically scattered by the target, and those scattered at large angles are detected. (The reasons for detection at large scattering angles will be discussed shortly.) A silicon surface-barrier detector is most often used to detect the scattered particles. The entire system, including beam, target, and detector, is kept under vacuum.

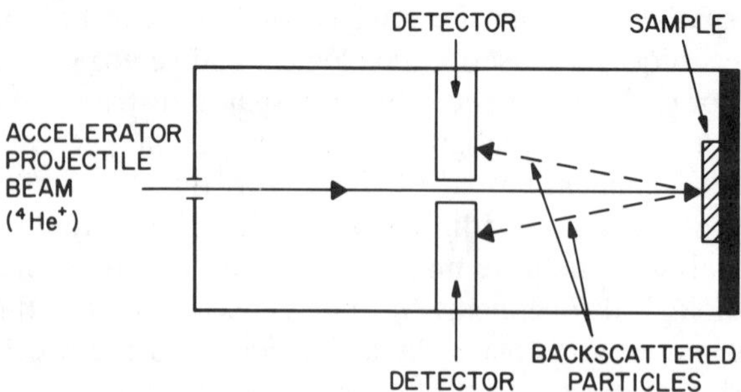

Figure 11.4. Experimental configuration for RBS analysis.

11.2.1. The Scattering Reaction

The He ions with mass M_1 and an initial energy, E_0, impinge on the target. In the scattering process, some kinetic energy is transferred to the target nuclei, so the scattered He ions have less energy, E_1, than they did originally. The ratio of the scattered energy to the initial energy, E_1/E_0, can be derived from classical concepts of energy and momentum conservation. The relationship is

$$K = \frac{E_1}{E_0} = \left[\frac{(M_2^2 - M_1^2 \sin^2 \theta)^{1/2} + M_1 \cos \theta}{M_2 + M_1}\right]^2 \tag{11.1}$$

where M_1 and M_2 refer to the masses of the projectile and target, respectively, and θ is the scattering angle. The ratio E_1/E_0 is called the **kinematic factor, *K*.** As can be seen in Eq. 11.1, the energy of a scattered projectile is related to its mass, the mass of the target atoms, and the scattering angle. Thus, if the mass of the projectile and the scattering angle are known, the masses of the target atoms can be determined. This is the basis for the qualitative identification of atoms in RBS. Tables giving K values for all the elements for typical scattering angles are found in the literature.

Equation 11.1 can be used to illustrate why large scattering angles are preferred for detection. The ability to determine the mass of the target atoms is related to the system's ability to detect energy differences among scattered particles. Therefore, the largest possible change in energy is desirable. This occurs when the ion is scattered directly back at 180°. For this angle, Eq. 11.1 simplifies to

$$\frac{E_1}{E_0} = \left(\frac{M_2 - M_1}{M_2 + M_1}\right)^2 \tag{11.2}$$

In practice, detection at 180° is not feasible, so an angle of around 170° is often used.

Equation 11.1 also shows that the kinematic factor increases as the mass of the target atoms increases and as the difference between the masses of two atoms increases. Thus, RBS is best used to determine heavy elements in a matrix of lower mass.

11.2.2. Surface Analysis Using RBS

Scattering may occur either from the surface atoms or from atoms some distance into the sample. In this section, only the scattering occurring from surface atoms is considered. Figure 11.5 shows a schematic diagram of the backscattering process for a sample consisting of equal numbers of Si, Cu, and Au atoms distributed on the surface of a sample, and the backscattering spectrum that would result. The beam of $^4He^+$ typically has an initial energy of around 1–3 MeV. Figure 11.5 assumed a beam of 2-MeV ions. The He ions are elastically scattered from the surface atoms and are detected. The signals are sent to a multichannel analyzer for display. Because the scattered energies (E_1) of the He ions are related to the masses of the target atoms, three different peaks are seen in the spectrum. The loss in energy is inversely related to the mass of the target atom, so the highest energy peaks will be seen for scatter from Au, the most massive atom in this example. The lowest energy peak would result from scatter from Si, which has the lowest mass of the three. Because the scattered energies are related to the target mass, the energy scale can also be converted into a mass scale.

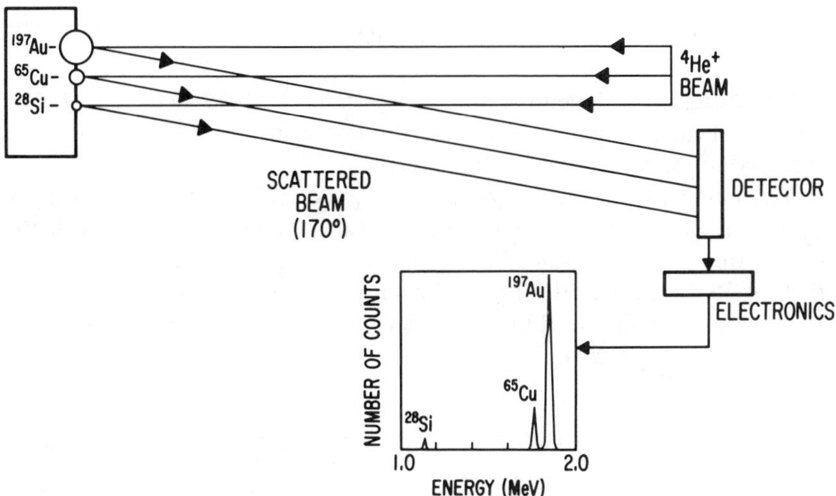

Figure 11.5. A representation of the RBS surface analysis of a sample containing equal numbers of Si, Cu, and Au atoms. The incoming beam consists of 2-MeV $^4He^+$ ions. Only surface scattering is represented.

Example. Calculate the energies of the three peaks in the RBS spectrum in Figure 11.5, assuming an incident He^+ ion of 2 MeV energy and a scattering angle of 170°.

Equation 11.1 is used to calculate the scattered energy:

$$\text{For silicon:} \quad E_0 = 2 \text{ MeV}, M_2 = 28, M_1 = 4, \theta = 170°$$

$$E_1 = (2 \text{ MeV})\left[\frac{(28^2 - 4^2 \sin^2 170°)^{1/2} + 4 \cos 170°}{28 + 4}\right]^2 = 1.14 \text{ MeV}$$

Similar calculations for Cu and Au would give energies of 1.56 and 1.84 MeV, respectively.

The ratios of the peak heights are directly related to the squares of the ratios of the atomic numbers, because the likelihood of Rutherford scattering is dependent on Z. The peak heights are therefore greater for Cu and Au. This points out again the greater sensitivity of RBS for heavier elements.

Qualitative analysis of a surface by RBS requires only a knowledge of the energies of the backscattered particles. For quantitative analysis, other parameters are needed, including the probability of scattering (i.e., the scattering cross section), the number of particles incident on the target, and the number of particles detected. The general form of the equation for calculating the number of target atoms, N, is

$$N = \frac{Q_D}{\sigma(\theta)\Omega Q} \tag{11.3}$$

where Q_D refers to the number of backscattered particles detected, Q is the number of incident particles, $\sigma(\theta)$ is the scattering cross section, and Ω is the solid angle subtended by the detector.

The number of incident particles may be determined by direct measurement of the beam current. The number of detected particles is represented by the area under the peaks in the backscattering spectrum. The cross section for Rutherford scattering can be calculated using classical concepts of electrostatic repulsion. Rutherford derived this relationship in his work with α-particle scattering:

$$\sigma(\theta) = \left(\frac{Z_1 Z_2 e^2}{4E}\right)^2 \frac{1}{\sin^4(\theta/2)} \tag{11.4}$$

where θ is the scattering angle, Z_1 and Z_2 are the atomic numbers of the projectile and target nuclei, e is the charge on the electron, and E is the energy of the projectile. For very low and very high energy projectiles, this cross section expression must be modified. At low energies, the screening effects of the electrons are significant and must be considered. At higher energies, nuclear interactions may also occur as the projectile approaches the nucleus more closely. The expression for the scattering cross section shows an inverse dependence on the energy of the incident particles. Therefore, the yield of backscattered particles is greater for lower-energy projectiles.

For quantitative work, RBS can be an absolute method in which the number of atoms is calculated directly from values of the parameters. Alternatively, standard materials can be used for comparator methods.

11.2.3. Depth Profiling Using RBS

In the preceding section, only those scattering interactions that occurred between the projectile and the atoms on the outermost surface layer were considered. In reality, the projectiles will also penetrate the target to some degree, and will lose energy by interactions with electrons. Therefore, two processes contribute to energy loss for the projectiles that penetrate the target: elastic scattering and electronic interactions as the projectile goes into, then out of, the target:

$$E_1 = E_0 - \Delta E_{\text{scatt}} - \Delta E_{\text{in}} - \Delta E_{\text{out}} \tag{11.5}$$

where E_1 and E_0 have the meanings defined earlier, E_{scatt} refers to the energy lost through scattering processes, E_{in} is the energy loss as the projectile travels into the target, and E_{out} is the energy loss as the projectile travels out of the target after scattering. Calculation of the actual energy losses incurred while the projectile travels through the target requires knowledge of the stopping power cross section, ϵ, of the target material. Values for these cross sections, for all the elements and for varying projectile energy, can be found in the literature. If the cross sections are known, the energy of the projectile can be related to the depth (x) by the following equation:

$$E(x) = x[K\epsilon_{\text{in}}(E_{\text{in}}) + (1/|\cos\theta|)\epsilon_{\text{out}}(E_{\text{out}})] \tag{11.6}$$

where K is the kinematic factor (Eq. 11.1), and ϵ's are the stopping cross sections as a function of the average energy of the projectile going into and out of the target. The yield of backscattered particles (Q_D) detected from a given layer of atoms is given by

$$Q_D = \Omega Q N \Delta x \left[\frac{Z_1 Z_2 e^2}{4E(x)}\right]^2 \tag{11.7}$$

where $N\Delta x$ is the areal density of atoms in the layer and the other symbols have the same meanings as defined previously.

Because the charged particles lose energy continuously as they travel through the target, the energy loss can be related to the total distance traveled. Therefore, measurement of the final energy (E_1) of the projectile can be used to provide information on the distance it has traveled through the target. This is the basis for the use of RBS as a depth profiling technique in thin films.

Figure 11.6 shows a backscatter spectrum from a thick silicon target that has a thin layer of gold on top of it. The particle beam is again composed of 2-MeV He^+. There are two features on the spectrum, corresponding to the Au layer and to the thick Si base. The incoming particles are scattered from the surface of the Au and these have the energy marked $E_{1,Au}$. For a scattering angle of 170°, the energy would be 1.84 MeV, as calculated above. The helium ions also penetrate the Au layer completely and are scattered throughout the thickness of the layer. This results in a continuum of energies that decreases with increased distance traveled through the target. In

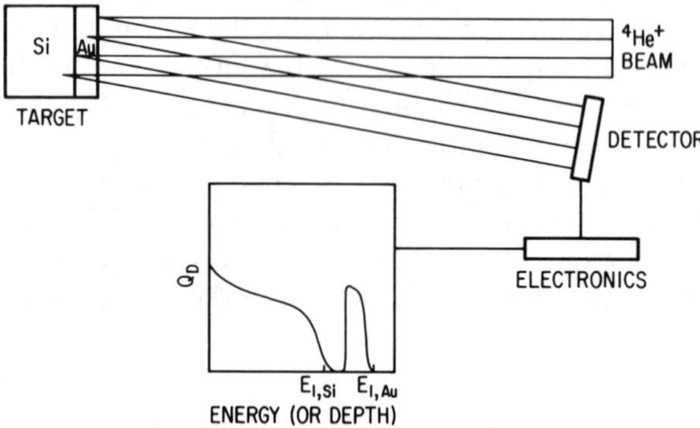

Figure 11.6. A representation of the RBS analysis of a thick silicon target which has a thin layer of Au on its surface. The particle beam is again 2-MeV $^4He^+$ ions. In this figure, scattered particles from the Au surface layer are easily distinguished from the continuum of scattered particle energies from various depths in the thick Si matrix.

this example, the layer is thin enough that the particles can penetrate it completely, and the thickness of the Au layer is related to the width of the peak.

After penetrating the Au layer, the particles will scatter from the thick Si base. The scattering from the first layer of Si at the interface will yield particles with the energy marked $E_{1,\text{Si}}$, and additional scattering throughout the thickness of the Si will yield a continuum of energies extending from $E_{1,\text{Si}}$ down to lower energies. Each point on the energy axis may be related to a certain depth.

Energy straggling effects account for the small tail at the high-energy end of the spectrum. These effects are due to the statistical nature of the energy loss process and are a limiting factor in the depth resolution attainable by RBS. For RBS with α particles, depths of a few micrometers may be effectively probed.

RBS is also used for analysis of more complex systems, such as those consisting of compounds or multiple impurities in a matrix. The RBS spectrum of a thin superconducting film is shown in Fig. 11.7. The stoichiometry and thickness of the film were determined by RBS analysis.

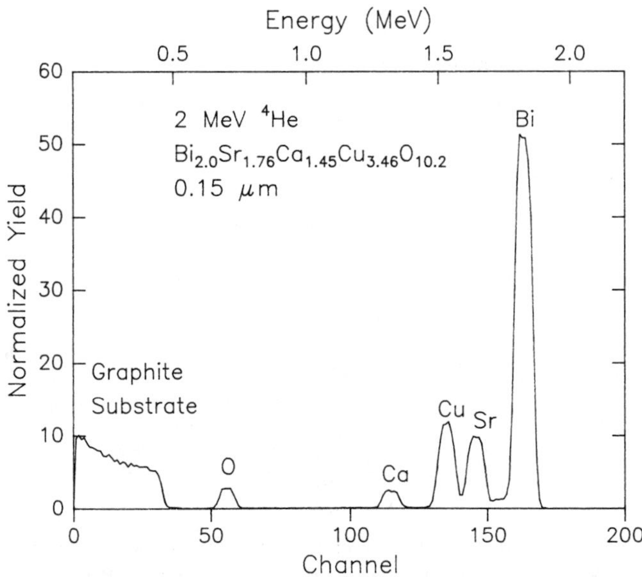

Figure 11.7. Alpha backscatter spectrum of a thin superconducting film containing five elements. The thickness of the film and its stoichiometry can be obtained from analysis of the RBS spectrum. (Courtesy of J. D. Robertson, University of Kentucky, Lexington.)

11.2.4. Channeling Effects

The phenomenon of **ion channeling** was postulated in the early 1900s, but was not observed and put to use until the 1960s. In recent years, channeling phenomena have become an important tool in crystal studies.

The RBS processes described above have assumed random scatter from a matrix, that is, that the matrix is amorphous. Crystals, of course, do not fit this description, because they consist of orderly arrays of rows or planes of atoms. These rows and planes may act as "guide tubes" for an incoming particle beam, directing the particles in the spaces between the rows of atoms. The confinement of a particle beam to a given trajectory within a crystal array is referred to as **channeling**. Figure 11.8 compares randomly scattered and channeled ion beams.

RBS techniques that use channeling are referred to as RBS/channeling, or **RBS-C**. The crystal to be examined is positioned so that the incoming ion beam is aligned with the "channels" formed by the atoms in the crystal. The ion beam is scattered in the usual way from the surface layers, but once inside a channel, the likelihood of the occurrence of Rutherford scattering is greatly reduced because the projectile cannot get close enough to

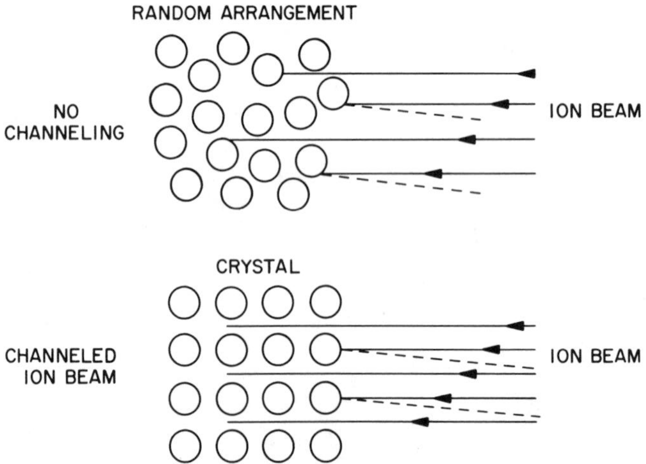

Figure 11.8. Representation of RBS channeling. In less-ordered solids, the spectrum may be dominated by scattering from matrix atoms or ions at various depths in the sample. In well-ordered crystals, incident particles may pass deep into the lattice. In this case, scattering from impurity atoms or ions lying in the channels and on the surface may contribute strongly to the RBS spectrum, while scattering from matrix atoms or ions at depth is minimal.

the nuclei to undergo scattering. RBS-C methods are thus particularly sensitive to surface atoms and to imperfections in the crystal lattice structure because these "out-of-position" atoms are the ones most likely to be encountered in the channels and thus to cause scattering.

Ion channeling methods have become useful in materials analysis. They have been used to determine foreign impurities in crystals, characterize films and layers, locate defects in crystals, and identify areas of radiation damage.

11.2.5. Applications of RBS

Rutherford backscattering spectrometry is a powerful tool for determining the stoichiometry, structure, thickness, and impurity concentration of surfaces and thin films. It has become a very widely used technique in materials science and solid-state physics. A selection of specific applications from current literature is given in Table 11.2.

Table 11.2. Examples of Recent RBS Applications

Area	References
Archaeology	
Mummy bones	Cholewa et al., *J. Nucl. Instrum. Methods Phys. Res.* **B22**(1–3):423 (1987)
Biology and medicine	
Lung	Paschoa et al., *J. Radioanal. Nucl. Chem.* **115**(2): 231 (1987) (RBS and microPIXE)
Geology	
Topaz stones	Rubel et al., *Nucl. Instrum. Methods Phys Res.* **B28**(2):284 (1987)
Zeolites	Baumann et al., *Anal. Chem.* **60**(10):1046 (1988)
Industrial	
Alloys	Le Boite et al., *Nucl. Instrum. Methods Phys. Res.* **B29**(4):653 (1988)
Catalysts	Pierson et al., *Anal. Chem.* **60**(24):2661 (1988)
Ceramics	Niiler et al., *Nucl. Instrum. Methods Phys. Res.* **B4041**(2):838 (1988/89)
Microelectronic circuits	Morris et al., ibid. **B15**(1–6):661 (1988)
Semiconductors	Ras et al., ibid. **B35**(3–4):488 (1988)
Superconductors	Boergesen et al., ibid. **B36**(1):1 (1989)

11.3. MÖSSBAUER SPECTROSCOPY

Mössbauer spectroscopy is a technique in which the recoilless resonance absorption of γ rays by nuclei is used to provide information about the chemical milieu of the nuclei. The concept of resonance absorption may be more familiar to most readers from atomic spectroscopy. A resonance wavelength refers to the wavelength of electromagnetic radiation that exactly corresponds to a particular electron energy level transition in an atom, usually that from the ground to the first excited state. In **resonance absorption,** a species emits and then reabsorbs radiation of the same energy. A specific example of this is atomic resonance fluorescence, in which the wavelengths of radiation emitted by the source and absorbed by the target atoms are identical. The nuclear counterpart of this phenomenon was predicted as early as the 1920s, but it was not until 1957 that R. Mössbauer experimentally observed the effect that now bears his name. Mössbauer was awarded the Nobel Prize in physics in 1961 for this discovery.

In atomic resonance fluorescence, the exciting radiation comes from a light source, and appropriate wavelengths are selected with a device called a monochromator. However, the wavelengths of energy that are needed to induce nuclear energy level changes are much higher and also much more restricted than those needed to induce electronic changes. There are no source/monochromator systems that could possibly produce a narrow enough band of radiation to exactly match a nuclear transition. Therefore, the exciting radiation must come from nuclei of the same type as are being studied. Mössbauer spectroscopy is based on the emission of γ rays by radionuclides and the subsequent absorption of those γ rays by other nuclei of the same identity. The actual energies at which the radiation is absorbed and emitted are affected by the chemical environment the atom is in, so determination of these very small energy variations is used to provide chemical information.

In discussions of radioactive decay in Chapter 2, it was noted that the energy of an emitted γ ray is not exactly equal to the energy of the nuclear transition because some energy is lost due to recoil effects in conserving momentum. Therefore, a γ ray emitted by one nucleus is not likely to be absorbed by another of the same type because of this small energy difference. However, there are some experimental ways to minimize this energy loss and increase the likelihood of resonance absorption. One is to minimize the loss due to recoil of the decaying nucleus. This can be accomplished by embedding the nucleus in a solid (crystalline) material, because then the recoil mass essentially becomes the entire mass of the solid and there is virtually no recoil. Cooling the solid can reduce the energy loss still further. In addition, the emitted γ rays do show some small spread in energy, due to Doppler effects resulting from thermal motions of the atoms in the crystal and uncertainty principle considerations. The γ rays at the higher end of this energy spectrum are likely to be of sufficient energy to be absorbed by another nucleus. These ideas are illustrated in Fig. 11.9.

In atomic resonance fluorescence, the energy of the exciting radiation can be varied by the monochromator. The energy of the γ rays emitted by nuclei cannot be directly controlled, but can be made to vary by small amounts by making use of the enhanced Doppler effect. The Doppler effect was explained in Section 5.4.4. It refers to the increase or decrease in the wavelength of radiation from a source as that source is moved toward or away from an observer. If a γ source were moved toward a target, the wavelength of the emitted gammas would decrease very slightly and the energy would increase slightly. This movement provides the experimenter with a way to vary the energy of the emitted γ rays so that they match perfectly the transition energy of the target. At that energy, the source will show a large increase in absorption. This is illustrated in Figure 11.10. To produce a Mössbauer spectrum, a radioactive source is mounted on a carrier which can move the source toward or away from the target. A detector is placed behind the target. As the carrier moves the γ-ray source, the energy of the incident radiation is varied, until at some point it matches the absorption energy needed for the target. At this energy, a decrease in the count rate will be observed because the incident γ rays are being absorbed by the target nuclei. This forms the Mössbauer spectrum.

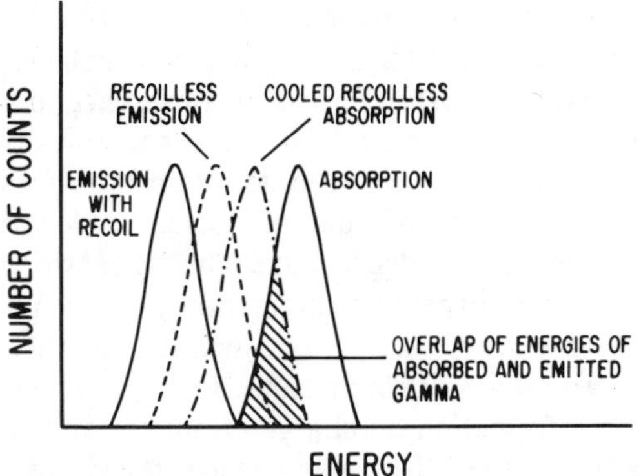

Figure 11.9. Representation of recoilless resonance absorption for γ rays. The solid curve at the far left represents the distribution of energies of γ rays emitted from a noncrystalline radioactive source. The two broken lines to the right of this represent the energies of γ rays emitted from a source that is embedded in a matrix and then cooled. Note how the distribution shifts position so that there is more and more overlap with the absorption energies shown by the solid curve at the far right.

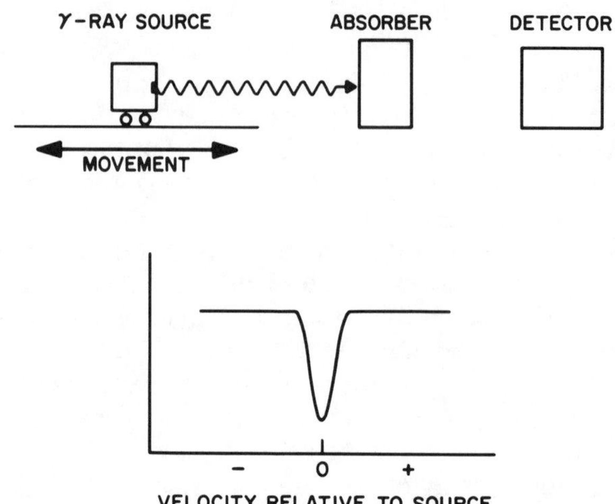

Figure 11.10. Experimental configuration for Mössbauer analysis. The radioactive source is moved toward the absorber with a velocity that just compensates for recoil energy loss, and a detector on the back side of the absorber will record a decreased counting rate due to resonance absorption of the incident γ rays by the absorber.

As noted previously, the energies of the nuclear absorption peaks are affected to a small but measurable extent by the electronic environment of the sample. These effects are referred to as the **hyperfine parameters,** and include the **isomer shift,** the **paramagnetic high-frequency (hf) interaction,** and the **quadrupole and magnetic hyperfine interaction.** They provide the main source of information from a Mössbauer spectrum. The isomer shift, or **chemical shift,** refers to the difference in transition energy between the source and the target. Differences in transition energies can arise because of differences in electron density between source and target. The isomer shift can provide information about the electron density at the nucleus, and therefore about valence state. The other two effects (paramagnetic hf and quadrupole hyperfine interactions) cause splitting of the nuclear energy levels due to interactions between the atomic electrons and the magnetic and electric fields of the nucleus. From these level splittings it is possible to obtain information on charge symmetry and atomic configurations in the vicinity of the Mössbauer nuclide, the degree of covalency in metal bonds to ligands, solvent–ion interactions, nuclear moments, and lattice vibrations in the source crystal. The Reading List at the end of this chapter contains references to more detailed discussion of the use of hyperfine parameters in Mössbauer spectroscopy.

Mössbauer spectroscopy is not equally applicable to all elements. Appropriate nuclei have low-level excited states because the maximum γ-ray

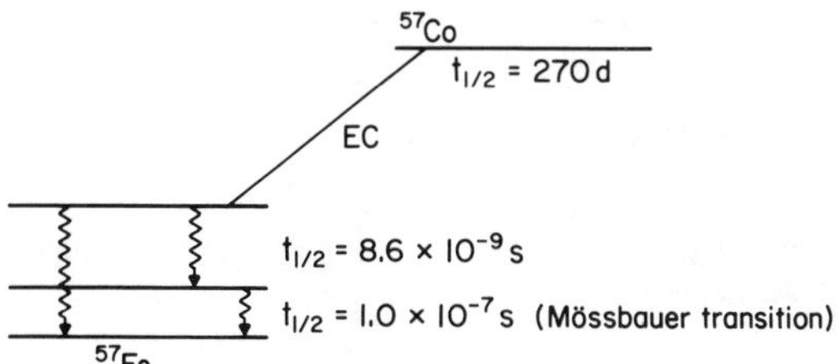

Figure 11.11. The decay scheme for ^{57}Co, a common radionuclide source for Mössbauer spectroscopy. The low-energy γ rays emitted and the short half-lives of the isomeric states of ^{57}Fe are ideal for this method of analysis.

energy that can be effectively used is around 160 keV. Also, the half-life of the excited state must be suitable, generally within the range of 10^{-6} to 10^{-10} s. By far the most used radionuclide has been ^{57}Co, which is the radioactive source for studies of absorbers containing different chemical forms of Fe. The decay scheme for ^{57}Co is shown in Figure 11.11. Mössbauer spectra of some coal samples are shown in Fig. 11.12. Other source–target Mössbauer nuclides that have received attention are ^{119m}Sn–Sn, ^{191}Os–Ir, ^{133}Ba–Cs, and ^{161}Tb–Dy. The radionuclide listed is the parent source material, but often it is an active daughter's γ rays that are used in the Mössbauer analysis. Many other Mössbauer pairs are now known.

Iron has a rich chemistry and is an element of great importance in a number of fields of study. In biology, iron is an essential element and is a cofactor in several enzyme systems and in hemoglobin, the molecule responsible for oxygen transport. Mössbauer spectroscopy has been used to help determine the reaction mechanisms for iron-containing enzymes by characterizing some of the intermediate states and structures of the systems. Iron chemistry has obvious importance for metallurgical studies of steel and of corrosion processes. Because Mössbauer spectroscopy is confined to solids, it provides a means for following the course of chemical reactions occurring in solids, which is usually a difficult task. Mössbauer spectroscopy has also been applied to the study of art and archaeological artifacts.

11.4. HOT-ATOM CHEMISTRY

During the course of an induced nuclear reaction or spontaneous radioactive decay, there are chemical phenomena that occur along with nuclear processes. Nuclear reactions often result in the formation of highly reactive chemical species having large kinetic energies and charges. Because the

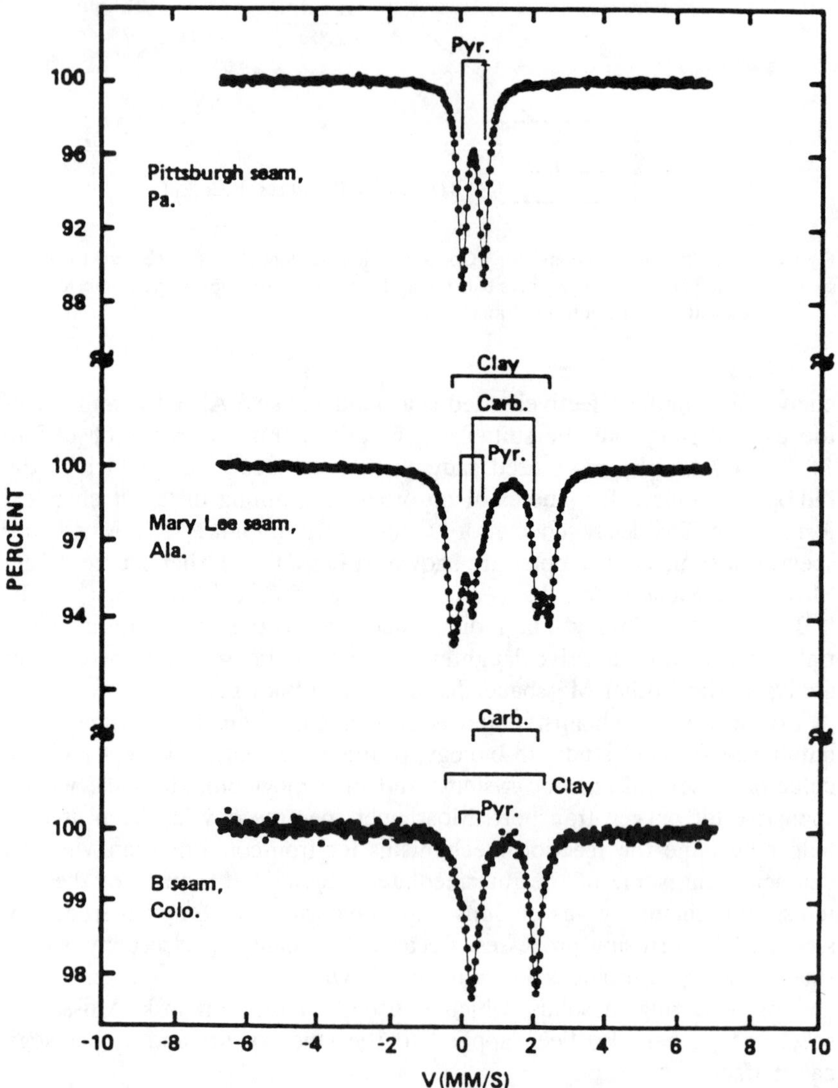

Figure 11.12. Mössbauer spectra of three coal samples containing in various amounts Fe as pyrite, in an Fe^{2+}-bearing clay, and as iron carbonate (siderite). Differences in the chemical shifts of the three forms of iron can be observed. (Courtesy of G. P. Huffman, University of Kentucky, Lexington.)

energies of these species are much greater than the thermal energy of the surrounding atoms and molecules, they are called **hot atoms,** and the study of their chemical effects is called **hot-atom chemistry.** Hot-atom chemistry should not be confused with radiation chemistry or photochemistry, which study the effects of electromagnetic radiation on complex molecules. In those studies, there are no nuclear reactions or decays taking place. Radiation from outside the system under study is used to induce the chemical changes, and it is the chemically active species produced by the interaction of radiation with the orbital electrons that is of interest. The energies involved are far less than those in hot-atom chemistry. Some applications of radiation chemistry are listed in Section 11.5.

11.4.1. Production of Hot Atoms

Hot atoms can be produced either by induced nuclear reactions or by radioactive decay. Both of these processes result in the recoil of the affected nucleus, due to conservation of momentum. Because of the large amounts of energy generated in nuclear processes, the recoiling nucleus almost always has more than enough energy to rupture the chemical bond that holds the atom in the molecule. Once the hot atom is separated from the molecule, it still has sufficient kinetic energy and a large enough charge to undergo a variety of chemical interactions with neighboring molecules to form radicals, ions, and excited states. The elucidation of these interactions is one area of research in hot-atom chemistry.

The nuclear reactions that have been most studied in hot-atom chemistry are those in which the product species is isotopic with the target atom. These are primarily the reactions involving neutron capture or loss, such as the (n, γ), $(n, 2n)$, and (γ, n) processes. Reactions that involve formation of products that are nonisotopic with the parent have not been studied as extensively. These include most charged-particle-induced reactions, such as (p, n) and (p, α) reactions. However, these reactions are now receiving more attention because of the increased use of cyclotron-produced radiopharmaceuticals for medical uses.

Radioactive decay processes can also form chemically excited species. Alpha decay, isomeric transitions, beta decay, and electron capture reactions are all capable of forming hot atoms.

11.4.2. Energy Calculations

As has already been noted, the energy released in nuclear processes is much greater than that involved in most chemical processes. The energy that is released in a nuclear reaction or decay is divided between the emitted particle or radiation and the nucleus that remains after the emission. The division of energy is governed by the law of conservation of momentum, which requires that the momentum of a γ ray or particle emitted from a nucleus be equal to the momentum of the nucleus remaining after the decay.

An equation to calculate the recoil energy of the nucleus may be simply derived. Reactions resulting in γ-ray emission will be discussed first. The energies of the recoil particle (E_r) and the γ ray (E_γ) can be given as

$$E_\gamma = mc^2 \qquad E_r = \tfrac{1}{2}m_r v^2 \tag{11.8}$$

Rearranging these equations for the particle and the γ ray gives

$$\frac{E_\gamma}{c} = mc = p_\gamma \qquad 2m_r(E_r) = (m_r v)^2 = p_r^2 \tag{11.9}$$

and

$$(p_\gamma)^2 = \frac{(E_\gamma)^2}{c^2} \qquad (p_r)^2 = 2m_r E_r \tag{11.10}$$

By the law of conservation of momentum,

$$(p_\gamma)^2 = (p_r)^2 \tag{11.11}$$

so

$$\frac{(E_\gamma)^2}{c^2} = 2m_r E_r \tag{11.12}$$

The energy of the recoil species is

$$E_r = \frac{(E_\gamma)^2}{2m_r c^2} \tag{11.13}$$

If energy is given in units of MeV and mass in daltons, Eq. 11.13 becomes

$$E_{r,\,\mathrm{MeV}} = \frac{(E_{\gamma,\,\mathrm{MeV}})^2}{1862 m_{r,\,\mathrm{daltons}}} \tag{11.14}$$

By using eV as energy units instead of MeV, the equation is

$$E_{r,\,\mathrm{eV}} = \frac{537(E_{\gamma,\,\mathrm{MeV}})^2}{m_{r,\,\mathrm{daltons}}} \tag{11.15}$$

Example. An organic bromide, R–Br, is irradiated with thermal neutrons. Calculate the energy, in eV and in kJ/mol, of the ^{82}Br hot atom produced.

$$\mathrm{R{-}^{81}Br + n \longrightarrow R^{\cdot} + (^{82}Br^{n+})^{*} + 7.0\ MeV}\ \gamma$$

The incident neutron imparts little energy to the Br atom, but the emitted γ ray causes the Br atom to recoil. The charge on the Br may be quite high (around 10+), as electrons are stripped off in the recoil process.

Equation 11.15 is used to solve the problem:

$$E_{\mathrm{r,eV}} = \frac{537(7.0\ \mathrm{MeV})^2}{82\ \mathrm{daltons}} = 321\ \mathrm{eV/atom}$$

Converting this result to kilojoules/mole gives 3.1×10^4 kJ/mol.

A normal chemical bond energy is 1–5 eV (tens to hundreds of kJ/mol). It is easy to see that the high energy imparted to the hot atom is far in excess of that required to break the bond. This is true even when taking into consideration that only the energy component in the direction of the bond is actually used in bond breaking.

The production of a hot atom by β decay, electron capture, or internal conversion produces lower recoil energies, sometimes even lower than the chemical bond energies. The equation used to calculate recoil energy for β decay is different from that used for γ decay:

$$E_{\mathrm{recoil,\,max}} = \frac{537E_{\beta}(E_{\beta} + 1.02)}{m_{\mathrm{recoil,\,daltons}}} \tag{11.16}$$

where E_{β} is in units of MeV and $E_{\mathrm{recoil,\,max}}$ is in units of eV. This energy would represent the maximum energy for the β particle.

11.4.3. Applications of Hot-Atom Reactions

Hot-atom chemistry has both practical and fundamental applications. The study of hot-atom chemistry has shed new light on both reaction kinetics and the mechanisms by which chemical reactions occur. A knowledge of hot-atom effects is important for many fields in which radioisotopes are used. For example, the biological effects of hot atoms from radiopharmaceuticals would be an important factor to consider for patient safety. In studies of materials for nuclear reactors, the behavior of structural components under the high irradiation conditions in the reactor must be known. Hot-atom processes are used to produce high specific activity isotopes and to label molecules with radioactive atoms.

Perhaps the best-known application of hot-atom chemistry is for the production and separation of high-specific-activity isotopes. These isotopes are desirable for many tracer applications because there is a smaller inactive isotope content of the tracer present, which, for example, could result in toxicity problems in living systems. The process of forming high-specific-activity isotopes via hot-atom reactions was first demonstrated by L. Szilard and T. A. Chalmers in 1934, and is often referred to as the **Szilard–Chalmers effect.** In the original experiment, ethyl iodide was irradiated with neutrons to produce $^{128}I^*$ hot atoms:

$$n + CH_3CH_2{}^{127}I \longrightarrow (^{128}I)^* + \gamma \text{ ray} + \text{organic fragment}$$

Szilard and Chalmers found that the inorganic iodine extracted from the irradiated organic iodide was essentially all ^{128}I (i.e., the radioisotope had a high specific activity). Thus, an isotopic separation had been effected by the hot-atom processes that took place subsequent to the irradiation. This separation effect can also be demonstrated with other halogens.

For the Szilard–Chalmers process to be effective, several conditions must be met. First, chemical bonds must be broken with high efficiency. This implies that the recoil energy of the nucleus must be large enough to break almost all the bonds holding the nuclei in the molecules. This is certainly the case for all recoils except those few directly along the bond axis. A second requirement is that the thermalized hot atom (i.e., the hot atom after it has "cooled down" to the same energy as the surrounding atoms) must not exchange rapidly with the parent molecule. If rapid exchange were to occur, the radioisotope would not remain separated. And finally, it must be possible to chemically separate the inorganic product from the organic parent fraction.

Nuclear isomers may also be separated by hot-atom processes. For example, an organic compound could be tagged with ^{80m}Br, which undergoes an isomeric transition to ^{80g}Br with a half-life of 4.42 h. The ground-state isomer may be isolated from the sample.

Another application of hot-atom chemistry is the labeling of organic compounds used as tracers in many investigations in both organic and biochemistry. One way to do this is by simple random labeling. A compound that is similar to the one desired is irradiated. There will be a variety of products formed, one of which will likely be the desired compound. If separation is possible, the desired compound can be removed from the mixture.

Table 11.3. Incorporation of Tritium in Organic Molecules by Use of Tritium Recoil Labeling[a]

Compound	% Entry Tritium into Parent Compound	% Survival of Parent	Specific Activity (dpm/mg)	Irradiation Time
Methane	47	≈100	10^4	30 s
Benzoic acid	25	66	2×10^6	150 h
Glucose	10	>95	6×10^3	3 d

[a]Irradiation in a thermal neutron flux density of approximately 2×10^{12} n cm^{-2} s^{-1}.

A method specific for tritium labeling is the **tritium-recoil method.** An organic compound may be randomly labeled with tritium by mixing the compound with a lithium salt (as a slurry) and irradiating the mixture in a reactor. The reaction that occurs is

$$^{6}\text{Li}(\text{n}, \alpha)^{3}\text{H}$$

In practice, particles of $LiCO_3$ or LiCl of 35–45 μm diameter are mixed in a 3–10% slurry with the organic compound to be tagged and then irradiated with thermal neutrons. The (n, α) reaction occurs within the particles. The range of the ^{3}H is about 50 μm, so the tritium escapes the $LiCO_3$ particles to act as hot atoms and tag the organic compound. The range of the α particles is less than that of the tritons (about 10 μm), so they do not escape the $LiCO_3$ particles to cause indiscriminate bond rupture. Use of this method will have variable results, depending on the parent molecule. Table 11.3 shows the percent incorporation into the parent for three organic molecules. The labeling that occurs in this process is nonspecific (any H may be replaced by a ^{3}H) and nonuniform (there may be more than one hydrogen tagged on a given molecule). Any position may be labeled in this process, not just the one that is most labile. The specific activity attainable with this method is moderately high, and it is a simple and cheap way to accomplish random labeling.

11.5. RADIATION CHEMISTRY

As mentioned at the start of Section 11.4, **radiation chemistry** refers to the study of the effects of externally applied radiation on chemical systems. Gamma rays and high-energy electrons are the most commonly used radiations in these studies. Electron beams provide high-intensity radiation over small areas with low penetration, while γ rays provide higher penetration over larger areas, but at a lower intensity. Charged particles, low-energy electrons, and neutrons are used less often.

Irradiation of a chemical system with γ rays or high-energy electrons results in the production of a variety of energetic, reactive chemical species. One of these species is the free radical, which is a highly reactive, short-lived reaction intermediate that is important in a variety of chemical reaction mechanisms. Free-radical reactions are involved in the generation of atmospheric pollutants and also take place in living organisms. With the current interest in environmental quality and the chemistry of biological processes, the importance of a better understanding of free-radical chemistry is clear.

The kinetic and mechanistic aspects of a chemical reaction can be studied using radiation chemistry. For example, the technique of **pulse radiolysis** can be used to help identify the reactive and very short-lived species produced during the course of a reaction. Several different mechanistic processes, including charge transfer, exciton transfer, and radical–molecule reactions have been examined by radiation chemistry.

The applications of radiation chemistry fall into a number of areas, including industrial process control, sterilization of medical supplies, treatment of food, and environmental improvements. In industrial processes, the largest use of radiation is for polymer modification, including cross-linking, scission, and induction of polymerization. **Cross-linking** refers to the process of linking two or more polymer chains by a side bridge. The radiation-induced cross-linking of polymer chains produces many changes in the properties of the polymers. For example, cross-linked polyethylene differs from normal polyethylene in its behavior upon heating. The cross-linked type changes to a rubbery material instead of a liquid when it is heated above the melting point. The cross-linked variety also can retain a "memory" of its molecular arrangement during irradiation, which has made it useful for shrink-wrapping. **Scission** refers to the fragmentation of a polymer. Radiation-induced scission of polymers can be used to control molecular mass. The production of Teflon is an example of the application of this process. Radiation can also be used to initiate the polymerization process itself.

A variation of the use of radiation on polymeric materials is for the curing of a variety of coatings applied to surfaces. In most cases, the curing of a coating, such as paint, is done by drying the substance. This results in the release of large amounts of organic solvents into the air and also requires a great deal of energy. Environmental regulations governing the release of chemical emissions are becoming increasingly stringent, so a process that did not release such large solvent amounts would be desirable. Thus, a useful alternative to conventional curing is the irradiation of the coating with electrons in a procedure known as **electron-beam curing.**

Polymeric materials, along with metals and ceramics, are widely used for medical purposes, so the field of **bioengineering** finds many uses for radiation chemistry. The polymeric materials that are used for medical procedures or implants may be created or modified using radiation and be sterilized in the process. A unique application of radiation chemistry in this field is the radiation-induced immobilization of a biologically active species on a polymer. These species could include antibiotics, anti-cancer agents, and enzymes, among many others. The materials may then be inserted or implanted into a target organ for therapeutic use.

Environmental applications of radiation chemistry are increasing. A major environmental concern is the addition of large amounts of sulfur dioxide and nitrogen oxides to the air by industrial processes and by automobile exhaust. Both of these oxides are washed out of the air by rainwater and are largely responsible for acid rain. The removal of these oxides from exhaust gases is necessary. One possible way to do this involves the irradiation of the gases with electron beams. A small amount of ammonia gas is added to the exhaust gases and the mixture irradiated with an electron beam. The result is the formation of ammonium salts by reaction of the SO_2 and NO_x with the ammonia. Not only can the process remove the harmful oxides from waste gases, but the product of this reaction can then be converted for use as an agricultural fertilizer. Waste water from industries can also be processed by irradiation with electron beams. This treatment results in the decomposition of a number of different kinds of organic waste material that are present. Another environmental use of radiation is the inactivation of bacteria and other microflora in sewage sludge. Conventional sewage treatment often does not reduce the pathogen level in the sludge to a sufficient degree for safe disposal of the sludge. Irradiation with electron beams can effectively sterilize this material.

The uses of radiation to sterilize food and to control insect populations through sterilization were mentioned in Chapter 10. Water may be sterilized by a γ-radiation dose of several kilograys. Doses of about 10 kGy can delay meat spoilage by killing microorganisms and can enhance the shelf-life of fruits such as strawberries. Doses up to 100 kGy are used to kill microorganisms and insects in spices.

TERMS TO KNOW

- Depth profiling
- Doppler effect
- Electron-beam curing
- Energy dispersive spectrometer
- Hot-atom chemistry
- Ion beam analysis
- Ion channeling
- Isomer shift
- Kinematic factor
- Micro-PIXE
- Mössbauer spectroscopy
- PIXE
- Polymer scission
- Pulse radiolysis
- Radiation chemistry

Radiation sterilization
RBS
Resonance absorption
Rutherford scattering
Szilard–Chalmers effect
Tritium-recoil labeling
Wavelength dispersive spectrometer
X-ray notation

READING LIST

Arnikar, H. J., *Isotopes in the Atomic Age.* New York: Wiley, 1989. [Szilard–Chalmers effect and applications]

Barnes, J. W., and W. H. Burgus, Chemical phenomena accompanying nuclear reactions (hot-atom chemistry). In *Radioactivity Applied to Chemistry,* A. C. Wahl and N. A. Bonner, eds. New York: Wiley, 1951, pp. 244–283. [hot-atom chemistry]

Bird, J. R., and J. S. Williams, *Ion Beams for Materials Analysis.* San Diego: Academic, 1989. [PIXE, RBS, channeling]

Cesareo, R., ed., *Nuclear Analytical Techniques in Medicine*. Amsterdam: Elsevier, 1988. [applications of PIXE in medicine]

Chu, W.-K., J. W. Mayer, and M-A. Nicolet, *Backscattering Spectrometry.* Orlando, FL: Academic, 1978. [RBS, channeling]

Cohen, R. L., ed., *Applications of Mössbauer Spectroscopy,* Vols. 1 and 2. New York: Academic, 1976-80. [Mössbauer applications]

Deconninck, G., *Introduction to Radioanalytical Physics.* Amsterdam: Elsevier Scientific, 1978. [PIXE and RBS]

Denaro, A. R., and G. G. Jayson, *Fundamentals of Radiation Chemistry.* Ann Arbor, MI: Ann Arbor Science, 1972. [principles of radiation chemistry]

Dole, M., ed., *The Radiation Chemistry of Macromolecules.* New York: Academic, 1972-73. [applications of radiation chemistry]

Feldman, L. C., and J. W. Mayer, *Fundamentals of Surface and Thin Film Analysis.* New York: North-Holland, 1986. [depth profiling]

Herber, R. H., ed., *Chemical Mössbauer Spectroscopy.* New York: Plenum, 1984. [proceedings of Mössbauer symposium]

IAEA, *Hot Atom Chemistry Status Report.* Panel Proceedings Series, Vienna, 1975. [hot-atom chemistry]

Johansson, S. A. E., and J. L. Campbell, *PIXE: A Novel Technique for Elemental Analysis.* New York: Wiley, 1988. [PIXE]

Khan, M. R., and D. Crumpton, Proton induced x ray emission analysis, Part I. *CRC Crit. Rev. Anal. Chem.* **11**(2):103 (1981). [principles of PIXE]

Khan, M. R., and D. Crumpton, Proton induced x ray emission analysis, Part II. *CRC Crit. Rev. Anal. Chem.* **11**(3):161 (1981). [PIXE]

Long, G. J., ed., *Mössbauer Spectroscopy Applied to Inorganic Chemistry.* New York: Plenum, 1984. [Mössbauer applications in inorganic chemistry]

Long, G. J., and J. G. Stevens, eds., *Industrial Applications of the Mössbauer Effect.* New York: Plenum, 1986. [industrial Mössbauer applications]

Perriere, J., Rutherford backscattering spectrometry. *Vacuum* **37**(5–6):429 (1987). [Rutherford backscattering]

Ranney, M. W., *Irradiation in Chemical Processes: Recent Developments.* Park Ridge, NJ: Noyes Data Corp., 1975. [industrial applications of radiation chemistry]

Urch, D. S., Nuclear recoil chemistry in gases and liquids. In *Radiochemistry,* Vol. 2. London: The Chemical Society, 1975, pp. 1–72. [hot-atom chemistry]

Wilson, J. E., *Radiation Chemistry of Monomers, Polymers, and Plastics.* New York: Dekker, 1974. [use of radiation chemistry in polymer processes]

Ziegler, J. F., P. J. Scanlon, W. A. Lanford, and J. L. Duggan, eds., *Ion Beam Analysis.* Amsterdam: North-Holland (Elsevier), 1990. [RBS, PIXE, micro-PIXE, depth profiling, ion channeling]

EXERCISES

1. Write the symbolic notation for x rays produced by the following electron transitions:

   (a) L-shell to K-shell transition in Fe

   (b) M-shell to L-shell transition in Pb

   (c) M-shell to K-shell transition in Sn

2. With reference to Fig. 11.5, calculate the scattered He ion energy from Au if the incident particle beam of He ions has an energy of 3.0 MeV and the scattering angle is only 120°.

3. Calculate the recoil energy of a hot atom that was formed by the emission of a 6.45-MeV γ ray from a nuclide with mass 46 which has captured a thermal neutron. Express your answer in units of eV/atom and kJ/mol.

4. By reference to the literature, identify five more nuclide pairs useful for Mössbauer spectroscopy, and the areas of study in which they are important.

5. By reference to the literature, give one application of PIGE in the following areas: biology, geology, and analysis of high-purity materials.

CHAPTER

12

NUCLEAR DATING METHODS

12.1. GENERAL PRINCIPLES OF NUCLEAR DATING METHODS

The phenomenon of radioactive decay has been widely used to determine time intervals in the history of living things, natural waters, rock systems, meteorites, and the evolution of the solar system. The decay of a radionuclide is a very precise metronome whose timing is essentially unaltered by chemical form, temperature, pressure, or other natural physical phenomena. The decay constant, and thus the half-life, of a radionuclide is known to vary slightly with chemical form for only a few radionuclides decaying by electron capture. The simple equations developed in Chapter 5 for the rates of radioactive decay are the basis of all the nuclear dating methods.

If at some time in the past, t_0, the quantity of a radionuclide present was N_0, and at some later time, t_1, (time is measured back from the present) the amount of the radionuclide remaining was N_1, the following decay law applies:

$$N_1 = N_0 e^{-\lambda(t_0 - t_1)} \tag{12.1}$$

where λ is the decay constant for the parent radionuclide. This equation can be solved for the decay interval $(t_0 - t_1)$ as follows:

$$t_0 - t_1 = \frac{1}{\lambda} \ln \frac{N_0}{N_1} \tag{12.2}$$

If the amounts of the radionuclide at the two times and the decay constant λ are known, the time interval $(t_0 - t_1)$ can be calculated.

If t_1 is the present time ($t_1 = 0$), it is clear that N_1 can be directly measured in a sample at hand. The problem lies in knowing the amount of the radionuclide, N_0, at the earlier time t_0. The two methods of determining N_0 serve to define the two basic types of nuclear dating methods: the **equilibrium decay clock** and the **accumulation clock.**

Figure 12.1*a* illustrates the principle of the equilibrium decay clock. The clock can be likened to a system in which the water in an elevated bucket is

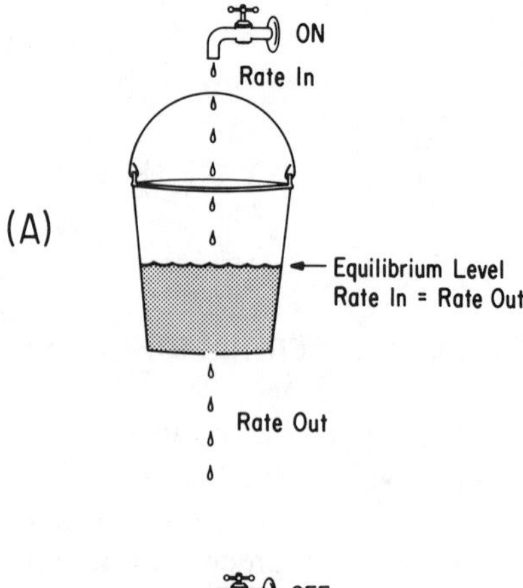

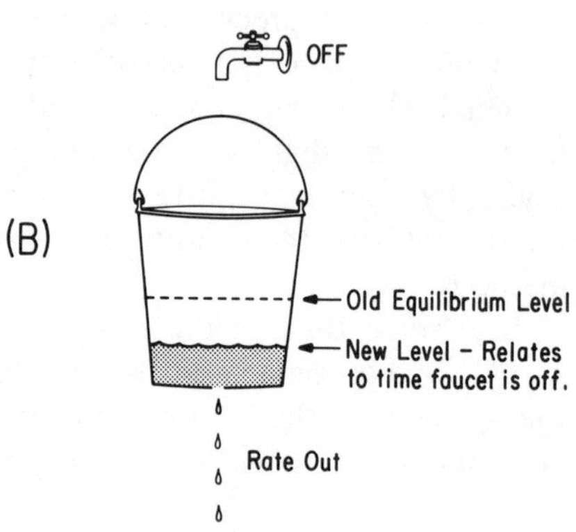

Figure 12.1. An illustration of the equilibrium decay clock. The water flowing into the top bucket (*A*) is analogous to the production rate of a parent radionuclide and the leak in the bucket represents the decay rate of the radionuclide. An equilibrium water level is maintained in the bucket as long as the input (faucet) is left on. In (*B*) the faucet has been turned off, and the amount of water in the bucket decreases as a function of the time since the input was stopped.

maintained at an equilibrium level, because the water lost through a leak in the bucket is replenished at a constant rate from a faucet for the time period in question. This equilibrium amount is analogous to the amount of a radionuclide, N_0 at t_0 in the discussion above. If the faucet is shut off and water input ceases, the amount of water in the bucket will decrease with time, as shown in Fig. 12.1*b*. In this simple illustration, if the rate of the leak R is known, the time t from faucet shutoff to when the amount of water in the bucket is again measured is calculated by the expression:

$$t = (1/R)\,(\text{Equilibrium amount of water} - \text{Residual amount of water}) \tag{12.3}$$

In the case of a natural radionuclide, if the balance between its production rate and its decay rate is terminated by removing the sample from the environment in which production occurs, the amount of the radionuclide, hence its activity, will begin to decrease with time according to the simple nuclear decay law. In radioactive decay the "leak rate" depends on the half-life of the radionuclide and the number of radioactive atoms present, and the basic decay equation $A_t = A_0 e^{-\lambda t}$ must be used to calculate t.

Radiocarbon dating and tritium dating are simple examples of the equilibrium decay clock. In these cases, equilibrium specific activity levels of ^{14}C and ^{3}H are established for the carbon and hydrogen in the atmosphere as a result of cosmic-ray interactions with the atmosphere. The equilibrium terminates when the sample containing carbon (in a living system) or hydrogen (usually in water) is removed from equilibrium contact with the atmosphere. For living systems, input of ^{14}C terminates at death. For ^{3}H in the form of atmospheric water, input stops when the water enters a deep aquifer, or when the water is bottled (as for wines), and atmospheric exchange cannot occur.

The second major type of nuclear dating method is illustrated in Fig. 12.2. If at t_0 the amount of water in the lower bucket is zero, or at least a known quantity, the added amount of water in the lower bucket at some later time, t_1, is a function of the time the upper bucket has been leaking. Therefore, the following expression can be used to determine the time since water was placed in the upper bucket and the leak started:

$$\text{Leak time} = \frac{\text{Accumulation in lower bucket corrected for } t_0 \text{ amount (mL)}}{\text{Leak rate (mL)/Unit time}} \tag{12.4}$$

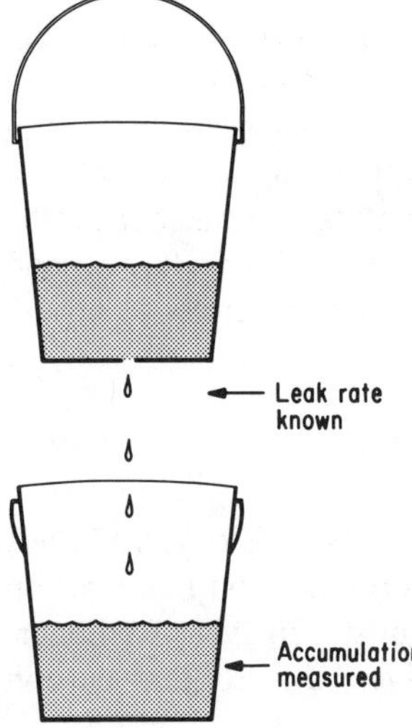

Figure 12.2. An illustration of the accumulation clock. If the leak rate of the top bucket and the amount of water in the bottom bucket (corrected for any amount present before the leak started) are determined, the time the bucket has been leaking is easily calculated. This is the basis of many of the geological dating methods, such as the K–Ar, U–Pb, and Rb–Sr methods.

In an analogous radionuclide decay system, the accumulation is the amount of the stable daughter product(s) formed, and the leak rate will depend on both the amount of the radioactive parent present at t_0 and the decay constant (or $t_{1/2}$) for the radionuclide. The amount of the parent radionuclide (N_0) present at t_0 is determined simply as the sum of the atoms of decay product(s) formed and the atoms of parent remaining at t_1.

In a natural system, the amount of the parent radionuclide (N_1) plus the amount of daughter product(s) present now ($t_1 = 0$) can be used to determine N_0, the amount of the parent radionuclide at t_0, only if the system is chemically closed. A **chemically closed system** is one in which there is no migration of the parent radionuclide or its daughter(s) in or out of the system for the time period measured. In rock systems, the chemical system is often closed for nonvolatile components when the rock solidifies from a melt, and for volatile components when cooling has proceeded to the point that gases are no longer evolved. Branching ratios in the decay scheme must be taken into account, because not all parent decays may yield the specific daughter nuclide measured. Remember, the half-life of a radionuclide which determines λ is related to the disappearance of the parent radionuclide, not to the appearance of a specific daughter nuclide when several decay paths are possible.

In the accumulation clock, amounts of the product daughter nuclide present at t_0 must be subtracted from the amount measured at t_1, so only the amount produced in the radioactive decay for the period of interest is used in the calculation (Eq. 12.2). Isotopic ratios in the daughter "accumulator" element can be measured with a mass spectrometer. It is likely that at least one of these isotopes will not be produced by natural radioactive decay processes. From the systematics of nucleosynthesis, or by analyses of certain phases of primitive meteorites that contain essentially no parent radionuclide, a **primordial ratio** for isotopes of the product element can be estimated. This ratio can then be used to correct the observed amount of the product nuclide for the "primordial" amount present at t_0. Examples of this correction will be given in later sections. Many of the nuclear dating methods applied to geological samples, such as the U–Pb, K–Ar, and Rb–Sr clocks, are examples of the accumulation clock principle.

12.2. RADIOCARBON DATING

Radiocarbon dating was developed by J. R. Arnold, W. F. Libby, and their collaborators at the University of Chicago. For his research on radiocarbon dating, Libby was awarded the Nobel Prize in chemistry in 1960. Radiocarbon is the term commonly used for the radioactive ^{14}C isotope of carbon that is produced in the earth's atmosphere by nuclear reactions induced by energetic neutrons. Primary cosmic rays are mostly high-energy protons, but in their interactions with atmospheric gases they produce a variety of nuclear fragments, including fast neutrons. These neutrons react with the most common isotope of nitrogen in the atmosphere, ^{14}N, as follows:

$$^{14}N(n, p)^{14}C$$

The flux of cosmic rays interacting with the earth's atmosphere is thought to have been almost constant for more than 70 000 years, so the production rate of ^{14}C has also been correspondingly constant. ^{14}C is a pure negatron emitter with a half-life of 5730 years. Because ^{14}C is both being produced by cosmic-ray reactions and undergoing radioactive decay, an equilibrium level is eventually established in the atmosphere.

The ^{14}C atoms are produced as hot atoms and promptly react chemically with the oxygen in the atmosphere to produce $^{14}CO_2$. The radioactive CO_2 is incorporated into plants by photosynthesis, into animals that eat the plants, and eventually into all living things. The equilibrium specific activity of ^{14}C measured in living plant tissues prior to the atmospheric testing of thermonuclear weapons after World War II was about 15 disintegrations min^{-1} g^{-1} of elemental carbon. If a living material is removed from atmospheric equilibrium by death, the ^{14}C activity level will decrease as a function of the time since death. If a bit of wood charcoal found in an archaeological site has a ^{14}C specific activity of 7.5 dpm/g C, one half-life (5730 years) must have passed since the time the tree that produced the charcoal died. In general, the basic decay equation is applied to calculate the radiocarbon age (t):

$$SA_{present} = SA_{equilibrium} e^{-\lambda t} \tag{12.5}$$

where SA represents the present and original equilibrium specific activities of ^{14}C. Thus, radiocarbon dating is an example of the equilibrium decay clock.

The radiocarbon method can be used to date once-living materials that died as long as 75 000 years B.P. (before present). The method is ordinarily limited to ages of less than this time because the specific activity of the ^{14}C will have decreased to $\approx 10^{-4}$ of its equilibrium value in 75 000 years and counting statistics after background correction would be poor.

As noted previously, fundamental assumptions in radiocarbon dating are that the cosmic-ray flux on the atmosphere is constant and that there are no other significant sources for ^{14}C input that would change its equilibrium level in the atmosphere. It was recognized early that neither of these assumptions is strictly true and that the method must be subjected to several small corrections to produce accurate ages. The primary cosmic-ray flux is modulated by changing solar activity and changes in the earth's magnetic field, so that there are time-dependent small changes in the equilibrium radiocarbon content of the atmosphere. Additionally, man-made alterations in the ^{14}C specific activity have also occurred. For example, H. Suess observed in 1955 that the atmospheric ^{14}C specific activity decreased in the 20th century as compared to the 19th century. He concluded that this was due to the introduction of massive amounts of nonactive fossil carbon into the atmosphere by coal and petroleum burning associated with the Industrial Revolution. This phenomenon is now called the **Suess effect.** Atmospheric thermonuclear weapon (H-bomb) tests after World War II had the opposite effect, raising equilibrium specific activities of ^{14}C by approximately a factor of two. It will probably take hundreds of years for the ^{14}C levels in living things to return to their pre-World War II value.

Fortunately, most of the above problems with radiocarbon dating can be resolved. To avoid problems associated with the man-made fossil fuel dilution or atmospheric thermonuclear production of ^{14}C, it is necessary to take care in the selection and sampling of objects to be dated. Samples contaminated with modern quantities of radiocarbon must be avoided. As an example, archaeological wood charcoal samples from buried campfire sites may be subjected to ground waters that carry fallout contamination. Sites in arid regions or dry caves would yield the best preserved material for dating. Also, rootlets from modern plants growing through the sample must be carefully removed, because they will carry modern levels of radiocarbon.

Calibrating the variations in the cosmic-ray production rate is a more difficult problem. The calibration is based on the radiocarbon dating of objects that have known ages and plotting variations between the radiocarbon age and the known age as a function of the known age. The known ages may be from historical records or samples of wood from interior tree rings of very old trees (e.g., bristlecone pines in the western United States and in Scandinavia). Dating by counting tree rings is called **dendrochronology.** Only the outer wood of the tree is in equilibrium with the atmosphere and by counting annual growth rings back toward the center of an old living tree, wood of defined ages up to 7000 years can be obtained. Analyses of organic matter deposited in sediments of glacial lakes of known geologic ages allow extension of the calibrations to earlier times. Cyclic variations in equilibrium ^{14}C specific activity levels have been observed using these approaches. Based on these data, it is likely that the equilibrium radiocarbon specific activities varied by at least $\pm 10\%$ over the past 30 000 years.

In addition to the previously mentioned factors, effects due to climatic variations, which alter the size of the CO_2 reservoir (C is found in CO_2, carbonate, and various types of organic carbon present in the earth's atmosphere, oceans, and biosphere), and isotopic fractionation of carbon by chemical, biological, and physical processes must be considered. For the interval from 1000 years B.P. to at least 10 000 years B.P. the required corrections are well understood and the method is considered to be very accurate. As an example, an experimental radiocarbon date of 1500 B.C. would be corrected by adding 250 years to the age to give a corrected date of 1750 B.C. It is interesting to note that age corrections for older samples are in the direction of the true age being greater than the uncorrected age. Dating of samples older than about 12 500 years depends mainly on extrapolation of established corrections for younger samples.

The methodology of radiocarbon dating is based on two quite different approaches: low-background counting and mass spectrometry. The original radiocarbon dates were based on conversion of the organic sample to CO_2 and its precipitation as $BaCO_3$. The solid samples were then counted with thin-window gas-filled counters. Because the ^{14}C negatron has an E_{max} of only 157 keV, self-absorption in the solid sample was a serious problem. Radiocarbon dates are now determined by converting the organic matter in the samples to gases such as acetylene (C_2H_2), methane (CH_4), and carbon dioxide (CO_2), or liquids such as benzene (C_6H_6). These compounds have a high stoichiometric percentage of carbon. The gas samples are used directly as counting gases in pressurized internal sample proportional counters that have counting efficiencies >90% for the ^{14}C negatrons. Liquid samples are counted with high-efficiency liquid-scintillation counting systems.

Detector backgrounds in ^{14}C dating are reduced by massive lead and steel shielding. The lead used for detector shielding is often recovered from 19th or early 20th century shipwrecks to avoid a background contribution from the negatron decay of ^{210}Pb ($t_{1/2}$ = 22.3 y) produced in lead ores by the decay of radionuclides in the ^{238}U natural decay series. Backgrounds for gas-filled counters are further reduced by surrounding the counter with an annular ring of cosmic-ray counters (cylindrical gas-filled counters, or large NaI scintillation detectors) operated in the anticoincidence mode to cancel cosmic-ray contributions to the central counter background. For liquid-scintillation counting, two detectors positioned at 180° are operated in the coincidence mode to avoid detection of spurious events occurring only in one detector.

Radiocarbon dating is also done by direct measurement of carbon isotopic ratios with a mass spectrometer. This approach is especially useful for very old samples or for very small samples where the counting rate of ^{14}C is low. The reason for the increased sensitivity is clear when it is noted that one gram of modern carbon contains about 6×10^{10} atoms of ^{14}C, but has an activity of only 15 dpm of ^{14}C. The most innovative approach uses a tandem Van de Graaff accelerator to introduce carbon ions into a mass spectrometer. Electron sputtering is used to produce negative carbon ions which are accelerated to as much as 8 MeV in the first stage of a tandem Van de Graaff accelerator. Ions of ^{14}N differ in mass only slightly from ^{14}C ions and would be a serious interference in mass spectrometry radiocarbon dating. Nitrogen atoms are abundant in living matter, while ^{14}C atoms are rare. However, negative nitrogen ions are very unstable and only a few will survive acceleration to 8 MeV. After the initial acceleration in the negative ion mode to eliminate nitrogen ions, the carbon ions are passed through a metal foil which strips off electrons and converts them to C^{4+} ions, which are further accelerated into a mass spectrometer for determination of $^{14}C/^{13}C$ or $^{14}C/^{12}C$ ratios. Special cyclotrons have also been designed to permit the mass analysis of the high-energy ions produced in this technique.

12.3. TRITIUM DATING

Tritium dating is an equilibrium decay method similar to radiocarbon dating. Tritium (^3H, $t_{1/2} = 12.3$ y) is produced in the atmosphere by the action of cosmic rays and reaches an equilibrium specific activity as the compound HTO in all atmospheric and surface waters. Tritium, like ^{14}C, is a weak negatron emitter and emits negatrons with an E_{max} of 18.6 keV. The tritium content of waters is measured in **tritium units (TU).** The TU is defined to be a tritium content equivalent to 1 atom of ^3H for every 10^{18} atoms of ^1H. Due to atmospheric thermonuclear tests the equilibrium levels of tritium in the hydrosphere have varied from pre-test levels of 1 TU (dependent in part on geographical location) to thousands of TU after the tests.

As in the case of radiocarbon dating, water in equilibrium with the atmosphere will exhibit the current “modern” TU level. However, water that is removed from atmospheric equilibrium while flowing in a deep aquifer will exhibit lower tritium levels related to its time of isolation. Water that has been isolated underground for several hundred years will exhibit essentially no tritium activity. Increased sensitivity for tritium dating can be achieved by partial evaporation of the water or electrolysis of alkaline solutions, processes that can enrich ^3H with respect to ^1H in the residual liquid water by as much as 10 000 times. The tritium dating method was originally calibrated by W. F. Libby and collaborators at the University of Chicago by use of dated French wines. Groundwater circulation and turnover rates of water in oceans and lakes have been studied using this technique.

12.4. U–Pb AND Th–Pb METHODS

The accumulation clock methods of nuclear dating that are based on measurement of Pb isotope decay products of the decay series originating from the three primary natural radionuclides ^{235}U, ^{238}U, and ^{232}Th are often referred to as **plumbology**. They are commonly used for dating solidification ages of geological samples. There are four stable isotopes of lead in nature: ^{204}Pb, ^{206}Pb, ^{207}Pb, and ^{208}Pb. ^{204}Pb is the rarest of these four isotopes and is the only one not produced by radioactive decay in nature. The other three are produced by U and Th decay (see Chapter 1). The U and Th decay chains also produce helium through α-particle decays, and the amount of He in a geological sample can be used for dating. The accumulation clock methods based on U and Th decay require a knowledge of present-day amounts of the parent primary natural radionuclide and either the amount of the corresponding Pb daughter isotope or the total amount of He in the sample. Several variations of the U and Th decay series methods are discussed below.

12.4.1. Simple He Accumulation Method

This method, which is often called the **helium clock,** is based on the fact that ^{235}U, ^{238}U, and ^{232}Th emit 7, 8 and 6 α particles, respectively, in their decay chains to isotopes of Pb. The amounts of U and Th in the sample can be determined chemically and the current production rate for He is then easily calculated. The sample is heated and the amount of radiogenic ^{4}He released is measured. Application of the simple decay equations can then yield a **helium-retention age.** Problems arise from the facts that He can be lost from rocks by heating and that U, which is often found along cracks and grain boundaries in rocks, is often mobilized by leaching and may not maintain a constant concentration. Because of these problems, the method is now rarely used.

12.4.2. Single-Decay-Chain Methods

Three decay chains are used in plumbology:

$$^{232}\text{Th}\ (t_{1/2} = 1.40 \times 10^{10}\ \text{y}) \longrightarrow\ ^{208}\text{Pb}$$

$$^{235}\text{U}\ (t_{1/2} = 7.04 \times 10^{8}\ \text{y}) \longrightarrow\ ^{207}\text{Pb}$$

$$^{238}\text{U}\ (t_{1/2} = 4.47 \times 10^{9}\ \text{y}) \longrightarrow\ ^{206}\text{Pb}$$

Both the amount of the parent radionuclide and the stable lead daughter that results from its decay must be determined. As an example, the method based on the ^{238}U decay series will be considered. Methods based on the other U and Th decay chains are similar.

The basic decay equation for the **^{238}U–Pb clock** can be written as follows:

$$^{238}\mathrm{U}_{\mathrm{now}} = (^{238}\mathrm{U}_{\mathrm{original}})e^{-\lambda t} \tag{12.6}$$

where λ is the decay constant for ^{238}U and t is the time since the system became closed. For a rock, this is usually the time that the rock solidified and no further migration of U or Pb into or out of the system occurred. It follows that

$$^{238}\mathrm{U}_{\mathrm{now}} = (^{238}\mathrm{U}_{\mathrm{now}} + {}^{206}\mathrm{Pb}_{\mathrm{now,\ by\ decay}})e^{-\lambda t} \tag{12.7}$$

After rearranging, the following working equation is obtained:

$$^{238}\mathrm{U}_{\mathrm{now}}(e^{+\lambda t} - 1) = {}^{206}\mathrm{Pb}_{\mathrm{now,\ by\ decay}} \tag{12.8}$$

Often, the equation is converted to one with isotopic ratios by dividing both sides by the atom amount of ^{204}Pb present. ^{204}Pb is not produced by natural radioactive decay and its content does not change with time. The U content of a sample can be measured chemically and isotopic ratios obtained by mass spectrometry for the calculation of $^{238}\mathrm{U}_{\mathrm{now}}$. The amount of Pb in the sample and the total amount of ^{206}Pb present are similarly determined. The ^{206}Pb in the sample is from two sources: the amount from the decay of ^{238}U and the nonradiogenic ^{206}Pb present in the sample due to its primordial Pb content. The troilite (FeS) phase in iron meteorites (see Chapter 13) contains Pb, but virtually no U. Because these meteorites are very old, the ratios of Pb isotopes in the troilite will approximate the primordial Pb ratios established during nucleosynthesis. Knowing the ^{206}Pb/^{204}Pb ratio in troilite and remembering that ^{204}Pb is not produced by natural decay processes, the following equation can be obtained:

$$\left[\left(\frac{^{206}\mathrm{Pb}}{^{204}\mathrm{Pb}}\right)_{\mathrm{primordial}}\right](^{204}\mathrm{Pb}_{\mathrm{sample}}) = {}^{206}\mathrm{Pb}_{\mathrm{primordial}} \tag{12.9}$$

and for the sample to be dated,

$$^{206}\mathrm{Pb}_{\mathrm{total}} - {}^{206}\mathrm{Pb}_{\mathrm{primordial}} = {}^{206}\mathrm{Pb}_{\mathrm{now,\ by\ decay}} \tag{12.10}$$

Solidification ages for rocks may be calculated by use of all three natural decay chains and the agreement among methods is usually excellent. These methods have been applied to a wide variety of terrestrial rock systems and yield solidification ages up to several aeons (one aeon = 10^9 y = 1 Gy or 1 Ga). Lunar rocks have been dated by these methods and provide evidence of a major lunar event resulting in the redistribution of Pb about 4 aeons ago.

12.4.3. Pb–Pb Method

The **Pb–Pb dating method** combines two of the single-chain U decay methods and permits the analyst to measure only isotopic ratios, rather than absolute amounts of the parent and daughter nuclides. If the working equation for the ^{235}U–Pb clock is divided by the similar working equation for the ^{238}U–Pb clock, the following expression is obtained:

$$\frac{[^{235}\mathrm{U}_{\mathrm{now}}(e^{+\lambda_{235}t} - 1)]}{[^{238}\mathrm{U}_{\mathrm{now}}(e^{+\lambda_{238}t} - 1)]} = \frac{(^{207}\mathrm{Pb}/^{204}\mathrm{Pb})_{\mathrm{total}} - (^{207}\mathrm{Pb}/^{204}\mathrm{Pb})_{\mathrm{primordial}}}{(^{206}\mathrm{Pb}/^{204}\mathrm{Pb})_{\mathrm{total}} - (^{206}\mathrm{Pb}/^{204}\mathrm{Pb})_{\mathrm{primordial}}} \tag{12.11}$$

where the right side of the equation has been normalized to the number of the nonradiogenic ^{204}Pb atoms present and has been corrected for primordial Pb content.

To use Eq. 12.11, it is only necessary to measure Pb isotopic ratios with a mass spectrometer. The absolute amounts of U and Pb in the sample need not be determined in Pb–Pb dating. The atomic ratio $(^{235}\mathrm{U}/^{238}\mathrm{U})_{\mathrm{now}}$ is a known quantity equal to 1/138. The primordial lead isotopic ratios, $^{207}\mathrm{Pb}/^{204}\mathrm{Pb}$ and $^{206}\mathrm{Pb}/^{204}\mathrm{Pb}$, are 10.3 and 9.31, respectively. The method represented by Eq. 12.11 is called the **Holmes–Houtermans model** for the lead isotopic composition in a single-stage radiogenic history. Normally, this equation is solved graphically, as shown in Fig. 12.3. The curved lines represent **growth curves** for separate U–Pb closed subsystems having different present day ratios (μ) of $^{238}\mathrm{U}/^{204}\mathrm{Pb}$. The straight lines are called **isochrons** and connect points on the growth curves that have the same ages for the isolation of their geological subsystems.

Not all geological systems remained closed throughout history. Sometimes ages obtained by the U–Pb, Th–Pb, and Pb–Pb methods do not agree with ages obtained by other nuclear dating methods. Often, these **discordant ages** are due to partial loss of Pb by diffusion from the otherwise closed system. If several related subsystems of equal solidification age suffer Pb loss at some later time due to a specific geologic episode, information on both the age of the system originally solidified, and when the

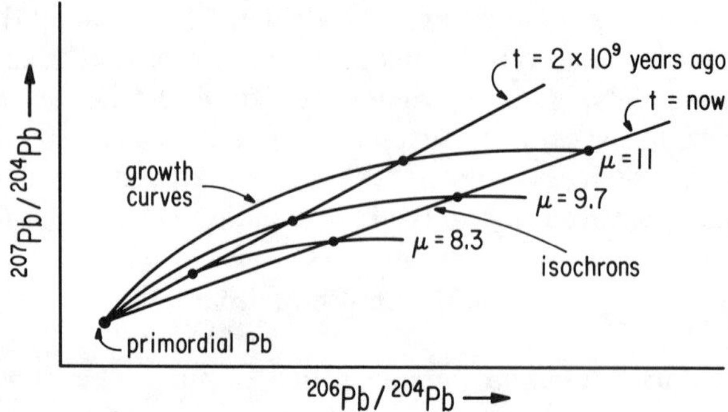

Figure 12.3. The Holmes–Houtermans model for lead isotopic evolution. The curved lines represent growth curves for the Pb isotopic ratios for rocks from a single original system that solidified with different relative amounts of U and Pb. Present $^{238}U/^{204}Pb$ ratios in the three rocks are given as values of μ. Straight lines joining points of equal age for the three rocks are called isochrons and their slope is related to the solidification age of the system. (Adapted from R. D. Russell and R. M. Farquhar, *Lead Isotopes in Geology.* New York: Interscience, 1960).

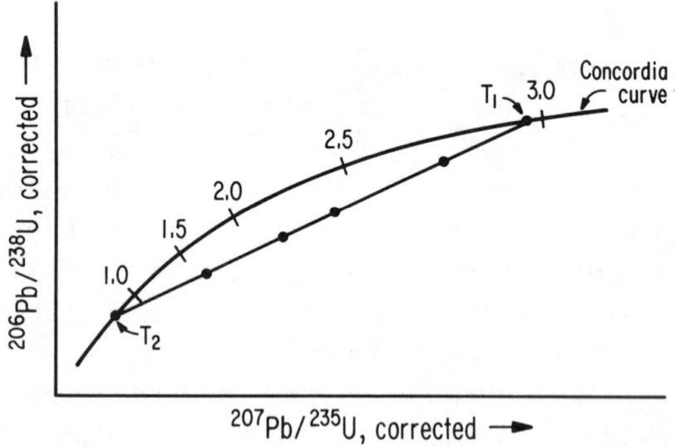

Figure 12.4. The Concordia curve. The curve is a plot of the changes in the $^{206}Pb/^{238}U$ and $^{207}Pb/^{235}U$ isotopic ratios in a closed system as a function of time. Numbers along the curve represent different ages (in aeons). The location of a point on the curve obtained by measurement of the two ratios in a rock determines its age. Because the half-life of ^{235}U is much shorter than that of ^{238}U, the $^{207}Pb/^{235}U$ ratio changes more rapidly than the $^{206}Pb/^{238}U$ ratio. If the system loses Pb at some time T_2, all Pb isotopes are lost proportionally. The various minerals in the rock after Pb loss would plot as points on the cord joining T_1 and T_2. Extrapolation of the cord to intersections with the Concordia curve yields the age of formation of the system, T_1, and the time of episodic Pb loss (or U gain).

episodic Pb loss occurred, can often be obtained. Loss of Pb by diffusion would not seriously alter Pb isotopic ratios, because of the small percentage differences in the masses of the Pb isotopes. If the ratio $^{206}Pb/^{238}U$ is plotted against the ratio $^{207}Pb/^{235}U$, a curve such as that shown in Fig. 12.4 is obtained. Only the radiogenic lead contents are used for the ratios in this plot. The curve is called the **Concordia curve** and points along it represent the Pb isotopic system after increasing times of decay in a closed system free of primordial Pb. If minerals in several subsystems with a general system age of T_1 suffered episodic loss of Pb (or, less likely, a gain of U) at T_2, data points for the subsystems would fall along a cord connecting T_1 and T_2 on the Concordia curve. G.W. Wetherill in 1956 showed that extrapolation of the discordant points along the cord toward the right to the intersection with the Corcordia curve would yield an age for the origin of the overall system. Extrapolation of the cord to the left would intersect the Concordia curve at the time of Pb loss. This approach is very useful in resolving discordant ages, but it is valid only for episodic rather than continuous modes of Pb loss. Overall, plumbology can be quite complex and other dating methods requiring fewer corrections are now more commonly used.

12.5. Rb–Sr METHOD

The **Rb–Sr dating method** is probably the most reliable, accurate, and widely used method to determine the solidification age of rocks. It is based on the following single-step decay:

$$^{87}\text{Rb (27.84\% of natural Rb)} \xrightarrow{t_{1/2}=4.8\times10^{10}\text{ y}} {}^{87}\text{Sr (7.00\% of modern Sr)}$$

The working equation for the method is

$$\frac{^{87}\text{Sr}}{^{86}\text{Sr}} = \frac{^{87}\text{Rb}}{^{86}\text{Sr}}(e^{\lambda t} - 1) \tag{12.12}$$

where the basic decay equation (see Eq. 12.8 for the U–Pb method) has been expressed in terms of isotopic ratios normalized to ^{86}Sr, a Sr isotope that is not produced by natural radioactive decay. Figure 12.5 shows mass spectra of Sr from both a terrestrial Sr ore and a mineral from a meteorite that has a high Rb content and has existed as a closed system for several aeons. It can be seen that in old rock systems that contain high levels of the radioactive parent nuclide ^{87}Rb, the amount of the stable daughter nuclide, ^{87}Sr, is increased in proportion to the time that the system has remained closed. The lowest values for the ^{87}Sr/^{86}Sr ratios are found in the achondrites, meteorites in which the ratio of Sr/Rb is very high (the alkaline earth elements are all high with respect to alkali metals). This ratio of Sr isotopes is known as **BABI (basaltic achondrite best initial)** and is equal to approximately 0.6988. This value is taken to be the primordial Sr isotope ratio. In the earth's crust, Sr is approximately four times more abundant

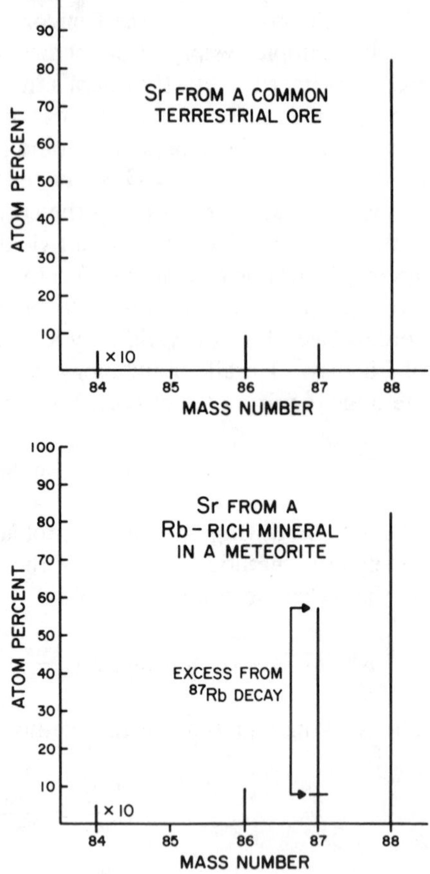

Figure 12.5. Mass spectra of Sr isotopes in a common alkaline earth ore and in an old mineral rich in Rb. The ^{87}Sr isotope is enriched in the old Rb-rich mineral due to the decay of ^{87}Rb.

than Rb. Because Sr is more abundant than Rb and the ^{87}Rb half-life is so long, the value for the ^{87}Sr/^{86}Sr atomic ratio in the oceans today, 0.708, is only slightly greater than the primordial ratio. Fortunately, mass spectrometry is able to measure this ratio to at least six decimal places and the small variations observed are the basis of the Rb–Sr dating method.

As was the case for Pb–Pb dating, graphical methods are often used to measure Rb–Sr ages. Figure 12.6 shows a Sr evolution diagram in which the two isotopic ratios in Eq. 12.12 are plotted. At the left of the diagram, R^0 is the primordial ^{87}Sr/^{86}Sr ratio. All minerals in a rock that solidified shortly after the formation of the earth will have a ^{87}Sr/^{86}Sr ratio close to this value at the time of the rock's crystallization, because ^{87}Rb decay has not yet altered the isotopic ratio. The points a–e in Fig. 12.6 reflect the

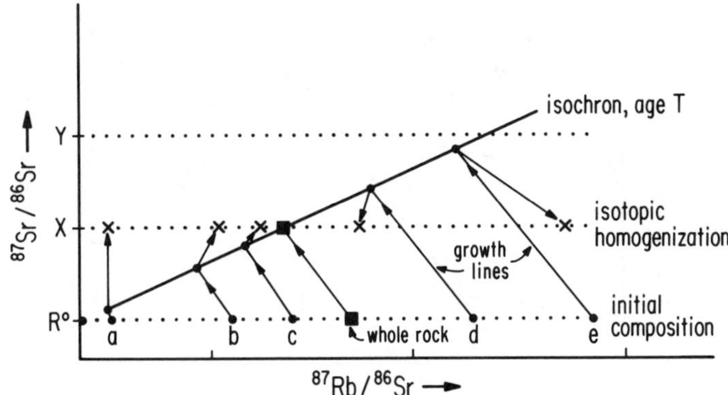

Figure 12.6. A Sr evolution diagram. R^0 corresponds to the primordial $^{87}Sr/^{86}Sr$ ratio. After the rock system is closed, minerals *a–e* will accumulate different amounts of radiogenic ^{87}Sr proportional to their Rb contents. At some time after the solidification of the system, *T*, the isotopic ratios for minerals in the rock will line along the diagonal line known as an isochron. Whole rock ratios are represented as solid squares. The slope of the isochron is related to the solidification age of the system. If the rock is melted at a later stage in its history, all minerals will have a new initial $^{87}Sr/^{86}Sr$ ratio represented by point *X* on the ordinate and a new isochron will develop from this intial value.

different Rb/Sr ratios in the different minerals in the rock and the solid square represents the average $^{87}Rb/^{86}Sr$ value for the whole rock. At some later geological time, the ^{87}Rb in the various minerals has undergone decay and produced additional amounts of ^{87}Sr in proportion to the relative amounts of Rb and Sr in the minerals. The new $^{87}Sr/^{86}Sr$ values now lie along a diagonal solid line called a **Rb–Sr isochron.** From Eq. 12.12 it can be seen that the slope of this line is equal to $(e^{+\lambda t} - 1)$ and can be used to calculate *t*, the Rb–Sr age for the crystallization of the rock.

If the rock dated in Fig. 12.6 is remelted and then recrystallized at a later time, the $^{87}Sr/^{86}Sr$ ratios in all the minerals will all again be the same and have a value corresponding to the point labeled *X* on the ordinate. Still another remelting and crystallization at a later time would result in a homogenization of the mineral $^{87}Sr/^{86}Sr$ ratios to a value of *Y* on the ordinate.

Figure 12.7 shows a Sr evolution diagram for two rocks with different initial Rb/Sr ratios that separated from a common geological closed system and then were remelted simultaneously at some later time. Points for individual minerals are shown as solid dots and the mean values for the whole rock are shown as solid squares. The point R^0 on the ordinate represents the $^{87}Sr/^{86}Sr$ value for the rocks when they were initially isolated at $(t_1 + t_2)$ years ago from the main system. After a passage of time, t_1, the rocks were remelted and all minerals in them were normalized to new $^{87}Sr/^{86}Sr$ ratios,

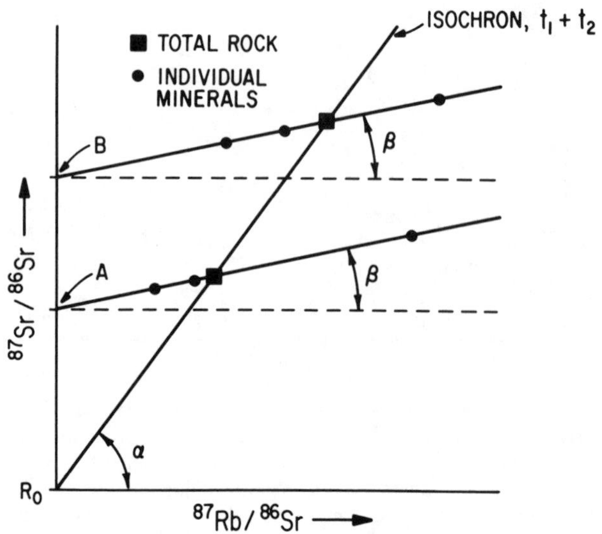

Figure 12.7. A Sr evolution diagram for two rocks with different initial Rb/Sr ratios that separated from a common closed geological system and were simultaneously remelted at a later time. Isotopic ratios for individual minerals can be used to determine the date of remelting, while total rock ratios can be used to determine the date of the separation of the rocks from a common geological system.

represented by points A and B on the ordinate. The lines with angle β intersecting points A and B are isochrons determined by plotting Sr isotopic ratios for the individual minerals in the two rocks. The tangent of angle β is used to calculate the time from the remelting to the present, t_2. A line drawn through the mean whole rock Sr isotopic ratios for the two rocks intersects the initial $^{87}Sr/^{86}Sr$ ratio for the system, R_0. Measurement of the the tangent of angle α permits calculation of the time $(t_1 + t_2)$ for the initial separation of the rocks from the common geological system. R^0 may be the primitive $^{87}Sr/^{86}Sr$ ratio if the system solidifed very early in the history of the solar system, or some higher value if the system has had a more complex geological history.

The Rb–Sr dating method is one of the best geological dating methods because the parent and daughter nuclides are not volatile or easily mobilized, and many rocks have an appreciable Rb content. An additional advantage is that the method is based on very accurate isotopic ratio measurements by mass spectrometry and low background counting equipment is not required.

12.6. K–Ar METHOD

12.6.1. Standard Method

The **K–Ar dating method** is another example of the accumulation clock. It is based on the following single-step decay process:

$$^{40}\text{K (0.0117\% of natural K)} \xrightarrow{t_{1/2}=1.28\times10^9\text{ y}} {}^{40}\text{Ar (11\% via E.C.)} + {}^{40}\text{Ca (89\% by } \beta^-)$$

The working equation for the method contains a branching ratio correction term (0.11) to correct for the fact that only 11% of ^{40}K decays proceed through a decay path to ^{40}Ar:

$$^{40}\text{K} = [^{40}\text{K} + (^{40}\text{Ar}/0.11)]e^{-\lambda t} \qquad (12.13)$$

where ^{40}K and ^{40}Ar represent the amounts (by atom) of these nuclides in the sample at the present time. ^{40}K and ^{40}Ar contents of rocks are measured by chemical and mass spectrometric techniques and used to calculate the K–Ar age, t.

Rock ages obtained by the K–Ar method may be discordant with ages obtained on the same rocks by the Pb–Pb or Rb–Sr dating methods. The reason for this is that Ar gas may be lost by slow diffusion in "loose lattice" minerals, by a partial melting of the rock, or by impact shock, as in the violent explosions associated with volcanism or meteorite impacts. Hence, the standard K–Ar age may at best be only a measure of the time since the rock was last heated to the point of Ar loss.

12.6.2. Incremental Heating Method

A clever approach to obtain better information from K–Ar dating was used by Merrihue and Turner in the dating of lunar rocks. This approach, called the **incremental heating** or **^{40}Ar/^{39}Ar method,** is based on first irradiating the rock in the fast neutron flux of a nuclear reactor. In the irradiation, some of the stable ^{39}K in the rock is converted to ^{39}Ar by the ^{39}K(n, p)^{39}Ar reaction. ^{39}Ar is a long-lived ($t_{1/2}$ = 269 y) negatron emitter and after the irradiation remains trapped in the same mineral sites as the parent K from which it was formed. The irradiated rock is then heated incrementally and the Ar gas released at each temperature is collected. The Ar samples are analyzed with a mass spectrometer to determine their ^{40}Ar/^{39}Ar ratios.

Some low-melting minerals may release their Ar at low temperatures, while higher-melting minerals will release their Ar at higher temperatures. Ideally, the $^{40}Ar/^{39}Ar$ ratios in the gas released at each temperature should be the same, because both the ^{40}Ar and the ^{39}Ar are derived from the same potassium-containing sites. However, it was found that the Ar released at lower temperatures from lunar rocks had a lower $^{40}Ar/^{39}Ar$ ratio than the Ar released at higher temperatures. This suggested that some of the ^{40}Ar produced by radioactive decay of ^{40}K over geologic time had been lost by diffusion or a related process. The present ^{40}K content of minerals releasing Ar at each temperature is determined by use of the activation analysis equation, the amount of ^{39}Ar produced from ^{39}K in the irradiation, and the present $^{40}K/^{39}K$ ratio. The apparent K–Ar age for the rock can then be determined from the calculated ^{40}K content and the amount of ^{40}Ar content released at each temperature. These apparent K–Ar ages may be plotted against the fraction of the total ^{39}Ar present in the irradiated sample that is released at each temperature. The total ^{39}Ar in the irradiated sample can be obtained by summation of the amounts released at each temperature up to the melting point of the rock.

An example of such an experiment is illustrated in Fig. 12.8. The Ar gas evolved at lower temperatures yields relatively low apparent K–Ar ages, most likely due to loss over geologic time of ^{40}Ar from certain low-temperature minerals in the rocks. At higher temperatures, the apparent K–Ar ages initially increase, because Ar is now being released from high-melting or tight-lattice minerals less susceptible to Ar loss by diffusion. Finally, at still higher temperatures, the Ar evolved attains a constant $^{40}Ar/^{39}Ar$ ratio and the apparent K–Ar ages obtained with further temperature increases no

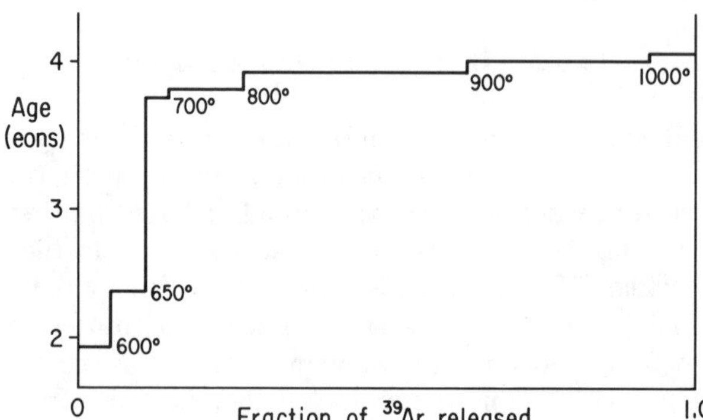

Figure 12.8. An example of the incremental heating approach in K–Ar dating (the $^{40}Ar/^{39}Ar$ dating method).

longer vary. It is assumed that any loss of ^{40}Ar and ^{39}Ar by diffusion will not alter the Ar isotopic ratios, because mass differences among the isotopes are small. The constant ^{40}Ar/^{39}Ar ratio is taken to mean that no ^{40}Ar loss by diffusion has occurred in these high-temperature minerals over geological time and that the apparent age obtained at the higher temperatures represents the true crystallization age of the rock.

In all K–Ar methods, care must be used to correct for contamination of the sample by atmospheric ^{40}Ar and other Ar isotopes that may be produced by the action of cosmic rays on meteorites and samples from the lunar surface. Levels of the nonradiogenic stable isotopes of Ar, ^{36}Ar (0.337%, natural abundance) and ^{38}Ar (0.063%, natural abundance), in the sample are used to provide contamination corrections.

12.7. PLEOCHROIC HALOS

The mineral mica often contains small inclusions of zircon or other minerals that have high contents of U and Th. Alpha particles emitted by the decay of U and Th and their daughters are sufficiently energetic to pass out of mineral inclusions of submicrometer size and penetrate a distance into the surrounding mica. The radiation damage caused by the α particles produces a dark halo in the mica surrounding the mineral inclusion. These **pleochroic halos,** so called because they exhibit color changes when rotated in polarized light, can be used for dating the mica. The degree of darkening is used as a measure of the age of the sample. Unfortunately, darkening of the halos is an inconsistent process and the method is not sufficiently reliable for absolute age determinations. It is more commonly used to estimate relative ages, or the heating history, of geologic subsystems having a common origin.

Halos with different radii are produced by α particles of different energies. This provides a means for the identification of the parent radionuclide. Extremely large halo radii observed in some old minerals and lunar samples were once believed to be the result of very high-energy α particles emitted by now extinct radioisotopes of super heavy elements ($Z \approx 110$–118). This interpretation of the data was later proved to be incorrect and evidence for the existence of these elements has yet to be discovered.

12.8. FISSION TRACKS

Fission events produce two energetic fission fragments that leave double radiation damage tracks in crystalline materials and glasses. The areal density of the fission tracks (tracks/cm^2) in a section of the sample is a function of the age of materials that have identical uranium contents and heating histories. The method is often used to date ancient geological and archaeological glasses.

The two natural isotopes of U differ in that ^{235}U undergoes thermal neutron-induced fission and has an extremely small decay branch via spontaneous fission, while ^{238}U undergoes spontaneous fission with one fission event for every 2.23×10^6 α-particle decays and does not undergo thermal neutron-induced fission. In the **fission track dating method** small mineral or glass grains are imbedded in a clear resin, polished to produce a flat surface, and etched with a chemical agent to reveal the activated (radiation-damaged) areas produced by the massive ($A \approx 100$), energetic ($\approx$200 MeV), and highly charged fission fragments. The fission tracks on the sample surface are counted both before and after irradiation of the sample with thermal neutrons. In old samples of natural glasses tens of thousands of spontaneous fission tracks can be counted per square centimeter of sample surface.

The areal density of ^{238}U spontaneous fission tracks on an etched surface is

$$d_s = \left(\frac{\lambda_{\text{fission}}}{\lambda_\alpha}\right) {}^{238}\text{U}(e^{\lambda_\alpha t} - 1)f \tag{12.14}$$

where d_s is the areal density of fission tracks on the etched surface and f is the fraction of total fission tracks produced that will cross the surface and will be counted after etching. Similarly, the density of induced-fission tracks from ^{235}U after irradiation of the polished section with thermal neutrons is

$$d_i = {}^{235}\text{U}\phi\sigma f \tag{12.15}$$

where ϕ is the dose of thermal neutrons, in n/cm^2, σ is the thermal neutron cross section for the induced fission of ^{235}U, and f is the same fraction of tracks produced that is used in Eq. 12.14. Dividing Eq. 12.14 by Eq. 12.15, and providing numerical values for the decay constants, the thermal neutron cross section of ^{235}U, and the atomic ratio ^{238}U/^{235}U, the following working equation is developed:

$$t = 6.446 \times 10^9 \ln\left(1 + 7.744 \times 10^{-18} \frac{d_s}{d_i}\phi\right) \tag{12.16}$$

The neutron dose ϕ is determined by irradiation of a monitor with the sample and the two areal track densities are determined by counting the tracks with the aid of a microscope. The technique has been described in detail in a book by Fleischer et al. (1975; see Reading List). Care must be taken to distinguish fission tracks from tracks caused by α particles and cosmic rays. A serious source of error results from the fact that fission tracks will fade with time, especially if the sample has been exposed to high temperatures, shear stress, or shock. Fading rates can be quite different for different matrices. It takes one hour of heating at over 1000 °C to completely erase fission tracks in quartz, while a temperature of only 300 °C for one hour will erase fission tracks completely in a glass derived from a volcanic basaltic glass. In general, the method is most useful in dating relatively young glass samples, especially archaeological samples that have been protected from heating since their production.

12.9. SPECIAL METHODS

12.9.1. Cosmic-Ray Exposure Ages of Meteorites

Radionuclides such as ^{3}H, ^{14}C, ^{26}Al, ^{36}Cl, ^{54}Mn, ^{60}Co and various stable and radioactive isotopes of the noble gases are produced in meteorites by spallation reactions induced by cosmic rays during the existence of the meteorites as small bodies in space. For example, ^{54}Mn ($t_{1/2}$ = 312.2 d) is formed by spallation reactions on ^{56}Fe in meteorites. For all practical purposes, the accumulation of the radionuclide or stable spallation product starts when the parent meteorite body breaks up into small pieces of a size that cosmic rays can penetrate the entire object. The **cosmic-ray exposure age** is defined as the time from the breakup of the large parent meteorite body to the time that the meteorite falls to earth, where it is shielded from further cosmic-ray interactions by the earth's atmosphere. It is assumed that in the large parent body of the meteorite, only a small surface layer would be exposed to the cosmic rays. The amount of exposed material would be small with respect to the total mass of the parent body. Thus, a random sample of a disrupted parent body immediately after fragmentation would have a low probability of exhibiting cosmic-ray induced radioactivities.

After the meteorite fragments begin their existence as small bodies in space, they start to accumulate cosmic-ray-induced radioactivities and some stable noble gas nuclides produced by spallation. For the shorter half-life radionuclides, activities will reach equilibrium levels quickly and production rates will be balanced by decay rates. With increasing exposure time, concentrations of their decay products build up in the meteorites. For example, the equilibrium production level of ^{3}H in a stony meteorite can be determined by extracting the hydrogen from a freshly fallen meteorite and counting the tritium activity. If it is assumed that the production of tritium (and thus also the cosmic-ray flux in space) has been relatively constant over the life history of the meteorite as a small body, the amount of ^{3}He in the meteorite can be measured and the exposure age of the meteorite calculated as follows:

$$\text{Exposure age (y)} = \frac{^{3}\text{He accumulated (atoms)}}{^{3}\text{H equilibrium activity (atoms/y)}} \tag{12.17}$$

In a related approach called the saturation method, activity levels of cosmic-ray-induced radioactivities with both moderate (^{54}Mn, $t_{1/2} = 312.2$ d) and very long (^{26}Al, $t_{1/2} = 7.3 \times 10^{5}$ y) half-lives are measured in the same meteorite. Measurement of the activity of the moderate half-life radionuclide which is assumed to have reached its equilibrium activity level, together with a knowledge of the chemical composition of the meteorite, the cosmic-ray energy spectrum, and spallation reaction cross sections will provide a measure of the cosmic-ray flux to which the meteorite was exposed. Using this value for the flux, the observed activity level of the longer-lived radionuclide, and the same chemical and nuclear parameters, the irradiation time required to produce the observed longer-lived activity (the cosmic-ray exposure age) may be calculated. The reader is referred to the Reading List at the end of this chapter for references to more advanced treatments of this topic.

Cosmic-ray exposure ages for meteorites are quite variable, which suggests that meteorites have a complex history and are not just fragments resulting from the breakup of a single large parent body in the asteroid belt between Mars and Jupiter. Stony meteorites commonly exhibit exposure ages of tens of millions of years, while those for iron meteorites are often hundreds of millions of years.

12.9.2. Terrestrial Ages of Meteorites

The **terrestrial age of a meteorite** is the time interval from when the meteorite fell to earth to the present. These ages are most often determined by the equilibrium decay clock method, similar to radiocarbon dating of once living materials. As noted in the previous section, equilibrium levels of relatively short-lived radionuclides are induced in meteorites during their existence as small bodies in space. Once the meteorite falls to earth, the production of the radionuclide nearly stops due to the shielding by the atmosphere, and the activity of the radionuclide will decrease with time. If equilibrium levels of several radionuclides, such as ^{3}H and ^{14}C, are first determined by measuring activity levels in a freshly fallen meteorite and then in a meteorite of unknown terrestrial age with similar chemistry (so amounts of target nuclei for spallation reactions are similar), the magnitude of the decrease in the activity level in the meteorite of unknown terrestrial age will be a function of the terrestrial age of the meteorite. Ages are calculated with the simple decay equation, using the equilibrium activity from freshly fallen meteorites, A_{eq}, as the initial activity and the present activity, A_{now}, in the older terrestrial age meteorite as the activity after a decay time, t, on the earth's surface:

$$A_{now} = A_{eq}e^{-\lambda t} \tag{12.18}$$

In a few cases, a fortunate accident has been used to date the fall (terrestrial age) of a large crater-forming meteorite. In forming a meteorite impact crater, ejecta may bury a tree standing nearby. In this case, simple radiocarbon dating of the death of the tree will also date the time of the meteorite fall and crater formation.

As noted above, the production of cosmic-ray-induced radioactivities in meteorites is assumed to stop when the meteorite falls to earth and is shielded from cosmic rays by the atmosphere, crater ejecta, ice in the polar regions, or water if the fall is in the oceans. Some cosmic rays do reach the earth's surface, however, and can induce very low levels of radioactivities to a depth of several inches in terrestrial rocks exposed at the surface for long periods of time. D. Lal and J. R. Arnold have developed a dating method based on cosmic-ray produced ^{10}Be and ^{26}Al in surface-exposed quartz. The activities of these two radionuclides are too low to be measured by direct counting techniques, but direct atom counting is possible by use of tandem Van de Graaff–cyclotron mass spectrometry. This is the same technique used to date very old samples by the radiocarbon method. The technique has been recently used to date the age of large terrestrial meteorite craters. In the meteorite impact event, a buried quartz vein originally well shielded from cosmic rays may be uncovered and thereafter exposed to cosmic rays that reach the earth's surface. Upon exposure, the production of ^{10}Be and ^{26}Al will start and the amounts of these long-lived radionuclides in the quartz, as measured by accelerator mass spectrometry, can be used to calculate the crater age.

12.9.3. Extinct Natural Radionuclides

Extinct natural radionuclides are radionuclides formed in stellar nucleosynthesis that have half-lives long enough to have survived through the early stages of planetary body formation in the solar system, but not long enough to have survived to the present. Radionuclides in this category generally have half-lives of millions to tens of millions of years. Those frequently studied are ^{26}Al ($t_{1/2}$ = 7.3 × 10^5 y), ^{41}Ca ($t_{1/2}$ = 1.03 × 10^5 y), ^{53}Mn ($t_{1/2}$ = 3.7 × 10^6 y), ^{107}Pd ($t_{1/2}$ = 6.5 × 10^6 y), ^{129}I ($t_{1/2}$ = 1.57 × 10^7 y), ^{146}Sm ($t_{1/2}$ = 1.03 × 10^8 y), and ^{244}Pu ($t_{1/2}$ = 8.0 × 10^7 y). Other candidates are ^{60}Fe, ^{135}Cs, ^{205}Pb, and ^{247}Cm. These radionuclides are primarily used to date the interval from the end of stellar nucleosynthesis (see Chapter 13) to the time of solidification of planetary bodies in the solar system. In some cases, other information on events in the very early history of the solar system can also be obtained.

The most used system employing an extinct natural radionuclide is that of **I–Xe dating.** The extinct natural radionuclide ^{129}I decays to Xe according to the following equation:

$$^{129}\text{I} \xrightarrow{t_{1/2}=1.57\times10^7\text{ y}} {}^{129}\text{Xe (26.4\% natural isotopic abundance)} + \beta^-$$

At some short time after the end of stellar nucleosynthesis, some ^{129}I will still remain to be trapped in halogen-rich minerals in matter condensing in the early solar system. In these halogen-rich minerals the ^{129}I will decay to form ^{129}Xe. This will result in a Xe isotopic anomaly in these minerals due to the addition of the radiogenic ^{129}Xe. Halogen-rich minerals that are in bodies that condensed several hundreds of millions of years after the end of stellar nucleosynthesis would not show this Xe isotopic anomaly, because nearly all the ^{129}I would have decayed to Xe prior to mineral condensation. Xenon gas would have little tendency to be trapped in the condensing rock. Theories of nucleosynthesis can be used to predict the amount of ^{129}I relative to the amount of stable ^{127}I produced in stellar nucleosynthesis (^{129}I/^{127}I ≈ 10^{-4}). Knowing the ^{127}I content of a sample (from NAA of ^{127}I to produce ^{128}I which decays to ^{128}Xe) and the above ratio, the maximum amount of radiogenic ^{129}Xe that could be produced in the sample can be calculated. For example, if the amount of ^{129}Xe is only one-half the calculated maximum amount that could be trapped in the rock from ^{129}I decay, one-half the ^{129}I must have decayed prior to the formation of the rock. In this example, the time interval between the end of stellar nucleosynthesis and the condensation of the rock is one half-life of ^{129}I, or 15.7 million years. As in all dating methods, corrections are necessary. In this case, corrections are required based on the small amount of primordial Xe trapped in the rock during condensation. In practice, an incremental heating method similar to that used in K–Ar dating is applied to neutron-irradiated samples, and the constant ^{129}Xe/^{128}Xe ratio reached at higher temperatures is used in the age calculation. The ^{128}Xe is produced by the negatron decay of ^{128}I produced by the ^{127}I(n, γ)^{128}I reaction during the thermal neutron irradiation of the sample. In effect, the ^{128}Xe serves as a measure of the iodine content of the sample. The I–Xe dating method is the only extinct natural radionuclide dating method for which an extensive data base exists.

12.9.4. Re–Os Method

The **Re–Os dating method** is another example of an accumulation clock. The equation for the decay process is

$$^{187}\text{Re (62.60\% natural abundance)} \xrightarrow{t_{1/2}=4.1\times10^{10}\ \text{y}}$$
$$^{187}\text{Os (1.6\% natural abundance)} + \beta^-$$

Rhenium is a dispersed element in nature and occurs most often in minerals of other elements. It is found in highest abundance in the mineral molybdenite (MoS_2) and in some rare-earth element ores. The working equation is similar to that for the Rb–Sr dating method, with the nonradiogenic isotope ^{186}Os being used for normalization. Isochron plots of ^{187}Os/^{186}Os against ^{187}Rb/^{186}Os are used for graphical computation of the ages. The Re–Os method is used primarily to date molybdenite-bearing vein deposits in terrestrial rock systems, Re-rich copper ores, and iron meteorites. The method is used for materials where U, K, and Rb contents are low, and methods based on the natural radionuclides of these elements cannot be effectively applied.

12.9.5. Lu–Hf Method

Lutetium is a rare-earth element that is concentrated in apatite, zircon, garnet, biotite, and various rare-earth minerals. The decay process is given by the following equation:

$$^{176}\text{Lu (2.59\% natural abundance)} \xrightarrow{t_{1/2}=3.6\times10^{10}\ \text{y}}$$
$$^{176}\text{Hf (5.206\% natural abundance)} + \beta^-$$

The working equation for the method is again similar to that for Rb–Sr dating, with the nonradiogenic nuclide ^{177}Hf being used for normalization. The technique is used to date rocks with high contents of zircons, but has not been widely applied because of difficulties in the separation of Hf from rocks for mass spectrometry.

12.9.6. K–Ca Method

In the discussion of the K–Ar dating method, it was noted that only 11% of the ^{40}K decays result in the formation of the ^{40}Ar daughter. The other decay branch produces stable ^{40}Ca. A working equation similar to that for K–Ar dating is used, with a fractional branching value of 0.89 for decay to ^{40}Ca and normalization to nonradiogenic ^{44}Ca. Because Ca is not volatile like Ar, this method can be used to date materials that have been reheated to the point of gas loss, or those with a crystal structure susceptible to Ar loss by slow diffusion at lower temperatures. Because the natural isotopic abundance of ^{40}Ca is 96.941%, samples selected for K–Ca dating must have very low Ca and high K contents. The method has been used to date K-rich and Ca-poor minerals such as micas and evaporite minerals. Evaporites often suffer loss of radiogenic ^{40}Ar and yield K–Ar ages that are too low as compared to stratigraphic ages. In these cases, the K–Ca method has yielded ages in good agreement with stratigraphic ages.

Other dating methods such as the ^{147}Sm–^{143}Nd clock have also been proposed, but the methods covered here represent the most widely used and best developed methods.

TERMS TO KNOW

- Accumulation clock
- $^{40}Ar/^{39}Ar$ dating
- BABI
- Closed geological system
- Concordia curve
- Cosmic-ray exposure age
- Dendrochronology
- Discordant ages
- Equilibrium decay clock
- Extinct natural radionuclide
- Fission track dating method
- Growth curves in Pb–Pb dating
- Helium clock
- I–Xe dating method
- Isochrons
- K–Ar dating method
- K–Ar incremental heating method
- K–Ca dating method
- Lu–Hf dating method
- Pb–Pb dating method
- Pleochroic halos
- Plumbology
- Primordial isotopic ratios
- Radiocarbon dating
- Rb–Sr dating method
- Re–Os dating method
- Suess effect
- Terrestrial ages of meteorites
- Tritium dating method
- Tritium unit
- U–Pb and Th–Pb dating methods

READING LIST

Berger, R., and H. E. Suess, eds., *Radiocarbon Dating: Proceedings of the 9th International Conference, Los Angeles and La Jolla, 1976.* Berkeley: University of California Press, 1979. [a good review of methodology and applications]

Boudin, A., and S. Deutsch, *Science* **168**:1219 (1970). [Lu–Hf dating]

Coleman, M. L., *Earth Planet. Sci. Lett.* **12**:399 (1971). [K–Ca dating]

Faul, H., *Ages of Rocks, Planets, and Stars.* New York: McGraw-Hill, 1966. [an excellent introduction to nuclear dating methods in geology and cosmochemistry]

Faure, G., *Principles of Isotope Geology.* New York: Wiley, 1977. [principles and applications of nuclear dating methods in geology]

Fleischer, R. L., P. B. Price, and R. M. Walker, *Nuclear Tracks in Solids.* Berkeley: University of California Press, 1975. [fission track dating]

Hirt, B., W. Herr, and W. Hoffmeister, in *Radioactive Dating.* Vienna: International Atomic Energy Agency, 1963, pp. 35–44. [introduction to the Re–Os method]

Kerridge, J. F., and M. S. Matthews, eds., *Meteorites and the Early Solar System.* Tucson: University of Arizona Press, 1988. [nuclear dating methods applied to meteorites, cosmic-ray exposure ages, etc.]

Laul, D., and J. R. Arnold, *Proc. Indian Acad. Sci. Earth Planet. Sci.* **94**:1 (1985). [^{10}Be and ^{26}Al dating of terrestrial surface rocks}

Libby, W. F., *Radiocarbon Dating,* 2d ed. Chicago: University of Chicago Press, 1955. [early development of the technique]

Luck, J. M., and C. J. Allègre, *Nature* **302**:130 (1983). [Re–Os dating]

McDougall, I., G.W. Lugmair, and J. F. Kerridge, *Geochronology and Thermochronology by the ^{40}Ar–^{39}Ar Method.* Cambridge: Cambridge University Press, 1987. [incremental K–Ar heating method]

Merrihue, C., and G. Turner, *J. Geophys. Res.* **71**:2852–2857 (1966). [incremental heating K–Ar method]

Minster, J. F., J. L. Birck, and C. J. Allègre, *Nature* **300**:414 (1982). [dating of meteorites by the Rb–Sr method]

Reedy, R. C., J. R. Arnold, and D. Lal, *Annu. Rev. Nucl. Part. Sci.* **33**:505 (1983). [cosmic-ray-induced radionuclides in meteorites]

Taylor, R. E., *Radiocarbon Dating: An Archaeological Perspective.* Orlando, FL: Academic, 1987. [radiocarbon dating in archeology]

Tilton, G. R., *Earth Planet. Sci. Lett.* **19**:321 (1973). [lead isotope dating of meteorites]

EXERCISES

1. A charcoal sample from an archaeological site was converted to methane and counted in an internal sample proportional counter that had a counting efficiency of 92.5%. The counter contained 1.35 L of gas at a pressure of 2.50 atm and a temperature of 27.0 °C. The ^{14}C counting rate after correction for background was 0.137 cpm. Assuming an equilibrium ^{14}C specific activity of 15.0 dpm/g C, calculate the radiocarbon age of the sample. What event does this date represent?

2. Assuming a pre-thermonuclear test equilibrium value of 1.3 TU for tritium in a specific grape-growing region, calculate the age of a wine produced there that has a $^{3}H/^{1}H$ ratio of 8.35×10^{-20}.

3. The isotopic atomic ratios in a uranium ore are determined by chemical and mass spectrometric measurements to be
$(^{238}U/^{204}Pb)_{now} = 143.0$
$(^{206}Pb/^{204}Pb)_{now} = 15.10$
After correcting for the primordial lead in the sample, calculate the solidification age of the ore. (*Hint:* Normalize the appropriate single-decay-chain equation to $^{204}Pb_{now}$, so only isotopic ratios given are required to solve for the U–Pb age.)

4. What volume of radiogenic ^{40}Ar gas at STP (0 °C, 1.0 atm pressure) would be produced in a 4.85-g sample of volcanic glass that contains 1.38% K, if the glass was last melted to the point of gas loss 570 000 years ago? How many atoms of radiogenic ^{40}Ca would be produced in the glass in the same time?

5. Based on the analogy to Rb–Sr dating, write the working equations and explain the methodology that would be used to graphically solve for rock ages by the following methods:
   (a) the Re–Os dating method
   (b) the K–Ca dating method

CHAPTER

13

THE ORIGIN OF THE CHEMICAL ELEMENTS

13.1. COSMOLOGY

Questions about the origin of the universe and its structure have been asked since earliest times, probably by all peoples on the earth. **Cosmology** is defined as the science or philosophy that studies the universe, including its form, nature, and origins. Some topics of interest to cosmologists are the origins of the universe and of space and time, the current structure of the universe, and its ultimate fate. These topics are fascinating, but are more appropriately dealt with in an astrophysics course. Much cosmological work is highly theoretical and speculative, partly because cosmologists are trying to explain phenomena that occur under conditions completely unlike anything that can be simulated in earthly laboratories. The reader who is interested in this aspect of cosmology should consult references such as those listed at the end of this chapter.

One arena of cosmological interest that is more closely allied to chemistry is the effort to describe the chemical makeup and evolution of the universe. This research area is called **cosmochemistry**. A major problem in this work is to gain an understanding of how the chemical elements came to be formed from primordial matter. **Nucleosynthesis** refers to the nuclear reactions that occurred to form the elements that we now observe in the universe. Nucleosynthesis occurred at two different times during the life of the universe. One phase took place at the time of the formation of the universe and was responsible for production of only a few of the lightest elements. This is referred to as **primordial**, or **"big bang," nucleosynthesis.** The other phase took place, and is still taking place, in the interiors of stars. This is called **stellar nucleosynthesis.** These two phases are discussed in separate sections later in the chapter.

There is a strong interplay between cosmology and elementary particle physics that is quite apparent in the attempt to describe the manner in which the chemical elements came to be. To postulate sensible reaction schemes, a theoretical knowledge of the rates of pertinent nuclear reactions and the cross sections for these reactions are required. The conditions existing at the location of the nucleosynthesis process will certainly affect the nuclear reactions that may occur, so good predictions of the temperature and composition of both the early universe and the interiors of stars are needed. To provide these facts, some understanding of the way in which stars and galaxies evolve is necessary. Correct theoretical understanding of nuclear forces and particle interactions, especially those occurring at very high temperatures and energies, is also needed. Information like this is generated by astrophysics and nuclear physics studies and will not be discussed here. However, an essential piece of this puzzle is the characterization of the actual elemental abundances in the universe. This is an area of more direct chemical interest, and is treated in the following section.

13.1.1. Elemental Abundances

An accurate description of the chemical composition of the universe is an obvious starting point for theories about nucleosynthesis. This characterization includes not only overall elemental abundances, but also isotopic abundances and various critical isotopic ratios. To ask for a chemical characterization of the universe is not a minor request! The compositions of the earth and its moon, the rest of the planets in our solar system, our sun, meteorites and micrometeorites from within and outside the solar system, interstellar matter (ISM), cosmic rays, stars, and nebulae must be known and all have been studied to some degree.

Determining the composition of the earth's crust is straightforward, because there is direct access to samples. Table 13.1 gives a list of the ten most abundant elements in the earth's crust. Only five elements (oxygen, silicon, aluminum, iron, and calcium) comprise more than 90% of the

Table 13.1. Most Abundant Elements in the Earth's Crust

Element	% Abundance by Mass	Isotopic Composition (% of total atoms)
Oxygen	49.20	^{16}O (99.762), ^{17}O (0.038), ^{18}O (0.200)
Silicon	25.67	^{28}Si (92.23), ^{29}Si (4.67), ^{30}Si (3.10)
Aluminum	7.50	^{27}Al (100)
Iron	4.71	^{54}Fe (5.8), ^{56}Fe (91.72), ^{57}Fe (2.2), ^{58}Fe (0.28)
Calcium	3.39	^{40}Ca (96.941), ^{42}Ca (0.647), ^{43}Ca (0.135), ^{44}Ca (2.086), ^{46}Ca (0.004), ^{48}Ca (0.187)
Sodium	2.63	^{23}Na (100)
Potassium	2.40	^{39}K (93.2581), ^{41}K (6.7302)
Magnesium	1.93	^{24}Mg (78.99), ^{25}Mg (10.00), ^{26}Mg (11.01)
Hydrogen	0.87	^{1}H (99.985), ^{2}H (0.015)
Titanium	0.58	^{46}Ti (8.0), ^{47}Ti (7.3), ^{48}Ti (73.8), ^{49}Ti (5.5), ^{50}Ti (5.4)

earth's crust. Of these, the isotopes ^{16}O, ^{28}Si, ^{27}Al, ^{56}Fe, and ^{40}Ca are the most common. In recent years, the composition of the moon's surface has also been determined directly through analysis of the rocks and soils brought back by the Apollo astronauts and unmanned Russian lunar missions. From these surface compositions and various physical properties of the moon, cosmochemists have calculated values for "whole moon" elemental abundances. Table 13.2 gives a list of calculated values of the most abundant elements in the whole moon, based on data derived from analyses of returned lunar samples.

Meteorites are objects that are believed to be mainly fragments of subplanetary-size bodies that existed in the present asteroid belt between Mars and Jupiter. Some meteorites have also been identified as originating on the moon and on Mars. These fragments are perturbed from their orbits and survive passage through the earth's atmosphere to reach the earth's surface. Meteorites that are collected after their fall has been observed are called *falls,* while those collected that are not associated with an observed fall are called *finds.* The term **meteors**, in contrast, refers to the visual phenomena produced in the atmosphere by small bits of extraterrestrial matter, most likely the residue of comets, that do not survive passage through the atmosphere to reach the earth's surface. Any relationship between periodic **meteor showers** and meteorites is unlikely. Meteorites are of interest to

Table 13.2. Most Abundant Elements in the Whole Moon (Calculated)

Element	% Abundance by mass
Oxygen	41.42
Silicon	18.62
Magnesium	17.37
Iron	9.00
Calcium	6.37
Aluminum	5.83
Nickel	0.51
Sulfur	0.39
Titanium	0.34
Chromium	0.12
Sodium	0.09
Phosphorus	0.05
Manganese	0.03
Cobalt	0.02

Source: Adapted from R. Ganapathy and E. Anders, *Geochim. Cosmochim. Acta.* **2** (Suppl. 5):1181 (1974).

cosmochemists because they were formed very early in the history of our solar system, and in the case of at least one type of meteorite, may have undergone very little chemical differentiation since they condensed from the primordial dust cloud from which the solar system formed. Hence, a study of these objects can provide valuable information about the chemical composition of the early solar nebula.

Meteorites can be divided into three major classes:

1. **Stony meteorites** (also called aerolites) consist largely of silicate minerals, often with small spherical mineral inclusions called **chondrules** and bits of an Fe–Ni–Co alloy. Stony meteorites containing chondrules are called **chondrites**. Those without chondrules are called **achondrites**. The chondrites are the most abundant (≈86%) of the meteorites collected after observed falls on the earth's surface (Kerridge and Matthews, 1988). Achondrites constitute ≈8% of all observed falls. One group of rare achondrites is believed to originate from Mars.
2. **Iron meteorites** (also called siderites) have an Fe–Ni–Co alloy matrix. These meteorites are among the most massive found on the earth's surface because they are able to survive passage through the atmosphere intact. Iron meteorites often contain inclusions of an iron sulfide mineral (troilite, FeS), phosphides (schreibersite, $(FeNi)_3P$), and have nickel contents of 2 to 35% by mass. Iron meteorites constitute ≈5% of all observed falls, but a higher percentage of finds.
3. **Stony-iron meteorites** (also called pallasites) consist of nearly equal amounts of metallic and stony (often the high-temperature mineral, olivine, a magnesium–iron silicate) matrix. This type of meteorite is much less common (≈1% of observed falls) than either of the previous two types. Only 9 falls have been collected.

A rare (only about 0.6% of all observed meteorite falls) subclass of chondrites called the **type I carbonaceous chondrites,** or **CI chondrites,** is of greatest interest to cosmochemists. This is because these chondrites exhibit little evidence of chemical differentiation, at least for metallic trace elements, since their condensation very early in the evolution of the solar system. This type of meteorite contains hydrated minerals, sulfur, magnetite, and a variety of organic materials. Included among the organic compounds are some rather complex molecules, such as amino acids. It is widely accepted that these meteorites represent our best sample of nonvolatile elemental abundances in the matter from which the solar system evolved. However, some CI chondrites do contain small bits of high-temperature

Table 13.3. Selected Cosmic Elemental Abundances and Concentrations in a CI Chondrite

Element	"Cosmic" Atomic Abundance Relative to Si = 10^6 atoms	Concentration in the Orgueil CI Chondrite (mass basis, unless noted)
H	2.72×10^{10}	2.02%
He	2.18×10^{9}	56 nL/g
Li	59.7	1.59 μg/g
C	1.21×10^{7}	3.45%
N	2.48×10^{6}	0.318%
O	2.01×10^{7}	46.4%
Na	5.70×10^{4}	0.483%
Mg	1.07×10^{6}	9.55%
Al	8.49×10^{4}	0.86%
Si	1.00×10^{6}	10.67%
P	1.04×10^{4}	0.12%
S	5.15×10^{5}	5.25%
Cl	5240	698 μg/g
K	3770	569 μg/g
Ca	6.11×10^{4}	0.90%
Ti	2400	436 μg/g
Cr	1.34×10^{4}	0.26%
Mn	9510	0.20%
Fe	9.00×10^{5}	18.51%
Co	2250	509 μg/g
Ni	4.93×10^{4}	1.1%
Cu	514	112 μg/g
Cd	1.69	0.67 μg/g
Sn	3.82	1.68 μg/g
Ir	0.66	0.47 μg/g
Au	0.186	0.14 μg/g
Pb	3.15	2.43 μg/g
U	0.009	0.008 μg/g

Source: Adapted from J. F. Kerridge and M. S. Matthews, *Meteorites and the Early Solar System*. Tucson: University of Arizona Press, 1988. Data derived largely from E. Anders and M. Ebihara, *Geochim. Cosmochim. Acta* **46:**2363 (1982).

minerals that cause abundances of some of the major elements, such as Mg, Si, and Fe, to vary between individual meteorites in this class. Selected elemental concentrations in a CI chondrite are given in Table 13.3.

The matrix of the CI chondrites has been altered by hydrolysis reactions and, while primordial elemental abundances may be preserved, the presence of actual primitive nebular particles in these meteorites is unlikely. Unaltered primitive particles may eventually be identified in the matrices of other types of relatively undifferentiated meteorites that have not been subject to aqueous alteration.

The composition of our sun and of other stars cannot be determined by direct sampling, but the use of spectroscopic techniques allows for the identification and approximate quantification of the elements in stars. Elemental abundances derived from stellar spectra are not as accurate as those derived from meteorite analyses, because the spectral line intensities depend on thermodynamic parameters, such as temperature and pressure, that can only be estimated. However, stellar spectra are utilized to assign relative abundances of hydrogen, helium, and other volatile elements.

Ringwood (1979) states that the initial bulk composition of the primordial solar nebula from which the planets formed was very similar to the present composition of the sun. The primordial solar nebula would have been composed of two major components: dust particles and a gas phase. The gas phase would consist of H and He and constitute 98% by mass of the nebula. The H/He atom ratio was probably about 5 or 6. The dust particles would be composed of two types of materials: ices and rock. The ices would contain C, N, O, Ne, S, Ar, and Cl (in some cases as hydrides) and constitute 1.5% by mass of the nebula. The remaining elements, principally metals such as Na, Mg, Al, Si, Ca, Fe, and Ni, would constitute only 0.5% by mass of the nebula and be present largely in the form of oxides.

Cosmic rays are radiations from space that constantly bombard the earth. Their sources include the sun in own own solar system, other stars in our galaxy (the Milky Way), or celestial objects, such as colliding galaxies, that are distant from the Milky Way. Cosmic rays consist mainly of high-energy protons and α particles (about 90% of the total), along with some light ($Z < 10$) nuclei that are completely stripped of their electrons. The H/He atomic ratio in cosmic rays is $\approx$12–14. Cosmic rays with energies greater than 10^9 eV (1 GeV) are probably extragalactic, while those with lower energies are more likely to have arisen within the Milky Way. The exact source of the cosmic rays and the ways in which they receive their energy are not completely understood. The composition of cosmic rays is an important clue in deciphering the origins of the light elements Li, Be, and B (see Section 13.6). These latter elements appear prominently in cosmic rays. Many nuclides with $Z > 6$ are an order of magnitude more abundant in cosmic rays than in the sun or similar stars, and the ratios of the volatile light elements (e.g., C/N/O) are different in cosmic rays than in the sun. Therefore, nuclidic abundances in cosmic rays are believed to vary considerably from their original abundances due to alterations caused by their acceleration and interaction with other particles in space prior to their arrival at an earth-based detector.

Characterization of the interstellar medium is difficult. Spectroscopic observations can be made on the luminous matter, but it is now thought that perhaps >90% of the matter in the universe is **dark matter,** that is, matter that is not luminous and is not detected using current observation methods. Elements heavier than H and He make up only 1–2% by mass of the interstellar medium and probably exist in dust grains between 0.001 and 0.1 μm in diameter. Some of these dust grains were probably trapped in the condensation of the early solar nebula. The nature and amounts of this dark matter are topics of great current interest in astronomy. It has many implications not only for nucleosynthesis, but for theories of the origin and fate of the universe as a whole.

13.1.2. Cosmic Abundance Curves

Hans Suess in 1947 attempted to measure the magnitude of possible cosmochemical fractionation processes and relate them to certain nuclear systematics, such as odd–even effects and magic numbers. In 1956, Hans Suess and Harold Urey compiled a very useful **abundance compilation** for many of the nonvolatile trace elements. These abundances were based largely on meteoritic data and certain terrestrial elemental ratios, and were supplemented by solar and stellar spectral data for light and volatile elements. These data represented the best estimates available for solar system atomic abundances. Although not strictly "cosmic," plots of nuclidic (or elemental) abundances versus A (or Z) based on these sources of data are commonly called **cosmic abundance curves.** A recent compilation of some cosmic atomic abundances is given in Table 13.3. In a 1957 publication, Burbidge et al. made use of these abundance patterns in proposing a variety of nuclear processes that must take place in the interiors of stars to produce the observed elemental and nuclidic abundances. This was the first time that trace element analyses of meteorites and some data for terrestrial rocks were definitively linked to primeval astrophysical processes. William A. Fowler (Fig. 13.1) received the Nobel Prize in physics in 1983 for his research on elemental synthesis in stars.

In Chapter 3 elemental abundances were discussed briefly. Figure 3.4 from that chapter shows a diagram of "cosmic" nuclidic abundances that have been derived largely from analyses of type I carbonaceous chondrites, stellar spectra, and some terrestrial nuclidic ratios. Certain features of this curve should be particularly noted. The most abundant elements are clearly H and He. There is a steep drop in elemental abundances at first, which slows for higher Z elements. Abundance peaks occur on the graph for the even–even nuclides, especially those that are either singly or doubly magic. There is much more ^{56}Fe, and much less Li, Be, and B, than is observed for their nearest neighbors. These features must be explained by any proposed reaction scheme.

Figure 13.1. William A. Fowler (*left*) and C. A. Barnes at the California Institute of Technology in 1982. (Courtesy of the University Library, California Institute of Technology, Pasadena, CA.)

13.2. PRIMORDIAL NUCLEOSYNTHESIS

The first era of element formation occurred at the time of the origin of the universe. The currently accepted model about the beginnings of the universe is the **big bang theory.** According to this model, the universe originated between 10 and 20 billion years ago. Matter, energy, space as we know it, and time were born in the cataclysmic explosion of an immensely dense and very small point. The complete details of what happened in the big bang are not all clear, but there is agreement among cosmologists regarding the general sequence of events that took place.

Figure 13.2 shows a diagram of some important stages in the universe's history. Physicists have been able to speculate on the events that occurred later than about 10^{-43} s after the big bang. This particular point is called **Planck time,** and the era between the instant of the big bang and 10^{-43} s is referred to as the **Planck era.** During this era, all four forces are thought to have been united, and no successful theory has yet described this state. For the next 10^{-8} s after the Planck era the **GUT era** existed. In this phase, three of the four basic forces (strong, weak, and electromagnetic) were unified. Energies were very high and radiation dominated the universe, but a

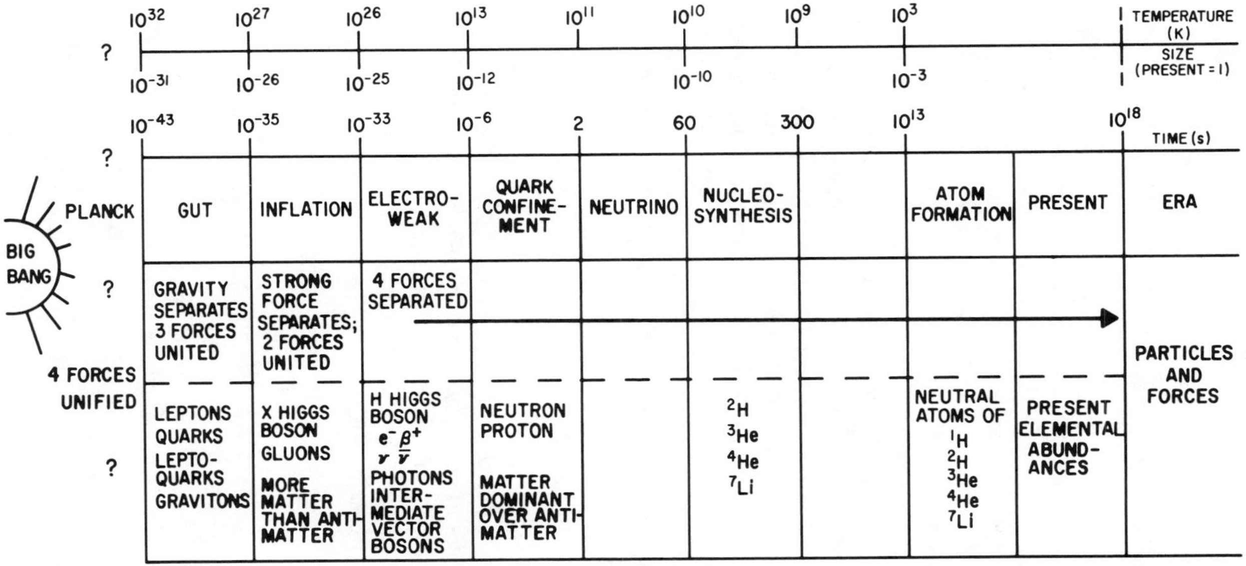

Figure 13.2. The early history of the universe. (Adapted from M. S. Turner, Cosmology and particle physics. In *Intersection Between Particle Physics and Cosmology,* T. Piram and S. Weinberg, eds. Singapore: World Scientific, 1986.)

few particles had come into existence, including leptons, quarks, leptoquarks (which did not survive past this era), and gravitons, the proposed gravity particle. At the end of the GUT era, the **inflation era** began. During this brief instant (from 10^{-35} to 10^{-33} s), the universe began to expand rapidly. The strong nuclear force separated from the electroweak, and the exchange particle for the strong force, the gluon, appeared. An important aspect of this era is that matter became more abundant than antimatter. During the **electroweak era,** the four basic forces separated, and their exchange particles came into permanent existence. In addition, electrons, positrons, neutrinos, and antineutrinos came into being. This phase lasted until about 10^{-6} s after the big bang, when the era of **quark confinement** began. At this time protons and neutrons appeared, and matter became dominant over antimatter.

The time of most interest for chemical evolution is the **nucleosynthesis era,** which lasted from about 1 to 5 min after the initial big bang. By this time, the universe had cooled to about 10^9 K, so nuclei could form and not be quickly broken apart by ambient high temperatures. The universe consisted mainly of photons, electrons and positrons, neutrinos and antineutrinos, and the neutrons and protons, which were continually interacting and interchanging with each other in a variety of reaction pathways.

The first nucleosynthetic step is the formation of deuterium from the reaction of a proton and a neutron:

$$\mathrm{p} + \mathrm{n} \longrightarrow \mathrm{d} + \gamma$$

Tritium could be formed by the d(n, γ)t or d(d, p)t reactions. Tritium is unstable with respect to negatron decay and would have quickly decayed, but it could have been involved in some nuclear reactions as an intermediate on the way to a stable product. The synthesis of helium isotopes could occur in a variety of ways, but these two reactions are most likely:

$$\mathrm{p} + \mathrm{d} \longrightarrow {}^{3}\mathrm{He} + \gamma$$

$${}^{3}\mathrm{He} + {}^{3}\mathrm{He} \longrightarrow {}^{4}\mathrm{He} + 2\mathrm{p}$$

After this series of reactions to form ^{2}H, ^{3}He, and ^{4}He, there are problems when trying to form any heavier elements. There are no stable nuclides with $A = 5$ or 8, so further simple binary synthesis reactions with the components available all lead to unstable products that decay before they can react further. For example, two ^{4}He particles might react to form ^{8}Be, but this nuclide quickly decays ($t_{1/2} = 7 \times 10^{-17}$ s) back to the two ^{4}He nuclei. The capture of a proton by ^{4}He to form ^{5}Li has a low probability and the ^{5}Li so formed would also decay quickly ($t_{1/2} = 3 \times 10^{-22}$ s) before additional reactions could take place. The capture of a neutron by ^{4}He yields ^{5}He, another short half-life radionuclide ($t_{1/2} = 7.6 \times 10^{-22}$ s). The only other element that was formed to any significant degree during primordial nucleosynthesis was ^{7}Li. Two possible reaction series could have occurred:

$$^4\text{He} + {}^3\text{H} \longrightarrow {}^7\text{Li}$$

or

$$^4\text{He} + {}^3\text{He} \longrightarrow {}^7\text{Be}$$

$$^7\text{Be} + \text{e}^- \longrightarrow {}^7\text{Li} + \nu$$

The end result of the nucleosynthetic era was the formation of nuclei of ^{1}H, ^{2}H, ^{3}He, ^{4}He, and ^{7}Li. By the time this era came to an end, the universe had expanded and cooled to such an extent that further kinds of simple capture or synthesis reactions were impossible.

The primordial nucleosynthesis that occurred shortly after the big bang is thus thought to have produced a universe consisting of about 76% H, 24% ^{4}He, and trace amounts of ^{3}He, ^{2}H, and ^{7}Li. At the present time, there is good agreement between these primordial elemental abundances predicted by theory and the actual relative abundances inferred from current data.

After the initial nucleosynthetic period, element synthesis stopped for more than a million years. The universe continued to expand and to cool, and eventually reached a point where matter became dominant over radiation. At this time, it became possible to form atoms by the combination of nuclei and electrons.

The universe has continued to expand and cool to the present. The remnants of the big bang explosion can be seen in the 3 K radiation that permeates the universe. The synthesis of the rest of the elements (except some Li, Be, and B) began taking place after the galaxies and stars had formed. This is the phase of stellar nucleosynthesis.

13.3. STELLAR EVOLUTION

Before discussing the production of elements in stars, it will be helpful to have a basic notion of how stars evolve, because the process of stellar evolution is related to the creation of the elements.

13.3.1. Star Populations

After the big bang, the material of the universe spewed out in all directions from the point of the explosion. Various instabilities caused inhomogeneities in the primeval matter, and the inexorable force of gravity pulled the more dense areas together into the beginnings of galaxies. Within the galaxies, further instabilities caused large clouds of H and He to begin to coalesce into stars. As the matter fell closer and closer toward the gravitational center of the **protostar**, heating occurred. Eventually, the matter became dense enough and the temperatures high enough to ignite the nuclear fusion reactions that produce the stars' energy (see Section 13.4.1).

The first generation of stars that formed is called **population III** stars. They were very massive, composed almost entirely of H and He, and had relatively short lifetimes. The nuclear reactions that took place in these stars formed new elements, thus providing fuel for further nucleosynthesis in later generations of stars. There are no population III stars existing now in our galaxy.

The next generation of stars, called **population II** stars, began life in the same way as those from population III, but with different material. Instead of just H and He, the population II stars also contained about 1% heavier elements, such as carbon and oxygen. The third generation of stars, called **population I** stars, consists of about 2–5% of elements heavier than H and He. Our sun is an example of a population I star.

13.3.2. Evolution of a Star

The evolution and fate of a single star depend on the original composition of the star and upon the mass of material present. A **Hertzsprung–Russell diagram (H-R diagram),** like that shown in Figure 13.3, may be used to discuss the evolution of a single star. An H-R diagram is a plot of stellar luminosity versus the surface temperature, or a color index related to temperature. After the first nuclear reactions are ignited in a star, it appears on the **main sequence** of the H-R diagram. The main sequence refers to the line of stars from the upper left to the lower right of the H-R diagram. Exactly where a star first appears on the main sequence will be governed by its initial mass. The most massive and brightest stars will lie in the upper left, while less massive and dimmer stars will lie in the lower right area of the main sequence. Our sun, a yellow star of relatively low mass, is marked on the diagram for reference. A star will stay on the main sequence during most of its life. For a star with the mass of our sun, this will be about 10 billion years. Less massive stars stay longer, because they burn their nuclear fuel at a slower rate. Very massive stars evolve more quickly. A star with a mass approximately 20 times that of our sun (described as "20 solar masses") would remain on the main sequence for only several million years.

There are two other areas on the H-R diagram that must be noted. In the upper right are stars that have large radii but are relatively cool. These

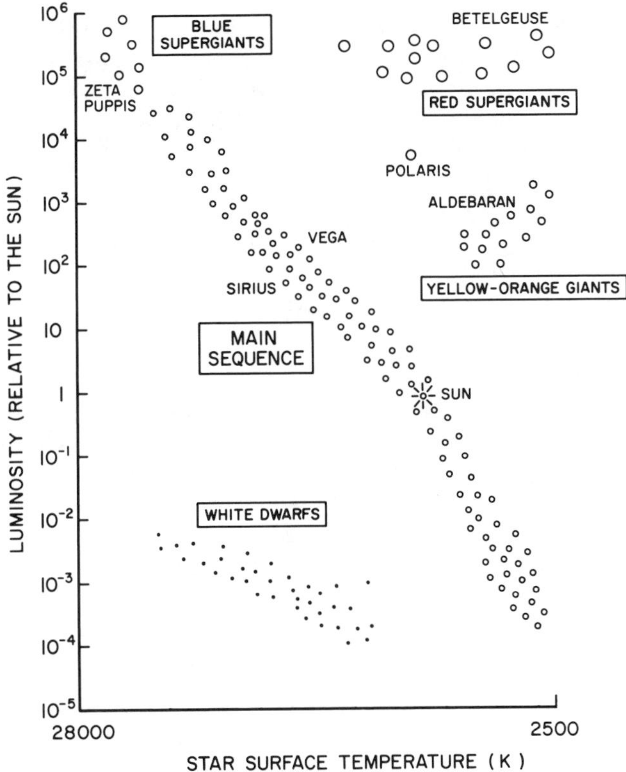

Figure 13.3. A Hertzsprung–Russell diagram. A large percentage of the stars, including our sun, lie near the line representing the main sequence of evolution. Star color, rather than surface temperature, is often plotted along the horizontal axis. Temperature values are not well determined. On the main sequence, the brightest stars observed have the highest surface temperatures and are blue in color. The dimmest stars are red. Our sun lies at approximately the middle of the main sequence line. Red giant stars lie off the main sequence and make up only a few percent of all stars observed. They are easily recognized by their high luminosities. White dwarfs are stars that are much smaller in volume than our sun, but may contain comparable masses. With their small surface areas, their luminosities are low. The white dwarfs represent the "graveyard" of stellar evolution.

stars are called the **red giants** and **supergiants**. Stars that are reaching the ends of their lives move off the main sequence into the red giant region. After this phase, the fate of the star depends on its initial mass. Lower mass stars (less than 1.4 solar masses) will move down toward the lower left corner of the H-R diagram and become **white dwarfs.** These stars are quite bright when first formed and are very small and dense. The white dwarfs gradually cool, become less bright, and die out quietly. More massive stars have a different and more spectacular fate in store. Instead of gradually cooling off, they will explode as a **nova** or **supernova**. The core that remains after the explosion will evolve into either a **neutron star** (for stars with 1.4–3 solar masses) or possibly a **black hole** (for stars with 3 or more solar masses).

13.4. STELLAR NUCLEOSYNTHESIS

13.4.1. Hydrogen Fusion

All stars begin life consisting primarily of H and ^{4}He, with some amount of heavier elements mixed in depending on the population the star belongs to. (These heavier elements are usually referred to as the "metals," although they are not all metals in the chemical sense.) The first series of reactions that occur in any star's evolution involves the fusion of hydrogen into helium.

When a new star's temperature has reached the point at which nuclear reactions can first begin ($\approx 10^8$ K), the reactions that occur are similar to the fusion reactions that took place in the big bang:

$$\mathrm{p} + \mathrm{p} \longrightarrow \mathrm{d} + \beta^+ + \nu$$

$$\mathrm{p} + \mathrm{p} + \mathrm{e}^- \longrightarrow \mathrm{d} + \nu$$

$$\mathrm{d} + \mathrm{p} \longrightarrow {}^{3}\mathrm{He} + \gamma$$

$${}^{3}\mathrm{He} + {}^{3}\mathrm{He} \longrightarrow {}^{4}\mathrm{He} + 2\mathrm{p}$$

This series of reactions is called a **proton–proton (PP)** chain, and the one above is specifically referred to as the **PPI** process. The net result is the fusion of 4 H nuclei into a nucleus of ^{4}He, with the release of energy. This sequence of reactions is not the only way to form a helium nucleus from 4 protons, however. If there is sufficient helium and a high enough temperature, the following series of reactions is also possible:

$${}^{3}\mathrm{He} + {}^{4}\mathrm{He} \longrightarrow {}^{7}\mathrm{Be} + \gamma$$

$${}^{7}\mathrm{Be} + \mathrm{e}^- \longrightarrow {}^{7}\mathrm{Li} + \nu$$

$${}^{7}\mathrm{Li} + \mathrm{p} \longrightarrow \gamma + {}^{8}\mathrm{Be} \longrightarrow {}^{4}\mathrm{He} + {}^{4}\mathrm{He}$$

This particular series of reactions is referred to as the **PPII** process. The ^{7}Be that is formed by the interaction of ^{3}He and ^{4}He may also experience these reactions:

$$^{7}\mathrm{Be} + \mathrm{p} \longrightarrow {}^{8}\mathrm{B} + \gamma$$

$$^{8}\mathrm{B} \longrightarrow {}^{8}\mathrm{Be} + \beta^{+} + \nu$$

$$^{8}\mathrm{Be} \longrightarrow 2\,{}^{4}\mathrm{He}$$

This is the **PPIII** chain. As can be seen, reactions of the PP series fail to effectively bridge the instability (short half-life) gaps at mass numbers 5 and 8. All three of the proton–proton chains will occur simultaneously in a single star. PPI will be the dominant pathway, but the actual percentages will vary depending upon the density, composition, and temperature of the star. Table 13.4 summarizes the PP chains, along with the CNO cycle to be discussed next. Detection of neutrinos emitted in these processes will aid in understanding the contributions of the various processes of energy production in the sun.

In stars that consisted only of H and He at the start, the proton–proton chains are the only mechanisms available to accomplish the fusion of 4 hydrogen nuclei into a helium nucleus. However, members of later generations of stars, which possess some carbon and nitrogen, can undergo another reaction cycle that also results in a He nucleus. This is the CNO bi-cycle.

13.4.2. The CNO Bi-cycle

In this cycle, isotopes of C and N serve as catalysts to bring about the fusion of 4 protons into a He nucleus. The CN reaction series, which is shown on the right side of Figure 13.4, is

$$^{12}\mathrm{C}(\mathrm{p},\gamma)^{13}\mathrm{N}(\mathrm{e}^{+} + \nu\ \mathrm{decay})^{13}\mathrm{C}(\mathrm{p},\gamma)^{14}\mathrm{N}(\mathrm{p},\gamma)^{15}\mathrm{O}(\mathrm{e}^{+} + \nu\ \mathrm{decay})^{15}\mathrm{N}(\mathrm{p},\alpha)^{12}\mathrm{C}$$

The net result of the CN cycle is the conversion of four protons into an α particle. A second cycle, shown on the left side of Figure 13.4, results from the (p, γ) reaction on ^{15}N to form ^{16}O:

$$^{15}\mathrm{N}(\mathrm{p},\gamma)^{16}\mathrm{O}(\mathrm{p},\gamma)^{17}\mathrm{F}(\mathrm{e}^{+} + \nu\ \mathrm{decay})^{17}\mathrm{O}(\mathrm{p},\alpha)^{14}\mathrm{N}$$

This cycle feeds back into the original CN cycle. The entire process is referred to as the **CNO bi-cycle.**

Table 13.4. Neutrino Producing Reactions in the Sun

Process	Neutrino Energies
PP I (≈91% of solar energy production)	
$^{1}H + ^{1}H \rightarrow ^{2}H + e^{+} + \nu$	0–0.42 MeV
$^{1}H + ^{1}H + e^{-} \rightarrow ^{2}H + \nu$ small but important	1.44 MeV
$^{2}H + ^{1}H \rightarrow ^{3}He + \gamma$	
$^{3}He + ^{3}He \rightarrow 2\ ^{1}H + ^{4}He$	
PP II (≈7% of solar energy production)	
$^{3}He + ^{4}He \rightarrow ^{7}Be + \gamma$	
$^{7}Be + e^{-} \rightarrow ^{7}Li + \nu$	0.861 MeV
$^{7}Li + ^{1}H \rightarrow \gamma + ^{8}Be \rightarrow 2\ ^{4}He$	
PP III (≈0.015% of solar energy production)	
$^{7}Be + ^{1}H \rightarrow ^{8}B + \gamma$	
$^{8}B \rightarrow ^{8}Be + e^{+} + \nu$	0–14 MeV
$^{8}Be \rightarrow 2\ ^{4}He$	
CN cycle (probably a very small % of solar energy production)	
$^{1}H + ^{12}C \rightarrow ^{13}N + \gamma$	
$^{13}N \rightarrow ^{13}C + e^{+} + \nu$	0–1.20 Mev
$^{13}C + ^{1}H \rightarrow ^{14}N + \gamma$	
$^{14}N + ^{1}H \rightarrow ^{15}O + \gamma$	
$^{15}O \rightarrow ^{15}N + \beta^{+} + \nu$	0–1.73 MeV
$^{15}N + ^{1}H \rightarrow ^{12}C + ^{4}He$	0–1.19 MeV

Detection

PP I = $^{71}Ga + \nu \rightarrow ^{71}Ge + e^{-}$: Use 15–25 tons of Ga. Dissolve Ga in HCl to form $GeCl_4$ gas, purge and clean up as GeH_4 with gas chromatography. Chemical yields of 100% possible. Count $^{71}GeH_4$ ($t_{1/2}$ = 11.4 d) in a proportional counter after a long exposure. Sensitive to neutrinos with energies >0.2 MeV. (PPI neutrinos are not detected in the Davis experiment.)

PP II = $^{81}Br + \nu \rightarrow ^{81m}Kr$ ($t_{1/2}$ = 13 s) $\rightarrow ^{81g}Kr$ ($t_{1/2} = 2.1 \times 10^{5}$ y): Sensitive to $^{7}Be\ \nu$. Use CH_2BrCH_2Br, or CH_2Br_2 as target. Use laser ionization mass spectrometry to detect ^{81g}Kr. Sensitive to neutrinos with energies >0.5 MeV. (PPII neutrinos are 15% of Davis experimental flux.)

PP III = $^{37}Cl + \nu \rightarrow ^{37}Ar$ (EC, $t_{1/2}$ = 35.0 d): Davis used 615 tons of C_2Cl_4 (100 000 gal = 2.2×10^{30} atoms of ^{37}Cl) in the Homestake gold mine. Expose target for 35–100 days, add ^{36}Ar as a carrier to determine the chemical yield by mass spectrometry), and count ^{37}Ar with a proportional counter. Sensitive to neutrinos with energies >0.8 MeV. (PPIII neutrinos are 85% of Davis experimental flux.) Kamiokande II experiment: Detects neutrinos >5 MeV by measuring the bremsstrahlung emitted when neutrinos pass through 3 kilotons of water and scatter-electrons in the forward direction.

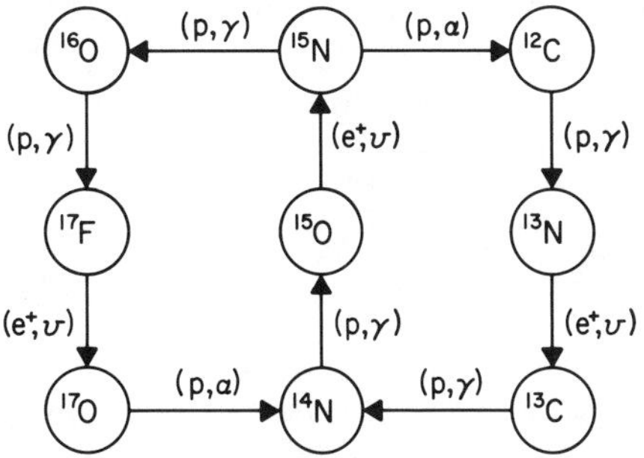

Figure 13.4. A representation of the C–N–O bi-cycle in stellar nucleosynthesis. (Adapted from C. E. Rolfs and W. S. Rodney, *Cauldrons in the Cosmos.* Chicago: University of Chicago Press, 1988.)

Whether it arises from the proton–proton chains or the CNO cycles, the energy released in the star by H fusion is sufficient to supply energy for the vast majority of a star's life. During this time, it remains on the main sequence of the H-R diagram. When the supply of H has been reduced sufficiently, the thermonuclear reactions will no longer produce enough energy to counteract the pull of gravity, the core of the star will begin to contract, and its outer regions will begin to expand. In this phase, the star moves off the main sequence into the red giant region. The contraction of the core causes more heating until, at a temperature of about 2×10^8 K, a new set of nuclear reactions involving helium fusion (commonly called **helium burning**) can begin.

13.4.3. Helium Burning

It was noted earlier in the discussion of primordial nucleosynthesis that there are no stable nuclei at either $A = 5$ or $A = 8$, and radionuclides with these mass numbers have very short half-lives. Under the big bang conditions, no further helium reactions were possible, because the ^{8}Be formed in the synthesis of two helium nuclei is very unstable. The nucleosynthesis era was very short, the universe was expanding, and there was no time for the ^{8}Be to even act as an intermediate in the formation of heavier elements. In the interior of a star, however, it is possible to build up a small concentration of ^{8}Be, and thus the following **triple alpha process** can occur:

$$^{4}\text{He} + {}^{4}\text{He} \longrightarrow {}^{8}\text{Be}$$

$$^{8}\text{Be} + {}^{4}\text{He} \longrightarrow {}^{12}\text{C} + \gamma$$

An excited state of ^{12}C that can decay either to three α particles, or to the ground state of ^{12}C, was discovered by W. A. Fowler in 1956. Although the branch to the ^{12}C ground state is small, this discovery meant that the essentially simultaneous collision of three α particles can also, through the ^{12}C excited state, bridge the mass 8 gap. The carbon thus formed can then react with another helium to produce ^{16}O:

$$^{12}C + {}^{4}He \longrightarrow {}^{16}O$$

These are the main reactions that occur during the helium burning phase. It appears that it should be possible for the ^{16}O to capture another helium nucleus to form ^{20}Ne, but this does not occur to any significant extent. At the end of helium burning, the major products are ^{12}C and ^{16}O. Helium burning lasts a much shorter time than hydrogen fusion did.

After the He is depleted, another period of core contraction and heating follows. For stars that have small masses (less than about 8 solar masses), the contraction and heating are not sufficient to start another round of nuclear reactions. The core will continue to contract until the matter becomes sufficiently dense to halt further contraction, and a stable state is reached. These stars are the white dwarfs, and no further elemental evolution occurs. However, in stars with greater masses the contraction and heating phases produce temperatures that are high enough to initiate further nuclear reactions. For stars that have 8–10 solar masses, a carbon–oxygen burning phase can occur next.

13.4.4. Heavier Element Burning

There are a number of reactions that could easily be imagined for ^{12}C nuclei. Two of them might fuse into a single ^{24}Mg nucleus, into a ^{23}Mg nucleus with neutron emission, or into an ^{16}O nucleus with emission of two helium nuclei. However, for **carbon burning,** the two reactions that turn out to be most important are

$$^{12}C + {}^{12}C \longrightarrow {}^{20}Ne + {}^{4}He$$

$$^{12}C + {}^{12}C \longrightarrow {}^{23}Na + {}^{1}H$$

Even though there is ^{16}O present, there is insufficient energy to initialize reactions between ^{12}C and ^{16}O. The carbon-burning phase lasts an even shorter time than helium burning.

As the carbon-burning phase ends, another type of reaction begins to occur. This is the **photodisintegration** reaction, in which photons induce nuclear changes via reaction pathways such as (γ, n), (γ, p), and (γ, α). The photodisintegration of ^{20}Ne produces more ^{16}O and α particles, which can react during a brief **neon-burning** phase to form more ^{24}Mg:

$$^{20}\text{Ne} + {}^{4}\text{He} \longrightarrow {}^{24}\text{Mg}$$

Another contraction–heating phase takes place that results in **oxygen burning** at a temperature of around 10^9 K. Oxygen burning results in the formation of ^{32}S, ^{31}P, and ^{28}Si by reactions such as

$$^{16}\text{O} + {}^{16}\text{O} \longrightarrow {}^{32}\text{S} + \gamma$$

$$^{16}\text{O} + {}^{16}\text{O} \longrightarrow {}^{31}\text{P} + {}^{1}\text{H}$$

$$^{16}\text{O} + {}^{16}\text{O} \longrightarrow {}^{28}\text{Si} + {}^{4}\text{He}$$

Other elements, including Ca, Ar, K, and Cl, are also produced. After oxygen burning, **silicon burning** starts. The reactions in this phase result in the production of the iron–nickel nuclei, and are often called the **equilibrium process** in nucleosynthesis, because a thermodynamic equilibrium is approached in the reactions. Recall from Chapter 3 that ^{56}Fe has the greatest binding energy per nucleon of any nuclide, and is therefore very stable. No additional charged-particle reactions will contribute significantly to element-building after formation of the iron–nickel region elements. Figure 13.5 shows the layered structure of a massive star after it has completed all of the above burning phases.

A detailed study of the sequence of events described above would show that the elements from H to the Fe region can be reasonably accounted for. However, the types of charged-particle-induced nuclear reactions and rearrangements that occur in the nucleosynthetic processes discussed so far cannot act to produce significant amounts of the heavier elements, in part because of the high Coulomb barriers for the reactions. Thus, the mechanisms by which these elements form are different from the ones that form the lighter elements. Heavier elements form via neutron capture reactions, designated as the slow (s) and rapid (r) neutron-capture processes.

The hydrogen fusion processes produce energy for the great majority of a star's lifetime. For a star of approximately 20 solar masses, the helium-burning phase would take only about a half-million years. The succeeding phases would be even shorter: a few hundred years for carbon burning, a year or so for neon burning, and a few months for oxygen burning. The formation of an iron core would require only a few days. The final collapse of the core, just before the star would explode as a supernova, would take place in only fractions of a second.

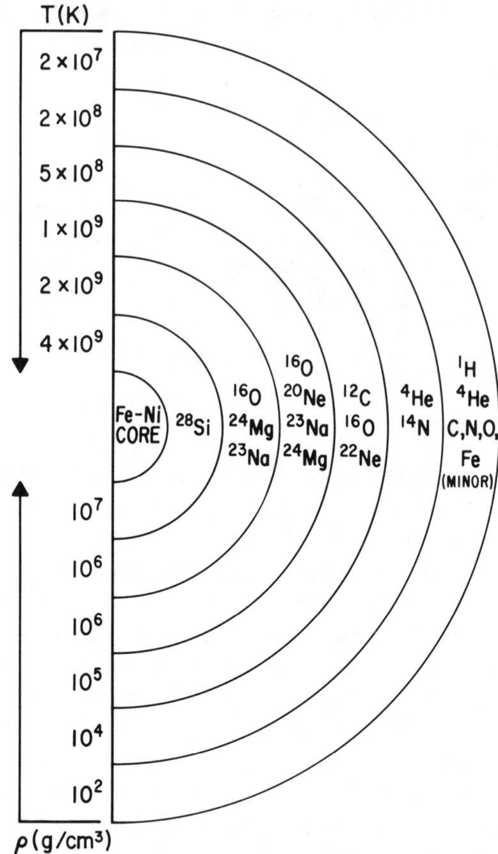

Figure 13.5. The layered structure of a star after nucleosynthesis has taken place through the equilibrium process, but prior to a supernova explosion. (Adapted from C. E. Rolfs and W. S. Rodney, *Cauldrons in the Cosmos.* Chicago: University of Chicago Press, 1988.)

13.4.5. The s-Process

The process in which neutron capture is slow relative to beta decay is called the **slow process,** or **s-process,** of heavy element formation. The nuclei that begin the neutron capture reactions in the s-process are mainly those of ^{56}Fe and its immediate neighbors. The following reaction is believed to be responsible for much of the neutron formation in stars:

$$^{21}\text{Ne} + {}^{4}\text{He} \longrightarrow {}^{24}\text{Mg} + \text{n}$$

Neutron production is favored when the amounts of hydrogen and nitrogen present are relatively low, because they would use up neutrons otherwise available for the s-process to form deuterium and ^{14}C, respectively.

The capture of a neutron by ^{56}Fe forms ^{57}Fe. This product nucleus may in turn capture another neutron to form ^{58}Fe, which captures still another neutron to form ^{59}Fe. A series of neutron captures like this will eventually form a nucleus (^{59}Fe in this example) that is too neutron-rich to be stable, and it will decay by β^- emission. In the stellar interior, the neutron flux is not very large, so capture events proceed at a slow rate (as long as 10^4 years/capture). Therefore, a nucleus that is radioactive is more likely to decay than it is to capture another neutron. Figure 13.6 shows how the s-process can serve to produce some of the elements in the region of $A = 115$ by successive (n, γ) reactions.

In regions of the chart of the nuclides away from the magic numbers it is found that the product of the capture cross section and the atomic

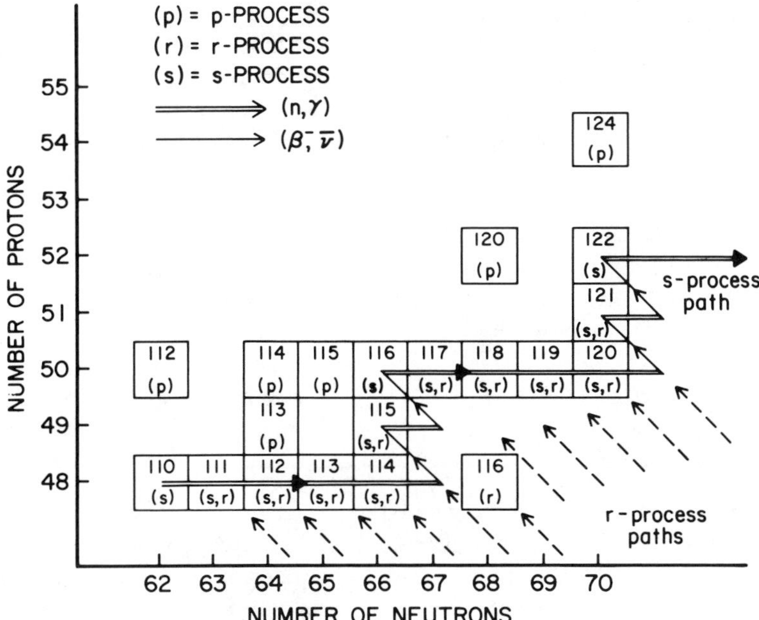

Figure 13.6. Paths of nucleosynthesis for stable nuclei in the region of Cd ($Z = 48$) to Xe ($Z = 54$). The r-process acting on much lower Z nuclides produces negatron emitters far to the right of the diagram. These decay along isobaric paths to form nuclides such as ^{116}Cd and also contribute to abundances of many unshielded nuclides formed by the s-process. (Adapted from C. E. Rolfs and W. S. Rodney, *Cauldrons in the Cosmos.* Chicago: University of Chicago Press, 1988.)

abundance of a nuclide produced by the s-process is approximately equal to the same product for an adjacent isotope of the same nuclide also produced by the s-process. Therefore, where a and b are adjacent s-process isotopes,

$$\sigma_a N_a \approx \sigma_b N_b \tag{13.1}$$

If neutron capture processes alone occur, nuclides with high cross sections would be expected to have lower abundances and nuclides with low cross sections would accumulate to higher abundance levels. This relationship for the s-process has been called the **local approximation.** In a given set of isotopes, those nuclides produced exclusively by the s-process should have approximately the same σN product. For example, the σN product for ^{148}Sm is 2930, and that for ^{150}Sm is 2770. Isotopes of Sm that are produced by other processes or multiple processes would have different (σN) products. Thus, for example, ^{144}Sm, produced by a p-process, has a σN product of 342, that for ^{149}Sm (r- and s-process) is 22 500, and the one for ^{154}Sm (r-process) is 7430. Over a large range of atomic masses, the local approximation fails. For the region between $A = 56$ and $A = 210$ where the s-process is operative, the (σN) product exhibits a smooth monotonic decrease with increasing atomic mass. The smoothness of the variation of the (σN) product is regarded as support for the correctness of the s-process in which each nucleus is produced by a succession of neutron-capture reactions on an initial pool of nuclei in the region of iron.

As noted above, there is a good deal of experimental evidence to attest to the success of the s-process model in accounting for many of the isotopes of the heavier elements up to bismuth ($Z = 83$). At Bi the process terminates in a cycle:

$$^{209}\text{Bi}(\text{n},\gamma)^{210}\text{Bi}(\beta^-, t_{1/2} = 5.01\ \text{d})^{210}\text{Po}(\alpha, t_{1/2} = 138\ \text{d})^{206}\text{Pb}$$
$$\times\ (\text{n},\gamma)\,(\text{n},\gamma)\,(\text{n},\gamma)^{209}\text{Pb}(\beta^-, t_{1/2} = 3.25\ \text{h})^{209}\text{Bi}$$

The s-process does not explain the presence of the small number of stable proton-rich nuclei, such as ^{124}Xe, nor can it explain the formation of elements heavier than bismuth. Thus, two other processes have been postulated to explain these elemental abundances, the r- and p-processes.

13.4.6. The r- and p-Processes

In the **rapid process,** or **r-process,** neutron capture reactions occur so rapidly (0.1–1 s per capture) that beta decay does not always have time to occur. As many as 20 to 50 neutrons may be captured before a negatron decay occurs. Therefore, the r-process results in the buildup of nuclei that are very neutron-rich. The neutron flux required for this process to occur is very large, much larger than that which exists normally in a star. It is thought that the extreme conditions of a supernova are needed to produce such massive neutron fluxes (see Section 13.4.7 for a discussion of supernovae). The r-process produces nuclei very far away from the valley of beta stability, which then decay by negatron emission to the nearest stable isobar in the valley of beta stability.

Fewer details of the r-process than of the s-process are known. The r-process is illustrated at the far right of Fig. 13.6, where the neutron-rich nuclei produced from lower-Z elements decay along isobaric negatron decay paths. In contrast to the s-process, it can be seen that the r-process does not "walk along" the line of beta stability, but rather results in a large jump in Z for every r-process capture sequence. "Bridging over" the s-process termination cycle at Bi by this mechanism results in the synthesis of the heaviest elements (e.g., Th, U, and the transuranium elements). Also, stable nuclides that are isolated from the s-process path by a short-lived radioisotope (e.g., the Cd isotope in Fig. 13.6, with $Z = 48$ and $A = 116$) can be produced by the r-process. Many stable medium mass nuclides that are not shielded by another stable nuclide that lies along the same isobaric decay path can be produced by both the s- and r-processes.

A magic number neutron configuration will be reached at a lower Z for r-process neutron capture than in the s-process. Therefore, in Fig. 3.4, two abundance peaks are seen for each neutron magic number in the s- and r-process region, the lower Z peak resulting from the r-process and the higher Z peak resulting from the s-process. Recall that these abundance peaks are related to the (σN) product relationships discussed previously. Magic neutron number nuclei have relatively low neutron-capture cross sections, hence abundances would be expected to be correspondingly high when these stable nuclear configurations are reached in the s- and r-processes.

In the mass region between $A = 70$ and $A = 200$ there are a number of proton-rich stable or very long half-life nuclei occurring in nature that cannot be formed by neutron capture reactions or the equilibrium process (e.g., ^{78}Kr, ^{112}Sn, ^{120}Te, ^{124}Xe, ^{144}Sm, ^{184}Os). These nuclides have very low abundances, often about 1% of their more neutron-rich neighbors. It is thought that these nuclides must be formed by photonuclear reactions in which neutrons are ejected following absorption of γ rays. Additionally, injection of small amounts of hydrogen during stages of the neutron capture processes may provide protons that can undergo capture reactions. The reactions leading to the formation of these low abundance proton-rich nuclides are called the **p-process.**

13.4.7. Supernovae

A **supernova** results from the explosion of a massive star (>10 solar masses). These are called **type II** supernovae. It is thought that supernovae may be the sites of a great deal of nucleosynthesis. They are certainly the sources of most of the heavier elements in the universe, because the stars that do not explode retain a large portion of their synthesized elements in their dying cores instead of releasing them out to the interstellar medium.

In a type II supernova, the massive star undergoes the series of burning reactions discussed in the preceding sections. Near the end of its life, the star is layered as a result of the different sets of reactions that have occurred, as shown in Figure 13.5. Eventually, the iron core becomes massive enough that it cannot support itself, and it will begin to collapse. The heat from the collapse results in photodisintegration reactions and the capture of electrons by protons to form neutrons. The collapse stops only when the core consists mainly of neutrons that have been compressed to nuclear density (around 10^{14} g/cm^3). When this density is attained, the core resists further collapse and there is a "rebound" of the core. The resulting shock wave is transmitted throughout the outer layers of the star, leading to the ejection of these outer layers in a supernova explosion. An enormous amount of energy is released in a supernova event—more than our sun would produce in its entire lifetime. In these extreme conditions, much nucleosynthesis occurs, and the products are cast out into space as material for the next generation of stars.

Astronomers do not often have the opportunity to observe a supernova. The last supernovae that occurred in our galaxy were in 1572 and 1604, and were observed by Tycho Brahe and Johannes Kepler. In February of 1987, however, astronomers were provided with a rare opportunity to observe a supernova and test some of their theories about them when the star designated Sanduleak −69°202 became a supernova. This star, containing about 20 solar masses, began life several million years ago in the Large Magellanic Cloud, a group of stars not far from our galaxy.

Theory predicted that a swarm of neutrinos should be emitted from the supernova a few seconds after core collapse. In fact, two different neutrino detectors simultaneously recorded bursts of neutrinos. The Irvine–Michigan–Brookhaven (IMB) detector in Ohio recorded 8 neutrinos in 5.5 s, and the Kamiokande-II detector in Japan recorded 11 neutrino interactions in 12.5 s. The detection of these neutrinos provided astronomers with experimental verification of the theories about neutrino production in a supernova, and let them calculate the energy carried away by the neutrinos (about 3×10^{53} ergs). The data may also enable physicists to obtain additional information about the neutrino itself, including its mass and lifetime.

Observations of many kinds continue to be made on supernova 1987A. The data obtained verified parts of the standard model of type II supernova, but also showed that some facets needed revision.

13.5. THE SOLAR NEUTRINO PROBLEM

As mentioned in Chapter 2, there is a great deal of interest in neutrino research in the fields of cosmology and astrophysics. Neutrinos may provide clues about the ultimate fate of the universe and the validity of current theories of stellar nuclear processes.

The mass of the neutrino is still not defined. The upper limit for the mass as determined by experiment is approximately equivalent to an energy of 18 eV (compare this to 5.11×10^5 eV for the energy equivalent of an electron mass). Some theoretical estimates of the mass of the electron neutrino are as low as 2×10^{-7} eV. The question of the neutrino mass has important implications for the Grand Unified Theories (GUTs) in physics and cosmology. For cosmology, neutrino mass may hold the key to the answer of the ultimate fate of the universe. If neutrinos have sufficient mass, they could provide the additional mass needed to eventually allow gravitational forces to halt the outward expansion of the universe and cause it to collapse back in on itself. In view of the recent limits placed on neutrino mass, such a collapse would require the presence of additional, as yet undetected, mass in the universe.

A greater knowledge of neutrinos can also provide insight into the actual processes taking place during stellar evolution. The nuclear fusion processes that have been described earlier are thought to be well understood, and consitute part of what is called a "standard model" of stellar nuclear processes. However, theoretical predictions about the neutrino flux from the sun based on the standard model have not been experimentally verified. The neutrino flux measured on earth is only one-third that predicted by theory. This discrepancy has been known for over 20 years and has come to be known as the **solar neutrino problem.**

Table 13.4 summarizes the equations of the proton–proton chains and the CN cycle. The table also gives the approximate abundances of each series of reactions, the energy of the neutrinos emitted, and the name of the experiment(s) presently attempting to detect neutrinos from that particular process.

The original solar neutrino detection experiment was conducted by Ray Davis in 1968. Some 100 000 gallons of liquid C_2Cl_4 were placed underground in a tank deep in the Homestake gold mine in South Dakota. The placement of the tank underground was to minimize background from cosmic rays. The reaction used to detect the neutrino was one of the so-called "inverse-beta" processes:

$$^{37}Cl + \nu \longrightarrow {}^{37}Ar^* + e^-$$

where the asterisk indicates that the argon is radioactive. After exposure for 35–100 days, the ^{37}Ar was purged from the tank and counted with an internal sample proportional counter. In the purging process, ^{36}Ar was added as a carrier and chemical yields were determined by mass spectrometry. The average experimental results over a period of many years show a neutrino capture rate of 2.07 $\pm$ 0.25 (1σ) SNU. SNU means a *solar neutrino unit,* and is equal to 10^{-36} interactions per second per atom. The value predicted from theory is 7.9 SNU. Thus, experimental results are only one-third of that predicted by the standard solar model.

The neutrinos to which this experiment is most sensitive are not the most abundant neutrinos from the solar reactions. Therefore, other detection systems are now being investigated. These new systems are either more sensitive to the neutrinos from the ^{8}B and ^{7}Be reactions, or can detect the more abundant neutrinos from the first two reactions in the PPI series.

One approach to neutrino detection uses the following reaction on gallium:

$$^{71}Ga + \nu \longrightarrow {}^{71}Ge + e^-$$

This reaction can detect the low-energy neutrinos from the first PPI reaction. There are two major experiments planned using this method: one by a European group (GALLEX) and one by a Soviet group. Large masses of gallium (30–60 tons) will be exposed to neutrinos, and the radioactive ^{71}Ge produced will be radiochemically separated and then counted in a proportional counter.

A geochemical experiment, using the reaction

$$^{98}Mo + \nu \longrightarrow {}^{98}Tc^* + e^-$$

is being conducted currently at Los Alamos National Laboratory. This experiment detects the neutrinos from the reactions in the PPII and PPIII chains, and can also evaluate the constancy of the neutrino flux over the past several million years.

The Kamiokande II experiment will detect high-energy (>5-MeV) neutrinos via Čerenkov radiation emitted by electrons scattered by incoming neutrinos. This experiment can provide unique information about the direction of incoming neutrinos and the time of the events.

Other detection schemes planned for the future include use of liquid argon, heavy water, a Si detector acting as a bolometer, a superfluid helium system, an ^{115}In scintillation detector, and an experiment analagous to the original ^{37}Cl detector, but using ^{81}Br.

A number of theories have been proposed to account for the differences between the theoretical predictions and the experimental results. These generally involve modifying current solar theories or theories about the neutrino's fate after it leaves the sun. Some of these suggestions include postulating the presence of WIMPs (weakly interacting massive particles) in the stellar interior that carry away heat and reduce neutrino flux, or the possible oscillation of the neutrino among three different forms (the electron, tauon, and muon neutrinos) while the neutrinos are in transit to earth. A recent review article by Bahcall et al. (1988) provides a more detailed discussion of these ideas.

13.6. SYNTHESIS OF Be, B, AND Li

The theories of primordial and stellar nucleosynthesis outlined in previous sections account reasonably well for the observed abundances of elements and isotopes in the observable universe, with three exceptions: the elements Be, B, and Li. No mechanism is able to successfully account for these elements, in either the big bang event or in stellar interiors. Because they cannot withstand high energies and would be broken down to α particles in stellar interiors, it was postulated long ago by Burbidge et al. (1957) that these elements must have been synthesized in a medium of lower density and energy, such as stellar atmospheres or the interstellar medium.

Several theories have been advanced to explain the observed abundances of these three elements. **Spallation reactions** (very high-energy nuclear reactions in which the target nucleus ends up as a product nucleus 10–20 mass units below its original mass) may account for some of the production of these elements. These spallation reactions can be initiated by particles accelerated to high energies in the large magnetic fields of some stars. It is also possible that the exposure of a shell of hydrogen to a very intense neutron flux during a supernova explosion of a star could build these light elements from hydrogen, or that they could be formed by high-energy neutron or proton spallation of C, N, and O in such an event. The most likely explanation to date is that the three elements are formed by interactions of cosmic rays with the nuclei found in the gas and dust clouds of the interstellar medium. This explanation ties in well with the observed high abundances of Li, B, and Be in cosmic rays.

TERMS TO KNOW

Big bang theory	Helium burning	Population I star
C–N–O bi-cycle	Inflation era	PP processes
Chondrites	Iron meteorites	Quark confinement
CI chondrites	Main sequence star	r-process
Cosmic abundances	Meteorites	Red giants
Cosmic rays	Meteors	s-process
Cosmochemistry	Nucleosynthesis	Spallation reactions
Cosmology	Nucleosynthesis era	Solar neutrino problem
Dark matter	Oxygen burning	Stony-iron meteorites
Electroweak era	p-process	Stony meteorites
Equilibrium process	Photonuclear disintegration	Supernovae
GUT theory	Planck era	White dwarfs
H-R diagram		x-process

READING LIST

Aller, L. H., *The Abundance of the Elements.* New York:Interscience, 1961. [fundamentals of nucleosynthesis and sources of cosmic abundance data]

Arnett, W. D., and J. W. Truran, eds., *Nucleosynthesis: Challenges and New Developments.* Chicago:University of Chicago Press, 1985. [nucleosynthesis]

Bahcall, J. H., R. Davis, Jr., and L. Wolfenstein, *Nature* **334**(6182):487 (1988). [the solar neutrino problem]

Bennett, G., *Astronomy.* **16**(8):18 (1988). [origin of the elements]

Burbidge, E. M., G. R. Burbidge, W. A. Fowler, and F. Hoyle, *Rev. Mod. Phys.* **21**:547 (1957). [processes of nucleosynthesis]

Contopoulos, G., and D. Kotsakis, *Cosmology: The Structure and Evolution of the Universe.* Berlin:Springer, 1987. [introduction to cosmology]

Clayton, D. D., *Principles of Stellar Evolution and Nucleosynthesis.* Chicago: University of Chicago Press, 1983. [nucleosynthesis]

Dominguex-Tenreiro, R., and M. Quiros, *An Introduction to Cosmology and Particle Physics.* Singapore:World Scientific, 1988. [introduction to cosmology]

Fowler, W. A., et al. *Bull. Am. Phys. Soc.* **1**:191(1956). [^{12}C excited state and the three-alpha reaction]

Hawking, S. W., *A Brief History of Time.* Toronto:Bantam, 1988. [early history of the universe]

Kerridge, J. F., and M. S. Matthews, eds., *Meteorites and the Early Solar System.* Tucson: University of Arizona Press, 1988. [a comprehensive review of research on the evolution of the solar system and a description of meteorite properties]

Macklin, R. L., J. H. Gibbons and T. Inada, *Nature* **197**:369 (1963). [nuclear data for s-process nuclides]

Ringwood, A. E., *Origin of the Earth and Moon.* New York:Springer, 1979. [theories of the origin of the earth and moon]

Rolfs, C. E., and W. S. Rodney, *Cauldrons in the Cosmos.* Chicago:University of Chicago Press, 1988. [big bang and stellar nucleosynthesis]

Suess, H. E., *Naturforsch. Teil A* **2a**:311, 604 (1947). [cosmochemical fractionation and abundance rules]

Suess, H. E., *Chemistry of the Solar System.* New York:Wiley, 1987. [cosmochemistry, nuclear systematics]

Suess, H. E., and H. C. Urey, *Rev. Mod. Phys.* **28**:53 (1956). [major early compilation of cosmic abundances]

Talcott, R., *Astronomy* **16**(2):6, (1988). [stellar evolution and death]

Viola, V. E., and G. J. Mathews, *Sci. Am.* **256**(3):39 (1987). [production of Li, Be and B]

EXERCISES

1. Based on the information presented in this chapter, predict which process(es) would be mainly responsible for the production of the following naturally occurring nuclides: ^{6}Li, ^{13}C, ^{16}O, ^{55}Mn, ^{84}Sr, ^{96}Mo, ^{94}Zr, ^{96}Zr, ^{204}Pb, ^{235}U.

2. Charged-particle reactions do not contribute significantly to nucleosynthesis above iron. Calculate the Coulomb barrier for the reaction of an α particle with a ^{56}Fe nucleus.

3. Suggest a series of neutron-induced nuclear reactions on the products of big-bang nucleosynthesis that might lead to the formation of ^{9}Be.

CHAPTER

14

PARTICLE GENERATORS

Many of the applications of radioactivity that have been discussed in previous chapters require a source of particles, either for activation or for production of specific radionuclides. In this chapter, the basic operating principles of the particle generators most used for these purposes are discussed. Many of these devices are better known for their application in nuclear physics studies, but these uses will be mentioned here only in passing.

From the standpoint of number of applications, the most important particle generator is the nuclear fission reactor, which is a source of neutrons for both analytical applications and for radionuclide production. Less often used are the accelerators that can serve as sources of energetic charged particles for charged-particle activation analysis, ion beam analysis, and production of radionuclides. Accelerators are also used to produce energetic neutrons for fast neutron activation analysis. These devices include Cockroft–Walton, Van de Graaff, and linear accelerators, cyclotrons, and synchrotrons. Some of the largest accelerators used for high-energy physics research are mentioned in the last section of the chapter. There are also natural sources of particles, which were historically very important, but are infrequently used now.

14.1. NATURAL PARTICLE SOURCES

The first man-made particle accelerators were not produced until the 1930s, so for the first 30 years after the discovery of radioactivity, natural sources of particles were the only ones available. Two of these are the naturally radioactive elements, especially α-particle emitters, and the cosmic radiation that constantly bombards the earth.

Alpha emitters were the most important sources for the early investigators, because neither β- nor γ-emitting nuclides induce nuclear changes in a target with high probability. The first experimentally induced nuclear transformation was brought about by Rutherford when he bombarded nitrogen with the α particles from radium (Chapter 1), and the Joliot–Curies formed the first artificial radionuclide by the reaction of Po α particles with

^{10}B (Chapter 1). The natural sources, essentially the α emitters, have limited flexibility with respect to energy, half-life, and beam intensity. Other than the isotopic neutron sources discussed in Chapter 9, naturally occurring radionuclides are not often used for activation or isotope production.

Cosmic rays (Chapter 13) consist mainly of very energetic protons with small amounts of α and heavier particles. As they enter earth's atmosphere, they can induce a variety of nuclear reactions and also generate showers of other kinds of particles (e.g., neutrons and mesons) that in turn induce further reactions. Cosmic rays are not directly used in analytical studies, but they are responsible for the production of radionuclides in the earth's atmosphere and in meteorites that are used in isotopic dating methods (Chapter 12). Cosmic-ray-induced nuclear reactions have been used for many years by particle physicists to study the mechanisms and products of very high-energy nuclear processes. These studies were originally conducted on mountain tops or by sending balloons carrying detecting devices (e.g., stacks of photographic plates) into the upper atmosphere. Mesons observed in early experiments like this provided evidence for H. Yukawa's hypothesis that the strong nuclear force resulted from the exchange of particles (mesons) among the nucleons. More recently these studies have profited by the use of space vehicles, very large shower detector telescopes mounted on the ground, and even large underground detectors used to detect neutrinos, which interact very weakly with any matter.

14.2. NUCLEAR REACTORS

Nuclear reactors are devices in which controlled nuclear fission occurs. Reactors are used for generation of electrical energy as well as for the production of neutrons for activation analysis, radionuclide formation, neutron radiography, and a multitude of neutron scattering experiments. The size and design of the reactor varies according to its function. There is a voluminous literature on the types, containment structures, operating characteristics, and economic and social implications of the nuclear reactors used for power production. In this section, only a basic overview of reactor components, operating characteristics, designs, and applications will be given.

14.2.1. The Fission Process

The fission that takes place in a reactor is largely not spontaneous, but is induced by neutrons. The process of neutron-induced fission is discussed in more detail in Section 4.5.1. A neutron is captured by a high-mass nucleus, which subsequently splits into two smaller nuclei with the concomitant release of about 200 MeV of energy and two or three additional neutrons. The probability of the occurrence of neutron-induced fission varies with the identity of the target nucleus and the energy of the incoming neutron. The potential of nuclear fission as a source of large quantities of energy was recognized soon after the discovery of fission in 1939.

The first controlled release of energy from nuclear fission occurred in December 1942. Enrico Fermi directed the construction and operation of a pile of graphite bricks (770 000 pounds), uranium (12 400 pounds), and uranium oxide fuel (80 590 pounds). This first "nuclear pile" was built under the stands of Stagg Field in Chicago and became the first operating nuclear reactor. An artist's representation of the scene as the control rods were withdrawn and the fission chain reaction became self-sustaining is shown in Fig. 14.1. No photographs were taken. Forty-two people were present. The distribution of the U fuel in the reactor was roughly in the shape of an oblate spheroid. The reactor was unshielded and uncooled, because Fermi had estimated that the total power output would be controlled at approximately 0.5 W. In fact it did operate at 0.5 W for 4.5 min that day. BF_3-filled neutron counter detectors were used to monitor the neutron output as the cadmium control rods were withdrawn. One control rod was attached to a rope that would be cut by personnel on top of the reactor if the rods driven by motors or the rod held by a solenoid failed. Several young scientists were stationed on top of the "pile," and were ready to pour solutions of cadmium sulfate into the reactor to terminate the chain reaction, if all attempts to insert control rods failed. The success of this first controlled self-sustaining chain reaction was announced to personnel in Washington and then to President Truman by the message, ". . . the Italian navigator has just landed in the new world." Although one fission event does release a large amount of energy (about 200 MeV), the real key to the usefulness of fission in energy production lies in the potential for a spontaneous continuation of the fission process in neighboring atoms. The neutrons that are released in a fission event play a key role here. If at least one neutron from every fission event causes another fission event, a **self-sustaining chain reaction** occurs. For this to take place, a minimum amount of material, referred to as a **critical mass,** must be present for the emitted neutrons to react with. If the chain reaction is not controlled and the density of fissile nuclei is very high, the number of atoms undergoing fission would increase exponentially in a short time with the release of tremendous amounts of energy. This is what occurs in an "atomic" bomb. In a nuclear bomb, ignition of the entire system must occur in a time less than it takes for a shock wave to propagate throughout the system. If the chain reaction is controlled and the density of fissile nuclei is low, then the energy and the neutrons that are released can be safely tapped for constructive purposes, such

Figure 14.1. Artist's rendition of the first self-sustaining controlled nuclear chain reaction, at the CP-1 reactor, University of Chicago, December 1942. (Courtesy of Community Affairs Office, Argonne National Laboratory.)

as the production of electricity or the induction of nuclear reactions for research or analysis. Commercial nuclear power reactors use uranium fuel enriched in ^{235}U to 1.7–3.2%, and, while they may undergo excursions of prompt criticality (as in the accident at Chernobyl in the USSR where a power excursion led to a fire and a release of radioactive materials), they cannot undergo a nuclear explosion as would occur with a bomb because their density of fissile nuclei is too low. This is also true for research reactors where fuel enrichment may be relatively high, but there is still insufficient fissile nuclei density for a nuclear explosion.

As noted above, the minimum condition necessary for a chain reaction to occur is that at least one neutron from each fission event induces one more fission event in a lattice of infinite size. This condition is specified by the **reproduction, or multiplication factor, *k*:**

$$k = \frac{\text{Number of neutrons in the } n + 1 \text{ generation}}{\text{Number of neutrons in the } n \text{ generation}} \tag{14.1}$$

If k is less than 1, it means that fewer neutrons are produced in the next generation of atoms. The chain reaction will not be able to sustain itself and will cease of its own accord. This state is referred to as **subcritical**. If k is greater than 1, there are larger numbers of neutrons released in the next generation and the series diverges to approach an infinite output of neutrons. This is referred to as a **supercritical** condition. If k equals 1, a steady-state chain reaction is sustained and the fissioning facility is said to be **critical**.

In practice, the fuel matrix is not infinite and k values will depend on the quantity and type of fuel, its physical arrangement, its purity, and the moderator used. More specifically, the value of k is affected by four important parameters, expressed by the **four factor formula:**

$$k_{\infty} = \epsilon p f \eta \tag{14.2}$$

where k = the multiplication factor

ϵ = fast fission factor

p = resonance escape probability

f = thermal utilization factor

η = eta, the number of fission neutrons produced for each neutron absorbed by the fuel

Descriptions of each of these factors are:

1. In a **thermal reactor,** the majority of fissions are induced by thermal neutrons, but it is also possible for a few of the fast neutrons to cause fission in ^{238}U. The **fast fission factor, ϵ,** is a measure of this. The presence of fast-neutron-induced fission results in more neutrons than would be expected from strictly thermal neutron induced fission. A typical value for ϵ in a reactor operating with natural uranium fuel would be around 1.03.
2. The concept of resonance capture of neutrons is discussed in Section 4.3.2. Although most fission is induced by thermal neutrons, the excitation function for neutron-induced fission also shows resonance peaks, so there is some probability that a neutron will be captured by a ^{238}U nucleus before it slows down to thermal energies. The probability that the neutron will escape capture in this region and actually be slowed to thermal energies is the **resonance escape probability, p.** A typical value for natural U could be 0.95.
3. The thermal neutrons that are present in a reactor may be absorbed not only by the fuel, but by all the surrounding components. Efforts are made to use materials that have very low neutron-capture cross sections, but there will still be some loss of neutrons to these other components. The **thermal utilization factor, f,** refers to the fraction of thermal neutrons that are actually absorbed by the fuel. For natural U, f could have a value of about 0.95.
4. The last factor in the formula is η, usually simply referred to as **eta**. This gives the average number of fission neutrons produced for every neutron that is actually absorbed by the fuel. A typical value would be 1.3.

The multiplication factor (k) for a reactor of infinite size is given the symbol k_∞. In an infinitely large reactor, no neutrons would be able to escape, and this assumption is used at first to simplify calculations. In a real reactor, of course, neutrons are lost from the system. For this case, the multiplication factor is referred to as k_{eff}, or an "effective" k. There are equations that can be used to relate these two multiplication factors. During normal operation, k_{eff} may be $\approx$1.001 and is controlled to within $\pm$0.001 quite easily. About 1% of the neutrons are those resulting from delayed neutron emission of the fission products. Because delayed neutron emission half-lives can be on the order of minutes, the response time of the reactor system is not faster than minutes and ample time is available to control against beginnings of unwanted power excursions.

A quantity that is defined in terms of k is the **excess reactivity,** δk:

$$\delta k = k_{\text{eff}} - 1 \tag{14.3}$$

For a reactor in operation, it is desirable that k be equal to one, but it is necessary that a reactor be designed in such a way that k may be made to exceed this value. There are several reasons for this. Excess reactivity would be needed during the start-up of the reactor to overcome capture events by residual fission products (reactor poisons), during changes in power levels, and as the fuel in the reactor is used up.

Reactor poisons are nuclides produced during the fission process that have a very high cross section for thermal neutrons. One of the most important poisons is ^{135}Xe, which has a σ_{th} of 2.6×10^6 b. As a reactor operates, ^{135}Xe forms as a fission product, and decays with a half-life of 9.10 h. The equilibrium concentration of ^{135}Xe is reached after about 30 h of reactor operation. The cross section of ^{135}Xe is so large that it absorbs a significant number of thermal neutrons, and reactivity must be increased during operation to overcome this loss of neutrons. The ^{135}Xe continues to be a problem even after reactor shutdown, because it is continuously being formed by the decay of ^{135}I, another fission product with a half-life of 6.57 h, but is not being removed by capture reactions that form ^{136}Xe. The amount of ^{135}Xe reaches a peak about 10 h after shutdown. At this time, an excess reactivity of 10–15% may be required to start the reactor again. If the reactor is not designed with this much excess reactivity, it would be impossible to start it up until a significant proportion of the ^{135}I and ^{135}Xe decay. Other fission products such as ^{149}Sm may also act as reactor poisons, although ^{135}Xe is the most important.

A concept that is important for safety in reactor design is the dependence of reactivity on temperature, referred to as the **temperature coefficient of reactivity.** Reactors can have either overall positive or negative temperature coefficients of reactivity. If the quantity is positive, it means that reactivity increases as temperature increases. It is easy to see that this kind of relationship could lead to a "runaway" situation, in which a temperature increase causes a reactivity increase, which would in turn cause still further increases in temperature. Eventually the reactor could get out of control, and the temperature could rise high enough to melt the elements of the core. This is what occurred in the Chernobyl nuclear accident in 1987. All power reactors using mixed uranium fuels have a negative *fuel* temperature coefficient, because as the temperature of the fuel rises, thermal vibration of fuel atoms increases and the range of neutron energies corresponding to the now Doppler-broadened resonance capture peaks in ^{238}U also increase. Due to this Doppler effect on the coefficient of reactivity, more neutrons are lost through resonance capture by ^{238}U at the higher temperatures, rather than being available to cause further fission of ^{235}U. However, there is also a *moderator* temperature coefficient effect that is not so simple. If the moderator temperature is increased, neutrons in the moderator will have higher average kinetic energies and will have a smaller probability of inducing another fission event. This is another negative temperature coefficient factor. However, if a significant amount of Pu has been produced in the reactor late in its fuel cycle, these higher-energy neutrons may be absorbed in a high cross-section resonance peak in Pu at 0.3 eV, and a positive temperature coefficient will result from enhanced Pu fission. A positive temperature reactivity effect can also result due to the coolant's neutron capture cross section decrease with increasing neutron thermal energy. At very high temperatures some moderator/coolants (as in water-moderated reactors) will actually vaporize, lowering the moderator density and thereby decreasing its moderating ability, leading to reduced reactivity. There are also negative reactivity effects connected with temperature-related changes in the structural members of the reactor.

For reactors that use water as a moderator, there is a reactivity void coefficient which is related to the temperature coefficient. If the temperature of the water moderator gets too high, cavitation voids appear (the water starts to boil), the moderator is correspondingly less effective, and the reactivity of the system will decrease. In all water-moderated power reactors this safety factor makes unwanted power excursions almost impossible.

Obviously, the most desirable situation for any reactor is one in which a temperature increase results in an overall reactivity decrease due to a combination of the factors discussed above. In this case, the temperature reactivity coefficient for the facility is always negative, and a rise in temperature would always lower the reactivity and enable the reactor to shut itself down instead of going out of control.

14.2.2. Major Components of Reactors

Although there are many types and designs of reactors, nearly all have similar components. These include the nuclear fuel itself, a covering for the fuel, a moderator to slow down the fission neutrons, a coolant to carry away the thermal energy emitted in the fission process, and control material to control the rate of fission. Each of these will be discussed in turn.

Nuclear Fuel

Nuclear reactors require fuel that is **fissile**, that is, fuel that can undergo neutron-induced fission. There are five fissile nuclides currently used in reactors: ^{232}Th, ^{233}U, ^{235}U, ^{238}U, and ^{239}Pu. Of these, ^{233}U, ^{235}U, and ^{239}Pu have

Table 14.1. Characteristics of Fissile Nuclides

Nuclide	σ_{th} (b)[a]	σ_f	Half-life or % Abundance
^{232}Th	7.37, 85	<2.5 μb	100%
			1.40×10^{10} y
^{233}U	46, 140	531 b	1.59×10^5 y
^{235}U	99, 140	585 b	0.72%
			7.04×10^8 y
^{238}U	2.68, 277	5 μb	99.28%
			4.47×10^9 y
^{239}Pu	271, 200	750 b	2.41×10^4 y

[a]The first cross section is to an isomeric state and the second is to the ground state.

high cross sections for fission induced by thermal neutrons (see Table 14.1). The other two nuclides, ^{238}U and ^{232}Th, do not experience significant fission with thermal neutrons, but can undergo fission with fast neutrons (1 MeV or greater). This distinction between fuel that fissions with either fast or thermal neutrons is one of the ways that reactors can be classified (see Table 14.2).

Besides its ability to fission, the fuel used in a nuclear reactor must meet other requirements. It should be mechanically strong, chemically stable, and resistant to radiation damage, so that it does not undergo significant physical or chemical changes in the extreme operating environments of a reactor. The thermal conductivity of the material should be high, so that heat may be easily removed. It is also desirable for the fuel to be easily obtainable, easily fabricated, relatively inexpensive, and not chemically hazardous.

The isotopes of uranium have found the greatest use as a nuclear fuel. Uranium ore is mined at a number of locations throughout the world. Natural U consists of 99.2745% ^{238}U and 0.720% ^{235}U. Another useful uranium isotope, ^{233}U, is radioactive, and is produced in reactors by irradiation of ^{232}Th, as discussed below. Depending on the design of the reactor, uranium fuel may consist only of the naturally occurring isotopes, or it may be **enriched** in the ^{235}U isotope. The enrichments may vary from a few percent up to the 90% range. The uranium may be used as a pure metal, as uranium dioxide (UO_2), or as uranium carbide (UC). The UO_2 has the best overall characteristics, and is the most commonly used form.

The ^{239}Pu that is used as a nuclear fuel is not naturally occurring, but comes from the chemical processing of irradiated U fuel rods. Plutonium is formed in these reactions:

$$^{238}\text{U} + \text{fast n} \longrightarrow {}^{239}\text{U} \xrightarrow[23.5\,\text{m}]{\beta^-} {}^{239}\text{Np} \xrightarrow[2.355\,\text{d}]{\beta^-} {}^{239}\text{Pu}$$

^{239}Pu has a half-life of 2.41×10^4 years. Plutonium metal is not used directly as a fuel because its mechanical and chemical properties are poor and it is toxic. Plutonium oxide (PuO_2) is a more convenient form for nuclear fuel. Sometimes, the PuO_2 is mixed with UO_2 to form a **mixed oxide fuel (MOX).**

^{232}Th is the only naturally occurring isotope of thorium. Like U, it is best used as a fuel in either the oxide (ThO_2) or carbide (ThC_2) form. Thorium is used primarily to produce the fissile ^{233}U isotope, as shown in these reactions:

$$^{232}\text{Th} + \text{fast n} \longrightarrow {}^{233}\text{Th} \xrightarrow[22.3\,\text{min}]{\beta^-} {}^{233}\text{Pa} \xrightarrow[27.0\,\text{d}]{\beta^-} {}^{233}\text{U}$$

Notice that both ^{232}Th and ^{238}U can undergo nuclear reactions that produce nuclides that are fissile with thermal neutrons. These two nuclides are said to be **fertile**, and the production of fissile material from them is referred to as **breeding**. This leads to the possibility that certain types of nuclear reactors could make as much or more fuel than they consume. These **breeder reactors** are discussed further in Section 14.2.3.

In summary, most types of nuclear reactors are based on the use of ^{235}U as the fissile material. The ^{235}U may be enriched or the naturally occurring amounts may be used. The chemical form of the fuel is usually uranium dioxide. Reactors using ^{235}U as fuel are relying on thermal neutron-induced fission to operate. The fuel may be fabricated into a variety of shapes and sizes.

Fuel Cladding

None of the varieties of fuel discussed in the preceding section are placed directly into a reactor. The fuel is always covered, and this covering is referred to as the **fuel cladding.** Figure 14.2 shows an arrangement of fuel and cladding, referred to as a **fuel element.** The cladding separates the fuel from the surrounding moderator and/or coolant. This protects the fuel from corrosion and prevents the spread of fission products into the surrounding material from the irradiated fuel. The cladding may also serve to provide structural support for the fuel and to aid in heat transfer.

The cladding material should, like the fuel, have good thermal and mechanical properties and be chemically stable to interactions with the fuel and surrounding materials. It should also have as low a cross section as possible for neutron-induced nuclear interactions and be radiation resistant. The most commonly used claddings are alloys of Zr with varying amounts of Ni, Cr, Fe, and Sn, referred to as **Zircaloys**. Stainless steel is used for cladding in some types of reactors, but the trace elements in steel may sometimes have high cross sections for neutron capture. Some alloys

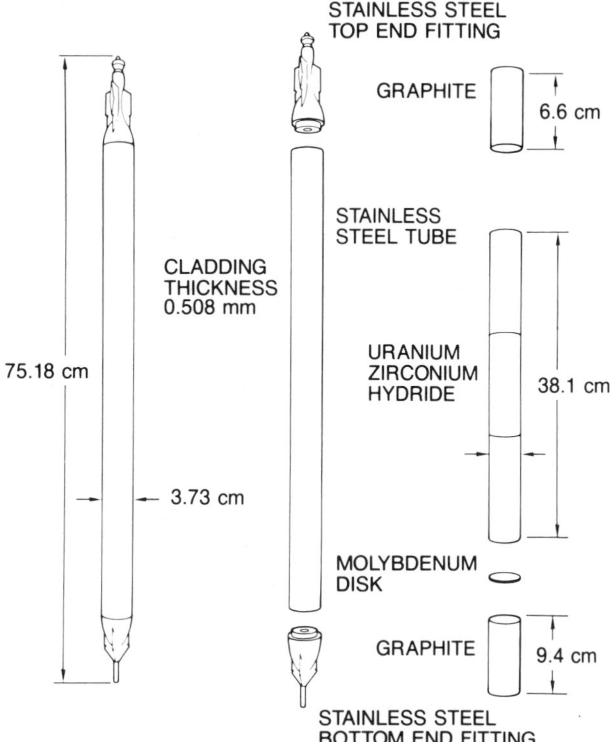

Figure 14.2. Fuel elements for the TRIGA® research reactor. (Courtesy of General Atomics, San Diego, CA.)

of Mg, Al, and Be (**Magnox**) can be used in reactors where temperatures are not too high.

Moderators

A **moderator** is a material that serves to slow down, or **thermalize**, fast neutrons. Most reactors operate with thermal neutron-induced fission, but the neutrons that are emitted in a fission event are fast neutrons. Therefore, these neutrons must be slowed to thermal energies before they are useful for inducing further fission events.

In Section 6.5, the process of neutron moderation was discussed, and it was shown that the best nuclei for slowing down neutrons are those whose atomic mass is close to that of the neutron itself. For reactor use, the moderator should also have a low neutron-capture cross section. These requirements limit the choice of moderator materials to a few, including hydrogen, deuterium, beryllium, and carbon. Because it is highly toxic, Be has so far found limited use as a reactor moderator. This leaves the two hydrogen isotopes, in the form of water and heavy water, and carbon, in the form of graphite, as moderating materials.

Ordinary water is the most common reactor-moderating material. It is usually referred to as **light water,** because it contains mostly the ^{1}H isotope. It has the obvious advantages of being plentiful, cheap, and easily purified. However, it has a fairly high neutron-capture cross section (0.333 b), so the fuel used for light-water-moderated reactors must contain enriched U for enhanced reactivity. In reactors used as power sources for electricity generation, water has some thermodynamic disadvantages that result in the need for high-pressure systems to prevent the water from boiling.

The primary advantage of **heavy water** (D_2O) as a moderator is its lower ^{2}H neutron-capture cross section (0.52 mb). Reactors moderated with heavy water may use natural U as a fuel. Heavy water does occur naturally, but it must be separated from ordinary water. This results in greatly increased costs for heavy water over light water.

Graphite (carbon) was the moderator used in the first nuclear reactors. Because of its higher mass, the moderating characteristics of graphite are not as good as those of water. However, it is easy to obtain in high purity and is fairly cheap. Very few U.S. reactors use graphite as a moderator now. The reactor at Chernobyl was graphite-moderated, and it was the burning of the graphite in the core that contributed to the spread of the fission products released in the accident.

Coolants

The heat released during fission must be removed from the fuel or it would ultimately become hot enough to melt the fuel assemblies. The heat that is removed from the fuel may be used to generate electricity, if the reactor is a power reactor, or it may simply be dissipated, as occurs in research reactors. The coolant material should have a high thermal conductivity, so that it is effective in removing heat, and should be chemically stable. It is also desirable that it have a low neutron-capture cross section, so that neutrons useful for fission are not wastefully absorbed by the coolant, and also so that the coolant itself does not become highly radioactive due to activities produced by (n, γ) reactions.

Both gases and liquids have been used as coolants. The two gases that are most commonly used are CO_2 and He. The properties of He are more ideal, but it is more expensive and difficult to obtain in large quantities. The liquid coolants include water, heavy water, and liquid metals. As noted earlier, water is a satisfactory but not ideal coolant, because of the high pressures necessary to prevent its boiling. The alkali metals are possible choices for liquid metal coolants, but only liquid sodium has been used to any significant extent. Liquid metals have better thermodynamic properties than water, but are not as easy to obtain nor are they as chemically stable. Sodium has a high cross section for thermal neutron capture via the ^{23}Na(n, γ)^{24}Na reaction, so it can become highly radioactive.

Control Materials

To achieve controlled fission, or to stop a fissioning system once it has started, there must be materials available that can absorb the excess neutrons to keep the multiplication factor near or below 1. These neutron-absorbing materials are called **control materials.** In contrast to all the other components mentioned above, where low neutron-capture cross section was desirable, control materials should have very high cross sections for neutron capture.

There are many nuclides that have sufficiently high cross sections to act as effective neutron absorbers, but only a few have mechanical and chemical characteristics that make them useful in this regard. Boron, with a capture cross section of 764 b, is one of the best choices. It can be fabricated into control rods as boron carbide, or its salts may be dissolved into solutions that can be used to flood a reactor. Cadmium and indium also have high capture cross sections. They are useful only as components of alloys, because of their low melting points.

Control of fission processes may also be achieved with the use of **burnable poisons.** These materials are mixed with the coolant at the initial start-up. The poison absorbs excess neutrons and as it does so it is used up, or **burned**. Thus, as the reactor fuel ages and produces fewer neutrons, there is also less control material in the coolant. The control material may be removed from the coolant by ion-exchange columns. An example of a burnable poison would be boric acid, a boron-containing compound that can be dissolved in water.

Another important factor in the control of reactors is the presence of delayed neutrons. (The concept of delayed neutrons was first discussed in Section 2.5.2.) Most ($\approx$99%) of the neutrons emitted in U fission are **prompt neutrons,** that is, they are released at the time of the fission event. **Delayed neutrons,** on the other hand, are emitted following the negatron decay of a fission product. This means that there is some time lag between the emission of the prompt and delayed neutrons. The significance of this time delay can be illustrated by considering its effect on the **period**, ***T***, of the reactor. The period gives a value for the rate of increase of the number of neutrons, and thus of the power increase, in a reactor. If values of T

are calculated based on the assumption that all the emitted neutrons are prompt neutrons, it is found that the reactor power increases so rapidly that control at start-up would be very difficult. For reactors based only on fast neutron fission, control would be impossible. However, if the calculations are modified to take into account the presence of the delayed neutrons, the period is longer and control of the reactor becomes a much easier problem to deal with.

The reactor components discussed above are all contained in the **core** of the reactor. The core, in turn, is placed into a thick-walled **reactor vessel.** Around the reactor vessel is a **containment building,** which should be strong enough to withstand either internal or external pressures or impacts.

14.2.3. Types of Reactors

Reactors can be classified in several ways, as shown in Table 14.2. Some of these groupings have been alluded to briefly in previous sections, including the distinctions according to neutron energy, fuel type, moderator, and coolant. The grouping based on phases of the fuel and moderator simply refers to whether the fuel and moderator are the same phase (**homogeneous**) or different phases (**heterogeneous**). Most reactors are heterogeneous, with the fuel in a solid phase and the moderator in a liquid phase, or with fuel and moderator in two distinct solid phases. An example of a homogenous reactor would be a solution of a uranium salt in water. Homogeneous reactors are uncommon.

The distinction among reactor types based on fuel consumption deserves further discussion. Recall from Section 14.2.2 that there are fertile nuclides present in reactor fuel, that is, nuclides that can undergo nuclear reactions to form fissile nuclides. In most cases, the amount of fissile material formed is very small compared to the amount used, so the reactor burns up more fuel than it creates. It is possible, however, to design reactors that create more fuel than they use, and these are called breeder reactors. The **breeding ratio,** ***B***, of a reactor is defined as the number of new fissile atoms produced per atom of fissile material present. Breeder reactors may be based on either thermal or fast-neutron induced fission.

Fast breeder reactors are more common than thermal breeders. The fast breeder requires the products from a conventional thermal reactor to begin its operation. The thermal reactor is fueled with uranium, either natural or ^{235}U-enriched. After the reactor has operated for some time, the composition of the fuel has changed. The ^{235}U is depleted somewhat, the ^{238}U is still present at virtually the same levels as at the start, and some fissile ^{239}Pu has been formed. The fuel rods can be chemically processed to separate the ^{238}U and the ^{239}Pu, which are then used as fuel for the fast breeder reactor.

Table 14.2. Classification of Reactors

According to type of fission process
Thermal
Fast
Intermediate
According to fuel consumption
Burner
Converter
Breeder
According to fuel type
Natural Uranium
^{235}U enriched
^{232}Th or $^{239}Pu/^{238}U$ mixtures
According to moderator
Heavy water
Light water
Graphite
Others (e.g., Be)
According to coolant
Gas (CO_2, He, air)
Liquid (water, metal)
According to phases of fuel/moderator
Homogeneous
Heterogeneous
According to application
Power
Nuclide production
Research

The ^{239}Pu is surrounded by the ^{238}U, and neutrons produced in the fission of ^{239}Pu will be captured by the ^{238}U to form more ^{239}Pu. Once begun, the reactor can continue to operate with the addition of fresh ^{238}U fuel, which is converted to fissile ^{239}Pu. Most of the work on development of the fast breeder reactor has taken place in France.

Thermal breeder reactors operate similarly, except that they are based on the production of ^{233}U from ^{232}Th. The thorium can be obtained from natural sources (100% abundant in ^{232}Th), but the ^{233}U, like the ^{239}Pu, must be obtained by chemically separating it from fuel used for conventional reactors. The thermal breeder is fueled with a ^{233}U core and surrounded by ^{232}Th. Fission neutrons from the ^{233}U are captured by the ^{232}Th to form more fissile ^{233}U. At the present, there are no operating thermal breeder reactors.

14.2.4. Applications of Reactors

Two major groups of nuclear reactors based on their uses are the **power** reactors and the **research** reactors. Power reactors generate electricity, while research reactors are used for basic nuclear research, applied analytical studies, and production of isotopes.

Power Generation

The first reactor to produce electrical power (EBR-I) was designed at Argonne National Laboratory and operated for the first time on December 20, 1951 in Arco, Idaho (Fig. 14.3). It produced 100 kW of electrical

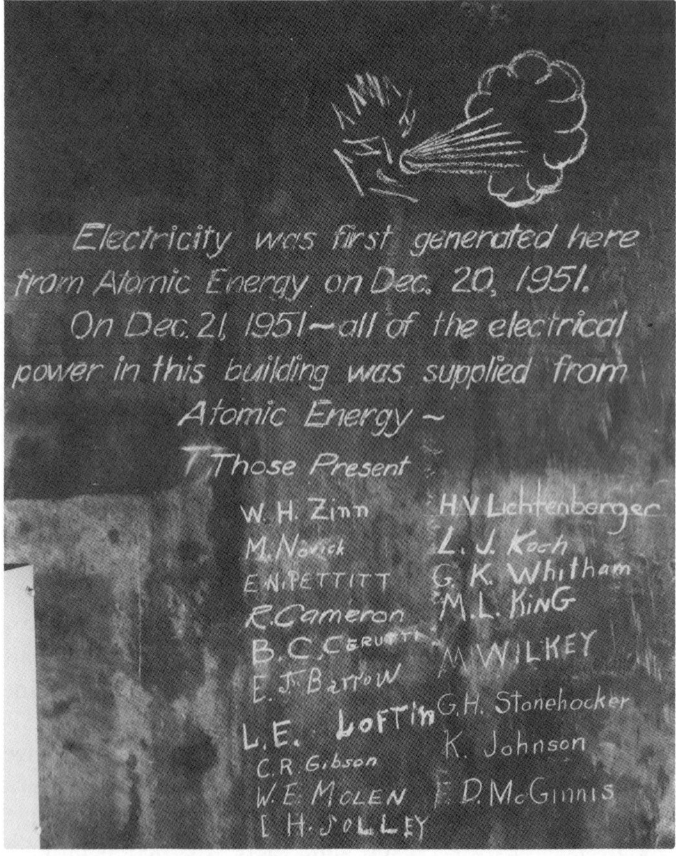

Figure 14.3. Graffiti at the site of the first nuclear reactor to produce electric power, Arco, Idaho, 1951. (Courtesy of Community Affairs Office, Argonne National Laboratory.)

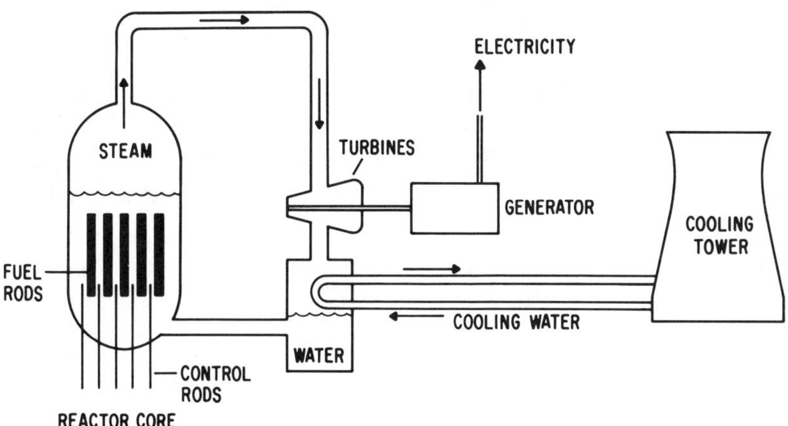

Figure 14.4. Representation of the components of a nuclear power reactor facility.

output on that date. The first nuclear reactor to produce substantial amounts of electric power was operated in the submarine U.S.S. *Nautilus* starting on May 31, 1953. A simple diagram of a reactor system that would be used for power production is shown in Fig. 14.4. Fission occurring in the fuel elements in the reactor core releases heat, which is absorbed by the coolant. The hot coolant produces steam, which drives turbines to produce electricity. Table 14.3 summarizes the location, number, and percentage of total power supplied by nuclear power plants all over the world. Altogether, about 17% of all electricity was generated by nuclear power plants in 1988. There are many designs for power reactors, but two or three account for the majority of the power generated. The graph in Fig. 14.5 shows the relative amounts of power generated by the different designs of reactor.

The largest number of power reactors are **light-water reactors (LWR),** so named because they use ordinary water as the moderator. Water also acts as the coolant in most of these reactors, except for the light-water graphite-moderated variety (**LWGR**). The largest proportion of the LWRs are **pressurized water reactors (PWRs).** The reason for the pressurization of the coolant was noted earlier, in that water has somewhat unfavorable thermodynamic properties as a coolant and must be kept under pressure to reach the higher temperatures that are desirable in power plant operation. The next largest group is the **boiling water reactors (BWRs).** This design allows the coolant water to boil, thus eliminating the need for high-pressure systems. The graphite-moderated light-water reactors are rarely used for power generation.

Heavy water is the second most commonly used moderator. In the **pressurized heavy-water reactor (PHWR),** heavy water is both the moderator

Table 14.3. Power Reactors Worldwide as of 31 December 1988

Country	Number of Units		% Electricity Supplied by Nuclear Power (1988)
	In Operation	Under Construction	
Argentina	2	1	11.2
Belgium	7	—	65.5
Brazil	1	1	0.3
Bulgaria	5	2	35.6
Canada	18	4	16.0
China	—	3	—
Cuba	—	2	—
Czechosolvakia	8	8	26.7
Finland	4	—	36.0
France	55	9	69.9
Germany (GDR)	5	6	9.9
Germany (FRG)	23	2	34.0
Hungary	4	—	48.9
India	6	8	3.0
Iran	—	2	—
Italy	2	—	—
Japan	38	12	23.4
Korea, Republic	8	1	46.9
Mexico	—	2	—
Netherlands	2	—	5.3
Pakistan	1	—	0.6
Poland	—	2	—
Romania	—	5	—
South Africa	2	—	7.3
Spain	10	—	36.1
Sweden	12	—	46.9
Switzerland	5	—	37.4
Taiwan	6	—	41.0
United Kingdom	40	2	19.3
USA	108	7	19.5
USSR	56	26	12.6
Yugoslavia	1	—	5.2
World total	429		

Source: Adapted from B. Semesnov, P. Dastidar, J. Kupitz, and A. Goodjohn, Growth projections and development trends for nuclear power, *IAEA Bull.* **31**(3):6–12 (1989).

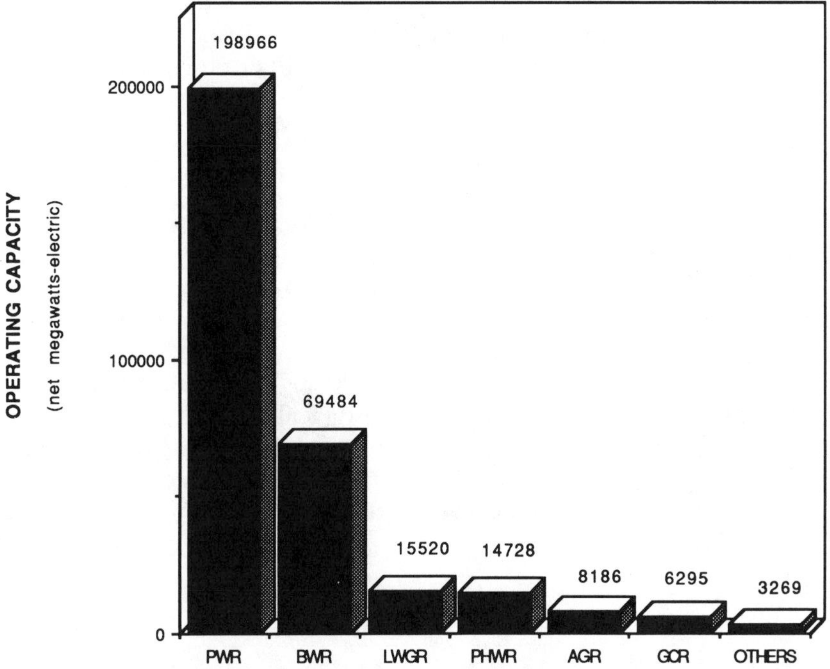

Figure 14.5. Operating capacities of different types of power reactors. (Adapted from *IAEA Bull.* **31:**63 (1989).)

and the coolant. Reactors using gas coolants, including the **advanced gas reactor (AGR)** and the **high-temperature gas-cooled reactor (HTGR),** are used by only a few power-generating stations. Even fewer **liquid-metal fast reactors (LMFR)** operate as power reactors. The few in operation use liquid sodium as a coolant.

The different designs of reactors can all be used effectively. In some cases, the variety of designs can be attributed to the availability of resources. For example, at the end of World War II, Canada was the only country that had the ability to produce large amounts of heavy water. Thus, virtually all of their reactors are of the heavy water variety, such as the CANDU reactor. In the United States, which had developed the capacity for enrichment of ^{235}U, light-water reactors became the norm. There is also a synergy among the various types of reactors. The spent fuel (consisting mostly of ^{238}U) from the LWRs can be used as fuel in the heavy-water reactors, and the ^{239}Pu formed during the operation of the LWRs can be used as fuel for the LMFR. The liquid-metal reactor can also use some of the transuranic (TRU) wastes produced by LWRs. This has the obvious benefit of providing a good use for nuclear waste materials.

The growth of nuclear power as an energy source is difficult to predict. Reactor developments now are concentrating on designing reactors that are inherently more safe, simple, and durable than previous generations of reactors.

Research Reactors

Table 14.4 shows the number of research reactors in countries around the world. These reactors are smaller than power reactors, and are often designed to maximize the neutron flux that is used for analysis, isotope production, and basic research. Most use light water as a moderator and coolant, and are **swimming pool** reactors, which means that the core is suspended in a pool of water. They usually use highly enriched uranium as fuel. Figure 14.6 shows a photograph of a typical research reactor facility. A diagram of the interior configuration of a swimming pool research reactor is given in Fig. 14.7. Figure 14.8 is a photograph looking down through the water at the core of an operating light-water swimming pool research reactor. The bright glow from the core region is due to Čerenkov radiation.

Table 14.4. Research Reactors in Operation Worldwide as of May 1989

Country	Number	Country	Number	Country	Number
Argentina	5	Greece	2	Phillippines	1
Australia	2	Hungary	3	Poland	3
Austria	3	India	5	Portugal	1
Bangladesh	1	Indonesia	3	Romania	2
Belgium	5	Iran	1	South Africa	1
Brazil	4	Iraq	2	Sweden	2
Bulgaria	1	Israel	2	Switzerland	4
Canada	14	Italy	6	Taiwan	5
Chile	1	Jamaica	1	Thailand	1
China	8	Japan	18	Turkey	2
Colombia	1	Korea, Dem. Peoples Rep.	1	United Kingdom	15
Czechoslovakia	3	Korea, Republic	3	USA	99
Denmark	2	Libya	1	USSR	24
Egypt	1	Malaysia	1	Venezuela	1
Finland	1	Mexico	3	Viet Nam	1
France	20	Netherlands	2	Yugoslavia	3
Germany (GDR)	5	Norway	2	Zaire	1
Germany (FRG)	21	Pakistan	1		
		Peru	2		
				World total =	325

Source: Adapted from "International Data File," *IAEA Bull.* **31**(3):62 (1989).

Figure 14.6. TRIGA® Mark II research reactor in Ljublijana. (Courtesy of General Atomics, San Diego, CA.)

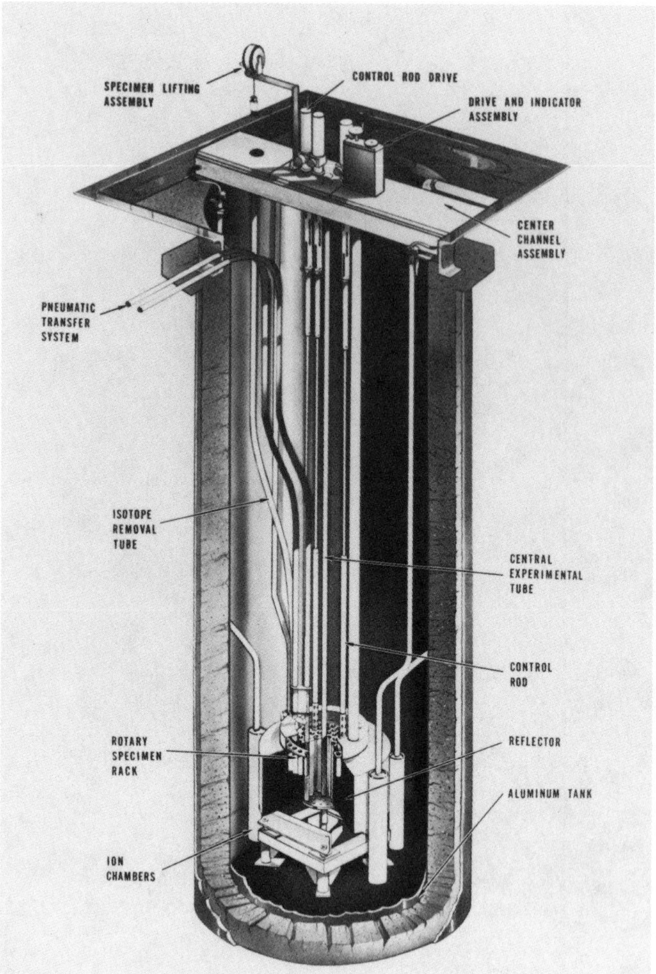

Figure 14.7. Cutaway diagram of the TRIGA® Mark I research reactor showing reactor components. (Courtesy of General Atomics, San Diego, CA.)

14.3. ACCELERATORS: BASIC COMPONENTS

Accelerators are devices whose purpose is to impart high energies to particles by accelerating them through electrical or magnetic fields. They were often referred to as "atom smashers" when first developed after World War II. These machines have enabled physicists to explore the world of subatomic particles, let chemists analyze the composition of substances,

Figure 14.8. Looking down through the water (note the bubbles) at the core of a TRIGA® swimming pool reactor. (Courtesy of General Atomics, San Diego, CA.)

and provided medical scientists with a new pharmacopoeia of substances and a weapon in their war against disease. Accelerators range in size and design from a small Cockroft–Walton-type neutron generator that could be carried in a person's arms to the proposed Superconducting Super Collider (SSC) that will have a circumference of about 54 miles.

Particle accelerators consist of four basic components. The first is a **source** of particles. The source produces electrically charged particles, because most accelerating devices use electrical and magnetic field effects for acceleration. The sources may provide negative ions, electrons, or positive

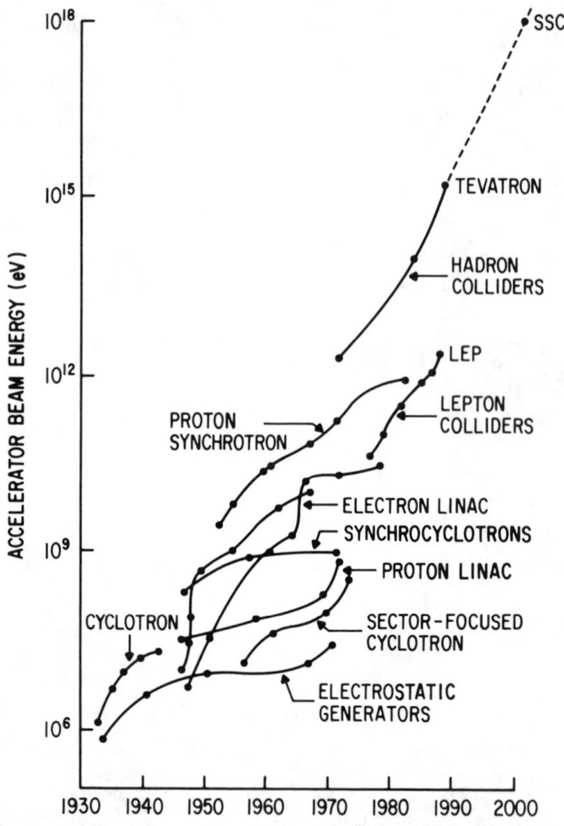

Figure 14.9. Livingston plot—the energy growth of accelerators with time. (Adapted from J. E. Leiss, Particle beams and twentieth century science and technology. In *Frontiers of Particle Beams: Proceedings of a Topical Course,* M. Month and S. Turner, eds. Berlin: Springer, 1988.)

ions. The latter are the most common, especially protons, deuterons, and alpha particles. After the ions are produced, they must be **injected** into the system. Sometimes this is a simple process, where the ions are attracted by simple electrostatics into the accelerating tube. In other cases, the injector is itself an accelerator that feeds a larger one. The **accelerating method** differs from one device to another, but relies on electromagnetic fields to accomplish the acceleration. Finally, the particles must be **extracted** from the accelerating device and allowed to strike the target. Figure 14.9 shows a *Livingston plot* of the types and energies of accelerators since their inception in the early 1930s.

In the sections below, the basic principles of the accelerating method are discussed, along with the analytical applications of the device.

14.4. COCKROFT–WALTON ACCELERATORS

The **Cockroft–Walton (CW) accelerator,** developed in the early 1930s by J. D. Cockroft and E. T. S. Walton, is one of the simplest types of accelerators. It uses a direct voltage applied between two terminals to accelerate charged particles toward a target. Cockroft–Walton accelerators are frequently used as injectors for the larger accelerator systems. Analytically, their greatest use is as a generator for 14-MeV neutrons.

A diagram of a Cockroft–Walton-type accelerator used for neutron generation is shown in Fig. 14.10. Deuterium ions are accelerated toward a tritium-containing target, and the following reaction occurs:

$$^{2}\text{H}(^{3}\text{H}, \text{n})^{4}\text{He}$$

The energy of the neutrons emitted in this reaction is 14.7 MeV. Neutron generators like this one can produce total neutron fluxes of $>10^{11}$ n/s, with an analytically useful flux density of about 10^9 n cm^{-2} s^{-1}. An interesting new adaption of the CW accelerator is the Karlsruhe ring ion source neutron generator (KARIN). Ions produced in a peripheral ring source are accelerated and focused to the center of the ring where they strike a coaxial water-cooled target of ScDT metal hydride. The generator is claimed to produce 5×10^{12} 14-MeV neutrons per second.

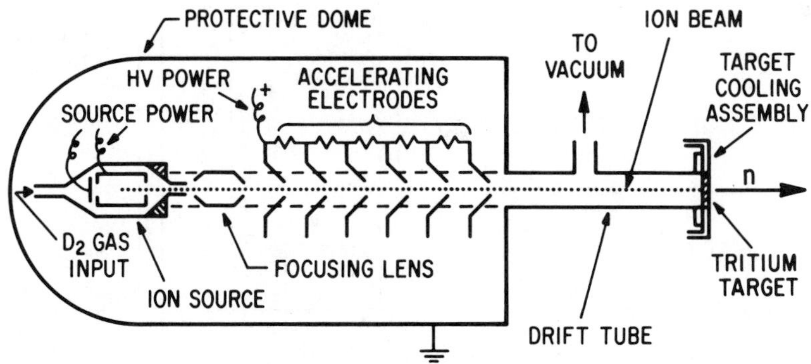

Figure 14.10. Schematic diagram of a Cockcroft–Walton 14-MeV neutron generator. The tritium is usually in the form of a solid metal hydride and the beam accelerated may be either pure deuterium or a mixture of deuterium and tritium.

14.5. VAN DE GRAAFF ACCELERATORS

Van de Graaff accelerators were developed by R. J. Van de Graaff in the early 1930s. A schematic diagram is shown in Fig. 14.11 and a photograph of a single-stage machine is shown in Fig. 14.12. A belt made of a nonconducting material is placed around two pulleys and made to rotate continuously. At one end, a voltage source sprays positive charge onto the belt. The positively charged particles are carried by the belt to the other end of the pulley system, which is inside a hollow metallic dome. Here, a series of fine points (a **comb**) connected to the dome removes the positive charges from the belt. The charges then distribute themselves around the sphere's surface. A significant amount of positive charge may be accumulated in

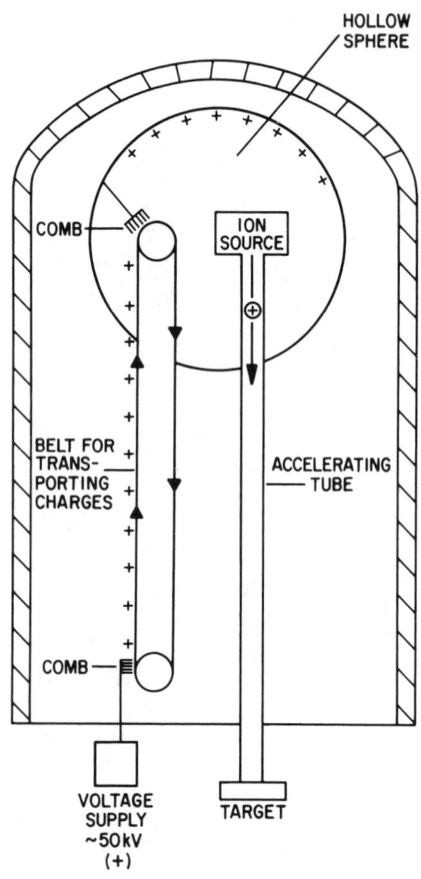

Figure 14.11. Schematic diagram of a Van de Graaff accelerator.

Figure 14.12. A 7-MeV Van de Graaff accelerator facility at the University of Kentucky, Department of Physics. The pressure dome has been removed for the photograph. (Courtesy of B. D. Kern, University of Kentucky, Lexington.)

this way, limited only by the insulating capacity of the surrounding atmosphere. An insulating gas such as SF_6 or N_2 fills the space around the dome so that a large charge can be accumulated.

Inside the positively charged hollow sphere is an **ion source** that can generate positive ions. The ions are repelled by the positive charge on the sphere, and are accelerated away from the source down an accelerating tube to ground potential. The target is at the end of this beam tube.

Many kinds of positive ions and even charged dust particles may be accelerated in a Van de Graaff, but protons are probably most common. A maximum energy of 7–8 MeV can be attained by protons in a single-stage Van de Graaff, and beam currents of approximately 1 mA can be achieved. A great advantage of the Van de Graaff accelerator is that the amount of

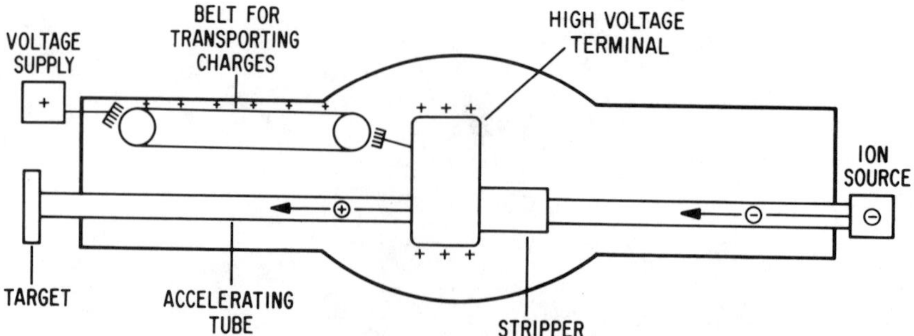

Figure 14.13. Schematic diagram of a tandem Van de Graaff.

charge on the sphere can be very precisely controlled, which means that the energy given to the particle beam can be regulated easily and precisely.

Particle energies greater than 8 MeV can be achieved with a variation of the basic Van de Graaff called a **tandem Van de Graaff.** The operation is similar to that of the single-stage device, except that there is an extra acceleration phase. A diagram of this accelerator is shown in Fig. 14.13. A belt and pulley system still provides positive voltage to a hollow metal sphere. However, the ion source is located outside the sphere some distance away, and produces negative ions instead of positive ions. These negative ions are attracted toward the positively charged dome, receiving the first acceleration. Once in the dome, the negative ions are stripped of some of their electrons to form positive ions, which are then repelled by the positively charged dome to receive the second acceleration.

The actual energy of the ion beam is dependent upon the number of electrons lost in the stripping step, which is not easily controlled. For example, consider the acceleration of oxygen ions in a tandem Van de Graaff with the dome at 8 MeV. If $^{16}O^-$ ions are formed, they would receive an energy of (8 MeV $\times$ 1) = 8 MeV energy in the first stage. If the $^{16}O^-$ ion is stripped to $^{16}O^{5+}$, it would receive (8 MeV $\times$ 5) = 40 MeV in the second accelerating stage. The total amount of energy possessed by the ions at the target would then be 48 MeV.

The tandem Van de Graaff can provide more energetic ions than the single-stage types, but the energies are not so readily controlled because the stripping step is not controlled. Beam currents from the tandem Van de Graaff are generally lower than those from the single-stage variety.

Van de Graaff accelerators are used analytically for charged-particle activation analysis, particle-induced x-ray emission, fast neutron activation analysis, and Rutherford backscattering spectrometry.

14.6. LINEAR ACCELERATORS

The first linear accelerator, or **linac**, was built by Wideröe in the late 1920s. The basic idea behind the linac is to give the particles many small accelerations, instead of one large one. A simple analogy to this process is pushing a child in a swing. By timing the push just right (i.e., at the height of the swing cycle), many small pushes can result in a big arc. This principle of *resonance acceleration* is followed in all of the accelerators discussed in succeeding sections.

In a linac, the particles are accelerated through a series of hollow tubes that are arranged in a straight line, as shown in Fig. 14.14. The ions from the source are attracted to the first tube, which has a charge opposite that of the particles from the ion source. The hollow tubes are called **drift tubes,** and while inside the drift tube the particle experiences no force or acceleration. Just as the particles reach the end of the first drift tube, a device called a **radiofrequency oscillator** alters the sign of the voltage of the drift tubes so that the particles are repelled by the tube they are exiting from and attracted across the gap to the next tube. Once inside the second tube, the particles drift toward the next gap, the voltage polarity is alternated, and they are accelerated again. This process continues until the particles have become highly energetic. The particle energy increases with passage through each tube, so to keep the time inside the tube constant, the tubes have to be made longer and longer. The ultimate energy attainable in a linac is limited by the length of the system.

Because the drift tube voltages must change sign very rapidly (every few microseconds), the linac did not become a practical accelerator until the required technology became available after World War II. The high-frequency generators developed for radar during the war proved useful for linac

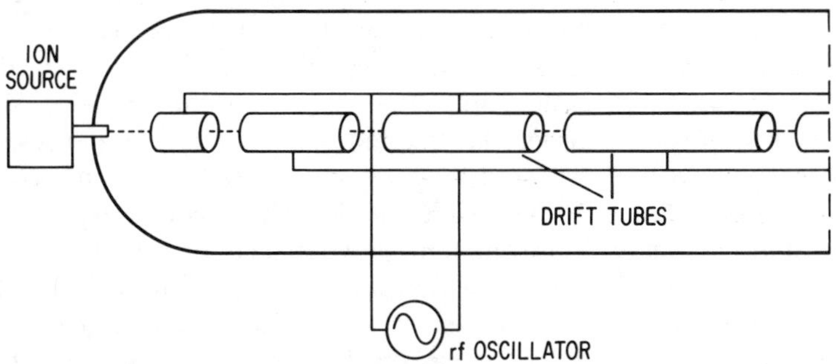

Figure 14.14. Schematic diagram of a linear accelerator.

construction. Linacs are used mainly for electron acceleration, because the low mass of the electron allows it to be accelerated to nearly the speed of light. Because the velocity of the electrons is essentially constant at high kinetic energies, the drift tubes can all be the same length. The largest electron linac is at Stanford University. The facility is referred to as **SLAC**, the Stanford Linear Accelerator Center. This accelerator is 1.8 miles (3 km) long, and can accelerate electrons to energies of 20 GeV. There are linacs that accelerate protons and heavy ions (**hilacs**), but generally heavy ions are more easily accelerated in the circular accelerators discussed in the following sections.

Electron linacs are not used by analytical chemists to any extent. They are used extensively in industrial radiation processing (Section 11.5), in physics research, and for medical radiation therapy (Section 10.5.5).

14.7. CYCLOTRONS

One of the physical drawbacks to the linear accelerator is the sheer physical dimensions of a device that can accelerate particles to high energies. In 1931, E. O. Lawrence first reported on the operation of a new kind of accelerator that used a spiral, semicircular orbit for the particles instead of a linear path. The device is the **cyclotron**, and Lawrence was awarded the Nobel Prize in 1939 for its development. The cyclotron and its later modifications quickly became favorite research tools for the nuclear physicist due to their ability to accelerate charged particles to energies of hundreds of MeV with beam currents of hundreds of microamperes. Figure 14.15 shows Niels Bohr and J. C. Jacobson with an early cyclotron in 1938.

A schematic diagram of a simple cyclotron is shown in Fig. 14.16. The cyclotron has an ion source placed between two **dees**, which are semicircular hollow disks. The gap between the dees is where the ion acceleration takes place. Above and below the dees and the ion source are the poles of a large magnet. It is the action of the magnetic field on the charged particles that causes them to move in a circular path. In the operation of a cyclotron, the ion source injects the charged (usually positive) particles into the space between the dees and they are accelerated to the inside of the oppositely charged dee. While inside the dee, the particles experience no electrical forces, but are constrained to move in a circular path by the magnetic field. At first, the energy of the particles is low and the path has a small radius. Just as the particles exit the first dee, a radiofrequency oscillator changes the signs of the voltage on the dees so that the particles are repelled by the first dee and accelerated across the gap into the second dee.

Figure 14.15. Niels Bohr (*right*) and J. C. Jacobson at an early cyclotron facility, 1938. (Courtesy of K. Heydorn, Risø National Laboratory, Roskilde, Denmark, and the Niels Bohr Archive, Copenhagen.)

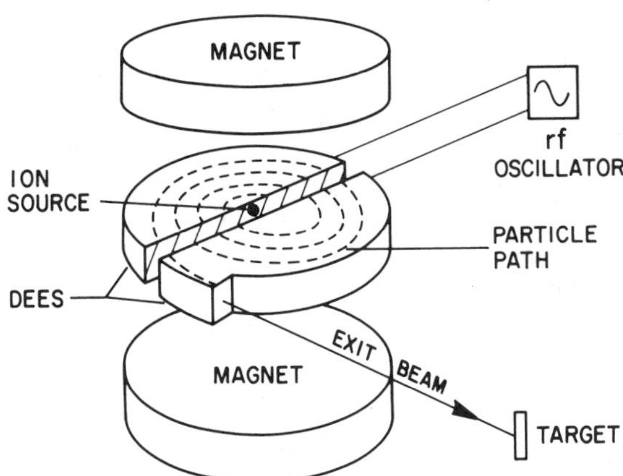

Figure 14.16. Schematic diagram of a simple cyclotron. (Adapted from R. Gouiran, *Particles and Accelerators.* New York: McGraw-Hill, 1967.)

This time the particles have a bit more kinetic energy, so they follow a curved path that has a larger radius, but the time that it takes the particle to come out of the dees will always be the same (their velocity is higher, but their path, which has a greater radius, is longer). The particles are accelerated in this way each time they cross the gap between the dees. Eventually, the radius of the spiral path becomes too large to be contained in the cyclotron in the next revolution, and the particles are electrically deflected out of the cyclotron toward a target.

The frequency of polarity change on the dees is constant in the simple cyclotron and is set for a specific charge/mass ratio (q/m) of the particle. $^{2}H^{+}$ and $^{4}He^{2+}$ have the same q/m ratio, while that of a proton, $^{1}H^{+}$, is different and would require a different frequency for the same particle energy. Most cyclotrons are used at a fixed frequency but accelerate particles to different energies for different q/m ratios. For a proton cyclotron the frequency is typically 23×10^{6} cps. For the motion of a charged particle in a magnetic field, the magnetic force on the particle is balanced by the centrifugal force.

$$\text{Magnetic force} = \frac{BqV}{c} \tag{14.4}$$

$$\text{Centrifugal force} = \frac{mV^2}{r} \tag{14.5}$$

where B = the magnetic field strength (G)

c = the speed of light in a vacuum (3.0×10^{10} cm/s)

q = electrical charge on the particle (esu)

V = velocity of the particle (cm/s)

r = radius of orbit (cm)

Because the two forces must be equal,

$$\frac{BqV}{c} = \frac{mV^2}{r} \tag{14.6}$$

and dividing by 2π

$$\frac{Bq}{2\pi c} = \frac{mV}{2\pi r} \tag{14.7}$$

Since frequency (ν, units of s^{-1}) of passage around a circular path is equal to the velocity of the particle divided by the circumference of the circle,

$$\nu = \frac{V}{2\pi r} \tag{14.8}$$

and

$$\nu = \frac{q}{m}\frac{B}{2\pi c} \tag{14.9}$$

Therefore, for a given magnetic field strength, the q/m ratio will determine the polarity change frequency used.

The first cyclotron built by Lawrence was only about 13 cm in diameter. With this device, he accelerated protons to an energy of 80 keV. Eventually he directed the construction of a 184-inch cyclotron, but larger sizes did not overcome a basic limitation of these instruments. The cyclotron requires that the circulating particles make the trips through the dees in a constant time. However, as the velocity of the particle increases, relativistic effects cause their masses to increase. This results in longer transit times through the dees, so eventually the particles get out of phase with the voltage polarity changes, and acceleration of the particle to higher energies is not effective regardless of the dimensions of the machine or the gap potential. For protons and α particles, the energy limits before relativistic effects are significant are around 25 and 50 MeV, respectively.

Electrons experience changes in mass at quite low energies, so the cyclotron is not a practical way to accelerate electrons to higher energies. In addition to this, electrons moving in circular paths continuously radiate large amounts of energy. This effect can be analytically useful (see Section 14.8), but is wasteful if the goal is to accelerate the particles to the highest possible energies. Heavier particles also radiate energy while traveling in curved paths, but not to such a large extent as electrons. Therefore, cyclotrons are used mainly for positive ion acceleration.

The problem of the increased time of flight for slower-moving particles was independently addressed by E. McMillan in the United States and V. Veksler in the USSR. They proposed changing the frequency of the voltage alterations to the dees to match the time of passage of the energetic ions through them. This variation is called a **synchrocyclotron** or a **frequency-modulated cyclotron,** and the first machine was operated successfully in 1946. In these machines, ions have to be carried through the entire acceleration path before a new "packet" of ions can be accelerated. Therefore the beams consist of a series of pulses with currents of several microamperes and pulse rates of several hundred pulses per second. The synchrocyclotron with its lower beam current is less useful than the simple cyclotron for isotope production. However, it can achieve proton energies of up to about 700 MeV, and is used extensively in high-energy particle physics research.

Another cyclotron variation aimed at overcoming the relativistic energy limitations was developed in the late 1950s. This goes by several names, including the **spiral ridge, isochronous**, **sector-focused,** or **azimuthally varying field (AVF)** cyclotron. The cyclotrons use fixed frequencies and constant travel times, but have specially shaped magnets to create variations in the magnetic fields that can allow the particles to maintain the same time of flight despite their increased relativistic mass. As can be seen in Eq. 14.9, if the mass of the particle increases through relativistic effects, this can be balanced by an increase in B, the magnetic field strength, as the radius of the path increases. These machines are used with a variety of both light and heavy ions and can reach energies greater than 500 MeV with continuous beam currents up to 1 mA of protons.

Simple cyclotrons and synchrocyclotrons are now used as injectors for the larger, new technology, accelerating systems. Cyclotrons are also heavily used for medical purposes, especially for the production of radionuclides used in medical imaging, and for radiation therapy.

14.8. SYNCHROTRONS

Even with the advances of the sector-focused cyclotron and the synchrocyclotron, there remained a limitation on the ultimate energy of the particles, and that was the size and strength of the magnets. There is a practical limit to the size and strength of a single large magnet. The next advance in particle acceleration was a device that could vary the magnetic field strength along with the frequency of the voltage alteration. If both the magnetic field and frequency can be altered, it is possible to constrain the particles to move in the same size orbit, instead of continuously spiraling into orbits of larger and larger radius. The largest modern accelerators are based on these principles, and are called **synchrotrons**.

Figure 14.17 shows a diagram of a synchrotron. Instead of two dees, there is only one closed curved tube that contains the particles. The large magnets above and below the dees have been replaced by a series of C-shaped magnets spaced at intervals along the length of the tube. Particles are injected into the ring by a smaller accelerator, such as a linac or cyclotron, and are held inside the tube by the electromagnets. Acceleration of the particles is accomplished by **accelerating cavities,** which act like the

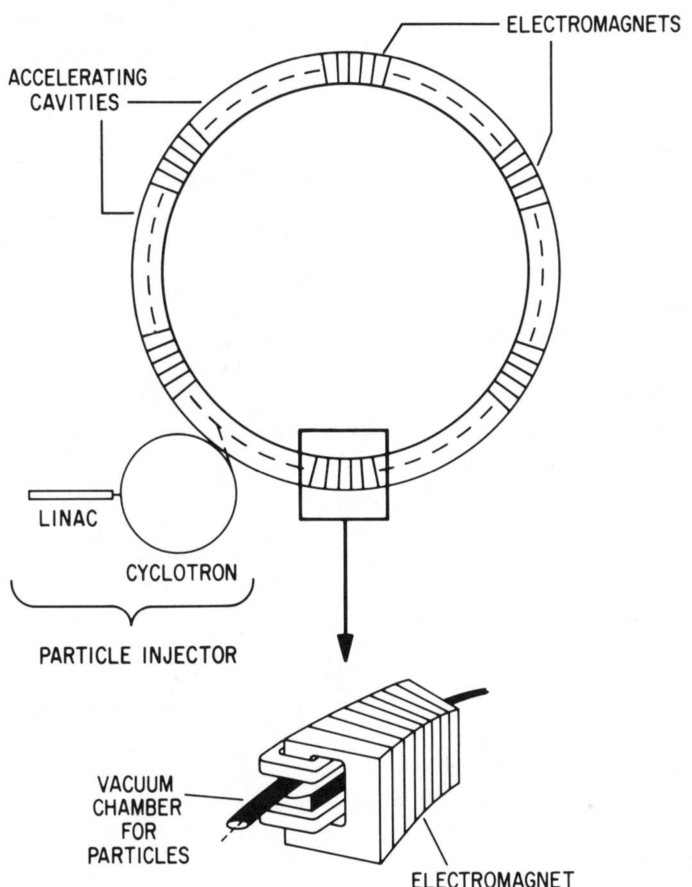

Figure 14.17. Schematic diagram of a synchrotron. (Adapted from R. Gouiran, *Particles and Accelerators.* New York: McGraw-Hill, 1967.)

drift tubes in a linear accelerator. Synchrotrons can be used to accelerate either electrons or positive ions.

The largest synchrotron in the world at this time is the Tevatron at the Fermi National Accelerator Laboratory near Chicago. The Tevatron has a diameter of 2 km, 2000 electromagnets, and can accelerate protons to an energy of 1 TeV (teraelectronvolts, 10^{12} eV). This accelerator can provide a burst of about 10^{13} protons every minute for high-energy physics research. The Tevatron uses several injector stages. First, a Cockroft–Walton acclerator feeds protons to a linac. The linac in turn feeds a small accelerator ring that finally injects protons into the main synchrotron.

The very large and powerful positive ion synchrotrons were not designed for analytical uses, but a by-product of the operation of electron synchrotrons has been used by analysts. This is the **synchrotron radiation** that is emitted as electrons circle the ring. Charged particles moving in a curved path will radiate energy (see bremsstrahlung radiation, Section 6.3.3). The National Synchrotron Light Source (NSLS) at Brookhaven National Laboratory on Long Island, New York, is a facility established to use the ultraviolet and x-ray region photons from an electron synchrotron. Electrons are first accelerated to 70 MeV with a linac and then to 750 MeV in a booster synchrotron. The booster then injects these electrons into either an ultraviolet photon or x-ray region photon storage ring where they can finally be accelerated to 2.5 GeV. The photon beams are pulsed and have an individual duration of less than 10^{-9} s. As many as 90 experiments can be conducted simultaneously. Typical analytical experiments at the facility include x-ray fluorescence analysis, gas phase spectroscopy, crystallography, and x-ray microscopy. Other major synchrotron light sources in the United States are located at Stanford University and the University of Wisconsin.

14.9. LARGE ACCELERATORS FOR NUCLEAR PHYSICS RESEARCH

In earlier sections of this chapter some of the large accelerators used primarily for biological, nuclear, and particle physics research have been mentioned, including the SLAC (for nuclear research) and the Tevatron (for particle physics). In the United States, the next large accelerator on the horizon is the **Superconducting Super Collider (SSC),** scheduled to be constructed in Texas in the 1990s. This advanced accelerator will use superconducting magnets instead of conventional magnets around its 54-mile ring. The SSC is expected to be able to accelerate protons to an energy of 20 trillion eV (10^{12} eV). The resulting proton collisions should help to unravel some of the unsolved problems in basic particle physics, including experimental proof for existence of the top quark, detection of Higgs bosons, and whether quarks, now postulated to be indivisible, really are. The most recently completed large European accelerator is the large electron–positron collider (LEP), built near Geneva, Switzerland, and operated by CERN (the European Center for Particle Physics). An early goal of work with this collider is to gain more information about the $Z°$ particle, which is one of the three that carries the weak nuclear force.

TERMS TO KNOW

Accelerators
AGR reactors
Breeder reactor
Breeding ratio, B
Burnable poisons
BWR reactors
Chain reaction
Control materials
Critical mass
Critical state
CW accelerator
Cyclotrons
Delayed neutron effects
Enriched fuels
Eta value
Excess reactivity
Fast fission factor
Fissile fuels
Four Factor Formula
Fuel assembly
Fuel cladding
Heavy-water reactor
Heterogeneous reactor
Hilacs
Homogeneous reactor
HTGR reactors
k_∞
Light water reactor
Linear accelerator
LMFR reactors
LWR reactors
Moderator
MOX
Multiplication factor
Nuclear reactor
PHWR reactors
Power reactor
PWR reactors
Radiofrequency oscillator
Reactor containment
Reactor coolants
Reactor core
Reactor period, T
Reactor poisons
Reactor vessel
Research reactor
Resonance escape probability
Sector-focused cyclotrons
Subcritical state
Superconducting supercollider (SSC)
Supercritical state
Swimming pool reactors
Synchrotron radiation
Synchrotrons
Temperature coefficient of reactivity
Tevatron
Thermal utilization factor
Van de Graaff accelerator
Zircaloys

READING LIST

Bennet, D. J., and J. R. Thomson, *The Elements of Nuclear Power,* 3d ed. New York: Wiley, 1989. [nuclear reactors and nuclear power]

Close, F., M. Marten, and C. Sutton. *The Particle Explosion.* New York: Oxford University Press, 1987. [accelerators]

Csikai, J., *Handbook of Fast Neutron Generators,* Vols. 1 and 2. Boca Raton, FL: CRC Press, 1987. [construction, properties, and operation of 14-MeV neutron generators]

Gouiran, R., *Particles and Accelerators.* New York: McGraw-Hill, 1967.

IAEA, *Multipurpose Research Reactors, Proceedings of an International Symposium on the Utilization of Multipurpose Research Reactors,* Grenoble, France, 19–23 Oct., 1987, Vienna, IAEA, 1988. [research reactor utilization]

Livingston, M. S., *Particle Accelerators: A Brief History.* Cambridge, MA: Harvard University Press, 1969. [a history of particle accelerators]

Schmidt, K. A., *Neutron Radiography.* EUR Report 11021, 1987, p. 239. [high-yield 14-MeV neutron generators]

EXERCISES

1. Research reactors are designed to maximize the neutron flux density available for experiments. With reference to the literature, discuss how this can be accomplished.

2. Calculate the magnetic field strength, in gauss, in a proton cyclotron that has a revolution frequency of 2.3×10^7 cps.

3. What types of background radiation would be expected at a counting station located near the containment vessel of a research nuclear reactor?

4. What types of shielding would be best to lower the background for the counting station in exercise 3?

5. What types of counting station shielding would be most appropriate for an electron synchrotron facility?

APPENDIX

A

STATISTICS FOR RADIOCHEMISTRY

STATISTICAL MODELS

To obtain useful information from determinations of the activity of a radioactive substance, some knowledge of counting statistics is necessary. This topic is normally covered in detail in the laboratory portion of a radiochemistry course, so only a short introduction is given here. The discussion is confined to the specific case of counting data from radioactive decay.

As noted in Chapter 5, radioactive decay is a statistically random process. The equations of decay from that chapter can be used to describe the behavior of a sample containing a large number of atoms. However, it is not possible to say with absolute certainty whether or not one specific nucleus out of the whole population of nuclei being observed will decay during a particular time interval. Only the *probability* of its decay can be discussed. For an individual nucleus, the probability of decaying during a time interval, t, is given by the expression $(1 - e^{-\lambda t})$.

The random nature of radioactive decay can be observed by measuring the activity of a sample several times in succession. For example, the data below are from 20 successive one-minute counts of a small ^{137}Cs sample:

Count No.	Counts	Count No.	Counts
1	2877	11	2763
2	2779	12	2956
3	2782	13	2842
4	2868	14	2950
5	2843	15	2824
6	2829	16	2812
7	2838	17	2913
8	2816	18	2989
9	2725	19	2835
10	2813	20	2828

Note that no two values are the same, even though the time of the counts, the counting geometry, and all ambient conditions remained constant for all counts. Over a 20-min period, ^{137}Cs, with its 30.17-year half-life, would not decay appreciably, so the differences cannot be accounted for by decay of the sample. It is not always possible to make multiple counts of a sample, as in the above example, so there are times when a sample is counted only once. In either case, a valid way to handle counting data from radioactive decay is needed. The two parameters needed are a "best value," (i.e., a value that closely approximates the true value, μ), and a way to estimate the uncertainty of this value.

For a data set like that shown above, the reader has probably already surmised that the "best value" that can be obtained will be the average, or **arithmetic mean ($\bar{x}$)** of the data points. The more observations that go into determining the mean, the closer $\bar{x}$ will be to μ. To obtain the mean, all the recorded counts are added, and then divided by the total number of individual measurements:

$$\bar{x} = \frac{\sum x}{n} \tag{A.1}$$

where $\sum x$ refers to the sum of all the individual measurements, and n refers to the number of measurements. For the data above, the mean is 2844 counts. The second piece of information desired is the uncertainty of this value. To obtain this uncertainty, some mathematical models and relationships must be considered.

If a very large number of counts of the ^{137}Cs sample were done and a graph was made of the frequency of occurrence of the values against the values themselves, a plot known as a **frequency distribution** is obtained. Figure A.1 shows an example of a frequency distribution plot. Notice that most values tend to cluster around the mean value. Frequency distributions for data that have resulted from statistically random processes can be mathematically described in several ways. For events that have only two possible outcomes, there are three common models. Radioactive decay fits into this class of events because there are only two possible outcomes for a nucleus during the time it is observed: either it decays or it doesn't. The probability that a recorded count rate of a radioactive sample is within a specified limit of the true or mean count rate is expressed accurately by a model using the **bionomial distribution law.** While this law is general, it is also difficult to use when the number of cases to be examined is very large as is the case for radioactive decay. Even a very small radioactive sample contains a very large number of atoms.

A mathematically simplifed model based on the **Poisson distribution** can be used in those cases where the probability of a single event is very

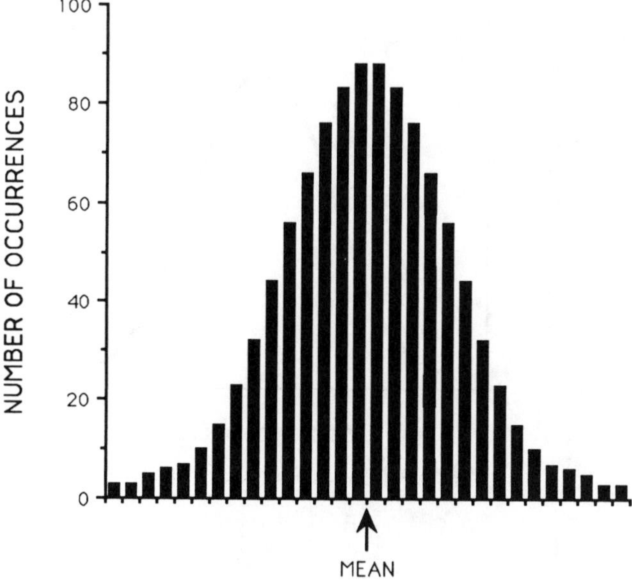

Figure A.1. A frequency distribution plot.

small compared to the total number of cases. This would be true for most observations of decay, because the likelihood of any one atom decaying out of a sample of 10^{18} or more atoms is small. The exception to this would be cases in which a count of a nuclide with a very short half-life was being done. The Poisson distribution is given by the expression:

$$P_n = \frac{\mu^x e^{-\mu}}{x!} \tag{A.2}$$

where μ is the true value which is approximated by the mean value ($\bar{x}$), and x is a measured value of counts. For low values, the Poisson distribution is not symmetrical about $\bar{x}$, in that values are skewed toward the right (high) side of the mean. If the number of events recorded is high (>100), as is usually the case in activity measurements in radiochemistry, a still simpler model, the **Gaussian distribution,** can be used with little error. The Gaussian distribution is given by the following equation:

$$G_n = \frac{1}{\sqrt{2\pi\mu}} e^{-(x-\mu)^2/2\mu} \tag{A.3}$$

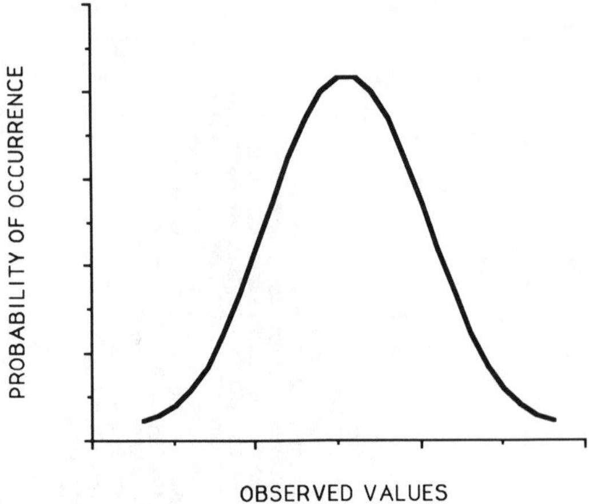

Figure A.2. A Gaussian distribution.

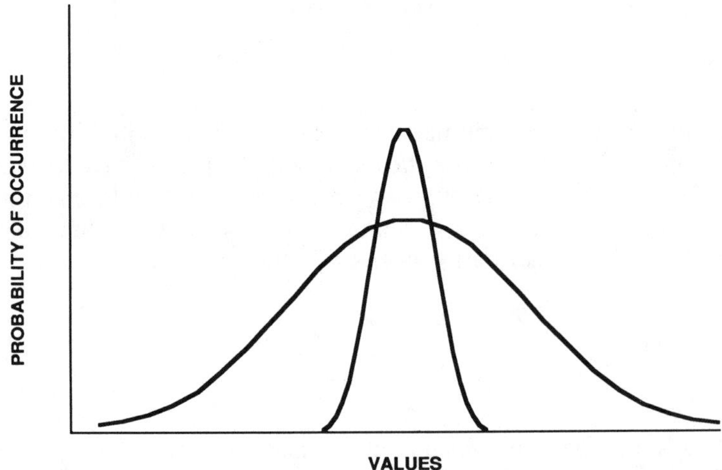

Figure A.3. Gaussian distributions with large and small variances (σ^2).

A plot of a Gaussian distribution is shown in Fig. A.2. This may be recognized by some readers as the familiar "normal curve." It is symmetrical about the mean. This is the model that is most often used to characterize data from radioactive decay.

The frequency distribution curves can be used to illustrate the concept of uncertainty of the mean. Figure A.3 shows two normal distributions, both with the same mean, but with an obvious difference between them. One curve is narrow and the other is broad. There is a greater uncertainty about the mean for the broad curve than for the narrow one. Uncertainty is most often expressed in statistics by a quantity called the **variance** (σ^2), or by the **standard deviation** (σ) which is the square root of the variance. It is an important mathematical property of the binomial and Poisson distributions that the variance, and thus the standard deviation, may be estimated from a knowledge of the mean value. This same estimate holds true for the Gaussian distribution, although it is not mathematically required. The equation used to estimate uncertainty in the case of the Poisson distribution where the number of measurements is large is

$$\sigma = (\bar{x})^{1/2} \tag{A.4}$$

In the above example, the standard deviation of the mean value would be $(2844)^{1/2} \approx 53$ counts. If a hand calculator using standard statistical formulae is used to calculate the standard deviation for the 20 successive counts in the example, a slightly different answer is obtained. This difference is due to the small number of measurements (20), and the possibility that the experimental distribution is not purely Poisson.

It also is true that just one measurement may be used to estimate the mean and standard deviation of the distribution by the equation above. The standard deviation for any single observation in a data set would be the square root of the number of counts for that specific measurement. So, for trial 10, the standard deviation would be $(2813)^{1/2} \approx 53$. It is important to note that this way of calculating the standard deviation is *not* universally applicable to all types of data. It does apply to the counting data from radioactive decay. It would not apply to data that do not originate from statistically random processes.

In the Gaussian distribution, 68.3% of the measurements will fall within $\pm 1\sigma$ of the mean value, which is our best estimate of the true counting rate, and 95.5% will fall within $\pm 2\sigma$.

ERROR PROPAGATION

The counting data obtained in an experiment will nearly always have to be mathematically manipulated to obtain a desired value. This presents no difficulty for the values themselves, but students are often unsure about what to do with the associated uncertainties. The following rules for **propagation of error** are useful for data processing in radiochemistry. *P* and

Q refer to the actual value of counts, and p and q refer to their respective uncertainties.

1. Addition and Subtraction. The error is equal to the square root of the sum of the squares of the errors:

$$(P \pm p) + (Q \pm q) = (P + Q) \pm \sqrt{p^2 + q^2} \tag{A.5}$$

$$(P \pm p) - (Q \pm q) = (P - Q) \pm \sqrt{p^2 + q^2} \tag{A.6}$$

2. Multiplication and Division.

$$(P \pm p)(Q \pm q) = PQ\left(1 \pm \sqrt{\frac{p^2}{P^2} + \frac{q^2}{Q^2}}\right) \tag{A.7}$$

$$\frac{P \pm p}{Q \pm q} = \frac{P}{Q}\left(1 \pm \sqrt{\frac{p^2}{P^2} + \frac{q^2}{Q^2}}\right) \tag{A.8}$$

3. Combining Numbers with Uncertainties with a Constant (k).

$$k(P \pm p) = kP \pm kp \tag{A.9}$$

$$\frac{P \pm p}{k} = \frac{P}{k} \pm \frac{p}{k} \tag{A.10}$$

BACKGROUND CORRECTIONS

1. Background Corrections. As noted in Chapter 7, radioactivity is a natural phenomenon, so there is always some low-level activity detected by the counting system. This is referred to as the **background**. The background counting rate must be subtracted from the gross counting rate level in order to obtain the net counting rate of the sample. The rules above for propagation of error should be followed.

Example. An electrodeposited sample of ^{60}Co for an isotope dilution analysis experiment has a counting rate of 1826 cpm. The background counting rate is found to be 128 cpm. What is the net counting rate of the ^{60}Co sample?

$$1826 \pm (1826)^{1/2} = 1826 \pm 43 \text{ cpm}$$

$$128 \pm (128)^{1/2} = 128 \pm 11 \text{ cpm}$$

Applying Eq. A.6, the net counting rate = 1698 ± 44 cpm

2. Calculation of Counting Rates. A count rate gives the number of counts per unit time. Because time can be measured very precisely compared to the uncertainty in the counts, time is usually considered to be a constant in calculating counting rates. Thus, to determine a counting rate from the total counts and total time, the rules of propagation of error involving a constant should be followed.

> **Example.** The formation constant for a calcium complex is determined using ^{45}Ca. A 2-min count of a sample produces 3137 counts. A background count for the same time produces 219 counts. Calculate the net counting rate for the sample. Using Eq. A.6,
>
> $$(3137 \pm 56) - (219 \pm 15) = 2918 \pm 58$$
>
> and
>
> $$(2918 \pm 58)/2 \text{ m} = 1459 \pm 29 \text{ cpm}$$
>
> The same result would be obtained if the individual counts were converted to counting rates first, and then subtracted.

DETECTION LIMITS IN RADIOCHEMISTRY

Sensitivity calculations are important in many radioanalytical methods. L. A. Currie (*Anal. Chem.* 586 (1968)) has presented an excellent review of statistical problems associated with qualitative and quantitative determinations in radiochemistry. He describes three specific quality levels for a measurement.

1. A **decision limit** (L_C) is the net instrument response above which an *observed* signal from a detector can be reliably recognized.
2. A **detection limit** (L_D) is the true net signal level that may be expected a priori to lead to detection.
3. A **determination limit** (L_Q) is the level at which there is adequate precision for a quantitative determination.

For equivalent observations of a sample and blank (B) pair, the levels for L_C, L_D, and L_Q are $2.33\sigma_B$, $4.65\sigma_B$, and $14.1\sigma_B$, respectively. If the blank, B, is very well known from a large number of observations, the levels are $1.64\ \sigma_B$, $3.29\ \sigma_B$, and $10\ \sigma_B$, respectively. For a more detailed treatment, the reader is referred to the original article by Currie.

APPENDIX

B

GENERAL REFERENCES AND DATA SOURCES

SELECTED MONOGRAPHS ON NUCLEAR AND RADIOCHEMISTRY

Arnikar, H. J., *Essentials of Nuclear Chemistry,* 2d ed. New York: Wiley, 1987.

Arnikar, H. J., *Isotopes in the Atomic Age.* New Delhi: Wiley Eastern, 1989.

Brune, D., B. Forkman, and B. Persson, *Nuclear Analytical Chemistry.* Deerfield Beach, FL: Verlag Chemie International, 1984.

Choppin, G. R., and J. Rydberg, *Nuclear Chemistry.* Oxford: Pergamon, 1980.

Friedlander, G., J.W. Kennedy, E. S. Macias, and J. M. Miller, *Nuclear and Radiochemistry,* 3d ed. New York: Wiley, 1981.

Geary, W., *Radiochemical Methods.* Chichester, England: Wiley, 1986.

Knoll, G. F., *Radiation Detection and Measurement,* 2d ed. New York: Wiley, 1989.

Nesmeyanov, A. N., *A Guide to Practical Radiochemistry,* Vols. 1 and 2. Moscow: Mir, 1984.

Tölgyessy, J. and E. Bujdosó, *CRC Handbook of Radioanalytical Chemistry,* Vols. I and II, Boca Raton, FL: CRC Press, 1991.

Vértes, A., and I. Kiss, *Nuclear Chemistry.* Amsterdam: Elsevier, 1987.

SELECTED DATA SOURCES

Browne, E., and R. B. Firestone, *Table of Radioactive Isotopes,* V. S. Shirley, ed. New York: Wiley, 1986.

Browne, E., J. M. Dairiki, and R. E. Doebler, *Table of Isotopes,* 7th ed., C. M. Lederer and V. S. Shirley, eds. New York: Wiley, 1978.

GE Nuclear Energy, *Nuclides and Isotopes,* 14th ed. (chart of the nuclides prepared by F.W. Walker, J. R. Parrington, and F. Feiner). San Jose, CA: General Electric Company, (Nuclear Energy Operations, 175 Curtner Ave.), 1989.

Tuli, J. K., *Nuclear Wallet Cards.* Upton, NY: National Nuclear Data Center, Brookhaven National Laboratory, 1990.

Note: On-line computer nuclear data bases are also available. Information may be obtained from the National Nuclear Data Center, Brookhaven National Laboratory, Upton, NY 11973 (Phone: 516-282-2902.)

SELECTED SOURCES FOR LABORATORY EXPERIMENTS IN NUCLEAR AND RADIOCHEMISTRY

Chase, G. C., *Experiments in Nuclear Science*. Minneapolis, MN: Burgess, 1971.

Choppin, G. R., *Experimental Nuclear Chemistry.* Englewood Cliffs, NJ: Prentice-Hall, 1961.

Duggan, J. L., *Laboratory Investigations in Nuclear Science.* Oak Ridge, TN: The Nucleus, Inc. and Tennelec, 1988.

Leo, W. R., *Techniques for Nuclear and Particle Physics Experiments.* Berlin: Springer, 1987.

Note: The *Journal of Chemical Education* has also published many radiochemistry experiments.

APPENDIX

C

NUCLIDIC PROPERTIES

The following compilation of nuclear data was provided by Jagdish K. Tuli of Brookhaven National Laboratory. The data appear in a document entitled *Nuclear Wallet Cards* published by the National Nuclear Data Center, Brookhaven National Laboratory, Upton, NY 11973. Research support for the compilation was provided by the Office of Basic Energy Sciences, U.S. Department of Energy. The compilation is dated July 1990. Sources of the data, acknowledgments, and additional tables may be found in *Nuclear Wallet Cards.* The authors appreciate the cooperation of Dr. Tuli in making his compilation available to us for inclusion in this book.

The majority of the nuclear data used in the example problems in this book were derived from *Nuclides and Isotopes,* 14th ed., a chart of the nuclides published in 1989 by the General Electric Co., San Jose, California. Dr. Tuli's compilation was received just as this book was going to press. Values in the text of this book are, therefore, based on somewhat older tables and may vary slightly from those on the following pages.

EXPLANATION OF THE TABLE

Column 1. Nuclides are listed in order of increasing atomic number (Z) and subordered by increasing mass number (A). Isomers with half-lives ≥ 1 s and a few other well-known isomers are listed. Isomeric states are denoted by the symbol "m" and are listed in order of increasing excitation energy. Significant ^{235}U fission products are italicized. Element symbols have not all been internationally accepted.

Column 2. Spin and parity assignments. Those judged not to be based on strong arguments are given in parentheses.

Column 3. Mass excess ($M - A$) values in MeV, relative to $\Delta^{12}C = 0$. For isomers, the values are obtained by adding the excitation energy to the ground-state values. An appended "s" denotes the value is from systematics and an appended "e" denotes the source as Reference 4 in *Nuclear Wallet Cards.*

Column 4. The half-life for radionuclides, or the natural abundance (atom %) for stable nuclides. For some very short half-life radionuclides, level widths rather than half-lives are given. The last number in italics is the uncertainty in the last significant figures given for the value.

Column 5. Decay mode. Standard designations for decay modes are given followed by the percent branching, if any exists and its value is known ("w" indicates a weak branch). The following special designations are used for less common decay modes:

$2\beta^-$ = double-negatron decay (and similarly for other multiple-particle decay modes)

β-n = negatron decay followed by delayed neutron emission (and similarly for other related delayed-particle modes of decay)

ϵp = β^+ or ϵ decay followed by delayed proton emission (and similarly for other related delayed-particle modes of decay)

Isotope Z	El	A	Jπ	Δ (MeV)	T1/2 or Abundance	Decay Mode
0	n	1	1/2+	8.071	10.4 m *2*	β-
1	H	1	1/2+	7.289	99.985% *1*	
		2	1+	13.136	0.015% *1*	
		3	1/2+	14.950	12.33 y *6*	β-
		4?	2-	25.840		
2	He	3	1/2+	14.931	0.000137% *3*	
		4	0+	2.424	99.999863% *3*	
		5	3/2-	11.390	0.60 MeV *2*	α, n
		6	0+	17.592	806.7 ms *15*	β-
		7	(3/2)-	26.110	160 keV *30*	n
		8	0+	31.598	119.0 ms *15*	β-, β-n 16%
		9		40.810	very short	n
3	Li	4	2-	25.120		
		5	3/2-	11.680	≈1.5 MeV	α, p
		6	1+	14.085	7.5% *2*	
		7	3/2-	14.907	92.5% *2*	
		8	2+	20.945	838 ms *6*	β-, β-2α
		9	3/2-	24.954	178.3 ms *4*	β-, β-n 49.5%, β-n2α
		10		33.840	1.2 MeV *3*	n
		11	(1/2-)	40.900	8.7 ms *1*	β-, β-n, β-nα 0.027%
4	Be	6	0+	18.374	92 keV *6*	α, 2p
		7	3/2-	15.768	53.29 d *7*	ε
		8	0+	4.941	6.8 eV *17*	2α
		9	3/2-	11.347	100%	
		10	0+	12.607	1.51×10^{6} y *6*	β-
		11	1/2+	20.174	13.81 s *8*	β-, β-α 3.1%
		12	0+	25.077	24.4 ms *30*	β-, β-n<1%
		13	(1/2,5/2)+	35.000		n
		14	0+	40.100	4.2 ms *7*	β-
5	B	7	(3/2-)	27.870	1.4 MeV *2*	p, α
		8	2+	22.920	770 ms *3*	ε, εα, ε2α
		9	3/2-	12.415	0.54 keV *21*	p, 2α
		10	3+	12.050	19.9% *2*	
		11	3/2-	8.668	80.1% *2*	
		12	1+	13.369	20.20 ms *2*	β-, β-3α 1.58%
		13	3/2-	16.562	17.36 ms *16*	β-, β-n 0.28%
		14	2-	23.664	16.1 ms *12*	β-
		15		28.970	8.8 ms *6*	β-, β-n
		16	(0-)	37.140s		n
		17	(3/2-)	43.310s		β-
		18		52.280s		
		19		59.360s		
6	C	8	0+	35.094	230 keV *50*	α, p
		9	(3/2-)	28.913	126.5 ms *9*	ε, εp, ε2α
		10	0+	15.699	19.255 s *53*	ε
		11	3/2-	10.650	20.385 m *20*	ε
		12	0+		98.90% *3*	
		13	1/2-	3.125	1.10% *3*	
		14	0+	3.020	5730 y *40*	β-
		15	1/2+	9.873	2.449 s *5*	β-
		16	0+	13.694	0.747 s *8*	β-, β-n≥98.8%

Isotope Z	El	A	Jπ	Δ (MeV)	T1/2 or Abundance	Decay Mode
6	C	17		21.035	202 ms *17*	β-
		18	0+	24.920	66 ms *20*	β-, β-n
		19		32.630s		
		20	0+	37.070s		β-
7	N	10		39.700s		
		11	1/2-	24.890	0.74 MeV *10*	p
		12	1+	17.338	11.000 ms *16*	ε, ε3α 3.44%
		13	1/2-	5.345	9.965 m *4*	ε
		14	1+	2.863	99.63% *2*	
		15	1/2-	0.101	0.37% *2*	
		16	2-	5.682	7.13 s *2*	β-, β-α 0.0012%
		17	1/2-	7.871	4.173 s *4*	β-, β-n 95%
		18	1-	13.117	624 ms *12*	β-, β-n, β-α
		19		15.871	290 ms *90*	β-, β-n
		20		21.880s	100 ms *25*	β-, β-n
		21		25.150s	95 ms *13*	β-, β-n 84%
		22		31.990e	24 ms *7*	β-, β-n 35%
8	O	12	0+	32.060	400 keV *250*	p
		13	(3/2-)	23.113	8.90 ms *20*	ε, εp
		14	0+	8.006	70.606 s *18*	ε
		15	1/2-	2.855	122.24 s *16*	ε
		16	0+	-4.737	99.76% *1*	
		17	5/2+	-0.809	0.038% *3*	
		18	0+	-0.782	0.20% *1*	
		19	5/2+	3.332	26.91 s *8*	β-
		20	0+	3.796	13.57 s *10*	β-
		21	(5/2+)	8.066	3.42 s *10*	β-
		22	0+	9.440	2.25 s *15*	β-
		23		14.540s	82 ms *37*	β-, β-n 31%
		24	0+	18.790s	61 ms *26*	β-, β-n 58%
9	F	14	(2-)	33.610s		p
		15	(1/2+)	16.770	1.0 MeV *2*	p
		16	0-	10.680	40 keV *20*	p
		17	5/2+	1.951	64.49 s *16*	ε
		18	1+	0.873	109.77 m *5*	ε
		19	1/2+	-1.487	100%	
		20	2+	-0.017	11.00 s *2*	β-
		21	5/2+	-0.047	4.158 s *20*	β-
		22	(3,4)+	2.789e	4.23 s *4*	β-
		23	(3/2,5/2)+	3.350	2.23 s *14*	β-
		24		7.640e	340 ms *80*	β-
		25		11.330e		
		26		18.460e		
		27		25.600e		
10	Ne	16	0+	23.989	122 keV *37*	2p
		17	1/2-	16.480	109.0 ms *10*	ε, εp
		18	0+	5.319	1.672 s *8*	ε
		19	1/2+	1.751	17.22 s *2*	ε
		20	0+	-7.047	90.48% *3*	
		21	3/2+	-5.737	0.27% *1*	
		22	0+	-8.027	9.25% *3*	
		23	5/2+	-5.156	37.24 s *12*	β-
		24	0+	-5.950	3.38 m *2*	β-

Isotope Z El	A	Jπ	Δ (MeV)	T1/2 or Abundance	Decay Mode
10 Ne	25	(1/2,3/2)+	-2.060	602 ms 8	β-
	26	0+	0.440	230 ms 60	β-
	27		6.960e		
	28	0+	11.000e		
11 Na	18		25.320s		p
	19		12.928	0.03 s ?	p
	20	2+	6.839	447.9 ms 23	ε, εα 21%
	21	3/2+	-2.189	22.48 s 3	ε
	22	3+	-5.185	2.6088 y 14	ε
	23	3/2+	-9.532	100%	
	24	4+	-8.420	14.9590 h 12	β-
	24m	1+	-7.948	20.20 ms 7	IT, β-≈0.05%
	25	5/2+	-9.360	59.1 s 6	β-
	26	3+	-6.904	1.072 s 9	β-
	27	5/2+	-5.650e	301 ms 6	β-, β-n 0.13%
	28	1+	-1.140	30.5 ms 4	β-, β-n 0.58%
	29	3/2	2.650	44.9 ms 12	β-, β-n 22%
	30	2+	8.330e	48 ms 2	β-, β-n 30%, β-2n 1.17%, β-α 0.0055%
	31	(3/2)	12.010e	17.0 ms 4	β-, β-n 37%, β-2n 0.9%
	32		16.550	13.2 ms 4	β-, β-n 39%, β-2n 1.2%
	33		21.470	8.2 ms 4	β-, β-n 52%, β-2n 12%
	34		26.650	5.5 ms 10	β-, β-n, β-2n
	35			1.5 ms 5	β-, β-n
12 Mg	20	0+	17.570	0.1 s	ε, εp
	21	(3/2,5/2)+	10.913	122 ms 3	ε, εp 32%
	22	0+	-0.397	3.857 s 9	ε
	23	3/2+	-5.473	11.317 s 11	ε
	24	0+	-13.933	78.99% 3	
	25	5/2+	-13.192	10.00% 1	
	26	0+	-16.214	11.01% 2	
	27	1/2+	-14.586	9.462 m 11	β-
	28	0+	-15.019	20.91 h 3	β-
	29	3/2+	-10.728e	1.30 s 12	β-
	30	0+	-9.070e	335 ms 17	β-
	31		-3.400e	0.23 s 2	β-
	32	0+	-0.790e	120 ms 20	β-, β-n 2.4%
	33		5.090e	90 ms 20	β-, β-n 17%
	34	0+	8.440s	20 ms 10	β-, β-n
	35		14.680s		
13 Al	22		18.040e	70 ms 45	ε, εp, ε2p
	23		6.767	0.47 s 3	ε, εp
	24	4+	-0.055	2.053 s 4	ε, εα 0.035%
	24m	1+	0.371	131.3 ms 25	IT 82%, ε 18%, εα 0.028%
	25	5/2+	-8.915	7.183 s 12	ε
	26	5+	-12.210	7.4×10^5 y 3	ε
	26m	0+	-11.982	6.3452 s 19	ε
	27	5/2+	-17.197	100%	
	28	3+	-16.851	2.2414 m 12	β-
13 Al	29	5/2+	-18.215	6.56 m 6	β-
	30	3+	-15.890	3.60 s 6	β-
	31	(3/2,5/2)+	-14.967e	0.644 s 25	β-
	32	1+	-11.190e	33 ms 4	β-
	33		-8.610e		
	34		-3.250e	60 ms 18	β-, β-n 27%
	35		-0.320e	0.15 s 5	β-, β-n 40%
	36	0+	5.050s		
	37				
14 Si	22	0+		6 ms 3	ε, εp
	23		23.530s		
	24	0+	10.755	102 ms 35	ε, εp
	25	5/2+	3.827	220 ms 3	ε, εp
	26	0+	-7.145	2.234 s 13	ε
	27	5/2+	-12.385	4.16 s 2	ε
	28	0+	-21.492	92.23% 1	
	29	1/2+	-21.895	4.67% 1	
	30	0+	-24.433	3.10% 1	
	31	3/2+	-22.950	157.3 m 3	β-
	32	0+	-24.081	172 y 4	β-
	33		-20.556e	6.18 s 18	β-
	34	0+	-19.992e	2.77 s 20	β-
	35		-14.390e	0.78 s 12	β-
	36	0+	-12.640e	0.45 s 6	β-, β-n < 10%
	37		-7.000s		
	38	0+	-5.360s		
15 P	25		22.080s		
	26		11.260s	≈20 ms	ε, εp, ε2p
	27	1/2+	-0.715e	0.26 s 8	ε, εp
	28	3+	-7.161	270.3 ms 5	ε
	29	1/2+	-16.951	4.140 s 14	ε
	30	1+	-20.207e	2.498 m 4	ε
	31	1/2+	-24.441	100%	
	32	1+	-24.305	14.262 d 14	β-
	33	1/2+	-26.338	25.34 d 12	β-
	34	1+	-24.557	12.43 s 8	β-
	35	1/2+	-24.857	47.3 s 7	β-
	36		-20.251	5.6 s 3	β-
	37		-18.930e	2.31 s 13	β-
	38		-13.430e	0.64 s 14	β-
	39		-12.500s	≈160 ms	β-, β-n 41%
	40		-7.020s	0.26 s 8	β-, β-n 30%
	41			0.12 s 2	β-, β-n 30%
	42			0.11 s 3	β-, β-n 50%
16 S	27		18.220s		
	28	0+	4.199e	125 ms 10	ε, εp
	29	5/2+	-3.115e	0.187 s 4	ε, εp 47%
	30	0+	-14.063	1.178 s 5	ε
	31	1/2+	-19.045	2.572 s 13	ε
	32	0+	-26.016	95.02% 9	
	33	3/2+	-26.586	0.75% 1	
	34	0+	-29.932	4.21% 8	
	35	3/2+	-28.846	87.51 d 12	β-
	36	0+	-30.664	0.02% 1	

Isotope Z	El	A	Jπ	Δ (MeV)	T1/2 or Abundance	Decay Mode
16	S	37	7/2-	-26.896	5.05 m 2	β-
		38	0+	-26.861	170.3 m 7	β-
		39	(7/2)-	-23.160e	11.5 s 5	β-
		40	0+	-22.520	8.8 s 22	β-
		41		-17.870s		
		42	0+	-16.420s		
17	Cl	29		15.050s		
		30		4.840s		
		31		-7.060	150 ms 25	ε, εp
		32	1+	-13.330	298 ms 1	ε, εα 0.09%, εp 0.026%
		33	3/2+	-21.003	2.511 s 3	ε
		34	0+	-24.440	1.5264 s 14	ε
		34m	3+	-24.294	32.00 m 4	ε 53.4%, IT 46.6%
		35	3/2+	-29.013	75.77% 5	
		36	2+	-29.522	3.01×10^5 y 2	β- 98.2%, ε 1.8%
		37	3/2+	-31.761	24.23% 5	
		38	2-	-29.798	37.24 m 5	β-
		38m	5-	-29.127	715 ms 3	IT
		39	3/2+	-29.802	55.6 m 2	β-
		40	2-	-27.530	1.35 m 2	β-
		41	(1/2,3/2)+	-27.400	38.4 s 8	β-
		42		-24.660e	6.8 s 3	β-
		43		-23.130	3.3 s 2	β-
		44		-20.010s		
18	Ar	32	0+	-2.172e	98 ms 2	ε, εp 43%
		33	1/2+	-9.398e	173 ms 2	ε, εp 31%
		34	0+	-18.379	844.5 ms 35	ε
		35	3/2+	-23.048	1.775 s 3	ε
		36	0+	-30.230	0.337% 3	
		37	3/2+	-30.948	35.04 d 4	ε
		38	0+	-34.715	0.063% 1	
		39	7/2-	-33.241	269 y 3	β-
		40	0+	-35.039	99.600% 3	
		41	7/2-	-33.066	1.822 h 2	β-
		42	0+	-34.420	32.9 y 11	β-
		43	(3/2,5/2)	-31.980	5.37 m 6	β-
		44	0+	-32.260	11.87 m 5	β-
		45	(7/2-)	-29.720	21.48 s 15	β-
		46	0+	-29.720	8.4 s 6	β-
		47		-25.910s		
		50	0+			
19	K	33		8.000s		
		34		-1.480s		
		35	3/2+	-11.196e	190 ms 30	ε, εp 0.37%
		36	2+	-17.425	342 ms 2	ε, εp 0.05%, εα 0.003%
		37	3/2+	-24.798	1.226 s 7	ε
		38	3+	-28.802	7.636 m 18	ε
		38m	0+	-28.802	923.9 ms 6	ε
		39	3/2+	-33.806	93.2581% 30	
		40	4-	-33.534	1.277×10^9 y 8, 0.0117% 1	β- 89.33%, ε 10.67%
		41	3/2+	-35.558	6.7302% 30	
		42	2-	-35.020	12.360 h 3	β-
		43	3/2+	-36.593	22.3 h 1	β-
		44	2-	-35.810	22.13 m 19	β-
		45	3/2+	-36.614	17.3 m 6	β-
		46	(2-)	-35.418	105 s 10	β-
		47	1/2+	-35.696	17.5 s 3	β-
		48	(2-)	-32.122	6.8 s 2	β-
		49	(1/2+,3/2+)	-30.770	1.26 s 5	β-, β-n 86%
		50	(0-,1,2-)	-25.520s	472 ms 4	β-, β-n 29%
		51	(1/2+,3/2+)		365 ms 5	β-, β-n 68%
		52			105 ms 5	β-, β-n > 88%
		53	(3/2+)		30 ms 5	β-, β-n 85%
		54			10 ms 5	β-, β-n
20	Ca	35		4.440e	0.05 s 3	ε, ε2p
		36	0+	-6.481e	100 ms 65	ε, εp
		37	3/2+	-13.159	175 ms 3	ε 24%, εp
		38	0+	-22.059	440 ms 8	ε
		39	3/2+	-27.275	859.6 ms 14	ε
		40	0+	-34.846	96.941% 13	
		41	7/2-	-35.137	1.03×10^5 y 4	ε
		42	0+	-38.547	0.647% 3	
		43	7/2-	-38.408	0.135% 3	
		44	0+	-41.469	2.086% 5	
		45	7/2-	-40.812	163.8 d 18	β-
		46	0+	-43.140	0.004% 3	
		47	7/2-	-42.345	4.536 d 2	β-
		48	0+	-44.214	$>6\times10^{18}$ y, 0.187% 3	
		49	3/2-	-41.289	8.715 m 23	β-
		50	0+	-39.570	13.9 s 6	β-
		51	(3/2-)	-35.010	10.0 s 8	β-
		51		-35.010	10 s	β-n ?
		52	0+	-32.460s	4.6 s 3	β-
		53	(3/2-,5/2-)		90 ms 15	β-, β-n > 30%
21	Sc	38		-4.460s		
		39		-14.172e		
		40	4-	-20.526	182.3 ms 7	ε, εp 0.44%, εα 0.017%
		41	7/2-	-28.643	596.3 ms 17	ε
		42	0+	-32.121	681.3 ms 7	ε
		42m	(7)+	-31.504	1.028 m 7	ε
		43	7/2-	-36.187	3.891 h 12	ε
		44	2+	-37.815	3.927 h 8	ε
		44m	6+	-37.544	2.442 d 4	IT 98.8%, ε 1.2%
		45	7/2-	-41.069	100%	
		45m	3/2+	-41.057	0.32 s 1	IT
		46	4+	-41.758	83.810 d 10	β-
		46m	1-	-41.615	18.75 s 4	IT
		47	7/2-	-44.330	3.345 d 3	β-
		48	6+	-44.492	43.7 h 1	β-
		49	7/2-	-46.558	57.2 m 2	β-
		50	5+	-44.537	102.5 s 5	β-
		50m	2+,3+	-44.280	0.35 s 4	IT > 97.5%, β- < 2.5%

Isotope Z El	A	Jπ	Δ (MeV)	T1/2 or Abundance	Decay Mode
21 Sc	51	(7/2)-	-43.218	12.4 s *1*	β-
	52	3+	-40.060s	8.2 s *2*	β-
	53		-38.230s		
22 Ti	40	0+	-9.063	50 ms *15*	ε, εp
	41	3/2+	-15.690	80 ms *2*	ε, εp
	42	0+	-25.121	199 ms *6*	ε
	43	7/2-	-29.320	509 ms *5*	ε
	44	0+	-37.548	49 y *3*	ε
	45	7/2-	-39.006	3.08 h *1*	ε
	46	0+	-44.125	8.0% *1*	
	47	5/2-	-44.931	7.3% *1*	
	48	0+	-48.487	73.8% *1*	
	49	7/2-	-48.558	5.5% *1*	
	50	0+	-51.426	5.4% *1*	
	51	3/2-	-49.726	5.76 m *1*	β-
	52	0+	-49.464	1.7 m *1*	β-
	53	(3/2)-	-46.830	32.7 s *9*	β-
	54	0+	-45.530s		
23 V	42		-8.220s		
	43		-17.920s		
	44		-23.800s	90 ms *25*	ε, εα
	45	7/2-	-31.875	539 ms *18*	ε
	46	0+	-37.075	422.37 ms *20*	ε
	47	3/2-	-42.004	32.6 m *3*	ε
	48	4+	-44.474	15.974 d *3*	ε
	49	7/2-	-47.956	338 d *5*	ε
	50	6+	-49.219	1.5×10^{17} y +3-7 0.250% *2*	ε 83%, β- 17%
	51	7/2-	-52.199	99.750% *2*	
	52	3+	-51.438	3.75 m *1*	β-
	53	7/2-	-51.846	1.61 m *4*	β-
	54	3+	-49.889	49.8 s *5*	β-
	55	(7/2-)	-49.150	6.54 s *15*	β-
	56		-46.110s		
24 Cr	44	0+	-13.450		
	45	(7/2-)	-19.410	50 ms *6*	ε, εp≈25%
	46	0+	-29.472	0.26 s *6*	ε
	47	3/2-	-34.553	508 ms *10*	ε
	48	0+	-42.818	21.56 h *3*	ε
	49	5/2-	-45.328	42.3 m *1*	ε
	50	0+	-50.257	>1.8×10^{17} y 4.345% *9*	2ε ?
	51	7/2-	-51.447	27.702 d *4*	ε
	52	0+	-55.414	83.79% *1*	
	53	3/2-	-55.282	9.50% *1*	
	54	0+	-56.930	2.365% *5*	
	55	3/2-	-55.105	3.497 m *3*	β-
	56	0+	-55.290	5.94 m *10*	β-
	57	3/2-to7/2-	-52.690s	21.1 s *10*	β-
	58	0+	-52.050s	7.0 s *3*	β-
	59			1.0 s *4*	β-
	60	0+		0.57 s *6*	β-
25 Mn	46		-12.470s		ε, εp
25 Mn	47		-22.650s		ε, εp
	48	4+	-29.211	0.15 s *1*	ε, εp
	49	5/2-	-37.611	384 ms *17*	ε
	50	0+	-42.625	283.07 ms *36*	ε
	50m	5+	-42.396	1.75 m *3*	ε>92.6%, IT<7.4%
	51	5/2-	-48.238	46.2 m *1*	ε
	52	6+	-50.702	5.591 d *3*	ε
	52m	2+	-50.324	21.1 m *2*	ε 98.25%, IT 1.75%
	53	7/2-	-54.686	3.7×10^{6} y *4*	ε
	54	3+	-55.553	312.12 d *10*	ε, β-<0:001%
	55	5/2-	-57.708	100%	
	56	3+	-56.907	2.5785 h *2*	β-
	57	5/2-	-57.487	87.2 s *8*	β-
	58	3+	-55.830	65.3 s *7*	β-
	58m	(0)+	-55.830	3.0 s *1*	β-
	59	3/2-,5/2-	-55.476	4.6 s *1*	β-
	60	0+	-52.900	51 s *6*	β-
	60m	3+	-52.900	1.77 s *2*	β-, IT
	61	(5/2)-		0.71 s *1*	β-
	62			0.88 s *15*	β-
	63			0.25 s *4*	β-
26 Fe	48		-18.130		ε, εp
	49	(7/2-)	-24.580	75 ms *10*	ε, εp≤60%
	50	0+	-34.470		ε, εp
	51	(5/2-)	-40.217	310 ms *5*	ε
	52	0+	-48.331	8.275 h *8*	ε
	52m	(12+)	-41.511	45.9 s *6*	ε
	53	7/2-	-50.943	8.51 m *2*	ε
	53m	19/2-	-47.903	2.58 m *4*	IT
	54	0+	-56.250	5.9% *2*	
	55	3/2-	-57.476	2.73 y *3*	ε
	56	0+	-60.603	91.72% *15*	
	57	1/2-	-60.178	2.1% *1*	
	58	0+	-62.151	0.28% *2*	
	59	3/2-	-60.661	44.496 d *7*	β-
	60	0+	-61.406	1.5×10^{6} y *3*	β-
	61	3/2-,5/2-	-58.919	5.98 m *6*	β-
	62	0+	-58.896	68 s *2*	β-
	63	5/2-	-55.190	6.1 s *6*	β-
	64			2.0 s *2*	β-
27 Co	50		-17.980s		ε, εp
	51		-27.420s		ε, εp
	52		-34.287		ε, εp
	53	(7/2-)	-42.639	240 ms *20*	ε
	53m	(19/2-)	-39.449	247 ms *12*	ε≈98.5%, p≈1.5%
	54	0+	-48.007	193.24 ms *14*	ε
	54m	(7)+	-47.808	1.48 m *2*	ε
	55	7/2-	-54.025	17.53 h *3*	ε
	56	4+	-56.037	77.12 d *7*	ε
	57	7/2-	-59.342	271.80 d *5*	ε
	58	2+	-59.844	70.82 d *3*	ε
	58m	5+	-59.819	9.15 h *10*	IT

Isotope Z El	A	Jπ	Δ (MeV)	T1/2 or Abundance	Decay Mode
27 Co	59	7/2-	-62.226	100%	
	60	5+	-61.646	5.2714 y 5	β-
	60m	2+	-61.587	10.47 m 4	IT 99.76%, β- 0.24%
	61	7/2-	-62.897	1.650 h 5	β-
	62	2+	-61.423	1.50 m 4	β-
	62m	5+	-61.401	13.91 m 5	β- >99%, IT <1%
	63	(7/2)-	-61.839	27.4 s 5	β-
	64	1+	-59.791	0.30 s 3	β-
	65	(7/2)-	-59.160	1.25 s 5	β-
	66			0.23 s 2	β-
	67			0.42 s 7	β-
28 Ni	51				ε, εp
	52		-22.640		ε, εp
	53	(7/2-)	-29.380	45 ms 15	ε, εp
	54	0+	-39.210		ε
	55	7/2-	-45.330	189 ms 5	ε
	56	0+	-53.901	6.10 d 2	ε
	57	3/2-	-56.077	35.65 h 5	ε
	58	0+	-60.225	68.077% 5	
	59	3/2-	-61.153	7.5×10^{4} y 13	ε
	60	0+	-64.470	26.223% 5	
	61	3/2-	-64.219	1.140% 1	
	62	0+	-66.745	3.634% 1	
	63	1/2-	-65.512	100.1 y 20	β-
	64	0+	-67.098	0.926% 1	
	65	5/2-	-65.124	2.520 h 1	β-
	66	0+	-66.029	54.6 h 4	β-
	67	1/2-	-63.743	21 s 1	β-
	68	0+	-63.483	19 s 5	β-
	69		-60.460	11.4 s 3	β-
29 Cu	55		-31.630s		ε, εp
	56		-38.584		ε, εp
	57	3/2-	-47.350	233 ms 16	ε
	58	1+	-51.662	3.204 s 7	ε
	59	3/2-	-56.353	81.5 s 5	ε
	60	2+	-58.344	23.7 m 4	ε
	61	3/2-	-61.982	3.347 h 12	ε
	62	1+	-62.797	9.74 m 2	ε
	63	3/2-	-65.578	69.17% 2	
	64	1+	-65.423	12.701 h 2	ε 62.9%, β- 37.1%
	65	3/2-	-67.262	30.83% 2	
	66	1+	-66.256	5.10 m 1	β-
	67	3/2-	-67.303	61.92 h 9	β-
	68	1+	-65.540	31.1 s 15	β-
	68m	(6-)	-64.818	3.75 m 5	IT 84%, β- 16%
	69	3/2-	-65.741	2.85 m 15	β-
	70	(1+)	-63.390	4.5 s 10	β-
	70m	(4)-	-63.250	47 s 5	β-
	71	(3/2-)	-62.920s	19.5 s 16	β-
	72			6.6 s 1	β-
	73			3.9 s 3	β-
	75			1.3 s 1	β-, β-n 3.5%
	76			0.61 s 10	β-, β-n
30 Zn	56		-26.130s		
	57	(7/2-)	-32.700	40 ms 10	ε, εp≥65%
	58	0+	-42.210		ε
	59	3/2-	-47.260	183.7 ms 23	ε, εp
	60	0+	-54.185	2.38 m 5	ε
	61	3/2-	-56.343	89.1 s 2	ε
	62	0+	-61.170	9.186 h 13	ε
	63	3/2-	-62.211	38.50 m 8	ε
	64	0+	-66.002	48.6% 3	
	65	5/2-	-65.910	243.9 d 1	ε
	66	0+	-68.898	27.9% 2	
	67	5/2-	-67.879	4.1% 1	
	68	0+	-70.006	18.8% 4	
	69	1/2-	-68.417	56.4 m 9	β-
	69m	9/2+	-67.978	13.76 h 2	IT 99.97%, β- 0.03%
	70	0+	-69.561	>5×10^{14} y 0.6% 1	2β- ?
	71	1/2-	-67.323	2.45 m 10	β-
	71m	9/2+	-67.165	3.96 h 5	β-, IT≤0.05%
	72	0+	-68.131	46.5 h 1	β-
	73	(1/2)-	-65.410	23.5 s 10	β-
	73m	(7/2+)	-65.215	5.8 s 8	β-, IT
	74	0+	-65.708	96 s 1	β-
	75	(7/2+)	-62.530	10.2 s 2	β-
	76	0+	-62.290	5.7 s 3	β-
	77	°(7/2+)	-58.820s	2.08 s 5	β-
	77m	(1/2-)	-58.048s	1.05 s 10	β- <50%, IT >50%
	78	0+	-57.660s	1.47 s 15	β-
	79		-53.820s	1.0 s 1	β-
	80	0+	-51.890	0.55 s 3	β-
31 Ga	61		-47.540s		
	62	0+	-51.999	116.12 ms 23	ε
	63	3/2-,5/2-	-56.690	32.4 s 5	ε
	64	0+	-58.837	2.630 m 11	ε
	65	3/2-	-62.654	15.2 m 2	ε
	66	0+	-63.724	9.49 h 7	ε
	67	3/2-	-66.878	3.261 d 1	ε
	68	1+	-67.085	67.629 m 24	ε
	69	3/2-	-69.322	60.108% 6	
	70	1+	-68.905	21.14 m 3	β- 99.59%, ε 0.41%
	71	3/2-	-70.139	39.892% 6	
	72	3-	-68.589	14.10 h 2	β-
	73	3/2-	-69.705	4.86 h 3	β-
	74	(3-)	-68.060	8.12 m 12	β-
	74m	(0)	-68.000	9.5 s 10	IT 75%, β- <50%
	75	3/2-	-68.466	126 s 2	β-
	76	(3-)	-66.440	29.1 s 7	β-
	77	(3/2-)	-66.320s	13.2 s 2	β-
	78	(3)	-63.560s	5.09 s 5	β-
	79	(3/2-)	-62.720	3.00 s 9	β-, β-n 0.1%
	80		-59.380	1.66 s 2	β-, β-n 0.84%
	81	(5/2-)	-57.990	1.23 s 1	β-, β-n 12%

Isotope Z El	A	Jπ	Δ (MeV)	T1/2 or Abundance	Decay Mode
31 Ga	82	(1,2,3)	-53.380s	0.602 s 6	β-, β-n 19.8%
	83			0.31 s 1	β-, β-n 43%
32 Ge	61			40 ms 15	ε, εp
	63		-47.310s		
	64	0+	-54.430	63.7 s 25	ε
	65	3/2-,5/2-	-56.410	30.9 s 7	ε, εp 0.013%
	66	0+	-61.620	2.26 h 5	ε
	67	(1/2-)	-62.656	18.7 m 5	ε
	68	0+	-66.978	270.82 d 27	ε
	69	5/2-	-67.097	39.05 h 10	ε
	70	0+	-70.561	21.23% 4	
	71	1/2-	-69.906	11.43 d 3	ε
	71m	9/2+	-69.708	20.40 ms 17	IT
	72	0+	-72.583	27.66% 3	
	73	9/2+	-71.295	7.73% 1	
	73m	1/2-	-71.228	0.499 s 11	IT
	74	0+	-73.423	35.94% 2	
	75	1/2-	-71.858	82.78 m 4	β-
	75m	7/2+	-71.718	47.7 s 5	IT 99.97%, β- 0.03%
	76	0+	-73.214	7.44% 2	
	77	7/2+	-71.216	11.30 h 1	β-
	77m	1/2-	-71.056	52.9 s 6	β- 79%, IT 21%
	78	0+	-71.863	88 m 1	β-
	79	(1/2-)	-69.490	19.1 s 3	β-
	79m	(7/2+)	-69.304	39.0 s 10	β- 96%, IT 4%
	80	0+	-69.380	29.5 s 4	β-
	81	(9/2+)	-66.310	7.6 s 6	β-
	81m	(1/2+)	-65.631	7.6 s 6	β-
	82	0+	-65.380	4.60 s 35	β-
	83		-61.140s	1.9 s 4	β-
	84	0+	-58.150s	1.2 s 3	β-
33 As	65		-47.510s		
	66		-52.070	95.8 ms 4	ε
	67	(3/2,5/2)	-56.650	42.5 s 12	ε
	68	3	-58.880	151.6 s 8	ε
	69	5/2-	-63.080	15.2 m 2	ε
	70	4(+)	-64.340	52.6 m 3	ε
	71	5/2-	-67.894	65.28 h 15	ε
	72	2-	-68.228	26.0 h 1	ε
	73	3/2-	-70.955	80.30 d 6	ε
	74	2-	-70.861	17.77 d 2	ε 66%, β- 34%
	75	3/2-	-73.035	100%	
	76	2-	-72.290	26.32 h 7	β-, ε<0.02%
	77	3/2-	-73.918	38.83 h 5	β-
	78	(2-)	-72.819	90.7 m 2	β-
	79	3/2-	-73.639	9.01 m 15	β-
	80	1+	-72.165	15.2 s 2	β-
	81	3/2-	-72.536	33.3 s 8	β-
	82	(1+)	-70.078	19.1 s 5	β-
	82m	(5-)	-70.078	13.6 s 4	β-
	83	(3/2-)	-69.880	13.4 s 3	β-
	84	0(-),1(-),2-	-66.080s	5.5 s 3	β-, β-n 0.08%
	84m		-66.080s	0.65 s	β-
33 As	85	(3/2-)	-63.510s	2.028 s 12	β-, β-n 23%
	86		-59.340s	0.9 s 2	β-, β-n 12%
	87	(3/2-)		0.73 s 6	β-, β-n 44%
34 Se	67		-46.860s		
	68	0+	-54.080s	1.6 m 4	ε
	69	(3/2-)	-56.300	27.4 s 2	ε, εp 0.05%
	70	0+	-61.540s	41.1 m 3	ε
	71	3/2-,5/2-	-63.090s	4.74 m 5	ε
	72	0+	-67.897	8.40 d 8	ε
	73	9/2+	-68.215	7.15 h 8	ε
	73m	3/2-	-68.189	39.8 m 13	IT 72.6%, ε 27.4%
	74	0+	-72.215	0.89% 2	
	75	5/2+	-72.171	119.779 d 4	ε
	76	0+	-75.254	9.36% 12	
	77	1/2-	-74.601	7.63% 5	
	77m	7/2+	-74.439	17.36 s 5	IT
	78	0+	-77.028	23.78% 15	
	79	7/2+	-75.920	≤6.5×10^{4} y	β-
	79m	1/2-	-75.824	3.91 m 5	IT
	80	0+	-77.762	49.61% 31	
	81	1/2-	-76.392	18.45 m 12	β-
	81m	7/2+	-76.289	57.28 m 5	IT 99.95%, β- 0.05%
	82	0+	-77.596	1.4×10^{20} y 4 8.73% 6	2β-
	83	9/2+	-75.343	22.3 m 3	β-
	83m	1/2-	-75.114	70.1 s 4	β-
	84	0+	-75.952	3.1 m 1	β-
	85	(5/2+)	-72.420	31.7 s 9	β-
	86	0+	-70.540	15.3 s 9	β-
	87	(5/2+)	-66.710s	5.85 s 15	β-, β-n 0.18%
	88	0+	-63.820s	1.52 s 3	β-, β-n 0.94%
	89	(5/2+)		0.41 s 4	β-, β-n 5%
	91			0.27 s 5	β-, β-n 21%
35 Br	69		-46.800s		
	70		-51.140s	80.2 ms 8	ε
	70m		-51.140s	2.2 s 2	ε
	71	1/2-to7/2-	-56.590s	21.4 s 6	ε
	72	(3)+	-59.000s	78.6 s 24	ε
	72m		-59.000s	7.2 s 5	εp
	72m	(1)-	-58.899s	10.6 s 3	IT, ε?
	73	(3/2)-	-63.600	3.4 m 3	ε
	74	(0-,1)	-65.301	25.4 m 3	ε
	74m	4(-)	-65.301	46 m 2	ε
	75	3/2-	-69.142	96.7 m 13	ε
	76	1-	-70.291	16.2 h 2	ε
	76m	(4)+	-70.188	1.31 s 2	IT>99.4%, ε<0.6%
	77	3/2-	-73.237	57.036 h 6	ε
	77m	9/2+	-73.131	4.28 m 10	IT
	78	1+	-73.455	6.46 m 4	ε≥99.99%, β-≤1×10^{-2}%
	79	3/2-	-76.070	50.69% 5	
	79m	9/2+	-75.863	4.86 s 4	IT

Isotope Z El	A	Jπ	Δ (MeV)	T1/2 or Abundance	Decay Mode
35 Br	80	1+	-75.891	17.68 m *2*	β- 91.7%, ε 8.3%
	80m	5-	-75.805	4.42 h *1*	IT
	81	3/2-	-77.978	49.31% *5*	
	82	5-	-77.499	35.30 h *2*	β-
	82m	2-	-77.453	6.13 m *5*	IT 97.6%, β- 2.4%
	83	(3/2)-	-79.010	2.40 h *2*	β-
	84	2-	-77.776	31.80 m *8*	β-
	*84*m	(5-,6-)	-77.456	6.0 m *2*	β-
	85	3/2-	-78.607	2.90 m *6*	β-
	86	(2-)	-75.640	55.1 s *4*	β-
	87	3/2-	-73.856	55.60 s *15*	β-, β-n 2.57%
	88	(1,2-)	-70.720	16.5 s *1*	β-, β-n 6.4%
	89	(3/2-,5/2-)	-68.420s	4.40 s *3*	β-, β-n 13%
	90	(2-)	-64.650	1.71 s *14*	β-, β-n 23%
	91			0.541 s *5*	β-, β-n 18.3%
	92			0.365 s *7*	β-, β-n 21%
	93	(5/2-)			β-, β-n
	94				β-, β-n
36 Kr	71		-46.490s	97 ms *9*	ε
	72	0+	-53.940s	17.2 s *3*	ε
	73		-56.890	27.0 s *12*	ε, εp 0.68%
	74	0+	-62.130	11.50 m *11*	ε
	75	(5/2+)	-64.214	4.3 m *2*	ε
	76	0+	-68.965	14.8 h *1*	ε
	77	5/2+	-70.194	74.4 m *6*	ε
	78	0+	-74.147	0.35% *2*	
	79	1/2-	-74.445	35.04 h *10*	ε
	79m	7/2+	-74.315	50 s *3*	IT
	80	0+	-77.894	2.25% *2*	
	81	7/2+	-77.697	2.13×10^5 y *21*	ε
	81m	1/2-	-77.506	13 s *1*	IT 99.99%, ε 0.01%
	82	0+	-80.592	11.6% *1*	
	83	9/2+	-79.982	11.5% *1*	
	*83*m	1/2-	-79.940	1.83 h *2*	IT
	84	0+	-82.430	57.0% *3*	
	85	9/2+	-81.477	10.756 y *18*	β-
	*85*m	1/2-	-81.172	4.480 h *8*	β- 78.6%, IT 21.4%
	86	0+	-83.262	17.3% *2*	
	87	5/2+	-80.706	76.3 m *6*	β-
	88	0+	-79.688	2.84 h *3*	β-
	89	(3/2+,5/2+)	-76.720	3.15 m *4*	β-
	90	0+	-74.947	32.32 s *9*	β-
	91	(5/2+)	-71.370	8.57 s *4*	β-
	92	0+	-68.650	1.85 s *1*	β-, β-n 0.03%
	93	(7/2+)	-64.160	1.29 s *1*	β-, β-n 3.2%
	94	0+		0.20 s *1*	β-, β-n 5.7%
	95			0.78 s *3*	β-, β-n
	97?			<0.1 s	β-
37 Rb	73		-46.590s		
	74	(0+)	-51.670	64.9 ms *5*	ε
	75	(3/2-,5/2-)	-57.210	19.0 s *12*	ε
	76	1	-60.530	39.1 s *6*	ε
37 Rb	77	3/2-	-64.917	3.75 m *8*	ε
	78	0(+)	-66.980	17.66 m *8*	ε
	78m	4(-)	-66.877	5.74 m *6*	ε 90%, IT 10%
	79	5/2+	-70.839	22.9 m *5*	ε
	80	1+	-72.176	34 s *4*	ε
	81	3/2-	-75.459	4.576 h *5*	ε
	81m	9/2+	-75.373	30.49 m *29*	IT 97.7%, ε 2.3%
	82	1+	-76.203	1.273 m *2*	ε
	82m	5-	-76.123	6.472 h *6*	ε, IT<0.33%
	83	5/2-	-79.049	86.2 d *1*	ε
	84	2-	-79.748	32.77 d *14*	ε 96.2%, β- 3.8%
	84m	6-	-79.284	20.26 m *4*	IT
	85	5/2-	-82.164	72.17% *1*	
	86	2-	-82.744	18.631 d *18*	β- 99.99%, ε 0.0052%
	86m	6-	-82.188	1.017 m *3*	IT
	87	3/2-	-84.593	4.75×10^{10} y *4* 27.83% *1*	β-
	88	2-	-82.601	17.78 m *11*	β-
	89	3/2-	-81.709	15.15 m *12*	β-
	90	(1-)	-79.350	153 s *3*	β-
	*90*m	(4-)	-79.243	258 s *5*	β- 95.7%, IT 4.3%
	91	3/2(-)	-77.786	58.4 s *4*	β-
	92	0(-)	-74.811	4.50 s *2*	β-, β-n 0.012%
	93	5/2-	-72.688	5.7 s *1*	β-, β-n 2.5%
	94	3(-)	-68.518	2.702 s *5*	β-, β-n 10.4%
	95	5/2	-65.813	384 ms *6*	β-, β-n 9.1%
	96	2+	-61.150	0.199 s *3*	β-, β-n 13%
	97	3/2	-58.290	171.8 ms *16*	β-, β-n 24.6%
	98	(0)	-54.090	0.114 s *5*	β-, β-n 15.9%, β-2n
	99	(5/2+)	-50.860	59 ms *1*	β-, β-n 15%
	100			51 ms *8*	β-, β-n 6%
	102?			90 ms *20*	β-, β-n
38 Sr	77	(5/2+,7/2+)	-57.880	9.0 s *2*	ε, εp<0.25%
	78	0+	-63.450s	2.5 m *3*	ε
	79	(3/2-)	-65.340s	2.25 m *10*	ε
	80	0+	-70.190	106.3 m *15*	ε
	81	(1/2-)	-71.470	22.3 m *4*	ε
	82	0+	-75.998	25.55 d *15*	ε
	83	(7/2+)	-76.781	32.41 h *3*	ε
	83m	(1/2-)	-76.522	4.95 s *12*	IT
	84	0+	-80.641	0.56% *1*	
	85	9/2+	-81.099	64.84 d *2*	ε
	85m	1/2-	-80.860	67.63 m *4*	IT 84.5%, ε 15.5%
	86	0+	-84.518	9.86% *1*	
	87	9/2+	-84.875	7.00% *1*	
	*87*m	1/2-	-84.486	2.804 h *3*	IT 99.7%, ε 0.3%
	88	0+	-87.916	82.58% *1*	
	89	5/2+	-86.211	50.53 d *7*	β-
	90	0+	-85.942	29.1 y *3*	β-
	91	5/2+	-83.652	9.63 h *5*	β-
	92	0+	-82.923	2.71 h *1*	β-
	93	5/2(+)	-80.160	7.423 m *24*	β-

Isotope Z El	A	Jπ	Δ (MeV)	T1/2 or Abundance	Decay Mode
38 Sr	94	0+	-78.836	75.1 s 7	β-
	95	(1/2+)	-75.050	25.1 s 2	β-
	96	0+	-72.880	1.06 s 4	β-
	97	(1/2+)	-68.810	420 ms 30	β-, β-n 0.27%
	98	0+	-66.380	0.65 s 3	β-, β-n 0.8%
	99	(3/2+)	-62.150	0.271 s 4	β-, β-n 0.32%
	100	0+	-60.200	202 ms 3	β-, β-n 0.73%
	101			115 ms 1	β-, β-n
	102	0+		68 ms 8	β-, β-n
39 Y	79		-58.140s		
	80	(4)	-61.190s	33.8 s 6	ε
	81	(1/2-)	-65.950	72.4 s 13	ε
	82	1+	-68.180	9.5 s 3	ε
	83	(9/2+)	-72.370	7.08 m 6	ε
	83m	(1/2-)	-72.160	2.85 m 2	ε
	84	1+	-74.230	4.6 s 2	ε
	84m	(5-)	-73.730	40 m 1	ε
	85	(1/2)-	-77.845	2.68 h 5	ε
	85m	(9/2)+	-77.825	4.86 h 13	ε, IT
	86	4-	-79.279	14.74 h 2	ε
	86m	(8+)	-79.061	48 m 1	IT 99.31%, ε 0.69%
	87	1/2-	-83.014	79.8 h 3	ε
	87m	9/2+	-82.633	13.37 h 3	IT 98.43%, ε 0.02%
	88	4-	-84.294	106.65 d 4	ε
	89	1/2-	-87.703	100%	
	89m	9/2+	-86.794	16.06 s 4	IT
	90	2-	-86.488	64.1 h 1	β-
	90m	7+	-85.806	3.19 h 1	IT, β- 0.002%
	91	1/2-	-86.349	58.51 d 6	β-
	91m	9/2+	-85.793	49.71 m 4	IT, β-<1.5%
	92	2-	-84.833	3.54 h 1	β-
	93	1/2-	-84.245	10.10 h 16	β-
	93m	(7/2)+	-83.486	0.82 s 4	IT
	94	2-	-82.348	18.7 m 1	β-
	95	1/2-	-81.214	10.3 m 2	β-
	96	0-	-78.300	6.2 s 2	β-
	96m	(3+)	-78.200	9.6 s 2	β-
	96m		-78.200	2.3 m 1	β-
	97	(1/2-)	-76.270	3.76 s 2	β-, β-n 0.06%
	97m	(9/2)+	-75.602	1.21 s 2	β-, IT<0.7%
	98	1+	-72.520	0.59 s 4	β-, β-n 0.3%
	98m	(4-)	-72.520	2.13 s 12	β-
	99	(5/2+)	-70.170	1.47 s 2	β-, β-n 0.96%
	100	1-,2-	-67.290	735 ms 7	β-, β-n 0.81%
	100m	(3,4,5)	-67.290	0.94 s 3	β-
	101	(5/2)	-64.650s	431 ms 7	β-, β-n
	102		-61.450s	0.36 s 4	β-, β-n
40 Zr	81		-58.790	15 s 5	ε, εp
	82	0+	-64.180	32 s 5	ε
	83	(1/2-)	-66.350	44 s 1	ε
	83m	(7/2+)	-66.350	7 s 2	ε
	84	0+	-71.430s	25.9 m 8	ε

Isotope Z El	A	Jπ	Δ (MeV)	T1/2 or Abundance	Decay Mode
40 Zr	85	(7/2)+	-73.150	7.86 m 4	ε
	85m	(1/2-)	-72.858	10.9 s 3	IT, ε>0.09%
	86	0+	-77.980s	16.5 h 1	ε
	87	(9/2)+	-79.348	1.68 h 1	ε
	87m	(1/2-)	-79.012	14.0 s 2	
	88	0+	-83.626	83.4 d 3	ε
	89	9/2+	-84.871	78.41 h 12	ε
	89m	1/2-	-84.283	4.18 m 1	IT 93.77%, ε 6.23%
	90	0+	-88.770	51.45% 2	
	90m	5-	-86.451	809.2 ms 20	IT
	91	5/2+	-87.893	11.22% 3	
	92	0+	-88.457	17.15% 2	
	93	5/2+	-87.120	1.53×10^{6} y 10	β-
	94	0+	-87.268	17.38% 3	
	95	5/2+	-85.659	64.02 d 4	β-
	96	0+	-85.442	$>3.56\times10^{17}$ y 2.80% 1	
	97	1/2+	-82.950	16.90 h 5	β-
	98	0+	-81.283	30.7 s 4	β-
	99	(1/2)+	-77.790	2.1 s 1	β-
	100	0+	-76.590	7.1 s 4	β-
	101	(3/2)	-73.380	2.1 s 3	β-
	102	0+	-71.770	2.9 s 2	β-
	103		-68.290	1.3 s 1	β-
	104	0+	-66.260s	1.2 s 1	β-
41 Nb	84	(3+)	-61.530s	12 s 3	ε, εp
	85	(9/2+)	-66.940s	20.9 s 7	ε
	86	(5+)	-69.580s	88 s 1	ε
	87	(9/2+)	-74.180	2.6 m 1	ε
	87m	(1/2-)	-74.180	3.7 m 1	ε
	88	(8+)	-76.430s	14.5 m 1	ε
	88m	(4-)	-76.430s	7.8 m 1	ε
	89	(1/2)-	-80.580	1.18 h 10	ε
	89m	(9/2+)	-80.580	1.9 h 2	ε
	90	8+	-82.659	14.60 h 5	ε
	90m	4-	-82.534	18.8 s 1	IT
	91	9/2+	-86.640	6.8×10^{2} y 13	ε
	91m	1/2-	-86.536	60.86 d 22	IT 95%, ε 5%
	92	(7)+	-86.451	3.5×10^{7} y 3	ε
	92m	(2)+	-86.315	10.15 d 2	ε
	93	9/2+	-87.210	100%	
	93m	1/2-	-87.179	16.1 y 2	IT
	94	(6)+	-86.368	2.03×10^{4} y 16	β-
	94m	3+	-86.327	6.26 m 1	IT 99.5%, β- 0.5%
	95	9/2+	-86.783	34.97 d 3	β-
	95m	1/2-	-86.547	3.61 d 3	IT 94.4%, β- 5.6%
	96	6+	-85.606	23.35 h 5	β-
	97	9/2+	-85.608	1.227 h 5	β-
	97m	1/2-	-84.865	58.1 s 5	IT
	98	1+	-83.528	2.86 s 6	β-
	98m	(5)+	-83.444	51.3 m 4	β-
	99	9/2+	-82.328	15.0 s 2	β-
	99m	1/2-	-81.963	2.6 m 2	β-, IT<2.5%

Isotope Z El	A	Jπ	Δ (MeV)	T1/2 or Abundance	Decay Mode
41 Nb	100	1+	-79.929	1.5 s 2	β-
	100m	(4+,5+)	-79.449	2.99 s 11	β-
	101		-78.950	7.1 s 3	β-
	102	low	-76.350	1.3 s 2	β-
	102	high	-76.350	4.3 s 4	β-
	103	(5/2+)	-75.240	1.5 s 2	β-
	104		-72.260	0.91 s 10	β-
	104		-72.260	4.8 s 4	β-
	105		-70.940	2.95 s 6	β-
	106		-67.290s	1.02 s 5	β-
42 Mo	86		-64.680s		
	87	(7/2+)	-67.440	13.4 s 4	ε, εp
	88	0+	-72.830s	8.0 m 2	ε
	89	(9/2+)	-75.005	2.04 m 11	ε
	89m	(1/2-)	-74.618	190 ms 15	IT
	90	0+	-80.170	5.67 h 5	ε
	91	9/2+	-82.208	15.49 m 1	ε
	91m	1/2-	-81.555	65.0 s 7	IT 50.1%, ε 49.9%
	92	0+	-86.809	14.84% 4	
	93	5/2+	-86.805	3.5×10^3 y 7	ε
	93m	21/2+	-84.380	6.85 h 7	IT 99.88%, ε 0.12%
	94	0+	-88.413	9.25% 2	
	95	5/2+	-87.709	15.92% 4	
	96	0+	-88.792	16.68% 4	
	97	5/2+	-87.542	9.55% 2	
	98	0+	-88.113	24.13% 6	
	99	1/2+	-85.967	65.94 h 1	β-
	100	0+	-86.186	9.63% 2	
	101	1/2+	-83.513	14.6 m 1	β-
	102	0+	-83.559	11.3 m 2	β-
	103	(3/2+)	-80.760	67.5 s 15	β-
	104	0+	-80.370	60 s 2	β-
	105	(3/2+)	-77.360	35.6 s 16	β-
	106	0+	-76.270	8.4 s 5	β-
	107		-72.910s	3.5 s 5	β-
	108	0+	-71.460s	1.5 s 4	β-
43 Tc	88		-62.330s		
	89		-68.000s		
	90	1+	-70.970s	8.3 s	ε
	90	high	-70.970s	49.2 s	ε
	91	(9/2)+	-75.990	3.14 m 2	ε
	91m	(1/2)-	-75.640	3.3 m 1	ε, IT<1%
	92	(8+)	-78.939	4.4 m 3	ε
	93	9/2+	-83.607	2.75 h 5	ε
	93m	1/2-	-83.216	43.5 m 10	IT 77.8%, ε 22.2%
	94	7+	-84.158	293 m 1	ε
	94m	(2)+	-84.082	52.0 m 10	ε, IT<0.1%
	95	9/2+	-86.018	20.0 h 1	ε
	95m	1/2-	-85.979	61 d 2	ε 96%, IT 4%
	96	7+	-85.819	4.28 d 6	ε
	96m	4+	-85.785	51.5 m 10	IT 98%, ε 2%
	97	9/2+	-87.222	2.6×10^6 y 4	ε
	97m	1/2-	-87.125	90.5 d 10	IT

Isotope Z El	A	Jπ	Δ (MeV)	T1/2 or Abundance	Decay Mode
43 Tc	98	(6)+	-86.429	4.2×10^6 y 3	β-
	99	9/2+	-87.324	2.111×10^5 y 12	β-
	99m	1/2-	-87.181	6.01 h 1	IT, β- 0.004%
	100	1+	-86.017	15.8 s 1	β-
	101	(9/2)+	-86.337	14.2 m 1	β-
	102	1+	-84.569	5.28 s 15	β-
	102m	(4,5)	-84.269	4.35 m 7	β-≈98%, IT≈2%
	103	5/2+	-84.601	54.2 s 8	β-
	104	(3+)	-82.490	18.3 m 3	β-
	105	(5/2+)	-82.350	7.6 m 1	β-
	106	(1,2)	-79.790	36 s 1	β-
	107m		-79.160s	21.2 s 2	β-
	108	(3)	-76.280s	5.17 s 7	β-
	109		-74.920s	1.4 s 4	β-
	110		-71.640s	0.83 s 4	β-
	111			0.30 s 3	β-
44 Ru	90		-65.470s		
	91	(9/2+)	-68.410s	9 s 1	ε
	91m	(1/2-)	-68.410s	7.6 s 8	ε, εp
	92	0+	-74.410s	3.65 m 5	ε
	93	(9/2)+	-77.270	59.7 s 6	ε
	93m	(1/2)-	-76.536	10.8 s 3	ε 77.8%, IT 22.2%, εp 0.03%
	94	0+	-82.569	51.8 m 6	ε
	95	5/2+	-83.451	1.64 h 1	ε
	96	0+	-86.073	5.54% 2	
	97	5/2+	-86.113	2.9 d 1	ε
	98	0+	-88.225	1.86% 2	
	99	5/2+	-87.617	12.7% 1	
	100	0+	-89.219	12.6% 1	
	101	5/2+	-87.950	17.1% 1	
	102	0+	-89.099	31.6% 2	
	103	(3/2)+	-87.260	39.26 d 2	β-
	104	0+	-88.093	18.6% 2	
	105	3/2+	-85.932	4.44 h 2	β-
	106	0+	-86.326	373.59 d 15	β-
	107	(5/2+)	-83.710	3.75 m 5	β-
	108	0+	-83.760	4.55 m 5	β-
	109	(5/2+)	-80.720s	34.5 s 10	β-
	110	0+	-80.240s	14.6 s 10	β-
	111		-77.030s	2.12 s 7	β-
	112	0+	-76.030s	1.75 s 7	β-
	113			0.80 s 5	β-
	114	0+		0.5 s	β-
45 Rh	92		-63.140s		
	93		-69.110s		
	94m	(8+)	-72.940	25.8 s 2	ε
	94m	(3+)	-72.940	70.6 s 6	ε
	95	(9/2)+	-78.340	5.02 m 10	ε
	95m	(1/2)-	-77.797	1.96 m 4	IT 88%, ε 12%
	96	5(+)	-79.626	9.6 m	ε
	96m	2(+)	-79.574	1.51 m 2	IT 60%, ε 40%
	97	(9/2)+	-82.590	31.1 m 8	ε
	97m	(1/2)-	-82.331	44.3 m 8	ε 95.1%, IT 4.9%

Isotope Z El	A	Jπ	Δ (MeV)	T1/2 or Abundance	Decay Mode
45 Rh	98	(2)+	-83.168	8.7 m 2	ε
	98m	(5+)	-83.118	3.5 m 3	ε
	99	(1/2-)	-85.519	16.1 d 2	ε
	99m	9/2+	-85.455	4.7 h 1	ε, IT<0.16%
	100	1-	-85.590	20.8 h 1	ε
	100m	(5+)	-85.590	4.6 m 2	IT≈98.3%, ε≈1.7%
	101	1/2-	-87.410	3.3 y 3	ε
	101m	9/2+	-87.253	4.34 d 1	ε 92.3%, IT 7.7%
	102	6(+)	-86.821	≈2.9 y	ε
	102m	(1-,2-)	-86.751	207 d 3	ε 75%, β- 20%, IT 5%
	103	1/2-	-88.024	100%	
	103m	7/2+	-87.984	56.12 m 1	IT
	104	1+	-86.952	42.3 s 4	β- 99.55%, ε 0.45%
	104m	5+	-86.823	4.34 m 5	IT 99.87%, β- 0.13%
	105	7/2+	-87.849	35.36 h 6	β-
	105m	1/2-	-87.719	≈40 s	IT
	106	1+	-86.365	29.80 s 8	β-
	106m	(6)+	-86.228	130 m 2	β-
	107	(7/2+)	-86.862	21.7 m 4	β-
	108m	1+	-85.080	16.8 s 5	β-
	108m	≥3+,≤6+	-85.080	6.0 m 3	β-
	109	7/2+	-85.021	80 s 2	β-
	110	1+	-82.940	3.2 s 2	β-
	110	(≥2)	-82.940	28.5 s 15	β-
	111		-82.330s	11 s 1	β-
	112m	1+	-79.730s	3.8 s 6	β-
	112m	≥4	-79.730s	6.8 s 2	β-
	113		-78.740s	2.72 s 22	β-
	114	(1+)	-75.960s	1.85 s 5	β-
	114m	(≥4)	-75.960s	1.85 s 5	β-
	115			0.99 s 5	β-
	116	1+		0.68 s 6	β-
	116	5,6,7		0.9 s 4	β-
46 Pd	94	0+	-66.270s	9.0 s 5	ε
	95m	(21/2+)	-68.150s	13.3 s 3	ε, εp>0.93%
	96	0+	-76.180	2.03 m 3	ε
	97	(5/2+)	-77.800	3.1 m 1	ε
	98	0+	-81.301	17.7 m 3	ε
	99	(5/2)+	-82.193	21.4 m 2	ε
	100	0+	-85.221	3.63 d 9	ε
	101	(5/2+)	-85.430	8.47 h 6	ε
	102	0+	-87.918	1.02% 1	
	103	5/2+	-87.471	16.991 d 19	ε
	104	0+	-89.393	11.14% 8	
	105	5/2+	-88.416	22.33% 8	
	106	0+	-89.907	27.33% 3	
	107	5/2+	-88.374	6.5×10^{6} y 3	β-
	107m	11/2-	-88.159	21.3 s 5	IT
	108	0+	-89.523	26.46% 9	
	109	5/2+	-87.605	13.7 h 1	β-

Isotope Z El	A	Jπ	Δ (MeV)	T1/2 or Abundance	Decay Mode
46 Pd	109m	11/2-	-87.416	4.69 m 1	IT
	110	0+	-88.345	11.72% 9	
	111	5/2+	-86.030	23.4 m 2	β-
	111m	11/2-	-85.858	5.5 h 1	IT 73%, β- 27%
	112	0+	-86.333	21.03 h 5	β-
	113	(5/2)+	-83.680	93 s 5	β-
	113m		-83.680	≥100 s	β-
	114	0+	-83.460	2.42 m 6	β-
	115		-80.590s	47 s 2	β-
	116	0+	-80.140	12.4 s 5	β-
	117			5.0 s +5-7	β-
	118	0+		2.4 s 4	β-
47 Ag	96	(8+,9+)	-64.430s	5.1 s 4	ε, εp 8%
	97		-70.790s	21 s 3	ε
	98	(7+)	-73.000s	47 s 1	ε, εp>0%
	99	(9/2)+	-76.760	124 s 3	ε
	99m	(1/2-)	-76.254	10.5 s 5	IT
	100	(5)+	-78.170	2.01 m 9	ε
	100m	(2)+	-78.154	2.24 m 13	ε, IT
	101	9/2+	-81.190	11.1 m 3	ε
	101m	1/2-	-80.916	3.10 s 10	IT
	102	5+	-82.080	12.9 m 3	ε
	102m	2+	-82.071	7.7 m 5	ε 51%, IT 49%
	103	7/2+	-84.787	65.7 m 7	ε
	103m	1/2-	-84.653	5.7 s 3	IT
	104	5+	-85.114	69.2 m 10	ε
	104m	2+	-85.107	33.5 m 20	ε 67%, IT 33%
	105	1/2-	-87.078	41.29 d 7	ε
	105m	7/2+	-87.053	7.23 m 16	IT 99.66%, ε 0.34%
	106	1+	-86.941	23.96 m 4	ε 99.5%, β-<1%
	106m	6+	-86.851	8.46 d 10	ε
	107	1/2-	-88.407	51.839% 5	
	107m	7/2+	-88.314	44.3 s 2	IT
	108	1+	-87.605	2.37 m 1	β- 97.15%, ε 2.85%
	108m	6+	-87.496	127 y 21	ε 91.3%, IT 8.7%
	109	1/2-	-88.721	48.161% 5	
	109m	7/2+	-88.633	39.6 s 2	IT
	110	1+	-87.459	24.6 s 2	β- 99.7%, ε 0.3%
	110m	6+	-87.341	249.76 d 4	β- 98.64%, IT 1.36%
	111	1/2-	-88.217	7.45 d 1	β-
	111m	7/2+	-88.157	64.8 s 8	IT 99.3%, β- 0.7%
	112	2(-)	-86.624	3.130 h 9	β-
	113	1/2-	-87.040	5.37 h 5	β-
	113m	7/2+	-86.997	68.7 s 16	IT≈80%, β-≈20%
	114	1+	-84.960	4.6 s 1	β-
	115	1/2-	-84.950	20.0 m 5	β-
	115m	(7/2+)	-84.950	18.0 s 7	β-, IT
	116	(2-)	-82.760	2.68 m 1	β-
	116m	(5+)	-82.679	10.4 s 8	β- 98%, IT 2%
	117m	(7/2+)	-82.250	5.34 s 5	β-
	117m	(1/2-)	-82.250	72.8 s +20-7	β-

Isotope Z El A	Jπ	Δ (MeV)	T1/2 or Abundance	Decay Mode
47 Ag 118	(1)	-79.580	3.76 s 15	β-
118m		-79.452	2.0 s 2	β- 59%, IT 41%
119	(7/2+)	-78.590	2.1 s 1	β-
120		-75.770	1.23 s 3	β-
120m		-75.567	0.32 s 4	β-≈63%, IT≈37%
121		-74.550	0.78 s 1	β-, β-n
122	(3+)		0.56 s 3	β-, β-n
122m			1.5 s 5	β-, β-n
123			0.31 s 2	β-, β-n
124	(1,2,3)+		0.22 s 3	β-, β-n
48 Cd 97			3 s +4-2	ε, εp
98	0+	-67.900s	≈8 s	ε, εp?
99	(5/2+)	-69.890s	16 s 3	ε, εp 0.17%, εα < 1×10^{-4}%
100	0+	-74.320s	49.1 s 5	ε
101	(5/2+)	-75.660	1.2 m 2	ε
102	0+	-79.720s	5.5 m 5	ε
103	(5/2+)	-80.650	7.3 m 1	ε
104	0+	-83.977	57.7 m 10	ε
105	5/2+	-84.339	55.5 m 4	ε
106	0+	-87.135	1.25% 4	
107	5/2+	-86.990	6.50 h 2	ε
108	0+	-89.253	0.89% 2	
109	5/2+	-88.507	462.0 d 6	ε
110	0+	-90.351	12.49% 12	
111	1/2+	-89.254	12.80% 8	
111m	11/2-	-88.858	48.54 m 5	IT
112	0+	-90.581	24.13% 14	
113	1/2+	-89.050	9.3×10^{15} y 19 12.22% 8	β-
113m	11/2-	-88.786	14.1 y 5	β- 99.86%, IT 0.14%
114	0+	-90.021	28.73% 28	
115	1/2+	-88.091	53.46 h 10	β-
115m	11/2-	-87.910	44.6 d 3	β-
116	0+	-88.720	7.49% 12	
117	1/2+	-86.416	2.49 h 4	β-
117m	(11/2)-	-86.280	3.36 h 5	β-
118	0+	-86.709	50.3 m 2	β-
119	(1/2+)	-83.940	2.69 m 2	β-
119m	(11/2-)	-83.793	2.20 m 2	β-
120	0+	-83.973	50.80 s 21	β-
121		-80.950	13.5 s 3	β-
121m		-80.950	8 s	β-
122	0+	-80.580s	5.3 s 1	β-
123	(3/2+)	-77.520s	2.09 s 3	β-
123m		-77.520s	1.9 s 1	β-
124	0+		1.24 s 5	β-
125	(3/2+)		0.68 s 5	β-
125m			0.66 s 3	β-
126	0+		0.52 s 4	β-
127	(3/2+)		0.4 s 1	β-
128	0+		0.28 s 4	β-
129			0.27 s 4	β-

Isotope Z El A	Jπ	Δ (MeV)	T1/2 or Abundance	Decay Mode
48 Cd 130	0+		0.20 s 4	β-, β-n≈4%
49 In 100		-63.870s		ε, εp
101		-68.360s		
102	(5)	-70.580s	23 s 4	ε
103	(9/2+)	-74.607	65 s 7	ε
104	5+	-76.080s	1.84 m 5	ε
104m		-76.080s	15.7 s 5	IT
105	(9/2+)	-79.493	5.07 m 7	ε
105m	(1/2-)	-78.819	48 s 6	IT
106	7+	-80.617	6.2 m 1	ε
106m	(3)+	-80.588	5.2 m 1	ε
107	9/2+	-83.568	32.4 m 3	ε
107m	1/2-	-82.890	50.4 s 6	IT
108	7+	-84.112	58.0 m 12	ε
108m	2+	-84.082	39.6 m 7	ε
109	9/2+	-86.487	4.2 h 1	ε
109m	1/2-	-85.837	1.34 m 7	IT
109m	(19/2+)	-84.377	0.21 s 1	IT
110	2+	-86.410	69.1 m 5	ε
110	7+	-86.410	4.9 h 1	ε
111	9/2+	-88.391	2.8049 d 1	ε
111m	1/2-	-87.854	7.7 m 2	IT
112	1+	-87.995	14.97 m 10	ε 56%, β- 44%
112m	4+	-87.838	20.56 m 6	IT
113	9/2+	-89.368	4.3% 2	
113m	1/2-	-88.976	1.6582 h 6	IT
114	1+	-88.571	71.9 s	β- 99.5%, ε 0.5%
114m	5+	-88.381	49.51 d 1	IT 95.6%, ε 4.4%
115	9/2+	-89.539	4.41×10^{14} y 25 95.7% 2	β-
115m	1/2-	-89.203	4.486 h 4	IT 95%, β- 5%
116	1+	-88.252	14.10 s 3	β- 99.94%, ε < 0.06%
116m	5+	-88.125	54.41 m 3	β-
116m	8-	-87.962	2.18 s 4	IT
117	9/2+	-88.945	43.8 m 7	β-
117m	1/2-	-88.630	116.5 m 7	β- 52.9%, IT 47.1%
118	1+	-87.232	5.0 s 5	β-
118m	5+	-87.172	4.45 m 5	β-
118m	8-	-87.032	8.5 s 3	IT 98.6%, β- 1.4%
119	9/2+	-87.733	2.4 m 1	β-
119m	1/2-	-87.422	18.0 m 3	β- 97.5%, IT 2.5%
120	1+	-85.800	3.08 s 8	β-
120	(3,4,5)+	-85.800	46.2 s 8	β-
120	(8-)	-85.800	47.3 s 5	β-
121	9/2+	-85.841	23.1 s 6	β-
121m	1/2-	-85.527	3.88 m 10	β- 98.8%, IT 1.2%
122	1+	-83.580	1.5 s 3	β-
122m	(4,5)+	-83.580	10.3 s 6	β-
122m	8-	-83.360	10.8 s 4	β-
123	(9/2)+	-83.420	5.98 s 6	β-
123m	(1/2)-	-83.100	47.8 s 5	β-
124	3+	-81.060	3.17 s 5	β-

Isotope Z El A	Jπ	Δ (MeV)	T1/2 or Abundance	Decay Mode
49 In 124m	(8-)	-80.870	3.4 s 5	β-
125	(9/2)+	-80.420	2.33 s 4	β-
125m	(1/2-)	-80.240	12.2 s 1	β-
126	(8-)	-77.810	1.63 s 5	β-
126m	3+	-77.660	1.5 s 2	β-
127	(9/2+)	-77.010	1.15 s 5	β-
127m	(1/2-)	-76.850	3.76 s 3	β-, β-n
128	3+	-74.020	0.80 s 1	β-, β-n
128m	8-	-73.940	0.7 s 1	β-, β-n
129	(9/2+)	-73.020	0.63 s 4	β-, β-n
129m	(1/2-)	-72.820	1.23 s 2	β-, β-n
130	1(-)	-70.010	0.32 s 2	β-, β-n 0.9%
130m	(5+)	-70.010	0.55 s 1	β-, β-n < 1.67%
130m	(10-)	-70.010	0.55 s 1	β-, β-n < 1.67%
131	(9/2+)	-68.490	0.27 s 2	β-, β-n
131m	(21/2+)	-68.490	0.32 s 6	β-
131m	1/2-	-68.490	0.35 s 5	β-
132		-63.210s	0.203 s 6	β-, β-n
133			180 ms 20	β-, β-n
50 Sn 102	0+	-65.020s		
103		-67.050s	7 s 3	ε, εp
104	0+	-71.680s	21.4 s 9	ε
105		-73.240	31 s 6	ε, εp
106	0+	-77.450	2.10 m 15	ε
107	5/2+	-78.470s	2.90 m 5	ε
108	0+	-82.050	10.30 m 8	ε
109	7/2(+)	-82.633	18.0 m 2	ε
110	0+	-85.834	4.11 h 10	ε
111	7/2+	-85.943	35.3 m 8	ε
112	0+	-88.658	0.97% 1	
113	1/2+	-88.330	115.09 d 4	ε
113m	7/2+	-88.253	21.4 m 4	IT 91.1%, ε 8.9%
114	0+	-90.560	0.65% 1	
115	1/2+	-90.034	0.36% 1	
116	0+	-91.526	14.53% 11	
117	1/2+	-90.399	7.68% 7	
117m	11/2-	-90.084	13.60 d 4	IT
118	0+	-91.654	24.22% 11	
119	1/2+	-90.068	8.58% 4	
119m	11/2-	-89.978	293.0 d 13	IT
120	0+	-91.103	32.59% 10	
121	3/2+	-89.203	27.06 h 4	β-
121m	11/2-	-89.197	55 y 5	IT 77.6%, β- 22.4%
122	0+	-89.946	4.63% 3	
123	11/2-	-87.820	129.2 d 4	β-
123m	3/2+	-87.795	40.08 m 7	β-
124	0+	-88.237	5.79% 5	
125	11/2-	-85.898	9.64 d 3	β-
125m	3/2+	-85.870	9.52 m 5	β-
126	0+	-86.021	≈1.0×10^5 y	β-
127	(11/2-)	-83.504	2.10 h 4	β-
127m	(3/2+)	-83.499	4.13 m 3	β-
128	0+	-83.330	59.1 m 5	β-

Isotope Z El A	Jπ	Δ (MeV)	T1/2 or Abundance	Decay Mode
50 Sn 128m	(7-)	-81.239	6.5 s 5	IT
129	(3/2+)	-80.620	2.4 m 1	β-
129m	(11/2-)	-80.585	6.9 m 1	β-, IT 0.0002%
130	0+	-80.130	3.72 m 4	β-
130m	(7-)	-78.183	1.7 m 1	β-
131		-77.380	39 s 2	β-
131m	(3/2+)	-77.380	61 s 2	β-
132	0+	-76.610	40 s 1	β-
133	(7/2-)	-71.190	1.44 s 4	β-, β-n 0.08%
134	0+	-67.230s	1.04 s 2	β-, β-n 17%
51 Sb 104		-59.380s		
105		-63.930s		
106		-66.520s		
107		-70.770s		
108		-72.510s	7.0 s 5	ε
109	(5/2+)	-76.253	17.0 s 7	ε
110	3+	-77.530s	24 s 1	ε
111	(5/2+)	-80.840s	75 s 1	ε
112	3+	-81.603	51.4 s 10	ε
113	5/2+	-84.424	6.67 m 7	ε
114	3+	-84.680	3.49 m 3	ε
115	5/2+	-87.004	32.1 m 3	ε
116	3+	-86.819	15.8 m 8	ε
116m	8-	-86.436	60.3 m 6	ε
117	5/2+	-88.644	2.80 h 1	ε
118	1+	-87.998	3.6 m 1	ε
118m	8-	-87.786	5.00 h 2	ε
119	5/2+	-89.475	38.1 h 2	ε
120	1+	-88.423	15.89 m 4	ε
120m	8-	-88.423	5.76 d 2	ε
121	5/2+	-89.591	57.36% 15	
122	2-	-88.327	2.70 d 1	β- 97.6%, ε 2.4%
122m	8-	-88.163	4.21 m 2	IT
123	7/2+	-89.223	42.64% 15	
124	3-	-87.619	60.20 d 3	β-
124m	5+	-87.608	93 s 5	IT 75%, β- 25%
124m	8-	-87.582	20.2 m 2	IT
125	7/2+	-88.258	2.73 y 3	β-
126	(8-)	-86.400	12.4 d 1	β-
126m	(5+)	-86.382	19.0 m 3	β- 86%, IT 14%
126m	(3-)	-86.360	≈11 s	IT
127	7/2+	-86.705	3.85 d 5	β-
128	8-	-84.610	9.01 h 3	β-
128m	5+	-84.590	10.4 m 2	β- 96.4%, IT 3.6%
129		-84.624	17.7 m	β-
129	7/2+	-84.624	4.40 h 1	β-
130	(8-)	-82.330	39.5 m 8	β-
130m	(5)+	-82.330	6.3 m 2	β-
131	(7/2+)	-82.020	23 m 2	β-
132	(8-)	-79.730	4.2 m 1	β-
132m	(4+)	-79.730	2.8 m 1	β-
133	(7/2+)	-79.020	2.5 m 1	β-
134	(0-)	-74.020	0.85 s 10	β-
134	(7-)	-74.020	10.43 s 14	β-, β-n 0.1%

Isotope Z El A	Jπ	Δ (MeV)	T1/2 or Abundance	Decay Mode
51 Sb 135	(7/2+)	-70.320s	1.71 s 2	β-, β-n 16.4%
136		-65.050s	0.82 s 2	β-, β-n 24%
52 Te 106	0+	-58.270s	70 μs 20	α
107		-60.640s	3.6 ms +6-4	α 70%, ε 30%
108	0+	-65.820s	2.1 s 1	α 68%, ε 32%
109		-67.620	4.6 s 3	ε 96%, α 4%, εp
110	0+	-72.300	18.6 s 8	ε, α
111		-73.470	19.3 s 4	ε, α, εp
112	0+	-77.270	2.0 m 2	ε
113	(7/2+)	-78.320s	1.7 m 2	ε
114	0+	-81.760s	15.2 m 7	ε
115	7/2+	-82.360	5.8 m 2	ε
115m	(1/2)+	-82.340	6.7 m 4	ε, IT
116	0+	-85.290	2.49 h 4	ε
117	1/2+	-85.110	62 m 2	ε
118	0+	-87.653	6.00 d 2	ε
119	1/2+	-87.182	16.05 h 5	ε
119m	11/2-	-86.882	4.69 d 4	ε
120	0+	-89.386	0.095% 5	
121	1/2+	-88.551	16.78 d 35	ε
121m	11/2-	-88.257	154 d 7	IT 88.6%, ε 11.4%
122	0+	-90.307	2.59% 1	
123	1/2+	-89.171	$1.3{\times}10^{13}$ y 0.905% 5	ε
123m	11/2-	-88.924	119.7 d 1	IT
124	0+	-90.525	4.79% 2	
125	1/2+	-89.024	7.12% 2	
125m	11/2-	-88.879	58 d 1	IT
126	0+	-90.067	18.93% 3	
127	3/2+	-88.286	9.35 h 7	β-
127m	11/2-	-88.198	109 d 2	IT 97.6%, β- 2.4%
128	0+	-88.992	$>8.{\times}10^{24}$ y 31.70% 2	2β-
129	3/2+	-87.006	69.6 m 2	β-
129m	11/2-	-86.901	33.6 d 1	IT 64%, β- 36%
130	0+	-87.348	$\leq1.25{\times}10^{21}$ y 33.87% 7	
131	3/2+	-85.206	25.0 m 1	β-
131m	11/2-	-85.024	30 h 2	β- 77.8%, IT 22.2%
132	0+	-85.222	78.2 h 8	β-
133	(3/2+)	-82.970	12.5 m 3	β-
133m	(11/2-)	-82.636	55.4 m 4	β- 82.5%, IT 17.5%
134	0+	-82.430	41.8 m 8	β-
135	(7/2-)	-77.870	19.0 s 2	β-
136	0+	-74.460	17.5 s 2	β-, β-n 1.1%
137	(7/2-)	-69.480	2.49 s 5	β-, β-n 2.7%
138	0+	-66.110s	1.4 s 4	β-, β-n 6.3%
53 I 108		-52.750s		
109		-57.710s	0.11 ms 2	p
110		-60.520s	0.65 s 2	ε 83%, α 17%, εα, εp
111		-65.070s	2.5 s 2	ε 99.9%, α≈0.1%

Isotope Z El A	Jπ	Δ (MeV)	T1/2 or Abundance	Decay Mode
53 I 112		-67.100s	3.42 s 11	ε, α≈0.0012%, εp, εα
113		-71.120	6.6 s 2	ε, εα, α $3.3{\times}10^{-7}$%
114		-72.760s	2.1 s 2	ε, εp
115	(5/2+)	-76.400s	1.3 m 2	ε
116	1+	-77.550	2.91 s 15	ε
117	(5/2)+	-80.600s	2.22 m 4	ε
118	2-	-81.050s	13.7 m 5	ε
118m	(7-)	-81.050s	8.5 m 5	ε
119	(5/2+)	-83.780	19.1 m 4	ε
120	2-	-83.771	81.0 m 6	ε
120m	>3	-83.771	53 m 4	ε
121	5/2+	-86.270	2.12 h 1	ε
122	1+	-86.073	3.63 m 6	ε
123	5/2+	-87.937	13.2 h 1	ε
124	2-	-87.368	4.18 d 2	ε
125	5/2+	-88.846	60.14 d 11	ε
126	2-	-87.916	13.02 d 7	ε 56.3%, β- 43.7%
127	5/2+	-88.982	100%	
128	1+	-87.736	24.99 m 2	β- 93.1%, ε 6.9%
129	7/2+	-88.507	$1.57{\times}10^{7}$ y 4	β-
130	5+	-86.897	12.36 h 3	β-
130m	2+	-86.857	9.0 m 1	IT 84%, β- 16%
131	7/2+	-87.457	8.04 d 1	β-
132	4+	-85.715	2.30 h 3	β-
132m	(8-)	-85.595	83.6 m 17	IT 86%, β- 14%
133	7/2+	-85.888	20.8 h 1	β-
133m	(19/2-)	-84.254	9 s 2	IT
134	4+	-83.990	52.6 m 4	β-
134m	8-	-83.674	3.69 m 7	IT 97.7%, β- 2.3%
135	7/2+	-83.821	6.57 h 2	β-
136	2-	-79.550	83.4 s 10	β-
136m	(6-)	-78.910	46.9 s 10	β-
137	(7/2+)	-76.507	24.5 s 2	β-, β-n 7.1%
138	(2-)	-72.290	6.49 s 7	β-, β-n 5.5%
139	(7/2+)	-68.880	2.29 s 2	β-, β-n 9.9%
140	(3)	-64.250s	0.86 s 4	β-, β-n 9.4%
141			0.43 s 2	β-, β-n 21.2%
142			≈0.2 s	β-
54 Xe 110	0+	-51.970s	≈0.2 s	ε, α
111		-54.510s	0.74 s 20	ε, α
111		-54.510s	0.9 s 2	ε, α
112	0+	-60.060s	2.7 s 8	ε 99.16%, α 0.84%
113		-62.090	2.74 s 8	ε 99.97%, εp 4.2%, α 0.03%, εα
114	0+	-67.180s	10.0 s 4	ε
115	(5/2+)	-68.670s	18 s 4	ε, εp, εα
116	0+	-73.050s	56 s 2	ε
117	(5/2+)	-74.200s	61 s 2	ε, εp 0.003%
118	0+	-77.950s	3.8 m 9	ε
119	(7/2+)	-78.750	5.8 m 3	ε
120	0+	-81.810	40 m 1	ε
121	(5/2+)	-82.510	39.0 m 5	ε

Isotope Z El	A	Jπ	Δ (MeV)	T1/2 or Abundance	Decay Mode
54 Xe	122	0+	-85.050	20.1 h 1	ε
	123	(1/2)+	-85.258	2.08 h 2	ε
	124	0+	-87.659	0.10% 1	
	125	(1/2)+	-87.191	16.9 h 2	ε
	125m	(9/2)-	-86.938	57 s 1	IT
	126	0+	-89.174	0.09% 1	
	127	(1/2+)	-88.319	36.4 d 1	ε
	127m	(9/2-)	-88.022	69.2 s 9	IT
	128	0+	-89.860	1.91% 3	
	129	1/2+	-88.698	26.4% 6	
	129m	11/2-	-88.462	8.89 d 2	IT
	130	0+	-89.881	4.1% 1	
	131	3/2+	-88.428	21.2% 4	
	131m	11/2-	-88.264	11.9 d 1	IT
	132	0+	-89.292	26.9% 5	
	133	3/2+	-87.659	5.243 d 1	β-
	133m	11/2-	-87.426	2.19 d 1	IT
	134	0+	-88.125	10.4% 2	
	134m	7-	-86.166	290 ms 17	IT
	135	3/2+	-86.506	9.14 h 2	β-
	135m	11/2-	-85.979	15.29 m 5	IT, β- 0.004%
	136	0+	-86.429	≥2.36×10^{21} y	2β-
				8.9% 1	
	137	(7/2)-	-82.383	3.818 m 13	β-
	138	0+	-80.110	14.08 m 8	β-, β-n
	139	3/2-	-75.690	39.68 s 14	β-
	140	0+	-72.990	13.60 s 10	β-
	141	5/2+	-68.320	1.73 s 1	β-, β-n 0.04%
	142	0+	-65.500	1.22 s 2	β-
	143	5/2-		0.30 s 3	β-
	144	0+		1.15 s 20	β-
	145			0.9 s 3	β-, β-n
	146	0+			
55 Cs	113		-51.810s	33 μs 7	p
	114	(1+)	-54.740s	0.57 s 2	ε, εp 7%, εα 0.16%,
					α 0.02%
	115		-59.650s	1.4 s 8	ε, εp
	116	>4+	-62.290	3.84 s 16	ε, εα, εp
	116	(1+)	-62.290	0.70 s 4	ε, εα, εp
	117m		-66.260	6.5 s 4	ε
	117m		-66.260	8.4 s 6	ε
	118m	2	-68.270	14 s 2	εα, ε, εp
	118m	8,7,6	-68.270	17 s 3	εα, ε, εp
	119	9/2(+)	-72.240	37.7 s 10	ε
	119	3/2	-72.240	28 s 1	
	120	2	-73.820	64 s 3	ε
	120	high	-73.820	57 s 6	ε, εp≤1.0×10^{-5}%
	121	3/2+	-77.110	2.27 m 5	ε
	121m	9/2(+)	-77.110	121 s 3	ε, IT
	122	1+	-78.140	21.0 s 7	ε
	122m		-78.140	0.36 s 2	IT
	122m	8-	-78.140	4.5 m 2	ε
	123	1/2+	-81.070	5.87 m 5	ε
	123m	(11/2-)	-80.911	1.60 s 15	IT

Isotope Z El	A	Jπ	Δ (MeV)	T1/2 or Abundance	Decay Mode
55 Cs	124	1+	-81.740	30.8 s 5	ε
	124m	(7)+	-81.277	6.3 s 2	IT
	125	1/2+	-84.113	45 m 1	ε
	126	1+	-84.347	1.64 m 2	ε
	127	1/2+	-86.243	6.25 h 10	ε
	128	1+	-85.928	3.62 m 2	ε
	129	1/2+	-87.506	32.06 h 6	ε
	130	1+	-86.853	29.21 m 4	ε 98.4%, β- 1.6%
	130m	5-	-86.690	3.46 m 6	IT 99.84%,
					ε 0.16%
	131	5/2+	-88.076	9.69 d 1	ε
	132	2(-)	-87.171	6.475 d 10	ε 98%, β- 2%
	133	7/2+	-88.086	100%	
	134	4+	-86.906	2.062 y 5	β-, ε 0.0003%
	134m	8-	-86.767	2.91 h 1	IT
	135	7/2+	-87.662	2.3×10^{6} y 3	β-
	135m	19/2-	-86.029	53 m 2	IT
	136	5+	-86.354	13.16 d 3	β-
	136m	8-	-86.354	19 s 2	IT, β-?
	137	7/2+	-86.556	30.1 y 2	β-
	138	3-	-82.896	32.2 m 1	β-
	138m	6-	-82.816	2.91 m 8	IT 81%, β- 19%
	139	7/2+	-80.710	9.27 m 5	β-
	140	1-	-77.053	63.7 s 3	β-
	141	7/2+	-74.472	24.94 s 6	β-, β-n 0.03%
	142	0-	-70.538	1.70 s 2	β-, β-n 0.28%
	143	3/2(+)	-67.745	1.78 s 1	β-, β-n 1.62%
	144	1	-63.370	1.01 s 1	β-, β-n 3.17%
	144m	(≥4)	-63.370	<1 s	β-
	145	3/2+	-60.210	0.594 s 13	β-, β-n 13.8%
	146	(2-)	-55.700	0.343 s 7	β-, β-n 14%
	147		-52.300	0.225 s 5	β- 57%, β-n 43%
	148		-47.580	158 ms 7	β-
56 Ba	117	(3/2)	-57.160s	1.8 s 1	ε, εα, εp
	118	0+	-62.350s		
	119		-64.460s	5.35 s 30	ε, εp
	120	0+	-69.020s	32 s 5	ε
	121		-70.420s	29.7 s 15	ε, εp 0.02%
	122	0+	-74.540s	1.95 m 15	ε
	123		-75.560s	2.7 m 4	ε
	124	0+	-79.140s	11.9 m 10	ε
	125	(1/2+)	-79.550	3.5 m 4	ε
	125		-79.550	8 m	ε
	126	0+	-82.770s	100 m 2	ε
	127	(1/2+)	-82.790	12.7 m 4	ε
	128	0+	-85.470	2.43 d 5	ε
	129	1/2+	-85.080	2.23 h 11	ε
	129m	(7/2)+	-85.072	2.17 h 4	ε
	130	0+	-87.291	0.106% 2	
	131	1/2+	-86.714	11.8 d 2	ε
	131m	9/2-	-86.526	14.6 m 2	IT
	132	0+	-88.447	0.101% 3	
	133	1/2+	-87.570	10.52 y 13	ε
	133m	11/2-	-87.282	38.9 h 1	IT 99.99%,
					ε 0.01%

Isotope Z El A	Jπ	Δ (MeV)	T1/2 or Abundance	Decay Mode
56 Ba 134	0+	-88.965	2.42% 4	
135	3/2+	-87.867	6.593% 24	
135m	11/2-	-87.599	28.7 h 2	IT
136	0+	-88.903	7.85% 5	
136m	7-	-86.872	0.3084 s 19	IT
137	3/2+	-87.732	11.23% 5	
137m	11/2-	-87.070	2.552 m 1	IT
138	0+	-88.272	71.70% 9	
139	7/2-	-84.924	83.06 m 28	β-
140	0+	-83.273	12.752 d 3	β-
141	3/2-	-79.732	18.27 m 7	β-
142	0+	-77.847	10.6 m 2	β-
143	5/2-	-73.979	14.33 s 8	β-
144	0+	-71.840	11.5 s 2	β-, β-n 3.6%
145	5/2-	-68.120	4.31 s 16	β-
146	0+	-65.060	2.20 s 3	β-
147		-61.500	0.893 s 1	β-, β-n 0.02%
148	0+	-58.130s	0.607 s 25	β-, β-n≤0.02%
149		-54.300s	0.356 s 8	β-n 0.43%
57 La 120			2.8 s 2	ε, εp
122			8.7 s 7	ε, εp
123			17 s 3	ε
124	(7/2+)	-70.240s	29 s 3	ε
125	(11/2-)	-73.810s	76 s 6	ε
126		-75.050s	1.0 m 3	ε
127	(3/2+)	-77.990s	3.8 m 5	ε
127m	(11/2-)	-77.990s	5.0 m 5	
128	4-,5-	-78.820	5.0 m 3	ε
129	(3/2+)	-81.360	11.6 m 2	ε
129m	11/2-	-81.188	0.56 s 5	IT
130	3(+)	-81.590s	8.7 m 1	ε
131	3/2+	-83.750	59 m 2	ε
132	2-	-83.740	4.8 h 2	ε
132m	6-	-83.551	24.3 m 5	IT 76%, ε 24%
133	5/2+	-85.520s	3.912 h 8	ε
134	1+	-85.252	6.45 m 16	ε
135	5/2+	-86.667	19.5 h 2	ε
136	1+	-86.030	9.87 m 3	ε
137	7/2+	-87.130	6×10⁴ y 2	ε
138	5+	-86.531	1.05×10¹¹ y 2 0.0902% 2	ε 66.4%, β- 33.6%
139	7/2+	-87.238	99.9098% 2	
140	3-	-84.327	1.6781 d 3	β-
141	7/2(+)	-82.983	3.92 h 3	β-
142	2-	-80.027	91.1 m 5	β-
143	7/2+	-78.200	14.2 m 1	β-
144	(3-)	-74.940	40.8 s 4	β-
145		-73.020	24.8 s 20	β-
146	(2-)	-69.200	6.27 s 10	β-
146m	(6)	-69.170	10.0 s 1	β-
147	(5/2+)	-67.250	4.015 s 8	β-, β-n 0.04%
148	(2-)	-63.810	1.05 s 1	β-, β-n 0.11%
149		-61.290s	1.2 s 4	β-, β-n

Isotope Z El A	Jπ	Δ (MeV)	T1/2 or Abundance	Decay Mode
57 La 150		-57.500s		
58 Ce 123			3.8 s 2	ε, εp
124	0+		6 s 2	ε
125	(5/2+)		10 s 1	ε, εp
126	0+	-71.070s	50 s 6	ε
127		-72.290s	32 s 4	ε
128	0+	-75.870s	6 m 2	ε
129		-76.480s	3.5 m 5	ε
130	0+	-79.590s	25 m 2	ε
131		-79.730	10 m 1	ε
131		-79.730	5 m 1	ε
132	0+	-82.440s	3.5 h 1	ε
133	9/2-	-82.470s	4.9 h 4	ε
133m	1/2+	-82.470s	97 m 4	ε
134	0+	-84.750	75.9 h 9	ε
135	1/2(+)	-84.641	17.7 h 2	ε
135m	11/2(-)	-84.195	20 s 1	IT
136	0+	-86.500	0.19% 1	
137	3/2+	-85.910	9.0 h 3	ε
137m	11/2-	-85.656	34.4 h 3	IT 99.22%, ε 0.78%
138	0+	-87.574	0.25% 1	
139	3/2+	-86.973	137.640 d 23	ε
139m	11/2-	-86.219	54.8 s 10	IT
140	0+	-88.088	88.43% 10	
141	7/2-	-85.445	32.501 d 5	β-
142	0+	-84.542	>5×10¹⁶ y 11.13% 10	
143	3/2-	-81.616	33.10 h 5	β-
144	0+	-80.441	284.893 d 8	β-
145	(3/2-)	-77.110	3.01 m 6	β-
146	0+	-75.730	13.52 m 13	β-
147	5/2-	-72.190	56.4 s 10	β-
148	0+	-70.430	56 s 1	β-
149		-66.800	5.2 s 3	β-
150	0+	-64.990	4.0 s 6	β-
151		-61.660s	1.02 s 6	β-
152	0+	-59.760s	3.1 s 3	β-
59 Pr 124			1.2 s 2	ε, εp
126			3.2 s 6	ε, εp
128		-66.320s	3.2 s 5	ε, εp
129		-70.060s	24 s 5	ε
130		-71.290s	40.0 s 4	ε
131		-74.450s	1.7 m 4	ε
132		-75.340s	1.6 m 3	ε
133	5/2(+)	-78.020s	6.5 m 3	ε
134	2-	-78.650s	17 m 2	ε
134m	(5-)	-78.650s	11 m	ε
135	3/2(+)	-80.920	24 m 2	ε
136	2+	-81.370	13.1 m 1	ε
137	5/2+	-83.200	1.28 h 3	ε
138	1+	-83.137	1.45 m 5	ε
138m	7-	-82.773	2.1 h 1	ε
139	5/2+	-84.844	4.41 h 4	ε

Isotope Z El	A	Jπ	Δ (MeV)	T1/2 or Abundance	Decay Mode
59 Pr	140	1+	-84.700	3.39 m 1	ε
	141	5/2+	-86.026	100%	
	142	2-	-83.798	19.12 h 4	β- 99.98%, ε 0.02%
	142m	5-	-83.794	14.6 m 5	IT
	143	7/2+	-83.078	13.57 d 2	β-
	144	0-	-80.760	17.28 m 5	β-
	144m	3-	-80.701	7.2 m 3	IT 99.93%, β- 0.07%
	145	7/2+	-79.636	5.984 h 10	β-
	146	(2)-	-76.760	24.15 m 18	β-
	147	(3/2+)	-75.470	13.6 m 5	β-
	148	1-	-72.490	2.27 m 4	β-
	148m	(4)	-72.400	2.0 m 1	β-
	149	(5/2+)	-70.988	2.26 m 7	β-
	150	1(-)	-68.000	6.19 s 16	β-
	151	1/2-to5/2-	-66.760s	18.90 s 7	β-
	152		-64.160s	3.24 s 19	β-
	153		-62.370s	4.3 s 2	β-
	154		-59.110s	2.3 s 1	β-
60 Nd	127	(5/2)		1.8 s 4	ε, εp
	128			4 s 2	ε, εp
	129	(5/2-)	-62.880s	4.9 s 2	ε, εp
	130	0+	-66.990s	28 s 3	ε
	131		-68.230s	24 s 3	ε, εp
	132	0+	-71.940s	1.8 m 2	ε
	133		-72.570s	70 s 10	ε
	133m	(9/2-)	-72.570s	<2 m	ε
	134	0+	-75.950s	8.5 m 15	ε
	135	9/2(-)	-76.220s	12.4 m 6	ε
	135m		-76.220s	5.5 m 5	ε
	136	0+	-79.160	50.65 m 33	ε
	137	1/2+	-79.700	38.5 m 15	ε
	137m	11/2-	-79.180	1.60 s 15	IT
	138	0+	-82.040s	5.04 h 9	ε
	139	3/2+	-82.060	29.7 m 5	ε
	139m	11/2-	-81.829	5.50 h 20	ε 88.2%, IT 11.8%
	140	0+	-84.471	3.37 d 2	ε
	141	3/2+	-84.203	2.49 h 3	ε
	141m	11/2-	-83.446	62.4 s 9	IT 99.97%, ε 0.03%
	142	0+	-85.960	27.13% 10	
	143	7/2-	-84.012	12.18% 5	
	144	0+	-83.758	2.29×10^{15} y 16 23.80% 10	α
	145	7/2-	-81.442	8.30% 5	
	146	0+	-80.935	17.19% 8	
	147	5/2-	-78.156	10.98 d 1	β-
	148	0+	-77.418	5.76% 3	
	149	5/2-	-74.385	1.72 h 1	β-
	150	0+	-73.693	$>1\times10^{18}$ y 5.64% 3	2β-
	151	(3/2)+	-70.956	12.44 m 7	β-
	152	0+	-70.160	11.4 m 2	β-

Isotope Z El	A	Jπ	Δ (MeV)	T1/2 or Abundance	Decay Mode
60 Nd	153		-67.170s	28.9 s 4	β-
	154	0+	-65.860s	25.9 s 2	β-
	155		-62.700s	8.9 s 2	β-
	156	0+	-60.570s	5.5 s 1	β-
61 Pm	130			2.2 s 5	ε, εp
	132		-61.940s	5.0 s 7	ε, εp
	133		-65.620s	12 s 3	ε
	134	(5+)	-67.050s	24 s 2	ε
	135	(11/2-)	-70.220s	49 s 7	ε
	136	(3+)	-71.300s	≈107 s	ε
	136	5(+),6-	-71.300s	107 s 6	ε
	137	(11/2)-	-74.020	2.4 m 1	ε
	138	1+	-75.140s	10 s 2	ε
	138m	(3+)	-75.140s	3.24 m 5	ε
	138m	(6-)	-75.140s	3.24 m	
	139	(5/2)+	-77.540	4.15 m 5	ε
	139m	(11/2)-	-77.351	180 ms 20	IT, ε?
	140	1+	-78.380	9.2 s 2	ε
	140m	7-	-78.380	5.95 m 5	ε
	141	5/2+	-80.472	20.90 m 5	ε
	142	1+	-81.090	40.5 s 5	ε
	143	5/2+	-82.970	265 d 7	ε
	144	5-	-81.425	363 d 14	ε
	145	5/2+	-81.278	17.7 y 4	ε, α 3×10^{-7}%
	146	3-	-79.458	5.53 y 5	ε 66.1%, β- 33.9%
	147°	7/2+	-79.052	2.6234 y 2	β-
	148	1-	-76.874	5.370 d 9	β-
	148m	6-	-76.736	41.29 d 11	β- 95%, IT 5%
	149	7/2+	-76.073	53.08 h 5	β-
	150	(1-)	-73.606	2.68 h 2	β-
	151	5/2+	-73.398	28.40 h 4	β-
	152	1+	-71.270	4.1 m 1	β-
	152m	4-	-71.100	7.52 m 8	β-
	152m	(8)	-71.100	13.8 m 2	β-, IT
	153	5/2-	-70.669	5.4 m 2	β-
	154	(0,1)	-68.410	1.73 m 10	β-
	154m	(3,4)	-68.410	2.68 m 7	β-
	155	(5/2-)	-67.100s	48 s 4	β-
	156		-64.370s	26.7 s 1	β-
	157		-62.370s	10.90 s 20	β-
	158		-59.410s	4.8 s 5	β-
62 Sm	131			1.2 s 2	ε, εp
	133	(5/2+)		2.9 s 2	ε, εp
	134	0+	-62.050s	11 s 2	ε
	135	(7/2+)	-63.520s	10 s 2	ε, εp
	136	0+	-67.260s	43 s 3	ε
	137	(9/2-)	-68.100s	45 s 1	ε
	138	0+	-71.540s	3.0 m 3	ε
	139	(1/2)+	-72.080	2.57 m 10	ε
	139m	(11/2)-	-71.622	10.7 s 6	IT 93.7%, ε 6.3%
	140	0+	-75.380s	14.82 m 10	ε
	141	1/2+	-75.943	10.2 m 2	ε
	141m	11/2-	-75.767	22.6 m 2	ε 99.69%, IT 0.31%

Isotope Z El A	Jπ	Δ (MeV)	T1/2 or Abundance	Decay Mode
62 Sm 142	0+	-78.986	72.49 m 5	ε
143	3/2+	-79.526	8.83 m 1	ε
143m	11/2-	-78.772	66 s 2	IT 99.66%, ε 0.34%
144	0+	-81.975	3.1% 1	
145	7/2-	-80.660	340 d 3	ε
146	0+	-81.000	10.31×10^{7} y 45	α
147	7/2-	-79.276	1.06×10^{11} y 2	α
			15.0% 2	
148	0+	-79.346	7×10^{15} y 3	α
			11.3% 1	
149	7/2-	-77.146	$>2\times10^{15}$ y	
			13.8% 1	
150	0+	-77.060	7.4% 1	
151	5/2-	-74.587	90 y 8	β-
152	0+	-74.773	26.7% 2	
153	3/2+	-72.569	46.27 h 1	β-
154	0+	-72.465	22.7% 2	
155	3/2-	-70.201	22.3 m 2	β-
156	0+	-69.374	9.4 h 2	β-
157		-66.870	8.07 m 12	β-
158	0+	-65.400s	5.51 m 9	β-
159		-62.370s	11.2 s 2	β-
160		-60.350s	9.6 s 3	β-
63 Eu 134			0.5 s 2	ε, εp
135			1.5 s 2	ε
136	(1+)	-57.000s	3.9 s 5	ε, εp
136m	(7+)	-57.000s	≈3.2 s	
137	(11/2-)	-60.720s	11 s 2	ε
138	(7+)	-62.340s	12.1 s 6	ε
139	(11/2)-	-65.630s	17.9 s 6	ε
140	1(-)	-66.980s	1.54 s 13	ε
140m		-66.980s	0.125 s 2	ε
141	5/2+	-69.980	40.0 s 7	ε
141m	11/2-	-69.884	2.7 s 3	IT 93%, ε 7%
142	1+	-71.590	2.4 s 2	ε
142m	8-	-71.410	1.22 m 2	ε
143	5/2+	-74.380	2.63 m 5	ε
144	1+	-75.646	10.2 s 1	ε
145	5/2+	-78.000	5.93 d 4	ε
146	4-	-77.125	4.59 d 3	ε
147	5/2+	-77.555	24.1 d 6	ε, α 0.0022%
148	5-	-76.239	54.5 d 5	ε, α 9.4×10^{-7}%
149	5/2+	-76.455	93.1 d 4	ε
150	5(-)	-74.800	35.8 y 10	ε
150m	0(-)	-74.758	12.8 h 1	β- 89%, ε 11%
151	5/2+	-74.663	47.8% 5	
152	3-	-72.899	13.542 y 10	ε 72.08%, β- 27.92%
152m	0-	-72.853	9.274 h 9	β- 72%, ε 28%
152m	8-	-72.751	96 m 1	IT
153	5/2+	-73.378	52.2% 5	
154	3-	-71.748	8.592 y 5	β- 99.98%, ε 0.02%

Isotope Z El A	Jπ	Δ (MeV)	T1/2 or Abundance	Decay Mode
63 Eu 154m	(8-)	-71.591	46.0 m 4	IT
155	5/2+	-71.829	4.68 y 5	β-
156	0+	-70.096	15.19 d 8	β-
157	5/2+	-69.472	15.18 h 3	β-
158	(1-)	-67.220	45.9 m 2	β-
159	5/2+	-66.058	18.1 m 1	β-
160	(0-)	-63.550s	38 s 4	β-
161		-61.770s	26 s 3	β-
162		-59.080s	10.6 s 10	β-
64 Gd 137			7 s 3	ε, εp
138		-56.640s		
139		-58.470s	4.9 s 10	ε, εp
140	0+	-62.480s	16 s 1	ε
141	1/2+	-63.540s	≈20 s	ε, εp 0.03%
141m	11/2-	-63.162s	24.5 s 9	ε 86%, IT 14%
142	0+	-67.390s	70.2 s 6	ε
143	(1/2)+	-68.470s	39 s 2	ε
143m	(11/2-)	-68.317s	112 s 2	ε
144	0+	-71.950s	4.5 m 1	ε
145	1/2+	-72.950	23.0 m 4	ε
145m	11/2-	-72.201	85 s 3	IT 94.3%, ε 5.7%
146	0+	-76.099	48.27 d 10	ε
147	7/2-	-75.367	38.06 h 12	ε
148	0+	-76.278	74.6 y 30	α
149	7/2(-)	-75.135	9.4 d 3	ε, α
150	0+	-75.771	1.79×10^{6} y 8	α
151	7/2-	-74.199	124 d 1	ε, α 1.0×10^{-6}%
152	0+	-74.718	1.08×10^{14} y 8	α
			0.20% 1	
153	3/2-	-72.893	241.6 d 2	ε
154	0+	-73.717	2.18% 3	
155	3/2-	-72.081	14.80% 5	
156	0+	-72.546	20.47% 4	
157	3/2-	-70.834	15.65% 3	
158	0+	-70.701	24.84% 12	
159	3/2-	-68.572	18.56 h 8	β-
160	0+	-67.953	21.86% 4	
161	5/2-	-65.517	3.66 m	β-
162	0+	-64.240	8.4 m 2	β-
163	(5/2-)	-61.590s	68 s 3	β-
164		-59.280s	45 s 3	β-
65 Tb 140		-51.780s	2.4 s 4	ε, εp
141	(11/2-)	-55.580s	3.5 s 2	ε
141m		-55.580s	7.9 s 6	ε
142	1+	-57.390s	597 ms 17	ε, εp≈3.0×10^{-7}%
142m	(5-)	-57.390s	303 ms 7	IT?, ε?, εp?
143	(11/2-)	-60.970s	12 s 1	ε
143m	(5/2+)	-60.970s	<17 s	
144	(1+)	-62.750s	≈1 s	ε
144m	(6-)	-62.353s	4.25 s 15	IT 66%, ε 34%
145	(1/2+)	-66.200s		
145m	(11/2-)	-66.200s	29.5 s 15	ε
146	1+	-67.860	8 s 4	ε
146m	5-	-67.860	23 s 2	ε

Isotope Z El A	Jπ	Δ (MeV)	T1/2 or Abundance	Decay Mode
65 Tb 147	1/2+	-70.880	1.7 h 1	ε
147m	11/2-	-70.829	1.83 m 6	ε
148	2-	-70.680	60 m 1	ε
148m	9+	-70.590	2.20 m 5	ε
149	1/2+	-71.499	4.13 h 2	ε 84.2%, α 15.8%
149m	11/2-	-71.463	4.16 m 4	ε, α
150m	(8+,9+)	-71.113	5.8 m 2	ε
150m	(2-)	-71.113	3.48 h 16	ε, α<0.05%
151	1/2(+)	-71.633	17.609 h 1	ε, α 0.0095%
151m	(11/2-)	-71.533	25 s 3	IT 93.8%, ε 6.2%
152	2-	-70.770	17.5 h 1	ε, α<7.0×10^{-7}%
152m	8+	-70.268	4.2 m 1	IT 78.9%, ε 21.1%
153	5/2+	-71.322	2.34 d 1	ε
154	0	-70.150	21.5 h 4	ε, β-<0.1%
154m	3-	-70.150	9.0 h 5	ε 78.2%, IT 21.8%, β-<0.1%
154m	7-	-70.150	22.7 h 5	ε 98.2%, IT 1.8%
155	3/2+	-71.261	5.32 d 6	ε
156	3-	-70.102	5.35 d 10	ε, β-?
156m	(7-)	-70.052	24.4 h 10	IT
156m	(0+)	-70.014	5.3 h 2	ε, IT
157	3/2+	-70.772	99 y 10	ε
158	3-	-69.480	180 y 11	ε 83.4%, β- 16.6%
158m	0-	-69.370	10.5 s 2	IT, β-<0.6%, ε<0.01%
159	3/2+	-69.542	100%	
160	3-	-67.846	72.3 d 2	β-
161	3/2+	-67.471	6.88 d 3	β-
162	1-	-65.680	7.76 m 10	β-
163	3/2+	-64.700	19.5 m 3	β-
164	(5+)	-62.090	3.0 m 1	β-
165	(3/2+)	-60.610s	2.11 m 10	β-
66 Dy 141			0.9 s 2	ε, εp
142	0+	-50.990s	2.3 s 3	ε, εp≈8.0×10^{-5}%
143		-52.870s	3.9 s 4	ε, εp
144	0+	-57.150s	9.1 s 4	ε, εp
145		-58.750s		
145m	(11/2-)	-58.750s	13.6 s 10	ε
146	0+	-62.860s	29 s 3	ε
146m	(10+)	-62.860s	150 ms 20	IT
147	1/2+	-64.330	40 s 10	ε, εp
147m	11/2-	-63.579	55.7 s 5	ε 67%, IT 33%
148	0+	-68.000	3.1 m 1	ε
149	(7/2-)	-67.900s	4.23 m 18	ε
150	0+	-69.324	7.17 m 5	ε 64%, α 36%
151	7/2(-)	-68.764	17.9 m 3	ε 94.4%, α 5.6%
152	0+	-70.127	2.38 h 2	ε 99.9%, α 0.1%
153	7/2(-)	-69.152	6.4 h 1	ε 99.99%, α 0.0094%
154	0+	-70.399	3.0×10^{6} y 15	α
155	3/2-	-69.166	10.0 h 3	ε
156	0+	-70.536	0.06% 1	
157	3/2-	-69.434	8.14 h 4	ε
158	0+	-70.418	0.10% 1	

Isotope Z El A	Jπ	Δ (MeV)	T1/2 or Abundance	Decay Mode
66 Dy 159	3/2-	-69.176	144.4 d 2	ε
160	0+	-69.682	2.34% 5	
161	5/2+	-68.064	18.9% 1	
162	0+	-68.189	25.5% 2	
163	5/2-	-66.389	24.9% 2	
164	0+	-65.976	28.2% 2	
165	7/2+	-63.621	2.334 h 6	β-
165m	1/2-	-63.513	1.257 m 6	IT 97.76%, β- 2.24%
166	0+	-62.593	81.6 h 1	β-
167	(1/2-)	-59.940	6.20 m 8	β-
168	0+	-58.500s	8.5 m 5	β-
67 Ho 144		-45.650s	0.7 s 1	ε, εp
145		-50.000s		
146	(10+)	-52.160s	3.6 s 3	ε, εp
147	(11/2)	-56.280s	5.8 s 4	ε, εp
148	1+	-58.380s	2.2 s 11	ε
148m	6-	-58.380s	9.59 s 15	ε, εp 0.08%
149	(1/2+)	-61.910s	>30 s	ε
149m	(11/2-)	-61.910s	21.4 s 18	ε
150	(9+)	-62.210	26 s 2	ε
150	(2-)	-62.210	72 s 4	ε
151	(11/2-)	-63.720	35.2 s 1	ε 78%, α 22%
151m	(1/2+)	-63.679	47.2 s 10	α>40%
152	2-	-63.750	161.8 s 3	ε 88%, α 12%
152m	9+	-63.590	49.5 s 3	ε 89.2%, α 10.8%
153	11/2(-)	-65.023	2.0 m 1	ε 99.95%, α 0.05%
153m	1/2(-)	-64.955	9.3 m 5	ε 99.82%, α 0.18%, IT
154	(1-,2,3+)	-64.647	11.8 m 5	ε 99.98%, α 0.02%
154m	8+	-64.647	3.25 m 10	ε, α<0.001%
155	5/2+	-66.064	48 m 1	ε, α
156	5+	-65.600s	56 m 1	ε
157	7/2-	-66.890	12.6 m 2	ε
158	5+	-66.200	11.3 m 4	ε
158m	2-	-66.133	27 m 2	IT>81%, ε<19%
158m	(9+)	-66.020	21.3 m 23	ε
159	7/2-	-67.338	33.05 m 1	ε
159m	1/2+	-67.132	8.30 s	IT
160	5+	-66.391	25.6 m 3	ε
160m	2-	-66.331	5.02 h 5	IT 65%, ε 35%
160m	(1+)	-66.330	3 s	
161	7/2-	-67.207	2.48 h 5	ε
161m	1/2+	-66.996	6.76 s 7	IT
162	1+	-66.050	15 m 1	ε
162m	6-	-65.944	67.0 m 10	IT 63%, ε 37%
163	7/2-	-66.386	4570 y 25	ε
163m	1/2+	-66.088	1.09 s 3	IT
164	1+	-64.990	29 m 1	ε 60%, β- 40%
164m	6-	-64.850	37.5 m +15-5	IT
165	7/2-	-64.907	100%	
166	0-	-63.079	26.80 h 2	β-
166m	(7)-	-63.073	1.20×10^{3} y 18	β-
167	7/2-	-62.291	3.1 h 1	β-

Isotope Z El A	Jπ	Δ (MeV)	T1/2 or Abundance	Decay Mode
67 Ho 168	3+	-60.260	2.99 m *7*	β-
169	7/2-	-58.805	4.7 m *1*	β-
170	(6+)	-56.250	2.76 m *5*	β-
170m	1(+)	-56.130	43 s *2*	β-
68 Er 146		-45.060s		
147	(11/2-)	-47.330s	2.5 s *2*	ε, εp
147m	(1/2+)	-47.330s	≈2.5 s	ε, εp
148	0+	-52.000s	4.6 s *2*	ε
149	(1/2+)	-54.950	10.7 s *4*	ε, εp
149m	(11/2-)	-54.208	10.8 s *6*	ε, εp, IT
150	0+	-58.120s	18.5 s *7*	ε
151	(7/2-)	-58.460s	23.5 s *13*	ε
152	0+	-60.640	10.3 s *1*	α 90%, ε 10%
153	(7/2-)	-60.670s	37.1 s *2*	α 53%, ε 47%
154	0+	-62.622	3.68 m *15*	ε 99.53%, α 0.47%
155	(7/2-)	-62.220	5.3 m *3*	ε 99.98%, α 0.02%
156	0+	-64.100s	19.5 m *10*	ε
157	3/2-	-63.420	18.65 m *10*	ε, α?
158	0+	-65.300s	2.24 h *7*	ε
159	3/2-	-64.570	36 m *1*	ε
160	0+	-66.063	28.58 h *9*	ε
161	3/2-	-65.203	3.21 h *3*	ε
162	0+	-66.346	0.14% *1*	
163	5/2-	-65.177	75.0 m *4*	ε
164	0+	-65.952	1.61% *1*	
165	5/2-	-64.530	10.36 h *4*	ε
166	0+	-64.933	33.6% *2*	
167	7/2+	-63.298	22.95% *13*	
167m	1/2-	-63.090	2.269 s *6*	IT
168	0+	-62.998	26.8% *2*	
169	1/2-	-60.930	9.40 d *2*	β-
170	0+	-60.117	14.9% *1*	
171	5/2-	-57.727	7.52 h *3*	β-
172	0+	-56.491	49.3 h *3*	β-
173	(7/2-)	-53.660s	1.4 m *1*	β-
174	0+		3.3 m *2*	β-
69 Tm 147	(11/2-)	-36.710s	0.56 s *4*	ε≈90%, p≈10%
148m	(10+)	-39.880s	0.7 s *2*	ε
149	(11/2-)	-44.510s	0.9 s *2*	ε
150	(6-)	-47.010s	2.3 s *4*	ε
151m	(11/2-)	-51.220s	4.13 s *11*	ε
151m	(1/2+)	-51.220s	5.2 s *20*	ε
152	(9+)	-51.850s	5.2 s *6*	ε
152m	(2-)	-51.850s	8.0 s *10*	ε
153	(11/2-)	-54.240s	1.48 s *1*	α 91%, ε 9%
153m	(1/2+)	-54.197s	2.5 s *2*	α 95%, ε 5%
154	(2-)	-54.700	8.1 s *3*	ε 56%, α 44%
154	(9+)	-54.700	3.30 s *7*	α 90%, ε 10%
155		-56.730	32 s *7*	ε >94%, α<6%
156	2-	-56.980	83.8 s *18*	ε 99.91%, α 0.09%
156m		-56.980	19 s *3*	α
157	1/2+	-58.890s	3.5 m *2*	ε
158	2-	-58.900s	4.02 m *10*	ε
158m	(5+)	-58.900s	≈20 s	
69 Tm 159	5/2(+)	-60.670s	9.15 m *17*	ε
160	1-	-60.460	9.4 m *3*	ε
160m	(5)	-60.360	74.5 s *15*	ε 15%
161	7/2+	-62.100	33 m *3*	ε
162	1-	-61.550	21.7 m *2*	ε
162m	5+	-61.358	24.3 s *17*	IT 82%, ε 18%
163	1/2+	-62.738	1.810 h *5*	ε
164	1+	-61.990	2.0 m *1*	ε, ε 39%
164	6-	-61.990	5.1 m *1*	IT≈80%, ε≈20%
165	1/2+	-62.938	30.06 h *3*	ε
166	2+	-61.894	7.70 h *3*	ε
167	1/2+	-62.550	9.25 d *2*	ε
168	3(+)	-61.319	93.1 d *2*	ε 99.99%, β- 0.01%
169	1/2+	-61.280	100%	
170	1-	-59.802	128.6 d *3*	β- 99.85%, ε 0.15%
171	1/2+	-59.217	1.92 y *1*	β-
172	2-	-57.382	63.6 h *2*	β-
173	(1/2+)	-56.265	8.24 h *8*	β-
174	(4)-	-53.870	5.4 m *1*	β-
175	(1/2+)	-52.300	15.2 m *5*	β-
176	(4+)	-49.700s	1.9 m *1*	β-
177			130 s *40*	
70 Yb 151	(1/2+)	-41.960s	≈1.6 s	ε, εp
151m	(11/2-)	-41.960s	≈1.6 s	ε, εp
152	0+	-46.640s	3.1 s *2*	ε
153		-47.270s	4.2 s *1*	α 50%, ε 50%
154	0+	-50.220s	0.402 s *17*	α≈98%, ε≈2%
155		-50.700s	1.72 s *12*	α 84%, ε 16%
156	0+	-53.410	26.1 s *7*	ε 90%, α 10%
157	(7/2-)	-53.630s	38.6 s *10*	ε 99.5%, α 0.5%
158	0+	-56.022	1.57 m *9*	ε, α≈0.003%
159	(5/2)	-55.900s	1.40 m *20*	ε
160	0+	-58.160s	4.8 m *2*	ε
161	3/2-	-57.900s	4.2 m *2*	ε
162	0+	-59.850s	18.87 m *19*	ε
163	3/2-	-59.370	11.05 m *25*	ε
164	0+	-60.990s	75.8 m *17*	ε
165	5/2-	-60.175	9.9 m *3*	ε
166	0+	-61.589	56.7 h *1*	ε
167	5/2-	-60.596	17.5 m *2*	ε
168	0+	-61.575	0.13% *1*	
169	7/2+	-60.371	32.022 d *8*	ε
169m	1/2-	-60.347	46 s *2*	IT
170	0+	-60.770	3.05% *5*	
171	1/2-	-59.314	14.3% *2*	
172	0+	-59.262	21.9% *3*	
173	5/2-	-57.558	16.12% *18*	
174	0+	-56.951	31.8% *4*	
175	7/2-	-54.702	4.19 d *1*	β-
176	0+	-53.501	12.7% *1*	
176m	(8-)	-52.451	11.4 s *3*	IT≥90%, β-<10%
177	9/2+	-50.996	1.9 h *1*	β-

Isotope Z El A	Jπ	Δ (MeV)	T1/2 or Abundance	Decay Mode
70 Yb 177m	1/2-	-50.664	6.41 s *2*	IT
178	0+	-49.705	74 m *3*	β-
179			8.1 m *8*	β-
180			2.4 m *5*	β-
71 Lu 150		-25.350s		
151		-31.000s	85 ms *10*	p
152	(6-,5-)	-34.050s	0.7 s *1*	ε
153		-38.840s		
154		-40.000s	0.96 s *10*	ε
155		-42.990s	70 ms *6*	α 79%, ε 21%
155m		-41.192s	2.60 ms *7*	α
156		-43.830s	≈0.5 s	α≈70%, ε
156m		-43.830s	0.18 s *2*	α≈95%, IT?, ε?
157		-46.690s	5.4 s *2*	ε 94%, α 6%
158		-47.490	10.4 s *1*	ε>98.5%, α<1.5%
159		-49.770	12.3 s *1*	ε, α 0.04%
160		-50.460s	35.5 s *8*	ε
161	(5/2+)	-52.600s	72 s	ε
162	(1-)	-52.860s	1.37 m *2*	ε
162m	(4-)	-52.860s	1.5 m	ε
162m		-52.860s	1.9 m	ε
163	(1/2-)	-54.770	238 s *8*	ε
164		-54.740s	3.14 m *3*	ε
165m	(7/2+)	-56.260	10.74 m *10*	ε
165m	1/2+	-56.260	12 m	
166	(6-)	-56.110	2.65 m *10*	ε
166m	(3-)	-56.076	1.41 m *10*	ε 58%, IT 42%
166m	(0-)	-56.067	2.12 m *10*	ε>80%, IT<20%
167	7/2+	-57.470	51.5 m *10*	ε
168	(6-)	-57.090	5.5 m *1*	ε
168m	3+	-56.870	6.7 m *4*	ε>95%, IT<5%
169	7/2+	-58.078	34.06 h *5*	ε
169m	1/2-	-58.049	160 s *10*	IT
170	0+	-57.311	2.00 d *3*	ε
170m	4-	-57.218	0.67 s *10*	IT
171	7/2+	-57.834	8.24 d *3*	ε
171m	1/2-	-57.763	79 s *2*	IT
172	4-	-56.741	6.70 d *3*	ε
172m	1-	-56.699	3.7 m *5*	IT
173	7/2+	-56.886	1.37 y *1*	ε
174	(1-)	-55.575	3.31 y *5*	ε
174m	(6-)	-55.404	142 d *2*	IT 99.38%, ε 0.62%
175	7/2+	-55.171	97.41% *2*	
176	7-	-53.394	3.78×10^{10} y *2* 2.59% *2*	β-
176m	1-	-53.271	3.635 h *3*	β- 99.9%, ε 0.1%
177	7/2+	-52.394	6.71 d *1*	β-
177m	23/2-	-51.424	160.9 d *3*	β- 79%, IT 21%
178	1(+)	-50.338	28.4 m *2*	β-
178m	(9-)	-50.118	23.1 m *3*	β-
179	(7/2+)	-49.110	4.59 h *6*	β-
180	3+,4+,5+	-46.690	5.7 m *1*	β-
181	(7/2+)		3.5 m *3*	β-

Isotope Z El A	Jπ	Δ (MeV)	T1/2 or Abundance	Decay Mode
71 Lu 182	(0,1,2)		2.0 m *2*	β-
183	(7/2+)		58 s *4*	β-
184			≈20 s	β-
72 Hf 154	0+	-33.420s	2 s *1*	ε, α?
155		-34.600s	0.89 s *12*	ε, α
156	0+	-38.180s	25 ms *4*	α
157		-38.960s	110 ms *6*	α 91%, ε 9%
158	0+	-42.400s	2.9 s *2*	ε 54%, α 46%
159		-43.050s	5.6 s *5*	ε 88%, α 12%
160	0+	-46.080	≈12 s	ε 97.7%, α 2.3%
161		-46.480s	17 s *2*	α, ε
162	0+	-49.178	37.6 s *8*	ε 99.99%, α 0.01%
163		-49.380s	40.0 s *6*	ε
164	0+	-51.790s	2.8 m *2*	ε
165	(11/2-)	-51.670s	1.7 m *1*	ε
166	0+	-53.790s	6.77 m *30*	ε
167	(5/2-)	-53.470s	2.05 m *5*	ε
168	0+	-55.290s	25.95 m *20*	ε
169	(5/2)-	-54.810	3.24 m *4*	ε
170	0+	-56.210s	16.01 h *13*	ε
171	(7/2+)	-55.430s	12.1 h *4*	ε
172	0+	-56.390	1.87 y *3*	ε
173	1/2-	-55.290s	23.6 h *1*	ε
174	0+	-55.851	2.0×10^{15} y *4* 0.162% *2*	α
175	5/2-	-54.488	70 d *2*	ε
176	0+	-54.582	5.206% *4*	
177	7/2-	-52.892	18.606% *3*	
177m	23/2+	-51.577	1.08 s *6*	IT
177m	37/2-	-50.152	51.4 m *5*	IT
178	0+	-52.446	27.297% *3*	
178m	8-	-51.299	4.0 s *2*	IT
178m	16+	-50.000	31 y *1*	IT
179	9/2+	-50.475	13.629% *5*	
179m	1/2-	-50.100	18.67 s *3*	IT
179m	25/2-	-49.369	25.1 d *3*	IT
180	0+	-49.791	35.100% *6*	
180m	8-	-48.649	5.5 h *1*	IT≥98.6%, β-<1.4%
181	1/2-	-47.416	42.39 d *6*	β-
182	0+	-46.062	9×10^{6} y *2*	β-
182m	8-	-44.889	61.5 m *15*	β- 58%, IT 42%
183	(3/2-)	-43.290	1.067 h *17*	β-
184	0+	-41.500	4.12 h *5*	β-
73 Ta 156		-26.230s		
157		-30.030s	5.3 ms *18*	α>77%
158		-31.370s	36.8 ms *16*	α 93%, ε 7%
159		-34.820s	0.57 s *18*	α 80%, ε 20%
160		-35.850s	1.4 s *2*	α
161		-38.980s	2.7 s *2*	ε≈95%, α≈5%
162		-40.060	3.52 s *12*	ε, α
163		-42.600	11.0 s *8*	ε≈99.8%, α≈0.2%
164		-43.320s	14.2 s *3*	ε 99.98%, α 0.02%
165		-45.850s	31.0 s *15*	ε

Isotope Z El	A	Jπ	Δ (MeV)	T1/2 or Abundance	Decay Mode
73 Ta	166	(2-)	-46.310s	34.4 s *5*	ε
	167		-48.470s	1.4 m *3*	ε
	168	(3+)	-48.590s	2.44 m *35*	ε
	169		-50.380s	4.9 m *4*	ε
	170	(3+)	-50.210s	6.76 m *6*	ε
	171	(5/2-)	-51.730s	23.3 m *3*	ε
	172	(3-)	-51.470	36.8 m *3*	ε
	173	5/2(-)	-52.490s	3.14 h *13*	ε
	174	3(-)	-51.850s	1.18 h *5*	ε
	175	7/2+	-52.490s	10.5 h *2*	ε
	176	(1)-	-51.470	8.09 h *5*	ε
	177	7/2+	-51.726	56.6 h *1*	ε
	178	1+	-50.530	9.31 m *3*	ε
	178	(7)-	-50.530	2.36 h *8*	ε
	179	7/2+	-50.365	1.79 y *8*	ε
	180	1+	-48.939	8.152 h *6*	ε 86%, β- 14%
	180m	9-	-48.864	$>1.2\times10^{15}$ y	
				0.012% *2*	
	181	7/2+	-48.444	99.988% *2*	
	182	3-	-46.436	114.43 d *3*	β-
	182m	5+	-46.420	283 ms *3*	IT
	182m	10-	-45.916	15.84 m *10*	IT
	183	7/2+	-45.299	5.1 d *1*	β-
	184	(5-)	-42.844	8.7 h *1*	β-
	185	(7/2+)	-41.402	49 m *2*	β-
	186	2,3	-38.620	10.5 m *5*	β-
74 W	158	0+	-24.380s	≈1.4 ms	α
	159		-25.720s	7.3 ms *27*	α
	160	0+	-29.690s	81 ms *15*	α
	161		-30.620s	410 ms *40*	α≈82%, ε≈18%
	162	0+	-34.300s	1.39 s *4*	ε 54%, α 46%
	163		-35.110s	2.75 s *25*	ε 59%, α 41%
	164	0+	-38.380	6.4 s *8*	ε 97.4%, α 2.6%
	165		-39.030s	5.1 s *5*	ε>98.5%, α<1.5%
	166	0+	-41.898	16 s *3*	ε 99.4%, α 0.6%
	167		-42.350s	19.9 s *5*	ε, α
	168	0+	-44.840s	53 s *2*	ε
	169		-44.940s	1.3 m *1*	ε
	170	0+	-47.240s	4 m *1*	ε
	171	(5/2-)	-47.080s	2.4 m *1*	ε, α
	172	0+	-48.970s	6.7 m *10*	ε
	173m	(5/2-)	-48.690s	7.97 m *27*	ε
	174	0+	-50.150s	31 m *1*	ε
	175	(1/2-)	-49.590s	34 m *1*	ε
	176	0+	-50.680s	2.5 h *1*	ε
	177	(1/2-)	-49.730s	135 m *3*	ε
	178	0+	-50.440	21.6 d *3*	ε
	179	(7/2-)	-49.306	37.5 m *5*	ε
	179m	(1/2-)	-49.084	6.4 m *1*	IT 99.72%,
					ε 0.28%
	180	0+	-49.647	0.12% *3*	
	181	9/2+	-48.256	121.2 d *2*	ε
	182	0+	-48.250	26.3% *2*	
	183	1/2-	-46.369	14.28% *5*	

Isotope Z El	A	Jπ	Δ (MeV)	T1/2 or Abundance	Decay Mode
74 W	183m	11/2+	-46.060	5.2 s *3*	IT
	184	0+	-45.709	$>3\times10^{17}$ y	
				30.7% *2*	
	185	3/2-	-43.393	75.1 d *3*	β-
	185m	11/2+	-43.196	1.67 m *3*	IT
	186	0+	-42.515	28.6% *2*	
	187	3/2-	-39.910	23.72 h *6*	β-
	188	0+	-38.673	69.4 d *5*	β-
	189	(3/2-)	-35.480	11.5 m *3*	β-
	190	0+	-34.310	30.0 m *15*	β-
75 Re	161		-21.170s	10 ms +*15*-*5*	α
	162		-22.670s	100 ms *30*	α
	163		-26.330s	260 ms *40*	α 64%, ε 36%
	164		-27.510s	0.88 s *24*	α 58%, ε 42%
	165		-30.910s	2.4 s *6*	ε 87%, α 13%
	166		-32.130	2.2 s *4*	α, ε
	167		-34.910	6.1 s *2*	ε≈99.3%, α≈0.7%
	168		-35.880s	6.9 s *8*	α, ε
	168m		-35.880s	6.6 s *15*	α, ε
	169m		-38.600s	12.9 s *11*	α
	170	(5)	-39.040s	8.0 s *5*	ε
	171	(9/2-)	-41.440s	15.2 s	α
	172	(5)	-41.660s	15 s *3*	ε, α
	172m	(2)	-41.660s	55 s *5*	ε, α, IT
	173		-43.650s	1.98 m *26*	ε
	174		-43.670s	2.4 m *1*	ε
	175		-45.280s	5.8 m	ε
	176	3(+)	-44.980s	5.3 m *3*	ε
	177	(5/2-)	-46.330s	14.0 m *10*	ε
	178	(3)	-45.780	13.2 m *2*	ε
	179	(5/2+)	-46.620	19.5 m *1*	ε
	180	(1)-	-45.840	2.44 m *6*	ε
	181	5/2+	-46.460s	19.9 h *7*	ε
	182	7+	-45.450	64.0 h *5*	ε
	182m	2+	-45.450	12.7 h *2*	ε
	183	5/2+	-45.813	70.0 h *14*	ε
	184	3(-)	-44.220	38.0 d *5*	ε
	184m	8(+)	-44.032	169 d *8*	IT 75.4%, ε 24.6%
	185	5/2+	-43.826	37.40% *2*	
	186	1-	-41.933	90.64 h *9*	β- 93.1%, ε 6.9%
	186m	(8+)	-41.784	2.0×10^{5} y *5*	IT, β-<10%
	187	5/2+	-41.222	4.35×10^{10} y *13*	β-,
				62.60% *2*	$\alpha<1.0\times10^{-4}$%
	188	1-	-39.022	16.98 h *2*	β-
	188m	(6)-	-38.850	18.6 m *1*	IT
	189	5/2+	-37.985	24.3 h	β-
	190	(2)-	-35.580	3.1 m *3*	β-
	190m	(6-)	-35.461	3.2 h *2*	β- 54.4%,
					IT 45.6%
	191	(3/2+,1/2+)	-34.360	9.8 m *5*	β-
	192		-31.790s	16 s *1*	β-
76 Os	163		-16.620s	?	ε, α
	164	0+	-20.780s	41 ms *20*	α, ε

Isotope Z El A	Jπ	Δ (MeV)	T1/2 or Abundance	Decay Mode
76 Os 165		-21.870s	65 ms +70-30	α≥60%, ε≤40%
166	0+	-25.740s	181 ms 38	α 72%, ε 28%
167		-26.710s	0.83 s 12	α 67%, ε 33%
168	0+	-30.130	2.2 s 1	ε 51%, α 49%
169		-30.880s	3.2 s 2	ε 84%, α 16%
170	0+	-33.933	7.1 s 2	ε 88%, α 12%
171		-34.550s	8.0 s 7	ε 98.3%, α 1.7%
172	0+	-37.190s	19 s 2	ε 99.8%, α 0.2%
173		-37.460s	16 s 5	ε 99.98%, α 0.02%
174	0+	-39.950s	44 s 4	ε 99.98%, α 0.02%
175		-39.920s	1.4 m 1	ε
176	0+	-42.080s	3.6 m 5	ε
177	1/2-	-41.870s	2.8 m	ε
178	0+	-43.540s	5.0 m 4	ε
179	(1/2-)	-42.970s	6.5 m 3	ε
180	0+	-44.380s	21.5 m 4	ε
181	(7/2)-	-43.530s	2.7 m 1	ε
181m	1/2-	-43.530s	105 m 3	
182	0+	-44.542	22.10 h 25	ε
183	9/2+	-43.510s	13.0 h 5	ε
183m	1/2-	-43.339s	9.9 h 3	ε 85%, IT 15%
184	0+	-44.259	$>5.6\times10^{13}$ y 0.02% 1	
185	1/2-	-42.813	93.6 d 5	ε
186	0+	-43.003	2.0×10^{15} y 11 1.58% 10	α
187	1/2-	-41.224	1.6% 1	
188	0+	-41.142	13.3% 2	
189	3/2-	-38.993	16.1% 3	
189m	9/2-	-38.962	5.8 h 1	IT
190	0+	-38.714	26.4% 4	
190m	(10)-	-37.009	9.9 m 1	IT
191	9/2-	-36.401	15.4 d 1	β-
191m	3/2-	-36.327	13.10 h 5	IT
192	0+	-35.892	41.0% 3	
192m	(10-)	-33.877	6.1 s	IT
193	3/2-	-33.405	30.5 h 4	β-
194	0+	-32.442	6.0 y 2	β-
195		-29.700	6.5 m	β-
196	0+	-28.300	34.9 m 2	β-
77 Ir 166		-13.540s	>5 ms	α
167		-17.360s	>5 ms	α
168		-18.670s	?	α
169		-22.210s	0.4 s 1	α
170		-23.530	1.05 s 15	α 75%, ε 25%
171		-26.420	1.5 s 1	α, ε?
172		-27.490s	2.1 s 1	ε≈97%, α≈3%
173		-30.230s	3.0 s 10	ε 97.98%, α 2.02%
174		-31.010s	4 s 1	ε 99.53%, α 0.47%
175		-33.490s	4.5 s 10	α
176		-34.000s	8 s 1	ε 97.9%, α 2.1%
177		-36.100s	21 s 2	α
178		-36.350s	12 s 2	ε
179		-38.050s	4 m 1	ε

Isotope Z El A	Jπ	Δ (MeV)	T1/2 or Abundance	Decay Mode
77 Ir 180		-37.840s	1.5 m 1	ε
181	(7/2+)	-39.360s	4.90 m 15	ε
182m	(5-)	-38.950s	15 m 1	ε
183	(7/2+)	-40.110s	57 m 4	ε
184	5-	-39.540	3.09 h 3	ε
185	5/2-	-40.210s	14.4 h 1	ε
186	5+	-39.172	16.64 h 3	ε
186m	2-	-39.172	2.0 h 1	ε, IT
187	3/2+	-39.720s	10.5 h 3	ε
188	1-	-38.333	41.5 h 5	ε
189	3/2+	-38.462	13.2 d 1	ε
190	(4)+	-36.710	11.78 d 10	ε
190m	(7)+	-36.684	1.2 h	IT
190m	(11)-	-36.535	3.25 h 20	ε 94.4%, IT 5.6%
191	3/2+	-36.715	37.3% 5	
191m	11/2-	-36.544	4.94 s 3	IT
191m		-34.668	5.5 s 7	IT
192	4(-)	-34.843	73.831 d 8	β- 95.4%, ε 4.6%
192m	1(+)	-34.785	1.45 m 5	IT 99.98%, β- 0.02%
192m	(9+)	-34.688	241 y 9	IT
193	3/2+	-34.544	62.7% 5	
193m	11/2-	-34.464	10.53 d 4	IT
194	1-	-32.539	19.15 h 3	β-
194m	(10,11)	-32.349	171 d 11	β-
195	3/2+	-31.700	2.5 h 2	β-
195m	11/2-	-31.600	3.8 h 2	β- 95%, IT 5%
196	(0-)	-29.460	52 s 2	β-
196m	(10,11-)	-29.050	1.40 h 2	β-
197	3/2+	-28.292	5.8 m 5	β-
197m	11/2-	-28.177	8.9 m 3	β-, IT
198		-25.830s	8 s 1	β-
78 Pt 168	0+	-11.370s	?	α
169		-12.610s	2.5 ms +25-1	α
170	0+	-16.610s	6 ms +5-2	α
171		-17.680s	25 ms 9	α
172	0+	-21.240	0.10 s 1	α 98%, ε 2%
173		-22.110s	342 ms 18	α 84%, ε 16%
174	0+	-25.324	0.90 s 1	α 83%, ε 17%
175		-25.950s	2.52 s 8	α 64%, ε
176	0+	-28.880s	6.33 s 15	ε 62%, α 38%
177		-29.390s	11 s 2	ε 91%, α 9%
178	0+	-31.950s	21.0 s 6	ε 92.3%, α 7.7%
179	1/2-	-32.200s	43 s 10	ε 99.76%, α 0.24%
180	0+	-34.400s	52 s 3	ε, α≈0.3%
181	(1/2-)	-34.310s	51 s 5	ε, α≈0.06%
182	0+	-36.170s	2.2 m 1	ε≈99.98%, α≈0.02%
183	1/2-	-35.700s	6.5 m 10	ε, α≈0.0013%
183m	(7/2-)	-35.665s	43 s 5	ε, IT
184	0+	-37.360s	17.3 m 2	ε, α≈0.001%
185	(9/2+)	-36.510s	70.9 m 24	ε
185m	(1/2-)	-36.407s	33.0 m 8	ε 99%, IT<2%
186	0+	-37.790	2.0 h 1	ε, α≈1.4×10^{-4}%

Isotope Z El A	Jπ	Δ (MeV)	T1/2 or Abundance	Decay Mode
78 Pt 187	3/2-	-36.820s	2.35 h 3	ε
188	0+	-37.827	10.2 d 3	ε, α 2.6×10^{-5}%
189	3/2-	-36.491	10.87 h 12	ε
190	0+	-37.331	6.5×10^{11} y 3	α
			0.01% 1	
191	3/2-	-35.701	2.9 d 1	ε
192	0+	-36.303	0.79% 5	
193	1/2-	-34.487	50 y 9	ε
193m	13/2+	-34.337	4.33 d 3	IT
194	0+	-34.787	32.9% 5	
195	1/2-	-32.821	33.8% 5	
195m	13/2+	-32.562	4.02 d 1	IT
196	0+	-32.671	25.3% 5	
197	1/2-	-30.446	18.3 h 3	β-
197m	13/2+	-30.046	95.41 m 18	IT 96.7%, β- 3.3%
198	0+	-29.932	7.2% 2	
199	5/2-	-27.432	30.8 m 4	β-
199m	(13/2)+	-27.008	13.6 s 4	IT
200	0+	-26.627	12.5 h 3	β-
201	(5/2-)	-23.750	2.5 m 1	β-
79 Au 173		-12.890s	59 ms +45-18	α
174		-14.330	120 ms 20	α
175		-17.210	0.20 s 2	α
176		-18.520s	1.25 s 30	α, ε
177		-21.370s	1.3 s 4	α
178		-22.530s	2.6 s 5	ε≤60%, α≥40%
179		-24.990s	7.5 s 4	ε 78%, α 22%
180		-25.750s	8.1 s 3	ε≤98.2%, α≥1.8%
181		-27.920s	11.4 s 5	ε 98.9%, α 1.1%
182		-28.390s	21 s 1	ε, α 0.038%
183	(5/2-)	-30.170s	42.0 s 12	ε 99.64%, α 0.36%
184	3+	-30.130s	53.0 s 14	ε, α 0.02%
185	5/2-	-31.750s	4.3 m 1	ε 99.9%, α 0.1%
185m		-31.750s	6.8 m 3	ε, IT
186	3-	-31.570s	10.7 m 5	ε
187	1/2+	-32.900s	8.4 m 3	ε, α
187m	9/2-	-32.779s	2.3 s 1	IT
188	1(-)	-32.530s	8.84 m 6	ε
189	1/2+	-33.640s	28.7 m 3	ε, α<3.0×10^{-5}%
189m	11/2-	-33.393s	4.59 m 11	ε, IT
190	1-	-32.889	42.8 m 10	ε, α<1.0×10^{-6}%
191	3/2+	-33.870	3.18 h 8	ε
191m	(11/2-)	-33.604	0.92 s 11	IT
192	1-	-32.787	4.94 h 9	ε
193	3/2+	-33.430s	17.65 h 15	ε
193m	11/2-	-33.140s	3.9 s 3	IT 99.97%,
				ε≈0.03%
194	1-	-32.295	38.02 h 10	ε
195	3/2+	-32.594	186.09 d 4	ε
195m	11/2-	-32.275	30.5 s 2	IT
196	2-	-31.166	6.183 d 10	ε 92.5%, β- 7.5%
196m	5+	-31.081	8.1 s 2	IT
196m	12-	-30.570	9.7 h 1	IT
197	3/2+	-31.165	100%	

Isotope Z El A	Jπ	Δ (MeV)	T1/2 or Abundance	Decay Mode
79 Au 197m	11/2-	-30.756	7.73 s 6	IT
198	2-	-29.606	2.6935 d 4	β-
198m	(12-)	-28.794	2.30 d 4	IT
199	3/2+	-29.119	3.139 d 7	β-
200	1(-)	-27.280	48.4 m 3	β-
200m	12-	-26.290	18.7 h 5	β- 82%, IT 18%
201	3/2+	-26.413	26 m 1	β-
202	(1-)	-24.420	28.8 s 19	β-
203	3/2+	-23.153	53 s 2	β-
204	(2-)	-20.720s	39.8 s 9	β-
80 Hg 175		-8.210s	≈20 ms	α
176	0+	-11.890	34 ms +18-9	α
177		-12.950s	0.17 s 5	α≈85%, ε≈15%
178	0+	-16.321	0.26 s 3	α≈50%, ε≈50%
179		-17.090s	1.09 s 4	α≈53%, ε≈47%,
				εp≈0.15%
180	0+	-20.200s	3.0 s 3	ε 51%, α 49%
181	1/2(-)	-20.680s	3.6 s 3	α 64%, ε 36%
182	0+	-23.530s	11.3 s 5	ε 84.8%, α 15.2%
183	1/2-	-23.740s	8.8 s 5	ε 74.5%, α 25.5%,
				εp 0.06%
184	0+	-26.310s	30.6 s 3	ε 98.89%, α 1.11%
185	1/2-	-26.110s	49 s 1	ε 94%, α 6%
185m	13/2+	-26.011s	21 s 1	IT 54%, ε 46%,
				α≈0.03%
186	0+	-28.540s	1.38 m 7	ε 99.98%, α 0.02%
187	13/2+	-28.130s	2.4 m 3	ε, α<1.2×10^{-4}%
187m	3/2-	-28.130s	1.9 m 3	ε, α 3.5×10^{-6}%
188	0+	-30.230s	3.25 m 15	ε, α 3.7×10^{-5}%
189	3/2-	-29.690s	7.6 m 1	ε, α<3.0×10^{-5}%
189m	13/2+	-29.690s	8.6 m 1	ε, α<3.0×10^{-5}%
190	0+	-31.410s	20.0 m 5	ε, α<5.0×10^{-5}%
191	(3/2-)	-30.690	49 m 10	ε
191m	13/2+	-30.690	50.8 m 15	ε
192	0+	-32.060s	4.85 h 20	ε
193	3/2-	-31.090s	3.80 h 15	ε
193m	13/2+	-30.949s	11.8 h 2	ε 92.9%, IT 7.1%
194	0+	-32.255	520 y 32	ε
195	1/2-	-31.070	9.9 h 5	ε
195m	13/2+	-30.894	41.6 h 8	IT 54.2%, ε 45.8%
196	0+	-31.852	0.15% 1	
197	1/2-	-30.566	64.14 h 5	ε
197m	13/2+	-30.267	23.8 h 1	IT 93%, ε 7%
198	0+	-30.979	9.97% 8	
199	1/2-	-29.572	16.87% 10	
199m	13/2+	-29.039	42.6 m 2	IT
200	0+	-29.529	23.10% 16	
201	3/2-	-27.688	13.10% 8	
202	0+	-27.370	29.86% 20	
203	5/2-	-25.292	46.612 d 18	β-
204	0+	-24.716	6.87% 4	
205	1/2-	-22.312	5.2 m 1	β-
206	0+	-20.969	8.15 m 10	β-
207	(9/2+)	-16.270	2.9 m 2	β-

Isotope Z El A	Jπ	Δ (MeV)	T1/2 or Abundance	Decay Mode
81 Tl 179		-8.020s	0.16 s +9-4	α
179m		-8.020s	1.4 ms 5	α
180		-9.300s		
181		-12.350s		
182		-13.500s		α
183	(1/2+)	-16.210s		
183m	(9/2-)	-16.210s	60 ms 15	α
184		-17.030s	11 s 1	ε 97.9%, α 2.1%
185	(1/2+)	-19.490s		
185m	(9/2-)	-19.036s	1.8 s 1	α, IT
186m	(7+)	-20.080s	27.5 s 10	ε, α 6.0×10^{-4}%
186m	(10-)	-19.706s	2.9 s 2	IT
187	(1/2+)	-22.200s	≈51 s	ε
187m	(9/2-)	-21.865s	15.60 s 12	ε, IT, α
188m	(2-)	-22.430s	71 s 2	ε
188m	(7+)	-22.430s	71 s 1	ε
189	(1/2+)	-24.450s	2.3 m 2	ε
189m	(9/2-)	-24.169s	1.4 m 1	ε, IT<4%
190m	(2)-	-24.490s	2.6 m 3	ε
190m	(7+)	-24.490s	3.7 m 3	ε
191	(1/2+)	-26.190s		
191m	9/2(-)	-25.891s	5.22 m 16	ε
192	(2-)	-25.950s	9.6 m 4	ε
192	(7+)	-25.950s	10.8 m 2	ε
193	1/2(+)	-27.450	21.6 m 8	ε
193m	(9/2-)	-27.085	2.11 m 15	IT 75%, ε 25%
194	2-	-27.070s	33.0 m 5	ε, α<1.0×10^{-7}%
194m	(7+)	-27.070s	32.8 m 2	ε
195	1/2+	-28.270	1.16 h 5	ε
195m	9/2-	-27.787	3.6 s 4	IT
196	2-	-27.500s	1.84 h 3	ε
196m	(7+)	-27.105s	1.41 h 2	ε 95.5%, IT 4.5%
197	1/2+	-28.400	2.84 h 4	ε
197m	9/2-	-27.792	0.54 s 1	IT
198	2-	-27.520	5.3 h 5	ε
198m	7+	-26.976	1.87 h 3	ε 54%, IT 46%
199	1/2+	-28.140	7.42 h 8	ε
200	2-	-27.073	26.1 h 1	ε
201	1/2+	-27.205	72.912 h 17	ε
202	2-	-26.006	12.23 d 2	ε
203	1/2+	-25.784	29.524% 9	
204	2-	-24.369	3.78 y 2	β- 97.43%, ε 2.57%
205	1/2+	-23.846	70.476% 9	
206	0-	-22.278	4.199 m 15	β-
206m	(12-)	-19.635	3.74 m 3	IT
207	1/2+	-21.049	4.77 m 2	β-
207m	11/2-	-19.701	1.33 s 11	IT, β-<0.1%
208	5(+)	-16.774	3.053 m 4	β-
209	(1/2+)	-13.652	2.20 m 7	β-
210	(5+)	-9.262	1.30 m 3	β-, β-n 0.007%
82 Pb 182	0+	-6.874	55 ms +40-35	α
183	(1/2-)	-7.720s	300 ms 80	α
184	0+	-11.000s	0.55 s 6	α

Isotope Z El A	Jπ	Δ (MeV)	T1/2 or Abundance	Decay Mode
82 Pb 185		-11.580s	4.1 s 3	α
186	0+	-14.630s	4.79 s 5	α
187m	(13/2+)	-14.920s	18.3 s 3	ε 98%, α 2%
187m		-14.920s	15.2 s 3	ε, α
188	0+	-17.780s	24.2 s 10	ε 78%, α 22%
189m		-17.820s	51 s 3	ε>99%, α≈0.4%
190	0+	-20.420s	1.2 m 1	ε 99.1%, α 0.9%
191		-20.300s	1.33 m 8	ε 99.99%, α 0.01%
191m	(13/2+)	-20.300s	2.18 m 8	ε
192	0+	-22.580s	3.5 m 1	ε 99.99%, α 0.01%
193	(3/2-)	-22.280s	?	ε
193m	(13/2+)	-22.180s	5.8 m 2	ε
194	0+	-24.250s	12.0 m 5	ε, α 7.3×10^{-6}%
195	3/2-	-23.780s	≈15 m	ε
195m	13/2+	-23.579s	15.0 m 12	ε
196	0+	-25.420s	37 m 3	ε, α<0.0001%
197	3/2-	-24.800s	8 m 2	ε
197m	13/2+	-24.481s	43 m 1	ε 81%, IT 19%, α<3.0×10^{-4}%
198	0+	-26.100s	2.40 h 10	ε
199	3/2-	-25.270	90 m 10	ε
199m	13/2+	-24.840	12.2 m 3	IT 93%, ε 7%
200	0+	-26.280s	21.5 h 4	ε
201	5/2-	-25.300	9.33 h 3	ε
201m	13/2+	-24.671	61 s 2	IT>99%, ε<1%
202	0+	-25.957	52.5×10^{3} y 28	ε, α<1%
202m	9-	-23.787	3.53 h 1	IT 90.5%, ε 9.5%
203	5/2-	-24.810	51.873 h 9	ε
203m	13/2+	-23.985	6.3 s 2	IT
203m	29/2-	-21.861	0.48 s 2	IT
204	0+	-25.132	≥1.4×10^{17} y 1.4% 1	
204m	9-	-22.946	67.2 m 3	IT
205	5/2-	-23.793	1.52×10^{7} y 7	ε
206	0+	-23.809	24.1% 1	
207	1/2-	-22.476	22.1% 1	
207m	13/2+	-20.843	0.805 s 10	IT
208	0+	-21.772	52.4% 1	
209	9/2+	-17.638	3.253 h 14	β-
210	0+	-14.752	22.3 y 2	β-, α 1.9×10^{-6}%
211	(9/2+)	-10.494	36.1 m 2	β-
212	0+	-7.571	10.64 h 1	β-
213		-3.240s	10.2 m 3	β-
214	0+	-0.188	26.8 m 9	β-
83 Bi 186		-3.380s		
187	(9/2-)	-6.100s	35 ms 4	α
187m	(1/2+)	-6.040s	8 ms 6	α
188m		-7.330s	0.21 s 9	α, ε
188m		-7.330s	44 ms 3	α, ε
189	(9/2-)	-9.800s	680 ms 30	α>50%, ε<50%
190m		-10.690s	6.2 s 1	α 68%, ε 32%
190m		-10.690s	6.3 s 1	α 82%, ε 18%
191	(9/2-)	-12.990s	12 s 1	α 60%, ε 40%
191m		-12.990s	20 s 15	α, ε

Isotope Z El A	Jπ	Δ (MeV)	T1/2 or Abundance	Decay Mode
83 Bi 192		-13.520s	37 s *3*	ε, α 18%
192		-13.520s	39.6 s *4*	ε, α 9.2%
193	(9/2-)	-15.720s	67 s *3*	ε 95%, α 5%
193m	(1/2+)	-15.413s	3.2 s *7*	α≈90%, ε≈10%
194	(2+,3+)	-16.040s	106 s *3*	ε
194m	(6+,7+)	-16.040s	92 s *5*	ε 99.93%, α 0.07%
194m	(10-)	-16.040s	125 s *2*	ε 99.79%, α 0.21%
195	(9/2-)	-17.930s	183 s *4*	ε 99.97%, α 0.03%
195m	(1/2+)	-17.529s	87 s *1*	ε 67%, α 33%
196m		-17.970	5 m	ε
196m	(10-)	-17.970	4.6 m *5*	ε, IT
197	(9/2-)	-19.640	?	ε, α≈1.0×10^{-4}%
197m	(1/2+)	-19.140	5.2 m *6*	α 55%, ε<45%, IT
198	(7+)	-19.540	11.85 m *18*	ε
198m	(10-)	-19.291	7.7 s *5*	IT
199	9/2-	-20.920	27 m *1*	ε
199m	(1/2+)	-20.240	24.70 m *15*	ε 99%, IT≤2%, α≈0.01%
200	7+	-20.400	36.4 m *5*	ε
200m	(2+)	-20.200	31 m *2*	ε>90%, IT<10%
200m	(10-)	-19.972	0.40 s *5*	IT
201	9/2-	-21.470	108 m *3*	ε, α<1×10^{-4}%
201m	1/2+	-20.624	59.1 m *6*	ε>93%, IT≤6.8%, α≈0.3%
202	5+	-20.800	1.72 h *5*	ε, α<1×10^{-5}%
203	9/2-	-21.580	11.76 h *5*	ε, α≈1.0×10^{-5}%
204	6+	-20.730	11.22 h *10*	ε
205	9/2-	-21.084	15.31 d *4*	ε
206	6(+)	-20.052	6.243 d *3*	ε
207	9/2-	-20.079	32.2 y *9*	ε
208	(5)+	-18.894	3.68×10^{5} y *4*	ε
209	9/2-	-18.282	100%	
210	1-	-14.815	5.013 d *5*	β-, α 1.3×10^{-4}%
210m	9-	-14.544	3.0×10^{6} y *1*	α
211	9/2-	-11.873	2.14 m *2*	α 99.72%, β- 0.28%
212	1(-)	-8.142	60.55 m *6*	β- 64.06%, α 35.94%, β-α 0.014%
212m	(9-)	-7.892	25 m	α≤93%, β-≥7%
212m	(15-)	-7.442	9 m	β-
213	9/2(-)	-5.244	45.59 m *6*	β- 97.84%, α 2.16%
214	1-	-1.218	19.9 m *4*	β- 99.98%, α 0.02%
215	(9/2-)	1.710	7.4 m *6*	β-
216		5.960s		
84 Po 192	0+	-8.030s	0.034 s *3*	α
193m		-8.280s	360 ms *50*	α
193m		-8.280s	260 ms *20*	α
194	0+	-11.010s	0.44 s *6*	α
195	(3/2-)	-11.120s	4.5 s *5*	α 75%, ε 25%
195m	(13/2+)	-10.890s	2.0 s *2*	α≈90%, ε≈10%, IT<0.01%

Isotope Z El A	Jπ	Δ (MeV)	T1/2 or Abundance	Decay Mode
84 Po 196	0+	-13.500s	5.5 s *5*	α, ε
197	(3/2-)	-13.450s	56 s *3*	ε 56%, α 44%
197m	(13/2+)	-13.246s	26 s *2*	α 84%, ε≤16%, IT
198	0+	-15.510s	1.76 m *3*	α 70%, ε 30%
199	3/2-	-15.280s	5.2 m *1*	ε 88%, α 12%
199m	13/2+	-14.970s	4.2 m *1*	ε 59%, α 39%, IT 2.1%
200	0+	-17.010s	11.5 m *1*	ε 85%, α 15%
201	3/2-	-16.570s	15.3 m *2*	ε 98.4%, α 1.6%
201m	13/2+	-16.146s	8.9 m *2*	ε 57%, IT 40%, α≈2.9%
202	0+	-17.970s	44.7 m *5*	ε 98%, α 2%
203	5/2-	-17.350	34.8 m *14*	ε 99.89%, α 0.11%
203m	13/2+	-16.709	1.2 m *2*	IT, ε 4.5%, α≈0.04%
204	0+	-18.370s	3.53 h *2*	ε 99.34%, α 0.66%
205	5/2-	-17.555	1.66 h *2*	ε 99.96%, α 0.04%
206	0+	-18.205	8.8 d *1*	ε 94.55%, α 5.45%
207	5/2-	-17.169	5.80 h *2*	ε 99.98%, α 0.02%
207m	19/2-	-15.786	2.8 s *2*	IT
208	0+	-17.492	2.898 y *2*	α, ε
209	1/2-	-16.390	102 y *5*	α 99.74%, ε 0.26%
210	0+	-15.977	138.376 d *2*	α
211	9/2+	-12.457	0.516 s *3*	α
211m	(25/2+)	-10.994	25.2 s *6*	α
212	0+	-10.394	0.298 μs *3*	α
212m	(16+)	-7.473	45.1 s *6*	α
213	9/2+	-6.676	4.2 μs *8*	α
214	0+	-4.493	164.3 μs *20*	α
215	(9/2+)	-0.542	1.780 ms *4*	α, β- 0.0002%
216	0+	1.760	0.145 s *2*	α
217		5.840s	<10 s	α>95%, β-<5%
218	0+	8.351	3.10 m *1*	α 99.98%, β- 0.02%
85 At 194		-0.760s	0.18 s *8*	α
195		-3.170s	?	α>75%, ε<25%
196		-3.890s	0.3 s *1*	α
197	(9/2-)	-6.190s	0.35 s *4*	α, ε
197m	(1/2+)	-6.138s	3.7 s *25*	α, ε, IT
198		-6.720s	4.9 s *5*	α, ε
198m		-6.620s	1.5 s *3*	α, ε, IT
199	(9/2-)	-8.730s	7.2 s *5*	α 90%, ε 10%
200	(5+)	-8.940	43 s *2*	ε 65%, α 35%
200m	(10-)	-8.650	4.3 s *3*	IT≈80%, α≈10%, ε≈10%
201	(9/2-)	-10.740	89 s *3*	α 71%, ε 29%
202	(5+)	-10.770	181 s *3*	ε 88%, α 12%
202m	(10-)	-10.379	≤1.5 s	IT
203	9/2-	-12.290	7.4 m *2*	ε 69%, α 31%
204	(7)+	-11.900	9.2 m *2*	ε 95.7%, α 4.3%
205	9/2-	-13.030	26.2 m *5*	ε 90%, α 10%
206	(5)	-12.490	30.0 m *6*	ε 99.04%, α 0.96%
207	9/2-	-13.290	1.80 h *4*	ε 91.3%, α 8.7%
208	6+	-12.560	1.63 h *3*	ε 99.45%, α 0.55%

Isotope Z El	A	Jπ	Δ (MeV)	T1/2 or Abundance	Decay Mode
85 At	209	9/2-	-12.902	5.41 h 5	ε 95.9%, α 4.1%
	210	5+	-11.995	8.1 h 4	ε 99.82%, α 0.18%
	211	9/2-	-11.674	7.214 h 7	ε 58.3%, α 41.7%
	212	(1-)	-8.640	0.314 s 2	α
	212m	(9-)	-8.415	0.119 s 3	α
	213	9/2-	-6.603	0.11 μs 2	α
	214	1-	-3.403	558 ns 10	α
	214m		-3.344	265 ns 30	α
	214m	9-	-3.171	760 ns 15	α
	215	(9/2-)	-1.269	0.10 ms 2	α
	216	1(-)	2.231	0.30 ms 3	α, ε<0.006%, β-<3×10^{-7}%
	217	9/2(-)	4.383	32.3 ms 4	α 99.99%, β- 0.01%
	218	(2-)	8.090	1.6 s 4	α 99.9%, β- 0.1%
	219		10.520	0.9 m 1	α 97%, β- 3%
	220		14.290s		
86 Rn	198	0+	-1.240s	?	α, ε
	198m	0+	-1.240s	50 ms 9	α, ε, IT
	199	(3/2-)	-1.560s	0.62 s 3	α 95%, ε 5%
	199m	(13/2+)	-1.560s	0.3 s 1	α, ε
	200	0+	-4.040s	1.06 s 2	α≈98%, ε≈2%
	201	(3/2-)	-4.160s	7.0 s 4	α≈80%, ε≈20%
	201m	(13/2+)	-3.880s	3.8 s 4	α≈90%, ε≈10%, IT
	202	0+	-6.320s	9.85 s 20	α, ε<30%
	203	(3/2,5/2)-	-6.230s	45 s 3	α 66%, ε 34%
	203m	(13/2+)	-5.869s	28 s 2	α≈80%, ε≈20%, IT<0.1%
	204	0+	-8.040s	1.24 m 3	α 68%, ε 32%
	205	(5/2)-	-7.760s	170 s 4	ε 77%, α 23%
	206	0+	-9.160s	5.67 m 17	α 62%, ε 38%
	207	5/2-	-8.670	9.3 m 2	ε 77%, α 23%
	208	0+	-9.690s	24.35 m 14	α 62%, ε 38%
	209	5/2-	-8.973	28.5 m 10	ε 83%, α 17%
	210	0+	-9.623	2.4 h 1	α 96%, ε 4%
	211	1/2-	-8.780	14.6 h 2	ε 74%, α 26%
	212	0+	-8.682	24 m 2	α
	213	(9/2+)	-5.722	25.0 ms 2	α
	214	0+	-4.343	0.27 μs 2	α
	214m	6+	-2.900	0.7 ns 3	α
	214m	8+	-2.718	6.5 ns 30	α
	215	(9/2+)	-1.193	2.30 μs 10	α
	216	0+	0.231	45 μs 5	α
	217	9/2+	3.634	0.54 ms 5	α
	218	0+	5.199	35 ms 5	α
	219	(5/2+)	8.828	3.96 s 1	α
	220	0+	10.590	55.6 s 1	α
	221	(7/2+,9/2+)	14.420s	25 m 2	β- 78%, α 22%
	222	0+	16.367	3.8235 d 3	α
	223			23 m 1	β-
	224	0+		107 m 3	β-
	225	7/2-		4.5 m 3	β-
	226	0+		6.0 m 5	β-
	227			23 s 1	β-

Isotope Z El	A	Jπ	Δ (MeV)	T1/2 or Abundance	Decay Mode
86 Rn	228	0+		65 s 2	β-
87 Fr	201	(9/2-)	3.770s	48 ms 15	α, ε<1%
	202		3.100s	0.34 s 4	α≈97%, ε≈3%
	203	(9/2-)	0.970s	0.55 s 2	α≈95%, ε≈5%
	204	(5+,6+)	0.650	2.1 s 2	α≈80%, ε≈20%
	205	(9/2-)	-1.270	3.85 s 10	α, ε<1%
	206	(5)	-1.420	15.9 s 2	α 88%, ε 12%
	206m		-0.889	0.7 s 1	IT, α
	207	9/2-	-2.960	14.8 s 1	α 95%, ε 5%
	208	7+	-2.710	59.1 s 3	α 90%, ε 10%
	209	9/2-	-3.830	50.0 s 3	α 89%, ε 11%
	210	6+	-3.400	3.18 m 6	α 60%, ε 40%
	211	9/2-	-4.200	3.10 m 2	α>70%, ε<30%
	212	5(+)	-3.600	20.0 m 6	ε 57%, α 43%
	213	9/2-	-3.572	34.6 s 3	α 99.45%, ε 0.55%
	214	(1-)	-0.983	5.0 ms 2	α
	214m	(9-)	-0.861	3.35 ms 5	α
	215	9/2(-)	0.292	0.12 μs	α
	216	(1-)	2.960	0.70 μs 2	α, ε<2×10^{-7}%
	217	9/2-	4.293	22 μs 5	α
	218	(1-)	7.036	1.0 ms 6	α
	219	(9/2-)	8.609	21 ms 1	α
	220	1+	11.456	27.4 s 3	α 99.65%, β- 0.35%
	221	5/2(-)	13.266	4.9 m 2	α
	222	2-	16.380	14.2 m 3	β-
	223	(3/2)	18.381	21.8 m 4	β- 99.99%, α 0.01%
	224	1(-)	21.620	3.30 m 10	β-
	225	3/2-	23.840	4.0 m 2	β-
	226	1	27.210	48 s 1	β-
	227	1/2+	29.590	2.48 m 3	β-
	228	2-	33.140s	39 s 1	β-
	229			50 s 20	β-
	230	(3)		19.1 s 5	β-
	231			17.5 s 8	β-
88 Ra	204	0+	5.990s		
	205		5.760s		
	206	0+	3.520s	0.24 s 2	α
	207	(5/2,3/2)-	3.470s	1.3 s 2	α≈90%, ε≈10%
	208	0+	1.660s	1.3 s 2	α 95%, ε 5%
	209		1.810s	4.6 s 2	α
	210	0+	0.420s	3.7 s 2	α 96%, ε 4%
	211	(5/2-)	0.800	13 s 2	α>93%, ε<7%
	212	0+	-0.230s	13.0 s 2	α≈94%, ε≈6%
	213	(1/2-)	0.311	2.74 m 6	α 80%, ε 20%
	213m		2.081	2.1 ms 1	IT 99%, α 1%
	214	0+	0.075	2.46 s 3	α 99.94%, ε 0.06%
	215	(9/2+)	2.509	1.59 ms 9	α
	216	0+	3.269	182 ns 10	α, ε
	217	(9/2+)	5.864	1.6 μs 2	α
	218	0+	6.627	25.6 μs 11	α
	219		9.363	10 ms 3	α
	220	0+	10.250	25 ms 5	α

Isotope Z El	A	Jπ	Δ (MeV)	T1/2 or Abundance	Decay Mode
88 Ra	221		12.938	29 s 2	α
	222	0+	14.303	38.0 s 5	α, ^{14}C 3×10^{-6}%
	223	1/2(+)	17.232	11.434 d 2	α, ^{14}C w
	224	0+	18.804	3.66 d 4	α, ^{12}C 4.3×10^{-9}%
	225	1/2+	21.988	14.9 d 2	β−
	226	0+	23.662	1600 y 7	α, ^{14}C 3×10^{-9}%
	227	(3/2+)	27.172	42.2 m 5	β−
	228	0+	28.936	5.75 y 3	β−
	229	5/2(+)	32.660	4.0 m 2	β−
	230	0+	34.660s	93 m 2	β−
	231			1.72 m 5	β−
	232	0+		250 s 50	β−
89 Ac	209		8.890	0.10 s 5	α
	210		8.620	0.35 s 5	α 96%, ε 4%
	211	(9/2−)	7.080	0.25 s 5	α>99.8%, ε<0.2%
	212		7.240	0.93 s 5	α≈98%, ε≈2%
	213	(9/2−)	6.100	0.80 s 5	α
	214		6.380	8.2 s 2	α≥89%, ε≤11%
	215	(9/2−)	5.970	0.17 s 1	α 99.91%, ε 0.09%
	216	(1−)	8.060	≈0.33 ms	α
	216m	(9−)	8.060	0.33 ms 2	α
	217	9/2−	8.685	0.07 μs 1	α
	217m		8.685	0.74 μs 4	α
	218		10.820	1.12 μs 11	α
	219	(9/2−)	11.540	7 μs 2	α
	220		13.730	26.1 ms 5	α, ε 5×10^{-4}%
	221°		14.500	52 ms 2	α
	222	(1−)	16.603	5.0 s 5	α 99%, ε≤2%
	222m		16.603	63 s 4	α≥88%, IT≤10%, ε≤2%
	223	(5/2−)	17.817	2.2 m 1	α 99%, ε 1%
	224	0−	20.204	2.9 h 2	ε 90.9%, α 9.1%, β−<1.6%
	225	(3/2−)	21.626	10.0 d 1	α
	226	(1)	24.303	29.4 h 1	β− 83%, ε 17%, α 0.006%
	227	3/2(−)	25.848	21.773 y 3	β− 98.62%, α 1.38%
	228	3(+)	28.890	6.15 h 2	β−, α 5×10^{-6}%
	229	(3/2+)	30.900	62.7 m 5	β−
	230	(1+)	33.760s	122 s 3	β−
	231	(1/2+)	35.910	7.5 m 1	β−
	232	(1+)	39.240s	119 s 5	β−
	233	(1/2+)		145 s 10	β−
	234	(1+)		44 s 7	β−
90 Th	212	0+	12.040s	30 ms 20	α
	213		12.080s	150 ms 25	α
	214	0+	10.670s	100 ms 25	α
	215	(1/2−)	10.890	1.2 s 2	α
	216	0+	10.270s	0.028 s 2	α, ε≈0.006%
	216	(8+,11−)	12.298s	0.18 ms 4	IT≈97%, α≈3%
	217		12.160	0.252 ms 7	α
	218	0+	12.348	109 ns 13	α
	219		14.450	1.05 μs 3	α
90 Th	220	0+	14.647	9.7 μs 6	α, ε 2×10^{-7}%
	221		16.917	1.68 ms 6	α
	222	0+	17.182	2.8 ms 3	α
	223		19.357	0.66 s 1	α
	224	0+	19.980	1.05 s 2	α
	225	(3/2)+	22.283	8.72 m 4	α≈90%, ε≈10%
	226	0+	23.183	30.6 m 1	α
	227	(3/2+)	25.803	18.718 d 5	α
	228	0+	26.749	1.9131 y 9	α
	229	5/2+	29.581	7340 y 160	α
	230	0+	30.858	7.538×10^{4} y 30	α, SF?
	231	5/2(+)	33.812	25.52 h 1	β−, α w
	232	0+	35.444	1.405×10^{10} y 6 100%	α, SF?
	233	1/2+	38.729	22.3 m 1	β−
	234	0+	40.607	24.10 d 3	β−
	235	(1/2+)	44.250	7.2 m 2	β−
	236	0+		37.5 m 2	β−
91 Pa	215		17.680	≈14 ms	α
	216		17.680	0.20 s 4	α≈80%, ε≈20%
	217		17.020	4.9 ms 6	α
	217m		17.020	1.6 ms 8	α
	218		18.600	0.12 ms +4−2	α
	219		18.500s		
	220		20.190s		
	221		20.310s	6 μs 3	α
	222		21.940	≈4.3 ms	α
	223		22.310	6.5 ms 10	α
	224		23.780	0.95 s 15	α 99.9%, ε 0.1%
	225		24.310	1.7 s 2	α
	226		26.015	1.8 m 2	α 74%, ε 26%
	227	(5/2−)	26.824	38.3 m 3	α≈85%, ε≈15%
	228	(3+)	28.856	22 h 1	ε 98.15%, α 1.85%
	229	(5/2+)	29.887	1.50 d 5	ε 99.52%, α 0.48%
	230	(2−)	32.168	17.4 d 5	ε 91.6%, β− 8.4%, α w
	231	3/2−	33.422	3.276×10^{4} y 11	α, SF?
	232	(2−)	35.924	1.31 d 2	β−, ε≈0.2%
	233	3/2−	37.485	26.967 d 2	β−
	234	4(+)	40.334	6.70 h 5	β−
	234m	(0−)	40.408	1.17 m 3	β− 99.87%, IT 0.13%
	235	(3/2−)	42.330	24.4 m 2	β−
	236	(1−)	45.340	9.1 m 2	β−, SF w
	237	(1/2+)	47.640	8.7 m 2	β−
	238	(3−)	50.910	2.3 m 1	β−
92 U	222	0+		1.0 μs +10−4	α
	225			0.05 s 3	α
	226	0+	27.170	0.5 s 2	α
	227		28.970s	1.1 m 3	α
	228	0+	29.209	9.1 m 2	α>95%, ε<5%
	229	(3/2+)	31.181	58 m 3	ε≈80%, α≈20%
	230	0+	31.600	20.8 d	α

Z	El	A	Jπ	Δ (MeV)	T1/2 or Abundance	Decay Mode
92	U	231	(5/2-)	33.780	4.2 d *1*	ε, α 0.006%
		232	0+	34.587	68.9 y *4*	α, SF w
		233	5/2+	36.915	1.592×10^{5} y *2*	α, ^{24}Ne w, SF < 6×10^{-9}%
		234	0+	38.141	2.45×10^{5} y *2* 0.0055% *5*	α, SF w
		235	7/2-	40.915	703.8×10^{6} y *5* 0.720% *1*	α, SF w
		235m	1/2+	40.916	≈25 m	IT
		236	0+	42.441	2.3415×10^{7} y *14*	α, SF w
		237	1/2+	45.387	6.75 d *1*	β-
		238	0+	47.305	4.468×10^{9} y *3* 99.2745% *15*	α, SF 0.0001%
		239	5/2+	50.570	23.50 m *5*	β-
		240	0+	52.711	14.1 h *1*	β-, α
		242	0+		16.8 m *5*	β-
93	Np	228			1.00 m *8*	SF?
		229		33.740	4.0 m *2*	α > 50%, ε < 50%
		230		35.220	4.6 m *3*	ε ≤ 97%, α ≥ 3%
		231	(5/2)	35.620	48.8 m *2*	ε 98%, α 2%
		232	(4+)	37.280s	14.7 m *3*	ε, α≈0.003%
		233	(5/2+)	38.010s	36.2 m *1*	ε, α ≤ 0.001%
		234	(0+)	39.952	4.4 d *1*	ε
		235	5/2+	41.039	396.2 d *12*	ε, α 0.0014%
		236	(6-)	43.370	115×10^{3} y *12*	ε 91%, β- 8.9%, α
		236m	1(-)	43.420	22.5 h *4*	ε 52%, β- 48%
		237	5/2+	44.868	2.14×10^{6} y *1*	α, SF ≤ 2×10^{-1}%
		238	2+	47.451	2.117 d *2*	β-
		239	5/2+	49.306	2.355 d *4*	β-
		240	(5+)	52.321	61.9 m *2*	β-
		240m	1(+)	52.321	7.22 m *2*	β- 99.89%, IT 0.11%
		241	(5/2+)	54.260	13.9 m *2*	β-
		242	(1+)	57.410	2.2 m *2*	β-
		242	(6)	57.410	5.5 m *1*	β-
		243		59.922		
94	Pu	231		38.390s		
		232	0+	38.349	34.1 m *7*	ε≈80%, α≈20%
		233		40.020	20.9 m *4*	ε 99.88%, α 0.12%
		234	0+	40.335	8.8 h *1*	ε 94%, α 6%
		235	(5/2+)	42.160	25.3 m *10*	ε, α 0.0027%
		236	0+	42.879	2.87 y *1*	α, SF w
		237	7/2-	45.090	45.2 d *1*	ε, α 0.004%
		237m	1/2+	45.236	0.18 s *2*	IT
		238	0+	46.160	87.74 y *4*	α, SF w
		239	1/2+	48.584	24119 y *26*	α, SF
		240	0+	50.122	6563 y *7*	α, SF 5.7×10^{-6}%
		241	5/2+	52.952	14.35 y *10*	β-, α
		242	0+	54.713	3.733×10^{5} y *12*	α, SF 5.5×10^{-4}%
		243	7/2+	57.751	4.956 h *3*	β-
		244	0+	59.802	8.08×10^{7} y *10*	α 99.88%, SF 0.12%
		245	(9/2-)	63.175	10.5 h *1*	β-
94	Pu	246	0+	65.391	10.84 d *2*	β-
		247			2.27 d *23*	β-
95	Am	232			55 s *7*	ε≈98%, α≈2%, εSF
		233		43.270s		
		234		44.340s	2.6 m *2*	ε, α?
		235		44.640s		
		236		46.010s		
		237	5/2(-)	46.640s	73.0 m *10*	ε 99.98%, α 0.02%
		238	1+	48.420	98 m *2*	ε > 99.99%, α 0.0001%
		239	(5/2)-	49.385	11.9 h *1*	ε 99.99%, α 0.01%
		240	(3-)	51.498	50.8 h *3*	ε, α 1.9×10^{-4}%
		241	5/2-	52.931	432.7 y *6*	α, SF
		242	1-	55.463	16.02 h *2*	β- 82.7%, ε 17.3%
		242m	5-	55.512	141 y *2*	IT 99.54%, α 0.46%, SF
		243	5/2-	57.169	7380 y *40*	α, SF w
		244	(6-)	59.877	10.1 h *1*	β-
		244m	1+	59.965	≈26 m	β- 99.96%, ε 0.04%
		245	(5/2)+	61.891	2.05 h *1*	β-
		246	(7-)	64.990	39 m *3*	β-
		246m	2(-)	64.990	25.0 m *2*	β-, IT < 0.01%
		247	(5/2)	67.230s	23.0 m *13*	β-
		248		70.590s	?	β-
96	Cm	235		48.020s		
		236	0+	47.870s		
		237		49.150s		
		238	0+	49.380	2.4 h *1*	ε ≥ 90%, α ≤ 10%
		239	(7/2-)	51.090s	≈2.9 h	ε, α < 0.1%
		240	0+	51.702	27 d *1*	α > 99.5%, ε < 0.5%, SF 3.9×10^{-6}%
		241	1/2+	53.700	32.8 d *2*	ε 99%, α 1%
		242	0+	54.800	162.79 d *9*	α, SF 6.2×10^{-6}%
		243	5/2+	57.177	29.1 y *1*	α 99.76%, ε 0.24%
		244	0+	58.449	18.10 y *2*	α, SF 1×10^{-4}%
		245	7/2+	60.998	8500 y *100*	α, SF w
		246	0+	62.614	4730 y *100*	α 99.97%, SF 0.03%
		247	9/2-	65.528	1.56×10^{7} y *5*	α
		248	0+	67.388	3.40×10^{5} y *4*	α 91.74%, SF 8.26%
		249	1/2(+)	70.746	64.15 m *3*	β-
		250	0+	72.985	9700 y	SF≈80%, α≈11%, β-≈9%
		251	(1/2+)	76.642	16.8 m *2*	β-
		252	0+		<2 d	β-
97	Bk	237		53.190s		
		238		54.070s		
		239		54.270s		
		240		55.600s	4.8 m *8*	ε, εSF w
		241		56.100s		
		242		57.800s	7.0 m *13*	ε
		243	(3/2-)	58.683	4.5 h *2*	ε 99.85%, α 0.15%

Isotope Z El A	Jπ	Δ (MeV)	T1/2 or Abundance	Decay Mode
97 Bk 244	(1-)	60.700	4.35 h *15*	ε 99.99%, α 0.006%
245	3/2-	61.809	4.94 d *3*	ε 99.88%, α 0.12%
246	2(-)	64.110s	1.80 d *2*	ε, α<0.2%
247	(3/2-)	65.484	1380 y *250*	α
248	1(-)	68.107	23.7 h *2*	β- 70%, ε 30%, α<0.001%
248	(6+)	68.107	>9 y	α>70%
249	7/2+	69.842	320 d *6*	β-, α 0.0014%, SF 4.7×10^{-8}%
250	2-	72.951	3.217 h *5*	β-
251	(3/2-)	75.222	55.6 m *11*	β-, α≈1.0×10^{-5}%
252		78.530s		
98 Cf 239		58.250s	42 s	α
240	0+	58.020s	1.06 m *15*	α
241		59.180s	3.78 m *70*	ε≈90%, α≈10%
242	0+	59.320	3.49 m *12*	α, ε?
243	(1/2+)	60.910s	10.7 m *5*	ε≈86%, α≈14%
244	0+	61.460	19.4 m *6*	α
245		63.380	43.6 m *8*	ε≈70%, α≈30%
246	0+	64.087	35.7 h *5*	α, ε<5.0×10^{-4}%, SF 2.0×10^{-4}%
247	(7/2+)	66.130	3.11 h *3*	ε 99.96%, α 0.03%
248	0+	67.237	333.5 d *28*	α, SF 0.0029%
249	9/2-	69.717	351 y *2*	α, SF 5.2×10^{-7}%
250	0+	71.167	13.08 y *9*	α 99.92%, SF 0.08%
251	1/2+	74.129	898 y *44*	α
252	0+	76.030	2.645 y *8*	α 96.91%, SF 3.09%
253	(7/2+)	79.296	17.81 d *8*	β- 99.69%, α 0.31%
254	0+	81.338	60.5 d *2*	SF 99.69%, α 0.31%
255	(9/2+)		85 m *18*	β-
256	0+		12.3 m *12*	SF, β-<1%, α≈1.0×10^{-6}%
99 Es 241		63.830s		
242		64.620s		
243		64.710s	21 s *2*	ε≤70%, α≥30%
244		65.970s	37 s *4*	ε 96%, α 4%
245		66.380s	1.33 m *15*	ε 60%, α 40%
246	(4-,6+)	67.940s	7.7 m *5*	ε 90.1%, α 9.9%
247		68.550	4.7 m *3*	ε≈93%, α≈7%
248	(2-,0+)	70.290	27 m *4*	ε>99%, α≈0.25%
249	7/2(+)	71.110	102.2 m *6*	ε 99.43%, α 0.57%
250	1(-)	73.270s	2.22 h *5*	ε≥99%, α≤1%
250	(6+)	73.270s	8.6 h *1*	ε>97%, α<3%
251	(3/2-)	74.506	33 h *1*	ε 99.51%, α 0.49%
252	(5-)	77.290	471.7 d *19*	α 76%, ε 24%, β-≈0.01%
253	7/2+	79.007	20.47 d *3*	α, SF 8.7×10^{-6}%
254	(7+)	81.994	275.7 d *5*	α, ε<1.0×10^{-4}%, SF<3.0×10^{-6}%, β- 1.7×10^{-6}%
99 Es 254m	2+	82.072	39.3 h *2*	β- 98%, IT<3%, α 0.33%, ε 0.08%, SF<0.05%
255	(7/2+)	84.083	39.8 d *12*	β- 92%, α 8%, SF 0.0041%
256	(1+)	87.160s	25.4 m *24*	β-
256	(8+)	87.160s	≈7.6 h	β-
100 Fm 242	0+		0.8 ms *2*	SF
243		69.360s	0.18 s *+8-4*	α 40%
244	0+	69.040s	3.7 ms *4*	SF
245		70.040s	4.2 s *13*	α
246	0+	70.120	1.1 s *2*	α 92%, SF 8%, ε≤1%
247		71.530s	35 s *4*	α 50%, ε 50%
247		71.530s	9.2 s *23*	α
248	0+	71.888	36 s *3*	α 99%, ε≈1%, SF≈0.05%
249	(7/2+)	73.510s	2.6 m *7*	ε≈85%, α≈15%
250	0+	74.060	30 m *3*	α>90%, ε<10%, SF≈6.0×10^{-4}%
250m		75.060	1.8 s *1*	IT>80%
251	(9/2-)	75.978	5.30 h *8*	ε 98.2%, α 1.8%
252	0+	76.814	25.39 h *5*	α, SF 0.0023%
253	1/2+	79.339	3.00 d *12*	ε 88%, α 12%
254	0+	80.900	3.240 h *2*	α 99.94%, SF 0.06%
255	7/2+	83.788	20.07 h *7*	α, SF 2.4×10^{-5}%
256	0+	85.482	157.6 m *13*	SF 91.9%, α 8.1%
257	(9/2+)	88.585	100.5 d *2*	α 99.79%, SF 0.21%
258	0+		370 μs *43*	SF
259			1.5 s *3*	SF
101 Md 247		76.060s	3 s	α
248		77.100s	7 s *3*	ε 80%, α 20%
249		77.270s	24 s *4*	α≈70%, ε≈30%
250		78.580s	52 s *6*	ε 93%, α 7%
251		79.050s	4.0 m *5*	ε≥90%, α≤10%
252		80.620s	2.3 m *8*	α<50%, ε>50%
253		81.240s	?	α, ε
254		83.490s	10 m *3*	ε
254		83.490s	28 m *8*	ε
255	(7/2-)	84.835	27 m *2*	ε 92%, α 8%
256	(0-,1-)	87.550	76 m *4*	ε 90.7%, α 9.3%, SF<3%
257	(7/2-)	89.010s	5.3 h *3*	ε 90%, α 10%, SF<4%
258	(1-)	91.840s	60 m *2*	ε
258	(8-)	91.840s	55 d *4*	α
259	(7/2-)		103 m *12*	SF>97%, α<3%
260			31.8 d *5*	SF≈70%, α≤25%, ε<15%, β-<10%
102 No 250	0+		0.25 ms *5*	SF, α≈0.05%
251		82.760s	0.8 s *3*	α, ε≈1%
252	0+	82.857	2.30 s *22*	α 73.1%, SF 26.9%

Isotope Z El	A	Jπ	Δ (MeV)	T1/2 or Abundance	Decay Mode
102 No	253	(9/2−)	84.330s	1.7 m *3*	α≈80%, ε≈20%
	254	0+	84.711	55 s *3*	α 90%, ε 10%, SF 0.25%
	254m		85.211	0.28 s *4*	IT>80%
	255	(1/2+)	86.848	3.1 m *2*	α 61.4%, ε 38.6%
	256	0+	87.793	3.3 s *2*	α 99.8%, SF≤0.25%
	257	(7/2+)	90.220	25 s *2*	α
	258	0+	91.430s	≈1.2 ms	SF, α 0.001%
	259	(9/2+)	94.018	58 m *5*	α 75%, ε 25%, SF<10%
	260	0+		106 ms *8*	SF
103 Lr	252			≈1 s	α≈90%, ε≈10%, SF<1%
	253		88.630s	1.3 s *+6−3*	α 90%, SF<20%, ε≈1%
	254		89.750s	13 s *2*	α 78%, ε 22%, SF<0.1%
	255		90.080s	22 s *4*	α 85%, ε<30%
	256		91.930s	28 s *3*	α>80%, ε<20%, SF<0.03%
	257	(9/2+)	92.670s	0.646 s *25*	α
	258		94.750s	4.3 s *5*	α>95%, ε<5%
	259		95.840	5.4 s *8*	α>50%, SF<50%, ε<0.5%
	260		98.130	180 s *30*	α 75%, ε≈15%, SF<10%
	261			39 m *12*	SF
	262			3.6 h *3*	ε
104 Rf	253			≈1.8 s	α≈50%, SF≈50%
	254	0+		0.5 ms *2*	SF, α≈0.3%
	255	(9/2−)	94.290s	1.5 s *2*	SF 52%, α 48%
	256	0+	94.234	6.7 ms *2*	SF 98%, α 2.2%
	257	(7/2+)	95.900s	4.7 s *3*	α 79.6%, ε 18%, SF 2.4%
	258	0+	96.340s	12 ms *2*	SF≈87%, α≈13%
	259		98.280	3.1 s *7*	α 93%, SF 7%, ε≈0.3%
	260	0+	99.020s	20.1 ms *7*	SF≈98%, α≈2%
	261		101.150s	65 s *10*	α>80%, ε≤10%, SF<10%
	262	0+		47 ms *5*	SF
105 Ha	255			1.6 s *+6−4*	α≈80%, SF≈20%
	256			2.6 s *+14−8*	α≤90%, SF≤40%, ε≈10%
	257		100.360s	1.3 s *+5−3*	α 82%, SF 17%, ε 1%
	258		101.620s	4.4 s *+9−6*	α 67%, ε 33%, SF<1%
	258		101.620s	20 s *10*	ε
	259		102.110s		
	260		103.620s	1.52 s *13*	α≥90%, SF≤9.6%, ε<2.5%
	261		104.170s	1.8 s *4*	α>50%, SF<50%
105 Ha	262		105.970s	34 s *4*	SF 71%, α 26%, ε≈3%
	263			26 m *2*	α
106	259	(1/2+)	106.590s	0.48 s *+28−13*	α 90%, SF<20%
	260	0+	106.580	3.6 ms *+9−6*	α 50%, SF 50%
	261		108.140s	0.23 s *3*	α 95%, SF<10%
	262	0+	108.460s		
	263		110.090	0.8 s *2*	SF≈70%, α≈30%
107	260				α
	261		113.330s	11.8 ms *+53−28*	α 95%, SF<10%
	262		114.650s	102 ms *26*	α≥80%, SF≤20%
	262m		114.965s	8.0 ms *21*	α>70%, SF<30%
	263		114.830s		
	264		116.150s		
108	263			?	α
	264	0+	119.710s	0.08 ms *+40−4*	α
	265		121.080s	1.8 ms *+22−7*	α
109	266		128.350s	3.4 ms *+16−13*	α

APPENDIX

D

USEFUL CONSTANTS AND CONVERSIONS*

h = Planck's constant = 6.626×10^{-34} J • s = 4.136×10^{-21} Mev • s
e = electronic charge = 1.602×10^{-19} C
c = speed of light = 2.998×10^{8} m/s
m_e = electron rest mass = 9.109×10^{-31} kg = 5.110×10^{-1} MeV
m_p = proton rest mass = 1.673×10^{-27} kg = 9.383×10^{2} MeV
m_n = neutron rest mass = 1.675×10^{-27} kg = 9.396×10^{2} MeV
m_d = deuteron rest mass = 3.344×10^{-27} kg = 1.876×10^{3} MeV
m_α = alpha rest mass = 6.645×10^{-27} kg = 3.727×10^{3} MeV

1 Ci = 3.700×10^{10} Bq
1 rad = 1.000×10^{-2} Gy
1 rem = 1.000×10^{-2} Sv
1 R = 2.580×10^{-4} C/kg
1 amu = 1 dalton = 1.661×10^{-27} kg = 9.315×10^{2} MeV
1 eV = 1.602×10^{-19} J
1 MeV = 1.602×10^{-13} J
1 cal = 4.184 J

*Taken from *Nuclides and Isotopes,* 14th ed. San Jose, CA: General Electric Company, 1989.

INDEX